STUDENT SOLUTIONS MANUAL

VOLUME 1

college physics

a strategic approach **4e**

knight • jones • field

P Pearson

Editor in Chief, Director Physical Science Courseware Portfolio: Jeanne Zalesky

Courseware Portfolio Manager: Darien Estes

Senior Content Producer: Martha Steele

Managing Producer: Kristen Flathman

Courseware Director, Content Development: Jennifer Hart

Senior Analyst, Content Development, Science: Suzanne Olivier

Courseware Editorial Assistant: Kristen Stephens

Rich Media Content Producer: Dustin Hennessey

Full-Service Vendor: Nesbitt Graphics/Cenveo

Compositor: Nesbitt Graphics/Cenveo

Art and Design Director: Mark Ong, Side By Side Studios

Interior/Cover Designer: tani hasegawa

Cover Printer: LSC Communications

Printer: LSC Communications

Manufacturing Buyer: Stacey J. Weinberger/LSC Communications

Product Marketing Manager, Physical Sciences: Elizabeth Bell

Cover Photo Credit: MirageC/Getty

ScoutAutomatedPrintCode

ISBN-10:   0-134-70419-3
ISBN-13: 978-0-134-70419-7

Pearson

www.pearson.com

Table of Contents

Preface

This *Student's Solutions Manual* is intended to provide you with examples of good problem-solving techniques and strategies. To achieve that, the solutions presented here attempt to:

- Follow, in detail, the problem-solving strategies presented in the text.
- Articulate the reasoning that must be done before computation.
- Illustrate how to use drawings effectively.
- Demonstrate how to utilize graphs, ratios, units, and the many other "tactics" that must be successfully mastered and marshaled if a problem-solving strategy is to be effective.
- Show examples of assessing the reasonableness of a solution.
- Comment on the significance of a solution or on its relationship to other problems.

We recommend you try to solve each problem on your own before you read the solution. Simply reading solutions, without first struggling with the issues, has limited educational value.

As you work through each solution, make sure you understand how and why each step is taken. See if you can understand which aspects of the problem made this solution strategy appropriate. You will be successful on exams not by memorizing solutions to particular problems buy by coming to recognize which kinds of problem-solving strategies go with which types of problems.

We have made every effort to be accurate and correct in these solutions. However, if you do find errors or ambiguities, we would be very grateful to hear from you.

Q1.1. Reason: The softball player starts with an initial velocity but as he slides he moves slower and slower until coming to rest at the base. The distance he travels in successive times will become smaller and smaller until he comes to a stop. See the figure below.

Assess: Compare to Figure 1.10 in the text.

Q1.3. Reason:

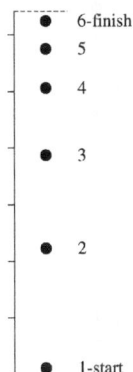

Assess: The spacing between dots is initially large, since the initial speed with which the bush baby leaves the ground is large. As the bush baby rises and gravity slows the ascent, the speed decreases and therefore the spacing between adjacent dots in the motion diagram decreases.

Q1.5. Reason: Position refers to a location in a coordinate frame. A displacement is the difference between two positions. In general, displacement is a vector and requires a direction. But in this one-dimensional case, we ignore that subtlety. The four miles Mark and Sofia walked definitely refer to a difference between their starting and ending positions. There is no information given about a reference frame. So it is more reasonable to associate the four miles with a displacement (magnitude) than with an absolute position.

Assess: Mark and Sofia's position could be specified as (for example) 100 m west of a particular intersection. But what is described is a difference between two positions. This is more like a displacement magnitude. Note that if their starting point had been the origin of a coordinate system, then both Mark and Sofia would be correct.

Q1.7. Reason: Both speed and velocity are ratios with a time interval in the denominator, but speed is a scalar because it is the ratio of the scalar distance over the time interval while velocity is a vector because it is the ratio of the vector displacement over the time interval. Speed and velocity have the same SI units, but one must specify the direction when giving a velocity.
An example of speed would be that your hair grows (the end of a strand of hair moves relative to your scalp) at a speed of about 0.75 in/month.
An example of velocity (where direction matters) would be when you spring off a diving board. Your velocity could initially be 2.0 m/s up, while later it could be 2.0 m/s down.
Assess: Saying that a velocity has both magnitude and direction does not mean that velocity is somehow "better" and that speeds are never useful. Sometimes the direction is unimportant and the concept of speed is useful. In other cases, the direction is important to the physics, and velocity should be cited. Each shows up in various physics equations.

Q1.9. Reason: If the position of the bicycle is negative it is to your left. The bicycle's velocity is positive, or to the right, so the bicycle is getting closer to you.
Assess: If the initial position had been positive and the velocity positive, the bicycle would be getting farther away from you.

Q1.11. Reason:

Start ●—▸●—▸●———▸●———▸●————▸●————————▸

Assess: The dots get farther apart and the velocity arrows get longer as she speeds up.

Q1.13. Reason: The initial velocity is zero. The velocity increases and the space between position markers increases until the chute is deployed. Once the chute is deployed, the velocity decreases and the spacing between the position markers decreases until a constant velocity is obtained. Once a constant velocity is obtained, the position markers are evenly spaced. See the following figure.

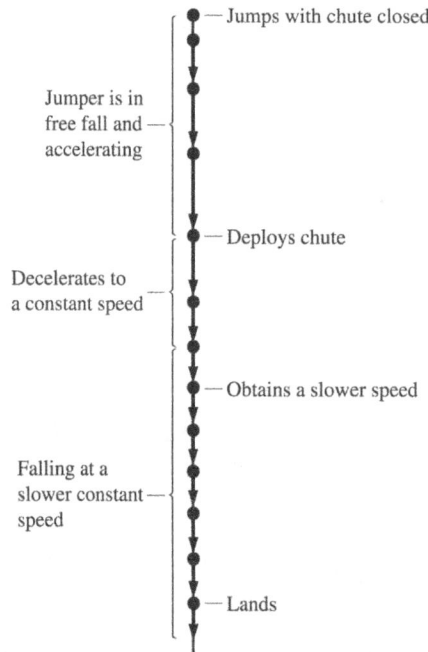

● — Jumps with chute closed

Jumper is in free fall and accelerating

● — Deploys chute

Decelerates to a constant speed

● — Obtains a slower speed

Falling at a slower constant speed

● — Lands

Assess: Knowing the velocity of the jumper will increase until the chute is deployed and then rapidly decrease until a constant descent velocity is obtained allows one to conclude that the figure is correct.

Q1.15. Reason:

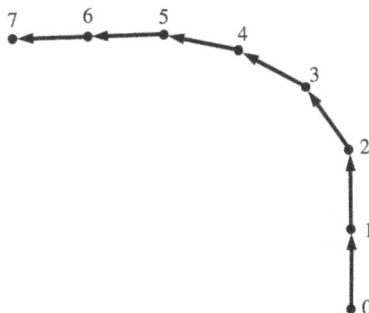

Assess: The car (particle) moves at a constant speed v so the distance between the dots is constant. While turning v remains constant, but the direction of $\vec{v}$ changes.

Q1.17. Reason: We do not believe Travis. 55 m is more than 165 ft. It is not reasonable to think that a unicycle could move that quickly.
Assess: The evaluation may be easier in different units. Consider converting to miles per hour:

$$\frac{55 \text{ m}}{\text{s}} \times \frac{1 \text{ mi}}{1610 \text{ m}} \times \frac{3600 \text{ s}}{1 \text{ hr}} = 120 \text{ mph}$$

This is twice the speed limit on many freeways, and not reasonable for a unicycle.

Q1.19. Reason: Since the rock is above the origin the position is positive; since it is still moving upward the velocity is also positive. Hence, the correct answer is A.
Assess: After it gets to the top and starts back down, the position will still be positive, but the velocity will be negative.

Q1.21. Reason: Because the dots are getting farther apart to the right (and the numbers are increasing to the right) we know that the object is speeding up. The choice that best fits that is a car pulling away (to the right) from a stop sign. So the correct choice is C.
Assess: An ice skater gliding (choice A) would likely have nearly constant velocity (constant spacing between dots). The motion diagram for a plane braking (choice B) might look like the given diagram with the dots numbered in reverse order. The pool ball reversing direction (choice D) would have dot numbers increasing in one direction at first but then going the other way.

Q1.23. Reason: The speed is the distance divided by the time.

$$\text{speed} = \frac{\text{distance}}{\text{time}} = \frac{0.30 \text{ km}}{5.0 \text{ min}} \left(\frac{1000 \text{ m}}{1 \text{ km}} \right) \left(\frac{1 \text{ min}}{60 \text{ s}} \right) = 1.0 \text{ m/s}$$

So the correct choice is B.
Assess: 1 m/s does seem like a reasonable speed for a seal in water.

Q1.25. Reason: This is a simple unit conversion problem:

$$\frac{400 \text{ m}}{51.9 \text{ s}} \times \frac{1 \text{ mi}}{1610 \text{ m}} \times \frac{3600 \text{ s}}{1 \text{ hr}} = 17.2 \text{ mph}$$

So the correct choice is C.
Assess: This is the speed of a slow-moving car, and is reasonable for a very fast-moving human.

Q1.27. Reason: When multiplying numbers, the correct number of significant digits is the smaller of the numbers of significant digits in the two numbers. That is, here we have four significant figures in 109.7 m and three in 48.8 m, and so our product must have three significant digits:

$$(109.7 \text{ m}) \times (48.8 \text{ m}) = 5.35 \times 10^3 \text{ m}^2$$

The correct answer is B.

Assess: Our final answer has three digits of precision, since we multiplied by something with only three digits of precision.

Q1.29. Reason: We are given an equation for density and are asked to calculate the density of the earth given its mass and volume. However, the units must be converted before the calculation is done since we're given volume in km^3 and the answer must be given in terms of m^3.

$$V = (1.08 \times 10^{12} \text{ km}^3) \left(\frac{10^3 \text{ m}}{1 \text{ km}} \right) \left(\frac{10^3 \text{ m}}{1 \text{ km}} \right) \left(\frac{10^3 \text{ m}}{1 \text{ km}} \right)$$

$$= (1.08 \times 10^{12} \text{ km}^3) \left(\frac{10^3 \text{ m}}{1 \text{ km}} \right)^3 = (1.08 \times 10^{12} \text{ km}^3) \left(\frac{10^9 \text{ m}^3}{1 \text{ km}^3} \right)$$

$$= 1.08 \times 10^{21} \text{ m}^3$$

Note carefully that we needed *three* conversion factors for the conversion from km to m here since we are dealing with cubic kilometers. Three factors are needed to cancel the factor of $\text{km}^3 = \text{km} \cdot \text{km} \cdot \text{km}$.

So the density is

$$\rho = \frac{M}{V} = \frac{5.94 \times 10^{24} \text{ kg}}{1.08 \times 10^{21} \text{ m}^3} = 5.50 \times 10^3 \text{ kg/m}^3$$

The correct choice is A.

Assess: For cubic and square units (or units to any power) you must include the correct number of conversion factors to convert every factor in the original quantity. Since the density of water is $1.0 \times 10^3 \text{ kg/m}^3$, it seems reasonable that the earth would be 5.5 times as dense.

Problems

P1.1. Strategize: A motion diagram consists of images or positions of an object shown with equal time intervals between each successive image. This is similar to images taken by a camera with fixed timing between images.

Prepare: Frames of the video are taken at equal intervals of time. As a result, we have a record of the position of the car at successive time equal intervals – this information allows us to construct a motion diagram.

Solve:

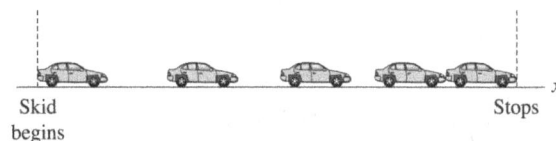

Assess: Once the brakes are applied, the car slows down and travels a smaller distance during each successive time interval until it stops. This is what the car in the figure is doing.

P1.3. Strategize: To draw an accurate motion diagram, we must consider when Amanda speeds up and when she slows down, and we must space the dots in our diagram accordingly.

Prepare: As the elevator begins to rise, its speed changes from zero to some other speed. An elevator typically maintains a steady speed for a while, and then slows as it reaches the desired floor. So our diagram should start with very small but increasing spacing between dots (getting started), then have wider spaced dots with constant spacing for some time (constant speed), and finally dots with decreasing spacing (slowing to a stop).

Solve:

```
                    ● 7-finish
                    ● 6

                    ● 5

                    ● 4

                    ● 3

                    ● 2
                    ● 1-start
```

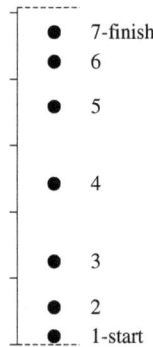

Assess: The speed (and therefore the spacing between dots) increases from 1 to 3, then stays constant from 3 to 5, and decreases from 5 to 7.

P1.5. Strategize: The displacement is the difference between two positions. It does not depend on the origin.
Prepare: To find Sue's displacement, we want to find the difference between her initial position between her home and the cinema, and her position at her home.
Solve: It is graphically clear that $\Delta x = x_f - x_i = 2$ mi. This is the answer for both parts (a) and (b).
Assess: Sue's position depends on where the origin is. But the displacement is the difference between two points (initial and final), and that is independent of what point in space we label as our origin.

P1.7. Strategize: Displacement is the difference between a final position x_f and an initial position x_i.
Prepare: The displacement can be written as $\Delta x = x_f - x_i$, and we are given that $x_i = 23$ m and that $\Delta x = -45$m.
Solve: $\Delta x = x_f - x_i$
Since we want to know the final position we solve this for x_f.

$$x_f = x_i + \Delta x$$
$$= 23 \text{ m} + (-45 \text{ m})$$
$$= -22 \text{ m}$$

Assess: A negative displacement means a movement to the left, and Keira has moved left from $x = 23$ m to $x = -22$ m.

P1.9. Strategize: We must combine three different displacements to determine the total displacement.
Prepare: We have been given three different displacements. The problem is straightforward since all the displacements are along a straight east-west line. All we have to do is add the displacements and see where we end up.
Solve: The first displacement is $\Delta \vec{x}_1 = 500$ m east, the second is $\Delta \vec{x}_2 = 400$ m west and the third displacement is $\Delta \vec{x}_3 = 700$ m east. These three displacements are added in the figure below.

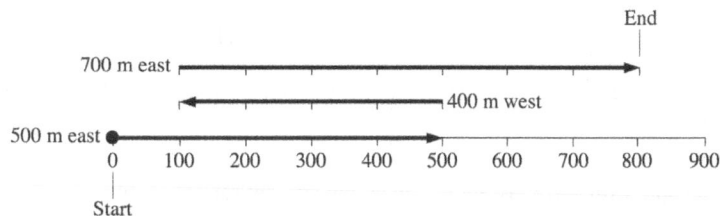

From the figure, note that the result of the sum of the three displacements puts the bee 800 m east of its starting point.
Assess: Knowing what a displacement is and how to add displacements, we are able to obtain the final position of the bee. Since the bee moved 1200 m to the east and 400 m to the west, it is reasonable that it would end up 800 m to the east of the starting point.

P1.11. Strategize: In all cases, objects are moving at a steady pace. So speed is just the distance divided by the time.

Prepare: We are asked to rank in order three different speeds, so we simply compute each one according to Equation 1.1:

$$\text{speed} = \frac{\text{distance traveled in a given time interval}}{\text{time interval}}$$

Solve: (i) Toy
$$\frac{0.15 \text{ m}}{2.5 \text{ s}} = 0.060 \text{ m/s}$$

(ii) Ball
$$\frac{2.3 \text{ m}}{0.55 \text{ s}} = 4.2 \text{ m/s}$$

(iii) Bicycle
$$\frac{0.60 \text{ m}}{0.075 \text{ s}} = 8.0 \text{ m/s}$$

(iv) Cat
$$\frac{8.0 \text{ m}}{2.0 \text{ s}} = 4.0 \text{ m/s}$$

So the order from fastest to slowest is bicycle, ball, cat, and toy car.

Assess: We reported all answers to two significant figures as we should according to the significant figure rules. The result is probably what we would have guessed before solving the problem, although the cat and ball are close. These numbers all seem reasonable for the respective objects.

P1.13. Strategize: Average velocity is defined as the displacement Δx divided by the time interval Δt.

Prepare: We are given $\Delta t = 35$ s, but we will do a preliminary calculation to find the displacement.

$$\Delta x = x_f - x_i = -47 \text{ m} - (-12 \text{ m}) = -35 \text{ m}$$

Solve:

$$v = \frac{\Delta x}{\Delta t} = \frac{-35 \text{ m}}{35 \text{ s}} = -1.0 \text{ m/s}$$

Assess: The answer is reasonable, and agrees with the approximate walking speed estimated in Example 1.5. The negative sign tells us that Harry is walking to the left.

P1.15. Strategize: We want to find the highest speed, meaning the highest value of: $\text{speed} = \dfrac{\Delta x}{\Delta t}$.

Prepare: Since we are told the times after intervals of 100 m, the Δx is the same over each interval. Only Δt changes. We need to find the shortest time interval, since that will correspond to the highest speed.

Solve: We find the duration of each of the four intervals:

$$\Delta t_1 = 11.20 \text{ s}$$
$$\Delta t_2 = (21.32 \text{ s}) - (11.20 \text{ s}) = 10.12 \text{ s}$$
$$\Delta t_3 = (31.76 \text{ s}) - (21.32 \text{ s}) = 10.44 \text{ s}$$
$$\Delta t_4 = (43.18 \text{ s}) - (31.76 \text{ s}) = 11.42 \text{ s}$$

(a) Clearly the second 100 m was done in the shortest amount of time. So the second 100 m was the fastest.

(b) $\text{speed} = \dfrac{\Delta x}{\Delta t} = \dfrac{100 \text{ m}}{10.12 \text{ s}} = 9.88 \text{ m/s}$

Assess: This is about 22 mph: extremely fast for a human, but still entirely plausible.

P1.17. Strategize: These are all simple unit conversions. SI units refer to the base units such as meters and seconds, with no scaling prefix. In particular, we do not want to use imperial units like feet or inches.

Prepare: We first collect the necessary conversion factors: 1 in = 2.54 cm; 1 cm = 10^{-2} m; 1 ft = 12 in; 39.37 in = 1 m; 1 mi = 1.609 km; 1 km = 10^3 m; 1 h = 3600 s.

Solve:

(a) $8.0 \text{ in} = 8.0 \text{ (in)} \left(\dfrac{2.54 \text{ cm}}{1 \text{ in}} \right) \left(\dfrac{10^{-2} \text{ m}}{1 \text{ cm}} \right) = 0.20 \text{ m}$

(b) $66 \text{ ft/s} = 66 \left(\dfrac{\text{ft}}{\text{s}} \right) \left(\dfrac{12 \text{ in}}{1 \text{ ft}} \right) \left(\dfrac{1 \text{ m}}{39.37 \text{ in}} \right) = 20 \text{ m/s}$

(c) $60 \text{ mph} = 60 \left(\dfrac{\text{mi}}{\text{h}} \right) \left(\dfrac{1.609 \text{ km}}{1 \text{ mi}} \right) \left(\dfrac{10^3 \text{ m}}{1 \text{ km}} \right) \left(\dfrac{1 \text{ h}}{3600 \text{ s}} \right) = 27 \text{ m/s}$

P1.19. Strategize: Review the rules for significant figures in Section 1.4 of the text.

Prepare: Pay particular attention to any zeros and whether or not they are significant. Note for example that the left-most zero in (c) is a place-holder and is not a significant digit, but the right-most zero in (c) specifies an additional digit of precision and is significant.

Solve: (a) The number 6.21 has three significant figures.

(b) The number 62.1 has three significant figures.

(c) The number 0.620 has three significant figures.

(d) The number 0.062 has two significant figures.

Assess: In part (c), the final zero is significant because it is expressed and in part (d), the second zero locates the decimal point but is not significant.

P1.21. Strategize: Review the rules for significant figures in Section 1.4 of the text.

Prepare: Pay particular attention to the rules for addition (and subtraction) and multiplication (and division).

Solve: (a) $33.3 \times 25.4 = 846$

(b) $33.3 - 25.4 = 7.9$

(c) $\sqrt{33.3} = 5.77$

(d) $333.3 \div 25.4 = 13.1$

Assess: In part (a) the two numbers multiplied each have three significant figures and the answer has three significant figures. In part (b), even though each number has three significant figures, no information is significant past the tenths column. As a result, the answer is expressed only to the tenths column. In part (c), the number and the answer both have three significant figures. In part (d) the answer is expressed to three significant figures since this is the least number of significant figures in either of the two numbers in the problem.

P1.23. Strategize: This is a simple unit conversion.

Prepare: To convert, we will use the fact that each foot contains exactly 12 inches, and each inch is 2.54 cm, or 0.0254 m.

Solve: We have: $29,029 \text{ ft} \left(\dfrac{12 \text{ in}}{1 \text{ ft}} \right) \left(\dfrac{2.54 \text{ cm}}{1 \text{ in}} \right) \left(\dfrac{1 \text{ m}}{100 \text{ cm}} \right) = 8.8480 \times 10^3 \text{ m}.$

Assess: Note that we have used exact conversions, such that there was no loss of significant digits in the conversion. The information was given to five significant digits, and we still have five significant digits.

P1.25. Strategize: This is an estimation problem. We will need to assemble estimates from life experience, and many answers are possible.

Prepare: My barber trims about an inch of hair when I visit him every month for a haircut. The rate of hair growth thus is one inch per month. We also need the conversions 1 in = 2.54 cm, 1 day = 24 h, 1 month = 30 days, 1 h = 3600 s, 1 cm = 10^{-2} m.

Solve: The rate of hair growth is

$$\left(\frac{1\,\text{in}}{\text{month}}\right)\left(\frac{2.54\,\text{cm}}{1\,\text{in}}\right)\left(\frac{10^{-2}\,\text{m}}{1\,\text{cm}}\right)\left(\frac{1\,\text{month}}{30\,\text{d}}\right)\left(\frac{1\,\text{d}}{24\,\text{h}}\right)\left(\frac{1\,\text{h}}{3600\,\text{s}}\right) = 9.8\times10^{-9}\ \text{m/s} \approx 35\ \mu\text{m/h}$$

Assess: Since we expect an extremely small number for the rate at which our hair grows per second, this figure is not unreasonable.

P1.27. Strategize: We are asked for the straight-line distance between two places. This is the same as the magnitude of the displacement between two locations, so this could be viewed as a displacement problem.
Prepare: In this problem we need to find the distance between Fort Collins and Greeley. We'll use the Pythagorean theorem.
Solve:

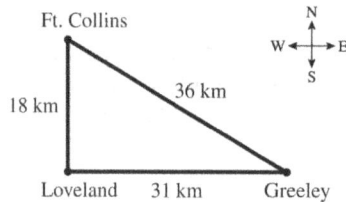

$$\sqrt{(18\ \text{km})^2 + (31\ \text{km})^2} = 36\ \text{km}$$

Assess: This is a reasonable distance between cities.

P1.29. Strategize: This problem is asking for the straight-line distance between initial and final points, given a number of smaller legs of the journey. A picture may be beneficial.
Prepare: We begin by sketching a drawing of the setup.

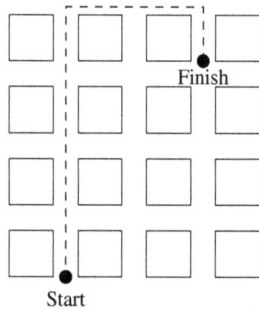

Clearly Veronica travels four blocks north, but then comes back one block south, such that she is finally three blocks north of her starting point. She also moves two blocks east. We can use the Pythagorean Theorem to combine these displacements in the east and north directions into one straight line distance.
Solve: The component of displacement eastward is $\Delta x = 2(400\ \text{ft}) = 800\ \text{ft}$, and the displacement northward is

$\Delta y = 3(280\ \text{ft}) = 840\ \text{ft}$. Then the total displacement is $d = \sqrt{\Delta x^2 + \Delta y^2} = \sqrt{(800\ \text{ft})^2 + (840\ \text{ft})^2} = 1.16\times10^3\ \text{ft}$.

Assess: This is a reasonable distance. One way of checking this would be to compare it to the distance required to walk two blocks east and then three blocks north, which is 1,640 ft. The straight-line distance is shorter, as it should be.

P1.31. Strategize: This is a displacement problem. The displacement is the difference between initial and final locations: $\Delta \vec{x} = \vec{x}_f - \vec{x}_i$. Because this is two-dimensional motion, the vector nature of the displacement is important; we must provide a magnitude and a direction.
Prepare: John's displacement is the vector from his starting point to his ending point.

Solve: Refer to the diagram below.

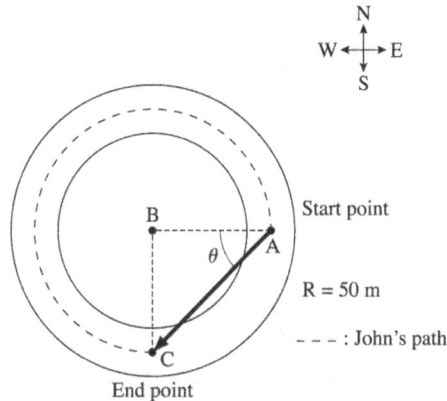

John stops at the southernmost end of the circle. His final position is 50 m west and 50 m south of his starting position since the radius of the circle is 50 m. We must find the displacement vector from the initial point to the final point. Point B is at the center of the circle.

The displacement vector has a length of

$$\overline{AC} = \sqrt{\overline{AB}^2 + \overline{BC}^2} = \sqrt{(50 \text{ m})^2 + (50 \text{ m})^2} = 71 \text{ m}$$

The angle in the diagram is

$$\theta = \arctan\left(\frac{1.0 \text{ m}}{1.0 \text{ m}}\right) = 45°$$

So the answer in (magnitude, direction) notation would be (71 m, 45° south of west).

Assess: Compare the solution to the solution for Problem 1.27. Here it would be difficult to sum his displacement vectors along the circle, but this is not necessary since the displacement vector is always the vector from the initial position to the final position.

P1.33. Strategize: We must consider two separate displacements and use them to find a total displacement.

Prepare: Knowing that the total trip consists of two displacements, we can add the two displacements to determine the total displacement and hence the distance of the goose from its original position. A quick sketch will help you visualize the two displacements and the total displacement.

Solve: The distance of the goose from its original position is the magnitude of the total displacement vector. This is determined as follows:

$$d = \sqrt{(32 \text{ km})^2 + (20 \text{ km})^2} = 38 \text{ km}$$

Assess: A quick look at your sketch shows that the total distance should be larger than the largest leg of the trip and this is the case.

P1.35. Strategize: This problem involves motion in two orthogonal directions: vertical and horizontal. We can use trigonometry to relate the distances and the given angle.

Prepare: It is helpful to draw a diagram.

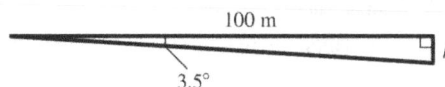

Solve: Use knowledge about right triangles from trigonometry.

$$\tan 3.5° = \frac{h}{100 \text{ m}} \Rightarrow h = (100 \text{ m})(\tan 3.5°) = 6.1 \text{ m}$$

The vulture loses 6.1 m of height as it flies 100 m horizontally.

Assess: This small loss of height would indeed allow a vulture to glide long distances.

P1.37. Strategize: We must use simple trigonometry to relate the height and distance along the slope of the Great Pyramid.

Prepare: We begin by making a sketch of the geometry described:

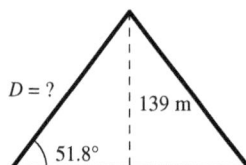

Clearly, we can relate the distance along the sloped side to the height by using the sine trigonometric function.

Solve: We know $\sin(\theta) = \frac{\text{opposite}}{\text{adjacent}} = \frac{h}{D}$, so $D = \frac{h}{\sin(\theta)} = \frac{139 \text{ m}}{\sin(51.8°)} = 177 \text{ m}$.

Assess: We know that the distance along the slope should be greater than the height, and we know from the given angle that it should not very much greater. Our answer of 177 m makes sense.

P1.39. Strategize: In this problem we have three displacements to add using the laws of vector addition.

Prepare: We will use the diagram below.

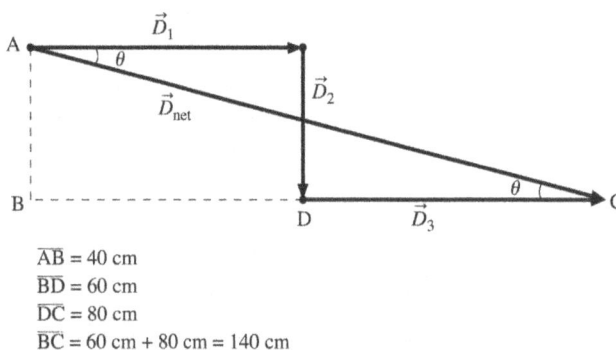

$\overline{AB} = 40$ cm
$\overline{BD} = 60$ cm
$\overline{DC} = 80$ cm
$\overline{BC} = 60$ cm $+ 80$ cm $= 140$ cm

Solve: A convenient place to place the origin is the origin of the motion. We could add the first two vectors and then add the third vector on to that result. However, by looking at the diagram, the total displacement in the x direction is $60 \text{ cm} + 80 \text{ cm} = 140 \text{ cm}$ to the right. The total displacement in the y direction is 40 cm downward. The net displacement will reflect these two displacements, so the result of the vector addition will be a vector pointing 140 cm to the right and 40 cm downward. Considering the right triangle ABC in the diagram, the magnitude of the displacement vector is then

$$\overline{AC} = \sqrt{\overline{AB}^2 + \overline{BC}^2} = \sqrt{(140 \text{ cm})^2 + (40 \text{ cm})^2} = 150 \text{ cm}$$

Where we have assumed that the measurements in the problem have been given to two significant figures.

The angle that the vector makes is $\theta = \arctan\left(\frac{40 \text{ cm}}{140 \text{ cm}}\right) = 16°$ below the positive x-axis.

Assess: When adding any number of displacements, the net displacement is always the vector between the initial and final point of the motion.

P1.41. Strategize: Because this problem involves motion in two orthogonal directions, it will involve a right triangle. We can use trigonometry to relate given distances and angles.
Prepare: Draw a diagram of the situation. Note the right triangle.

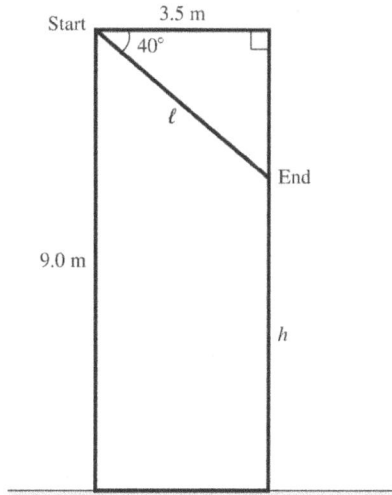

Solve: Use knowledge about right triangles from trigonometry.
(a) The length of the path is

$$\ell = \frac{3.5 \text{ m}}{\cos 40°} = 4.6 \text{ m}$$

(b) The height h above the ground is the vertical side of the triangle subtracted from 9.0 m.

$$h = 9.0 \text{ m} - (3.5 \text{ m})(\tan 40°) = 6.1 \text{ m}$$

Assess: Because of the angle we expect the squirrel to lose less height than the horizontal distance traveled.

P1.43. We interpret the described motion in a motion diagram. We assume equal time intervals between dots, and we draw velocity vectors with their length proportional to the dog's speed.
Prepare: Since the dog accelerates for 30 m, the position dots will be successively farther apart and the velocity vectors will increase for this part of the race. After the dog reaches its top speed (at 30 m), the position dots are uniformly spaced and the velocity vectors are all the same length.
Solve:

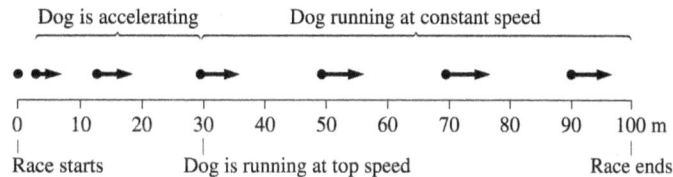

Assess: While the dog is accelerating, the dot spacing and the velocity vector must increase, and they do. While the dog is traveling at a constant speed the dot spacing and the velocity vector must remain the same, and they do.

P1.45. Strategize: We will draw the motion diagram with velocity vectors initially pointing down an incline and then leveling off to travel over flat ground.
Prepare: The length of velocity arrows and spacing between dots will increase as the skater rolls down the incline. Once the skater reaches the end of the incline, the skater will move at a constant speed. So on level ground the arrows should be a constant length and the dots should be equally-spaced.

Solve:

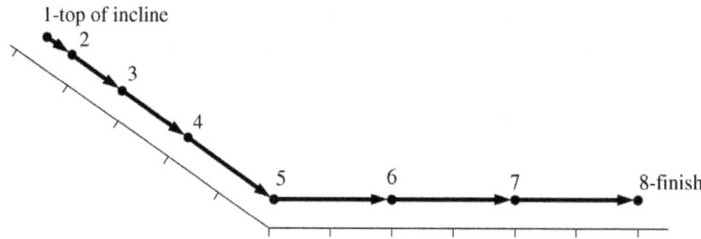

Assess: We expect the speed to increase as the skater moves down the incline, and then remain steady. This is reflected in the diagram above.

P1.47. Strategize: We interpret the described motion in a motion diagram. We assume equal time intervals between dots, and we draw velocity vectors with their length proportional to the eland's speed.
Prepare: Since the eland has a positive velocity but is slowing down, the velocity will decrease to zero and the spacing between the position dots will decrease. The velocity vector at each position on the way up has the same magnitude but opposite direction as the velocity at each position on the way down.
Solve:

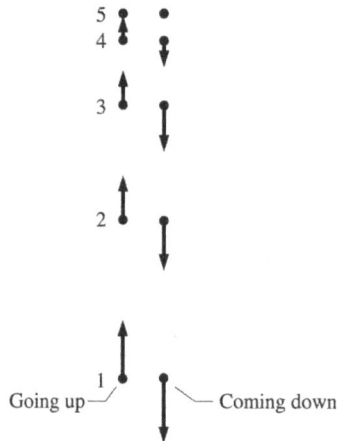

Assess: On the way up, the velocity vector decreases to zero as it should and the spacing between the position dots decreases as it should. The magnitude of the velocity vector at any position is the same on the way up as it is on the way down. This allows us to conclude that the figure is correct.

P1.49. Strategize: We interpret the described motion in a motion diagram. We assume equal time intervals between dots, and we draw velocity vectors with their length proportional to the car's speed.
Prepare: The motorist, represented as a particle, is moving along the x-axis. He slows down during braking. During his reaction time his velocity doesn't change.
Solve:

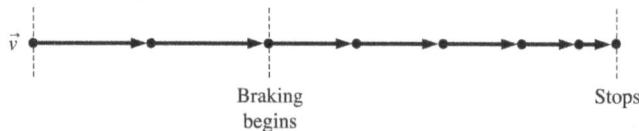

Assess: During the reaction time the velocity vector is constant in length and the position dots are uniformly spaced. During the braking process, the velocity vector decreases in length and the position dots get closer together.

P1.51. Strategize: Note any changes in velocity vectors, and make a physically plausible explanation for the changes.
Prepare: Keep in mind that the dots represent the position of an object at equal time intervals and the vectors represent the velocity of the object at these times.
Solve: Reema passes 3rd Street doing 40 mph, slows steadily to the stop sign at 4th Street, stops for 1 s, then speeds up and reaches her original speed as she passes 5th Street. If the blocks are 50 m long, how long does it take Reema to drive from 3rd Street to 5th Street?
Assess: The statement that Reema slows to a stop in one block and regains her initial velocity in one block is consistent with the symmetry of the position dots and the velocity vectors about the stop position.

P1.53. Strategize: Note any changes in velocity vectors, and make a physically plausible explanation for the changes.
Prepare: Keep in mind that the dots represent the position of an object at equal time intervals and the vectors represent the velocity of the object at these times. We can construct a situation to match the motion diagram.
Solve: A bowling ball is at rest at the top of an incline. You nudge the ball giving it an initial velocity and causing it to roll down an incline. At the bottom of the incline it bounces off a sponge and travels back up the incline until it stops.
Assess: The statement that you give the ball an initial velocity is consistent with the fact that the start position dot has a velocity vector. The statement that the ball rolls down the incline is consistent with the fact that the dots are getting farther apart and the velocity vectors are increasing in length. The statement that the ball bounces off a sponge is consistent with the fact that ball does not bounce back to its original position.

P1.55. Strategize: This is an estimation problem, so many different answers are possible.
Prepare: We must first estimate a human lifespan, and then do a series of unit conversions. A human lifespan depends heavily on where that human lives and on risk factors. For humans with ready access to medicines and without serious risk factors, at lifespan of 80 years is reasonable. We use this as our estimate.
Solve: Converting 80 years into units of seconds, we find

$$80 \text{ yr} \left(\frac{365 \text{ days}}{1 \text{ yr}} \right) \left(\frac{24 \text{ hrs}}{1 \text{ day}} \right) \left(\frac{3600 \text{ s}}{1 \text{ hr}} \right) = 2.5 \times 10^9 \text{ s}$$

Assess: We expect an extremely large number, since there are already thousands of seconds in just a single hour. This answer is reasonable.

P1.57. Strategize: We will use the relationship $v_{av} = \dfrac{d}{\Delta t}$ to determine the new speed required.

Prepare: We know that the distance between Evan and his grandmother's house does not change. But his speed and the duration of the trip will both be different today than on a normal day. Thus we will write out $v_{av,norm} = \dfrac{d}{\Delta t_{norm}}$

and $v_{av,today} = \dfrac{d}{\Delta t_{today}}$ separately and then relate the two.

Solve: From the information about ordinary days, we know

$$d = v_{av,norm} \Delta t_{norm} = \left(55 \text{ mi/hr} \right) \left(25 \text{ min} \right) \left(\frac{1 \text{ hr}}{60 \text{ min}} \right) = 22.9 \text{ mi}$$

Now we can use this distance, and the fact that Evan must make the trip in 5 minutes less than usual to write

$$v_{av,today} = \frac{d}{\Delta t_{today}} = \frac{\left(22.9 \text{ mi} \right)}{\left(20 \text{ min} \right)} \left(\frac{60 \text{ min}}{1 \text{ hr}} \right) = 69 \text{ mph.}$$

Assess: Since Evan has to make the trip in 25% less time than usual, it makes perfect sense that his average speed must be 25% greater than usual.

P1.59. Strategize: We will use the expression for average speed $v_{av} = \dfrac{d}{\Delta t}$ and apply it to each leg of the swim (against the current, and with the current).

Prepare: In order to calculate anything about the time, we must consider how the required time will be affected when we swim with the current: $\Delta t_{with} = \dfrac{d}{v + v_w}$, and against the current: $\Delta t_{against} = \dfrac{d}{v - v_w}$. The distance is the same either way. But we also know that we want to compare the total time we get from this to the time required in still water: $\Delta t_{still} = \dfrac{2d}{v}$. Here the factor of two came from the fact that we don't need to consider two separate legs of the swim in this case. We can just write down the total two-way time.

Solve: We want to compare $\Delta t_{with} + \Delta t_{against}$ to Δt_{still}, so let us calculate the ratio

$$\frac{\Delta t_{with} + \Delta t_{against}}{\Delta t_{still}} = \frac{\dfrac{d}{v + v_w} + \dfrac{d}{v - v_w}}{2d/v} = \frac{1}{2}\left(\frac{1}{1 + v_w/v} + \frac{1}{1 - v_w/v}\right)$$

$$= \frac{1}{2}\left(\frac{1}{1 + (0.52\text{ m/s})/(1.78\text{ m/s})} + \frac{1}{1 - (0.52\text{ m/s})/(1.78\text{ m/s})}\right) = 1.093$$

This means the trip takes 9.3% longer taking the current into account, than it would in still water.

Assess: This is a reasonable fractional difference, given that the speed of the water is significantly smaller than the swimming speed in still water.

P1.61. Strategize:

Prepare: Knowing that speed is distance divided by time, the distance is the circumference of a circle of radius 93,000,000 miles and the time is one year, we can determine the speed of the earth orbiting the sun. In order to get an answer in m/s, some unit conversion will be required.

Solve: The speed of the earth in its orbit about the sun may be determined by

$$v = \frac{\text{distance}}{\text{time}} = \frac{2\pi r}{t} = \frac{2\pi(9.3 \times 10^7\text{ mi})}{1\text{ yr}}\left(\frac{1\text{ yr}}{3.16 \times 10^7\text{ s}}\right)\left(\frac{1.61 \times 10^3\text{ m}}{1\text{ mi}}\right) = 3.0 \times 10^4\text{ m/s}$$

Assess: All of the unit conversions are correct, after units are canceled we obtain the desired units (m/s) and we are expecting a large number.

P1.63. Strategize: We will use the equation $v_{av} = \dfrac{d}{\Delta t}$ over any period when the average speed is given.

Prepare: We can apply the above expression to the uphill segment, the downhill segment.

$$v_{av,uphill} = \frac{d_{uphill}}{\Delta t_{uphill}}, \qquad v_{av,downhill} = \frac{d_{downhill}}{\Delta t_{downhill}}, \text{ and we also note } \Delta t_{total} = \Delta t_{downhill} + \Delta t_{uphill}$$

We know both distances, the total time and the average speed for the uphill leg. We can relate the speeds, distances and times for each segment and solve for the unknown average speed on the downhill leg.

Solve: Starting with $\Delta t_{total} = \Delta t_{downhill} + \Delta t_{uphill}$, and inserting $\Delta t_{uphill} = \dfrac{d_{uphill}}{v_{av,uphill}}$ we can obtain

$$\Delta t_{downhill} = \Delta t_{total} - \frac{d_{uphill}}{v_{av,uphill}} = (2{,}845\text{ s}) - \frac{(4.6\text{ mi})}{(8.75\text{ mi/hr})}\left(\frac{3600\text{ s}}{1\text{ hr}}\right) = 952\text{ s}$$

Now that we know the time, we simply use

$$v_{av,downhill} = \frac{d_{downhill}}{\Delta t_{downhill}} = \frac{6.9\text{ mi}}{952\text{ s}}\left(\frac{3600\text{ s}}{1\text{ hr}}\right) = 26\text{ mph}$$

Assess: This is much faster than the uphill leg of the trip, which is to be expected.

P1.65. Strategize: Distance is the quantity measured by the odometer and depends on the path taken. The displacement is the difference between two locations, and is independent of the path taken between the two endpoints. The speed is the distance divided by the time interval.
Prepare: Knowing that speed is distance divided by time, the distance is the speed multiplied by the time. 15 min = ¼ h.
Solve:
(a) The distance traveled during the ¼ hour is

$$\text{distance} = \text{speed} \times \text{time} = (100 \text{ km/h})(0.25 \text{ h}) = 25 \text{ km}$$

(b) Since the circumference of the track is 12.5 km, then the car goes completely around the track exactly twice in covering 25 km. Hence, the displacement from the initial position is 0 km.
(c) The speed of the car is

$$v = \left(100 \frac{\text{km}}{\text{h}}\right)\left(\frac{1000 \text{ m}}{1 \text{ km}}\right)\left(\frac{1 \text{ h}}{60 \text{ min}}\right)\left(\frac{1 \text{ min}}{60 \text{ s}}\right) = 28 \text{ m/s}$$

Assess: 28 m/s seems like a reasonable speed for a fast car.

P1.67. Strategize: This is a straightforward application of the definition of speed.
Prepare: Knowing the distance and time to travel that distance, we can determine the speed. Since we lack detailed information about the flight, and knowing that the flight is made by a bird, it is conceivable that its actual path darted back and forth and maybe even backtracked at times. If that is the case its actual distance of travel and hence average speed would be larger.
Solve: **(a)** The minimum average speed of the albatross may be determined by

$$v = \frac{\text{distance}}{\text{time}} = \left(\frac{1.2 \times 10^3 \text{ km}}{1.4 \text{ day}}\right)\left(\frac{1 \text{ day}}{24 \text{ hr}}\right)\left(\frac{0.621 \text{ mi}}{1 \text{ km}}\right) = 22 \text{ mph}$$

(b) The average speed of the bird is 22 mph if it flies in a straight line between the end points. If the bird deviates from this line, the average speed will have to be greater than 22 mph in order to have the 1200 km displacement in 1.4 days.
Assess: You can ride a bike 10 to 15 mph and while you are riding your bike, birds easily fly past you. In light of that this is a reasonable answer.

P1.69. Strategize: This problem involves displacement, which is the difference between initial and final positions. We also distinguish between speed and velocity. Since velocity has a direction associated with it, in order for two segments to describe the same velocity, they would need to be the same length and the same direction. For them to describe the same speed, they only need to have the same length (since time intervals are fixed).
Prepare: Assume that the bacterium moves along the path to consecutive letters.
Solve: **(a)** The displacements in segments AB and CD are the same (five right and one up). No other pairs appear to be the same.
(b) The problem explicitly stated that the bacteria move at a constant speed, so the answer is all of the segments.
(c) Since the displacements in segments AB and CD are the same (and the bacterium had the same speed in both segments, i.e., Δt is the same for both segments), then the velocity is the same in those two segments. Since no other pairs of segments have the same direction, they can't have the same velocity.
Assess: Remember that both displacement and velocity are vectors, so the direction matters. Only the length (magnitude) matters with speed since it is a scalar.

P1.71. Strategize: We can relate the distances given to the angle using trigonometry.
Prepare: Draw a diagram of the situation. Note the right triangle.

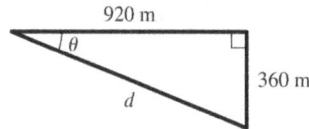

Solve: Use knowledge about right triangles from trigonometry.
(a) The angle of the path below the horizontal is

$$\theta = \arctan\left(\frac{360 \text{ m}}{920 \text{ m}}\right) = 21°$$

(b) The distance d covered is

$$d = \sqrt{(360 \text{ m})^2 + (920 \text{ m})^2} = 988 \text{ m}$$

which should be reported as 990 m to two significant figures. We keep one more significant figure to use in the next step.
(c) The seal's speed is

$$v = \frac{\text{distance}}{\text{time}} = \left(\frac{988 \text{ m}}{4.0 \text{ min}}\right)\left(\frac{1 \text{ min}}{60 \text{ s}}\right) = 4.1 \text{ m/s}$$

Assess: Because the seal descended less than the horizontal distance we expect the angle to be less than 45 degrees.

P1.73. Strategize: We can relate the given distance and angle to the required distance using trigonometry.
Prepare: Draw a diagram of the situation. Note the right triangle.

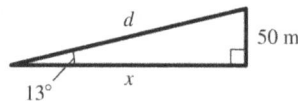

Solve: Use knowledge about right triangles from trigonometry.
(a) The horizontal distance x is

$$x = \frac{50 \text{ m}}{\tan 13°} = 216.6 \text{ m}$$

This should be reported as 220 m to two significant figures.
(b) The distance d covered is

$$d = \sqrt{(50 \text{ m})^2 + (216.6 \text{ m})^2} = 222.3 \text{ m}$$

which should be reported as 220 m to two significant figures. We keep two more significant figures to use in the next step.
(c) The time it takes the shark is

$$\text{time} = \frac{\text{distance}}{\text{speed}} = \left(\frac{222.3 \text{ m}}{0.85 \text{ m/s}}\right) = 260 \text{ s}$$

Assess: 260 s is just over 4 min, which is impressive but reasonable.

P1.75. Strategize: We can relate the speed to the displacement, easily. The two legs of the displacement are in orthogonal directions, such that we can use the Pythagorean Theorem.
Prepare: Since during part of the motion John is traveling north and then turns east, the rules of vector addition will be used to determine the net displacement.

Solve: **(a)** The vectors form a right triangle. See the following vector diagram.

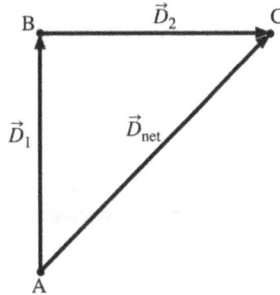

The length of the net displacement vector is

$$\overline{AC} = \sqrt{\overline{AB}^2 + \overline{BC}^2} = \sqrt{(1.00 \text{ km})^2 + (1.00 \text{ km})^2} = 1.41 \text{ km}$$

(b) Jane walks along John's net displacement vector, so she only travels 1.41 km, while John travels a total distance of 2.00 km. Since he travels at 1.50 m/s during the entire stroll, the time John takes to get to his destination is

$$\Delta t_{\text{John}} = \frac{2000 \text{ m}}{1.50 \text{ m/s}} = 1.33 \times 10^3 \text{s}$$

For Jane to walk 1.41 km in this time, her velocity would need to be

$$v_{\text{Jane}} = \frac{1410 \text{ m}}{1.33 \times 10^3 \text{ s}} = 1.06 \text{ m/s}$$

Assess: Jane must walk slower than John to walk the shorter distance in the same time, so the answer makes sense. For displacements in different directions you must use the law of vector addition.

P1.77. Strategize: This is a simple unit conversion problem.
Prepare: We must convert ft/yr to m/s.
Solve: The tree grows at 9 ft/yr, so

$$9\frac{\text{ft}}{\text{yr}} = \left(\frac{9 \text{ ft}}{\text{yr}}\right)\left(\frac{0.305 \text{ m}}{1 \text{ ft}}\right)\left(\frac{1 \text{ yr}}{365 \text{ days}}\right)\left(\frac{1 \text{ day}}{24 \text{ h}}\right)\left(\frac{1 \text{ h}}{60 \text{ min}}\right)\left(\frac{1 \text{ min}}{60 \text{ s}}\right) = 9 \times 10^{-8} \text{ m/s}$$

The correct answer is A.
Assess: Use the method of multiplying by one to keep track of conversion factors.

Q2.1. Reason: The elevator must speed up from rest to cruising velocity. In the middle will be a period of constant velocity, and at the end a period of slowing to a rest.

The graph must match this description. The value of the velocity is zero at the beginning, then it increases, then, during the time interval when the velocity is constant, the graph will be a horizontal line. Near the end the graph will decrease and end at zero.

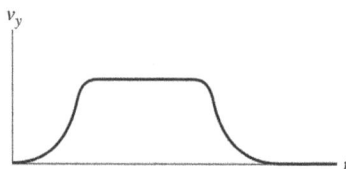

Assess: After drawing velocity-versus-time graphs (as well as others), stop and think if it matches the physical situation, especially by checking end points, maximum values, places where the slope is zero, etc. This one passes those tests.

Q2.3. Reason: Where the rings are far apart the tree is growing rapidly. It appears that the rings are quite far apart near the center (the origin of the graph), then get closer together, then farther apart again.

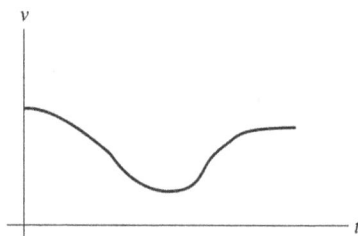

Assess: After drawing velocity-versus-time graphs (as well as others), stop and think if it matches the physical situation, especially by checking end points, maximum values, places where the slope is zero, etc. This one passes those tests.

Q2.5. Reason: Let $t_0 = 0$ be when you pass the origin. The other car will pass the origin at a later time t_1 and passes you at time t_2.

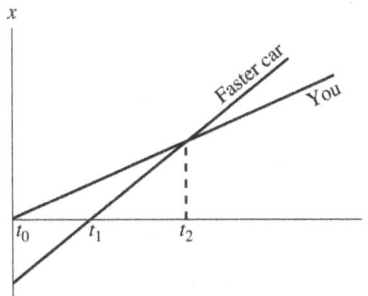

Assess: The slope of the position graph is the velocity, and the slope for the faster car is steeper.

Q2.7. Reason: A predator capable of running at a great speed while not being capable of large accelerations could overtake slower prey that were capable of large accelerations, given enough time. However, it may not be as effective as surprising and grabbing prey that are capable of higher acceleration. For example, prey could escape if the safety of a burrow were nearby. If a predator were capable of larger accelerations than its prey, while being slower in speed than the prey, it would have a greater chance of surprising and grabbing prey, quickly, though prey might outrun it if given enough warning.
Assess: Consider the horse-man race discussed in the text.

Q2.9. Reason: Consider the ball thrown upward. The path from Janelle's hand to its peak is symmetric to the path back from the peak to Janelle's hand. That means that whatever the initial upward speed was, when the ball returns and passes Janelle on its way down, it will have that same speed, just directed downward now. From that moment on, the trip down to Michael is exactly the same as for the ball thrown downward. Thus the two balls will be moving at the same speed when they reach Michael.
Assess: The ball initially thrown downward will certainly reach Michael first. But we are not asked about the time required or the average velocity. We are only asked about the speed at the moment the balls reach Michael. So this makes sense.

Q2.11. Reason: There are five different segments of the motion, since the lines on the position-versus-time graph have different slopes between five different time periods.
(a) A fencer is initially still. To avoid his opponent's lunge, the fencer jumps backwards very quickly. He remains still for a few seconds. The fencer then begins to advance slowly on his opponent.
(b) Referring to the velocities obtained in part (a), the velocity-versus-time graph would look like the following diagram.

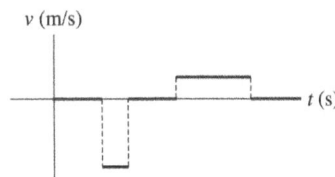

Assess: Velocity is given by the slope of lines on position-versus-time graphs. See Conceptual Example 2.1 and the discussion that follows.

Q2.13. Reason: (a) D. The steepness of the tangent line is greatest at D.
(b) C, D, E. Motion to the left is indicated by a decreasing segment on the graph.
(c) C. The speed corresponds to the steepness of the tangent line, so the question can be re-cast as "Where is the tangent line getting steeper (either positive or negative slope, but getting steeper)?" The slope at B is zero and is greatest at D, so it must be getting steeper at C.
(d) A, E. The speed corresponds to the steepness of the tangent line, so the question can be re-cast as "Where is the tangent line getting less steep (either positive or negative slope, but getting less steep)?"
(e) B. Before B the object is moving right and after B it is moving left.
Assess: It is amazing that we can get so much information about the velocity (and even about the acceleration) from a position-versus-time graph. Think about this carefully. Notice also that the object is at rest (to the left of the origin) at point F.

Q2.15. Reason: This graph shows a curved position-versus-time line. Since the graph is curved the motion is *not* uniform. The instantaneous velocity, or the velocity at any given instant of time, is the slope of a line tangent to the graph at that point in time. Consider the graph below, where tangents have been drawn at each labeled time.

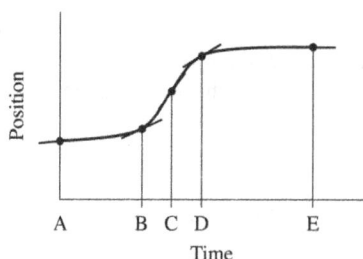

Comparing the slope of the tangents at each time in the figure above, the speed of the car is greatest at time C.
Assess: Instantaneous velocity is given by the slope of a line tangent to a position-versus-time curve at a given instant of time. This is also demonstrated in Conceptual Example 2.4.

Q2.17. Reason: The velocity of an object is given by the physical slope of the line on the position-versus-time graph. Since the graph has constant slope, the velocity is constant. We can calculate the slope by using Equation 2.1, choosing any two points on the line since the velocity is constant. In particular, at $t = 0$ s the position is $x = 5$ m. At time $t = 3$ s the position is $x = 15$ m. The points on the line can be read to two significant figures.
The velocity is

$$v = \frac{\Delta x}{\Delta t} = \frac{x_2 - x_1}{t_2 - t_1} = \frac{15 \text{ m} - 5 \text{ m}}{3 \text{ s} - 0 \text{ s}} = \frac{10 \text{ m}}{3 \text{ s}} = +3.3 \text{ m/s}$$

The correct choice is C.
Assess: Since the slope is positive, the value of the position is increasing with time, as can be seen from the graph.

Q2.19. Reason: The initial velocity is 20 m/s. Since the car comes to a stop, the final velocity is 0 m/s. We are given the acceleration of the car, and need to find the stopping distance. See the pictorial representation, which includes a list of values below.

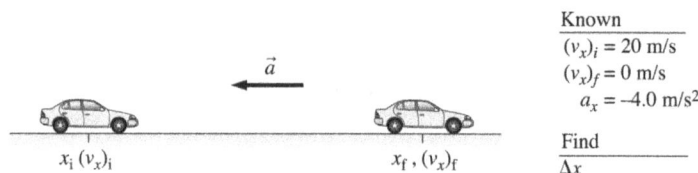

An equation that relates acceleration, initial velocity, final velocity, and distance is Equation 2.13.

$$(v_x)_f^2 = (v_x)_i^2 + 2a_x \Delta x$$

Solving for Δx,

$$\Delta x = \frac{(v_x)_f^2 - (v_x)_i^2}{2a_x} = \frac{(0 \text{ m/s})^2 - (20 \text{ m/s})^2}{2(-4.0 \text{ m/s}^2)} = 50 \text{ m}$$

The correct choice is D.

Assess: We are given initial and final velocities and acceleration. We are asked to find a displacement, so Equation 2.13 is an appropriate equation to use.

Q2.21. Reason: The slope of the tangent to the velocity-versus-time graph gives the acceleration of each car. At time $t = 0$ s the slope of the tangent to Andy's velocity-versus-time graph is very small. The slope of the tangent to the graph at the same time for Carl is larger. However, the slope of the tangent in Betty's case is the largest of the three. So Betty had the greatest acceleration at $t = 0$ s. See the figure below.

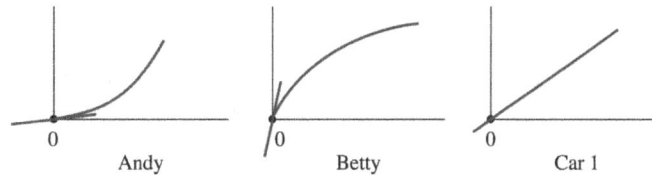

The correct choice is B.

Assess: Acceleration is given by the slope of the tangent to the curve in a velocity-versus-time graph at a given time.

Q2.23. Reason: There are two ways to approach this problem, and both are educational. Using algebra, first calculate the acceleration of the larger plane.

$$a = \frac{\Delta v}{\Delta t} = \frac{80 \text{ m/s}}{30 \text{ s}} = 2.667 \text{ m/s}^2$$

Then use that acceleration to figure how far the smaller plane goes before reading 40 m/s.

$$(v_x)_f^2 = (v_x)_i^2 + 2a_x\Delta x \Rightarrow \Delta x = \frac{(v_x)_f^2 - (v_x)_i^2}{2a_x} = \frac{(40 \text{ m/s})^2}{2(2.667 \text{ m/s}^2)} = 300 \text{ m}$$

So choice A is correct.

The second method is graphical. Make a velocity vs. time graph; the slope of the straight line is the same for both planes. We see that the smaller plane reaches 40 m/s in half the time that the larger plane took to reach 80 m/s. And we see that the area under the smaller triangle is ¼ the area under the larger triangle. Since the area under the velocity vs. time graph is the distance then the distance the small plane needs is ¼ the distance the large plane needs.

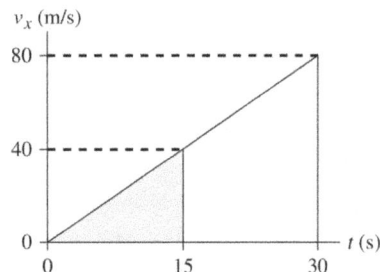

Assess: It seems reasonable that a smaller plane would need only ¼ the distance to take off as a large plane.

Q2.25. Reason: Let us call the vertically upward direction $+y$. We can determine the initial velocity by using Equation 2.13, with the minor adjustment of changing the subscripts from x to y. Then we have

$$\left(v_y\right)_f^2 = \left(v_y\right)_i^2 + 2a_y\Delta y \Rightarrow \left(v_y\right)_i = \pm\sqrt{\left(v_y\right)_f^2 - 2a_y\Delta y} = \pm\sqrt{(2.8\ \text{m/s}) - 2(-9.8\ \text{m/s})(3.8\ \text{m})} = \pm9.1\ \text{m/s}$$

Clearly, since the ball was thrown upward, we want the positive answer. So the correct answer is C.
Assess: It makes sense that the initial upward component of velocity would have to be greater in magnitude than the final.

Q2.27. Reason: By definition the acceleration in the x direction is the rate of change of the velocity in the x direction. This means

$$a_x = \Delta v_x / \Delta t = \frac{\left(v_x\right)_f - \left(v_x\right)_i}{t_f - t_i} = \frac{(15\ \text{m/s}) - (5\ \text{m/s})}{(4\ \text{s}) - (0\ \text{s})} = 2.5\ \text{m/s}^2.$$

The correct answer is B.

Assess: Since the slope is positive, we expect a positive acceleration. The magnitude is also reasonable for the components of velocity given.

Problems

P2.1. Strategize: The dots represent positions after fixed time intervals.
Prepare: The car is traveling to the left toward the origin, so its position decreases with increase in time.
Solve: (a)

Time t (s)	Position x (m)
0	1200
1	975
2	825
3	750
4	700
5	650
6	600
7	500
8	300
9	0

(b)

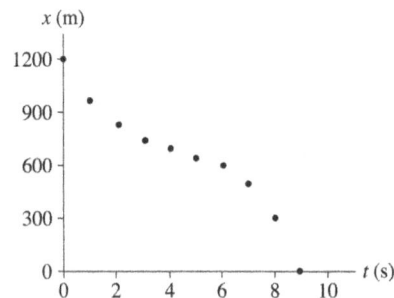

Assess: A car's motion traveling down a street can be represented at least three ways: a motion diagram, position-versus-time data presented in a table (part (a)), and a position-versus-time graph (part (b)).

P2.3. Strategize: This is a position vs. time graph, so the x component of velocity is given by the slope.
Prepare: The position graph has a shallow (negative) slope for the first 8 s, and then the slope increases.
Solve:
(a) The change in slope comes at 8 s, so that is how long the dog moved at the slower speed.
(b)

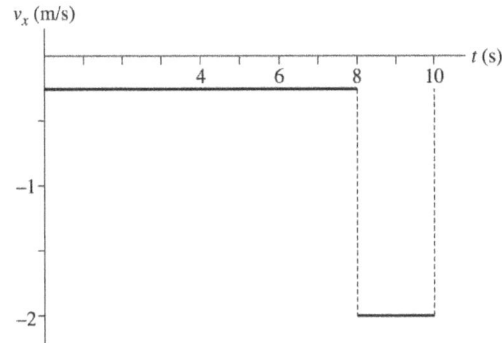

Assess: We expect the sneaking up phase to be longer than the spring phase, so this looks like a realistic situation.

P2.5. Strategize: We want to produce a position vs. time plot in which the slope of x vs. t yields the v_x vs. t plot we are given.
Prepare: To get a position from a velocity graph we count the area under the curve.
Solve:
(a)

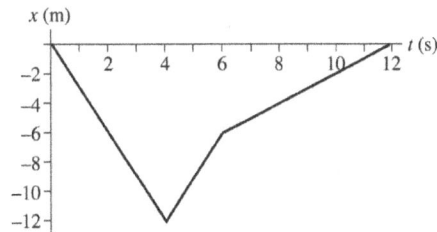

(b) We need to count the area under the velocity graph (area below the x-axis is subtracted). There are 12 m of area below the axis and 12 m of area above. $12 \text{ m} - 12 \text{ m} = 0 \text{ m}$.

(c) A football player runs left at 3 m/s for 4 s, then cuts back to the right at 3 m/s for 2 s, then walks (continuing to the right) back to the starting position.
Assess: We note an abrupt change of velocity from 3 m/s left to 3 m/s right at 4 s. It is also important that the problem state what the position is at $t = 0$, or we wouldn't know how high to draw the position graph.

P2.7. Strategize: We want to indicate position relative to the ground on the vertical axis of one plot, and time on the horizontal axis.
Prepare: We assume the speed is roughly constant when the elevator is moving. We start with the position plot, and then we can determine the components of the velocity from the slope.
Solve: The entire trip takes 24 s and 10 s are spent stopped. So the motion takes a total of 14 s. If we assume the elevator has the same speed when going upward as it does downward, then the total distance of 7 floors (5 up and then 2 down) corresponds to 2 s per floor. Thus, we expect it took 10 s to get up to the fifth floor, and then 4 s to go back down to the third floor. Clearly, the slope of this plot is either ± 2 m/s or 0 m/s. This allows us to complete the velocity vs. time plot as well.

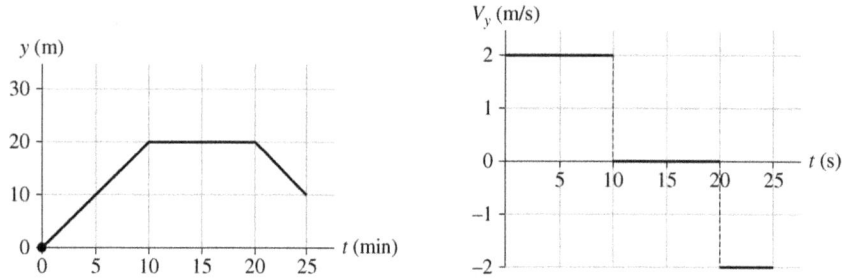

Assess: Note that the sign of the velocity is accurately reflected in the slope of the position vs. time plot.

P2.9. Strategize: Ignoring air resistance, the horizontal component of the velocity should be constant.
Prepare: Assume that the ball travels in a horizontal line at a constant v_x. It doesn't really, but if it is a line drive then it is a fair approximation.

$$\text{time} = \frac{\text{distance}}{\text{speed}} = \frac{60\ \text{ft}}{95\ \frac{\text{mi}}{\text{h}}} \left(\frac{1\ \text{mi}}{5280\ \text{ft}} \right) \left(\frac{60\ \text{min}}{1\ \text{h}} \right) \left(\frac{60\ \text{s}}{1\ \text{min}} \right) = 0.43\ \text{s}$$

Assess: Just under a half second is reasonable for a major league pitch.

P2.11. Strategize: We assume both drivers maintain a steady speed.
Prepare: A visual overview of Alan's and Beth's motion that includes a pictorial representation, a motion diagram, and a list of values is shown below. Our strategy is to calculate and compare Alan's and Beth's time of travel from Los Angeles to San Francisco.

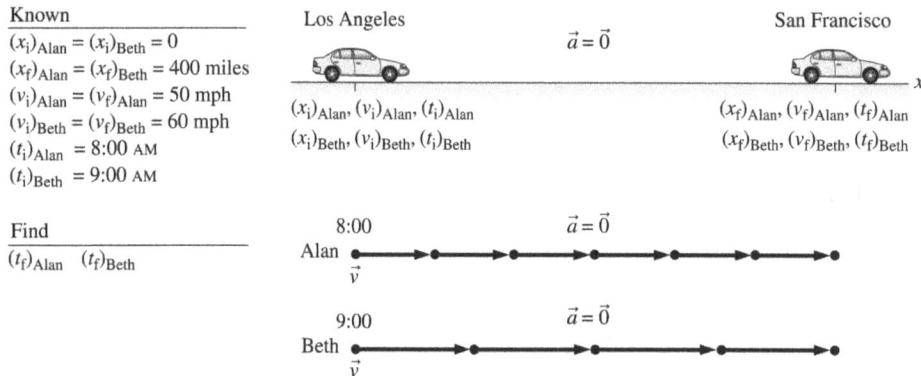

Solve: Beth and Alan are moving at a constant speed, so we can calculate the time of arrival as follows:

$$v = \frac{\Delta x}{\Delta t} = \frac{x_f - x_i}{t_f - t_i} \Rightarrow t_f = t_i + \frac{x_f - x_i}{v}$$

Using the known values identified in the pictorial representation, we find

$$\left(t_f \right)_{\text{Alan}} = \left(t_i \right)_{\text{Alan}} + \frac{\left(x_f \right)_{\text{Alan}} - \left(x_i \right)_{\text{Alan}}}{v} = 8{:}00\ \text{AM} + \frac{400\ \text{mile}}{50\ \text{miles/hour}} = 8{:}00\ \text{AM} + 8\ \text{hr} = 4{:}00\ \text{PM}$$

$$\left(t_f \right)_{\text{Beth}} = \left(t_i \right)_{\text{Beth}} + \frac{\left(x_f \right)_{\text{Beth}} - \left(x_i \right)_{\text{Beth}}}{v} = 9{:}00\ \text{AM} + \frac{400\ \text{mile}}{60\ \text{miles/hour}} = 9{:}00\ \text{AM} + 6.67\ \text{hr} = 3{:}40\ \text{PM}$$

(a) Beth arrives first.
(b) Beth has to wait 20 minutes for Alan.
Assess: Times of the order of 7 or 8 hours are reasonable in the present problem.

P2.13. Strategize: Since each runner is running at a steady pace, they both are traveling with a constant speed.
Prepare: Each runner must travel the same distance to finish the race. We assume they are traveling uniformly. We can calculate the time it takes each runner to finish using Equation 2.1.
Solve: The first runner finishes in

$$\Delta t_1 = \frac{\Delta x}{(v_x)_1} = \frac{5.00 \text{ km}}{12.0 \text{ km/h}} = 0.417 \text{ h}$$

Converting to minutes, this is $(0.417 \text{ h})\left(\frac{60 \text{ min}}{1 \text{ h}}\right) = 25.0 \text{ min}$

For the second runner

$$\Delta t_2 = \frac{\Delta x}{(v_x)_2} = \frac{5.00 \text{ km}}{14.5 \text{ km/h}} = 0.345 \text{ h}$$

Converting to seconds, this is

$$(0.345 \text{ h})\left(\frac{60 \text{ min}}{1 \text{ h}}\right) = 20.7 \text{ min}$$

The time the second runner waits is 25.0 min – 20.7 min = 4.3 min
Assess: For uniform motion, velocity is given by Equation 2.1.

P2.15. Strategize: This is a position vs. time plot, and we are asked for the top speed. So we are interested in the maximum slope in the given plot.
Prepare: The slope of the path traced by the dots in the position vs. time plot is initially somewhat inclined, but then increases after the first two seconds. After that point near the two second mark (which we approximate as 1.9 s), the slope appears fairly constant. We can take this larger, sustained slope as the maximum speed.
Solve: Since the speed after the dot at 1.9 s mark appears roughly constant, we can use the expression for the average speed over that interval:

$$v_{x,\text{av}} = \frac{\Delta x}{\Delta t} = \frac{x_f - x_i}{t_f - t_i} = \frac{(100 \text{ m}) - (10 \text{ m})}{(9.5 \text{ s}) - (1.9 \text{ s})} = 12 \text{ m/s}$$

Assess: This is extremely fast. One can perform a simple check by noting that the average time for the entire run is $(100 \text{ m})/(9.5 \text{ s}) = 11$ m/s. So the fact that we got a slightly higher speed for Usain Bolt's maximum speed is reasonable.

P2.17. Strategize: In this problem the velocity is changing, but we can determine v_x at a given time by looking at the slope of the x vs. t graph.
Prepare: The graph in Figure P2.17 shows distinct slopes in the time intervals: 0 – 1 s, 1 s – 2 s, and 2 s – 4 s. We can thus obtain the velocity values from this graph using $v = \Delta x/\Delta t$.
Solve: (a)

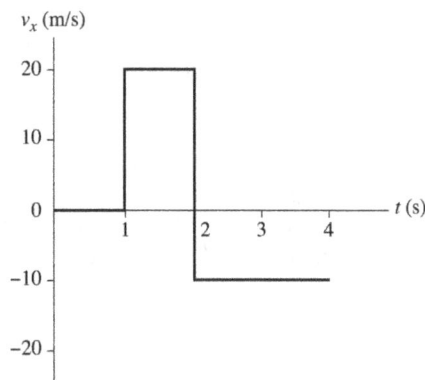

(b) There is only one turning point. At $t = 2$ s the velocity changes from $+20$ m/s to -10 m/s, thus reversing the direction of motion. At $t = 1$ s, there is an abrupt change in motion from rest to $+20$ m/s, but there is no reversal in motion.

Assess: As shown above in (a), a positive slope must give a positive velocity and a negative slope must yield a negative velocity.

P2.19. Strategize: Displacement is given by the area under the a velocity vs. time graph.

Prepare: In this case, the displacement is equal to the area under the velocity graph between t and t. We can find the car's final position from its initial position and the area.

Solve: (a) Using the equation $x = x +$ area of the velocity graph between t and t,

$$x_{2\,s} = 10 \text{ m} + \text{area of trapezoid between } 0 \text{ s and } 2 \text{ s}$$

$$= 10 \text{ m} + \frac{1}{2}(12 \text{ m/s} + 4 \text{ m/s})(2 \text{ s}) = 26 \text{ m}$$

$$x_{3\,s} = 10 \text{ m} + \text{area of triangle between } 0 \text{ s and } 3 \text{ s}$$

$$= 10 \text{ m} + \frac{1}{2}(12 \text{ m/s})(3 \text{ s}) = 28 \text{ m}$$

$$x_{4\,s} = x_{3\,s} + \text{area between } 3 \text{ s and } 4 \text{ s}$$

$$= 28 \text{ m} + \frac{1}{2}(-4 \text{ m/s})(1 \text{ s}) = 26 \text{ m}$$

(b) The car reverses direction at $t = 3$ s, because its velocity becomes negative.

Assess: The car starts at $x = 10$ m at $t = 0$. Its velocity decreases as time increases, is zero at $t = 3$ s, and then becomes negative. The slope of the velocity-versus-time graph is negative which means the car's acceleration is negative and a constant. From the acceleration thus obtained and given velocities on the graph, we can also use kinematic equations to find the car's position at various times.

P2.21. Strategize: The graph in Figure P2.21 shows the horizontal component of velocity as a function of time. We know the acceleration is the rate of change of the velocity. So we can determine the acceleration using the slope of this graph.

Prepare: We will use $a = \Delta v / \Delta t$. A linear decrease in velocity from $t = 0$ s to $t = 2$ s implies a constant negative acceleration. On the other hand, a constant velocity between $t = 2$ s and $t = 4$ s means zero acceleration.

Solve:

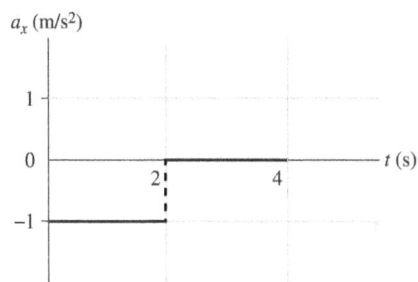

P2.23. Strategize: Acceleration is the rate of change of velocity. We must draw a velocity vs. time graph in which the slope at a given point is equal to the value of the acceleration in the plot above.

Prepare: We are told the initial speed of the object is 2.0 m/s. We can simply start drawing a line from that point with the appropriate slope, changing slopes at the appropriate times.

Solve:

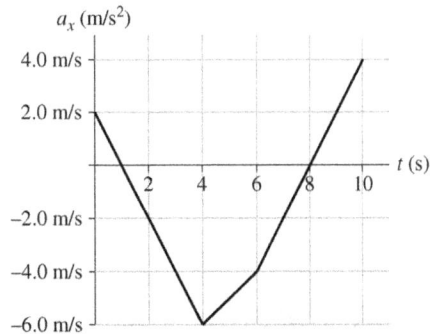

$a_x \; (\text{m/s}^2)$

Assess: We can check our answer by calculating the velocity after a certain time and seeing if it matches the graph. Let us check the lowest point, which on our graph is –6.0 m/s and occurs at 4 s. Using Equation 2.11, we have $(v_x)_f = (v_x)_i + a_x\Delta t = (2.0 \text{ m/s}) + (-2.0 \text{ m/s}^2)(4 \text{ s}) = -6.0 \text{ m/s}$, which is consistent.

P2.25. Strategize: From a velocity-versus-time graph we find the acceleration by computing the slope.
Prepare: We will compute the slope of each straight-line segment in the graph.

$$a_x = \frac{(v_x)_f - (v_x)_i}{t_f - t_i}$$

The trickiest part is reading the values off of the graph.
Solve: (a)

$$a_x = \frac{5.5 \text{ m/s} - 0.0 \text{ m/s}}{0.9 \text{ s} - 0.0 \text{ s}} = 6.1 \text{ m/s}^2$$

(b)

$$a_x = \frac{9.3 \text{ m/s} - 5.5 \text{ m/s}}{2.4 \text{ s} - 0.9 \text{ s}} = 2.5 \text{ m/s}^2$$

(c)

$$a_x = \frac{10.9 \text{ m/s} - 9.3 \text{ m/s}}{3.5 \text{ s} - 2.4 \text{ s}} = 1.5 \text{ m/s}^2$$

Assess: This graph is difficult to read to more than one significant figure. I did my best to read a second significant figure but there is some estimation in the second significant figure.
It takes Carl Lewis almost 10 s to run 100 m, so this graph covers only the first third of the race. Were the graph to continue, the slope would continue to decrease until the slope is zero as he reaches his (fastest) cruising speed.
Also, if the graph were continued out to the end of the race, the area under the curve should total 100 m.

P2.27. Strategize: We will assume constant accelerations for both animals.
Prepare: We can calculate acceleration from Equation 2.8:
Solve: For the gazelle:

$$(a_x) = \left(\frac{\Delta v_x}{\Delta t}\right) = \frac{13 \text{ m/s}}{3.0 \text{ s}} = 4.3 \text{ m/s}^2$$

For the lion:

$$(a_x) = \left(\frac{\Delta v_x}{\Delta t}\right) = \frac{9.5 \text{ m/s}}{1.0 \text{ s}} = 9.5 \text{ m/s}^2$$

For the trout:

$$(a_x) = \left(\frac{\Delta v_x}{\Delta t}\right) = \frac{2.8 \text{ m/s}}{0.12 \text{ s}} = 23 \text{ m/s}^2$$

The trout is the animal with the largest acceleration.
Assess: A lion would have an easier time snatching a gazelle than a trout.

P2.29. Strategize: This problem consists of unit conversion, and application of the definition of acceleration. Note that since the acceleration is constant, we are also free to use kinematic equations.
Prepare: First, we will convert units:

$$60 \frac{\text{miles}}{\text{hour}} \times \frac{1 \text{ hour}}{3600 \text{ s}} \times \frac{1609 \text{ m}}{1 \text{ mile}} = 26.8 \text{ m/s}$$

We also note that $g = 9.8$ m/s . Because the car has constant acceleration, we can use kinematic equations.
Solve: (a) For initial velocity $v = 0$, final velocity $v = 26.8$ m/s, and $\Delta t = 10$ s, we can find the acceleration using

$$v_f = v_i + a\Delta t \Rightarrow a = \frac{v_f - v_i}{\Delta t} = \frac{(26.8 \text{ m/s} - 0 \text{ m/s})}{10 \text{ s}} = 2.68 \text{ m/s}^2 \approx 2.7 \text{ m/s}^2$$

(b) The fraction is $a/g = 2.68/9.8 = 0.273$. So a is 27% of g, or 0.27 g.
(c) The displacement is calculated as follows:

$$x_f - x_i = v_i\Delta t + \frac{1}{2}a(\Delta t)^2 = \frac{1}{2}a(\Delta t)^2 = 134 \text{ m} = 440 \text{ feet}$$

Assess: A little over tenth of a mile displacement in 10 s is physically reasonable.

P2.31. Strategize: This is a question about acceleration and how it relates to other kinematic quantities. We will assume the large acceleration is constant, such that we can make use of kinematic equations.
Reason: We can use Equation 2.11 to relate acceleration to initial and final speeds. To relate the acceleration and time to distance covered, we can use the initial velocity from part (a) and Equation 2.13. Let us call the initial direction of motion the $+x$ direction, such that the acceleration will be in the $-x$ direction.
Solve: The maximum initial speed would require the maximum allowed time to stop. So we assume $\Delta t = 30$ ms. Then

$$(v_x)_f = (v_x)_i + a_x\Delta t \Rightarrow (v_x)_i = (v_x)_f - a_x\Delta t = (0 \text{ m/s}) - \left(-(50)(9.8 \text{ m/s}^2)\right)(30\times10^{-3} \text{ s}) = 14.7 \text{ m/s}$$

We would report our answer to part (a) as 15 m/s. We have kept an extra digit above for use in part (b).
To determine the minimum distance, we again assume that all 30 ms of allowable time are used in the stopping process, and we use the initial velocity from part (a), such that we can write

$$(v_x)_f^2 = (v_x)_i^2 + 2a_x\Delta x \Rightarrow \Delta x = \frac{(v_x)_f^2 - (v_x)_i^2}{2a_x} = \frac{(0 \text{ m/s})^2 - (14.7 \text{ m/s})^2}{2\left(-(50)(9.8 \text{ m/s}^2)\right)} = 0.22 \text{ m}$$

Assess: The maximum initial speed we found is around 33 mph. This means that going from full speed to a full stop in 30 ms could be fatal if the initial speed is greater than around 33 mph. Of course, seatbelts, airbags, and crumple zones in cars are designed to increase the distance over which the humans in the car stop to considerably more than 0.22 m. This way humans can survive head-on collisions starting from even greater speeds.

P2.33. Strategize: Because the skier slows steadily, her acceleration is a constant during the glide and we can use the kinematic equations.

Prepare: We can use Equation 2.13 to determine the unknown acceleration.

Solve: Since we know the skier's initial and final speeds and the width of the patch over which she decelerates, we will use

$$v_f^2 = v_i^2 + 2a(x_f - x_i)$$

$$\Rightarrow a = \frac{v_f^2 - v_i^2}{2(x_f - x_i)} = \frac{(6.0 \text{ m/s})^2 - (8.0 \text{ m/s})^2}{2(5.0 \text{ m})} = -2.8 \text{ m/s}^2$$

The magnitude of this acceleration is 2.8 m/s .

Assess: A deceleration of 2.8 m/s or 6.3 mph/s is reasonable.

P2.35. Strategize: Because the car slows steadily, the deceleration is a constant and we can use the kinematic equations of motion under constant acceleration.

Prepare: We look for an equation in which we know all but one variable, and find that we can solve this using Equation 2.13.

Solve: Since we know the car's initial and final speeds and the width of the patch over which she decelerates, we will use

$$v_f^2 = v_i^2 + 2a(x_f - x_i)$$

$$\Rightarrow a = \frac{v_f^2 - v_i^2}{2(x_f - x_i)} = \frac{(0 \text{ m/s})^2 - (90 \text{ m/s})^2}{2(110 \text{ m})} = -37 \text{ m/s}^2$$

The magnitude of this acceleration is 37 m/s .

Assess: A deceleration of 37 m/s is impressive; it is almost 4 *g*s.

P2.37. Strategize: Let us assume the acceleration of the car is constant.

Prepare: A visual overview of the car's motion that includes a pictorial representation, a motion diagram, and a list of values is shown below. We label the car's motion along the *x*-axis. For the driver's maximum (constant) deceleration, kinematic equations are applicable. This is a two-part problem. We will first find the car's displacement during the driver's reaction time when the car's deceleration is zero. Then we will find the displacement as the car is brought to rest with maximum deceleration.

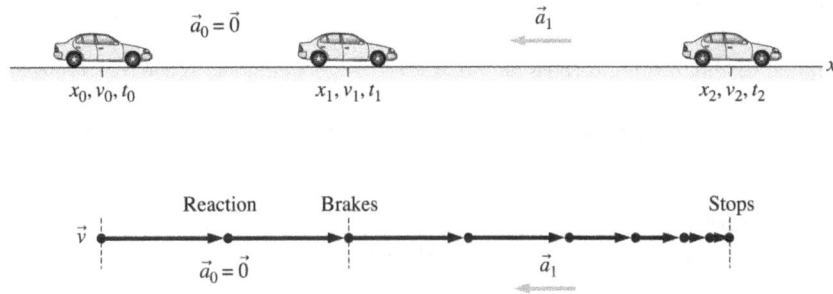

Solve: During the reaction time when *a* = 0, we can use

$$x_1 = x_0 + v_0(t_1 - t_0) + \frac{1}{2}a_0(t_1 - t_0)^2$$

$$= 0 \text{ m} + (20 \text{ m/s})(0.50 \text{ s} - 0 \text{ s}) + 0 \text{ m} = 10 \text{ m}$$

During deceleration,

$$v_2^2 = v_1^2 + 2a_1(x_2 - x_1) \qquad 0 = (20 \text{ m/s}) + 2(-6.0 \text{ m/s})(x - 10 \text{ m}) \Rightarrow x = 43 \text{ m}$$

She has 50 m to stop, so she can stop in time.

Assess: While driving at 20 m/s or 45 mph, a reaction time of 0.5 s corresponds to a distance of 33 feet or only two lengths of a typical car. Keep a safe distance while driving!

P2.39. Strategize: We will assume that you achieve the maximum magnitude of acceleration possible, and that this acceleration is constant.

Prepare: A visual overview of your car's motion that includes a pictorial representation, a motion diagram, and a list of values is shown below. We label the car's motion along the x-axis. For maximum (constant) deceleration of your car, kinematic equations hold. This is a two-part problem. We will first find the car's displacement during your reaction time when the car's deceleration is zero. Then we will find the displacement as you bring the car to rest with maximum deceleration.

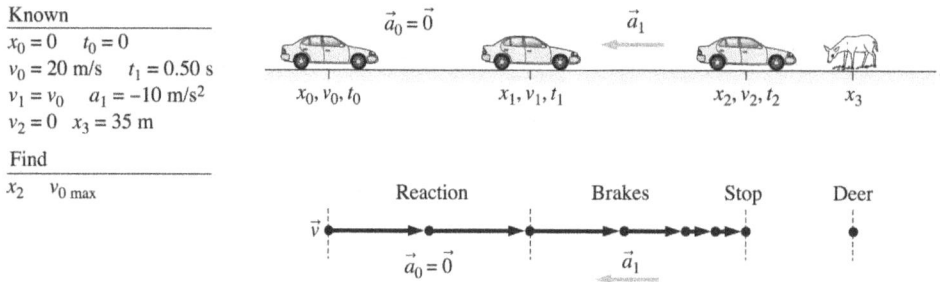

Known
$x_0 = 0$ $t_0 = 0$
$v_0 = 20$ m/s $t_1 = 0.50$ s
$v_1 = v_0$ $a_1 = -10$ m/s^2
$v_2 = 0$ $x_3 = 35$ m

Find
x_2 $v_{0\ max}$

Solve: (a) To find x_2, we first need to determine x_1. Using $x_1 = x_0 + v_0(t_1 - t_0)$, we get $x_1 = 0$ m $+ (20$ m/s$)(0.50$ s $- 0$ s$)$ $= 10$ m. Now, with $a_1 = 10$ m/s^2, $v_2 = 0$ and $v_1 = 20$ m/s, we can use

$$v_2^2 = v_1^2 + 2a_1(x_2 - x_1) \Rightarrow 0 \text{ m}^2/\text{s}^2 = (20 \text{ m/s})^2 + 2(-10 \text{ m/s}^2)(x_2 - 10 \text{ m}) \Rightarrow x_2 = 30 \text{ m}$$

The distance between you and the deer is $(x_3 - x_2)$ or $(35$ m $- 30$ m$) = 5$ m.

(b) Let us find $v_{0\ max}$ such that $v_2 = 0$ m/s at $x_2 = x_3 = 35$ m. Using the following equation,

$$v_2^2 - v_{0\ max}^2 = 2a_1(x_2 - x_1) \Rightarrow 0 \text{ m}^2/\text{s}^2 - v_{0\ max}^2 = 2(-10 \text{ m/s}^2)(35 \text{ m} - x_1)$$

Also, $x_1 = x_0 + v_{0\ max}(t_1 - t_0) = v_{0\ max}(0.50$ s $- 0$ s$) = (0.50$ s$)v_{0\ max}$. Substituting this expression for x_1 in the above equation yields

$$-v_{0\ max}^2 = (-20 \text{ m/s}^2)[35 \text{ m} - (0.50 \text{ s}) v_{0\ max}] \Rightarrow v_{0\ max}^2 + (10 \text{ m/s})v_{0\ max} - 700 \text{ m}^2/\text{s}^2 = 0$$

The solution of this quadratic equation yields $v_{0\ max} = 22$ m/s. (The other root is negative and unphysical for the present situation.)

Assess: An increase of speed from 20 m/s to 22 m/s is very reasonable for the car to cover an additional distance of 5 m with a reaction time of 0.50 s and a deceleration of 10 m/s^2.

P2.41. Strategize: We will assume constant acceleration, such that we can use kinematic equations.
Prepare: Call the point where the motorcycle started the origin.
Solve:
(a)

$$a = \frac{\Delta v}{\Delta t} \Rightarrow \Delta t = \frac{\Delta v}{a} = \frac{80 \text{ km/h}}{8.0 \text{ m/s}^2}\left(\frac{1 \text{ h}}{3600 \text{ s}}\right)\left(\frac{1000 \text{ m}}{1 \text{ km}}\right) = 2.78 \text{ s} \approx 2.8 \text{ s}$$

(b) Compute the distance traveled in 10 s for each vehicle.

For the car: $\Delta x = v\Delta t = (80 \text{ km/h})(2.78 \text{ s})\left(\frac{1 \text{ h}}{3600 \text{ s}}\right)\left(\frac{1000 \text{ m}}{1 \text{ km}}\right) = 61.7 \text{ m}$

For the motorcycle: $\Delta x = \frac{1}{2}a(\Delta t)^2 = \frac{1}{2}(8.0 \text{ m/s}^2)(2.78 \text{ s})^2 = 30.7 \text{ m}$

The difference is the distance between the motorcycle and the car at that time. 61.7 m $- 30.7$ m $= 31$ m
Assess: The motorcycle will never catch up if it never exceeds the speed of the car.

P2.43. Strategize: During the acceleration phase, acceleration is constant, and we can use kinematic equations. Once the acceleration drops to zero, acceleration will once again be constant, and we can apply kinematic equations over the phase of constant velocity. But we cannot apply kinematic equations from the beginning of the dash to the end, since acceleration changes in between.
Prepare: Use Equation 2.11 to find the acceleration.

$$v_x = a_x t_1 \qquad \text{where } v_0 = 0 \text{ and } t_0 = 0$$

$$a_x = \frac{v_x}{t_1} = \frac{11.2 \text{ m/s}}{2.14 \text{ s}} = 5.23 \text{ m/s}^2$$

Solve: The distance traveled during the acceleration phase will be

$$\Delta x = \frac{1}{2} a_x (\Delta t)^2$$
$$= \frac{1}{2} (5.23 \text{ m/s}^2)(2.14 \text{ s})^2$$
$$= 12.0 \text{ m}$$

The distance left to go at constant velocity is $100 \text{ m} - 12.0 \text{ m} = 88.0 \text{ m}$. The time this takes at the top speed of 11.2 m/s is

$$\Delta t = \frac{\Delta x}{v_x} = \frac{88.0 \text{ m}}{11.2 \text{ m/s}} = 7.86 \text{ s}$$

The total time is $2.14 \text{ s} + 7.86 \text{ s} = 10.0 \text{ s}$.
Assess: This is indeed about the time it takes a world-class sprinter to run 100 m (the world record is a bit under 9.8 s). Compare the answer to this problem with the accelerations given in Problem 2.25 for Carl Lewis.

P2.45. Strategize: This problem involves freefall, in which the acceleration is constant. Thus, we can use the kinematic equations.
Prepare: The bill must drop its own length in freefall.
Solve:

$$\Delta y = \frac{1}{2} g (\Delta t)^2 \Rightarrow \Delta t = \sqrt{\frac{2\Delta y}{g}} = \sqrt{\frac{2(0.16 \text{ m})}{9.8 \text{ m/s}^2}} = 0.18 \text{ s}$$

Assess: This is less than the typical 0.25 s reaction time, so most people miss the bill.

P2.47. Strategize: This problem involves freefall, in which the acceleration is constant. Thus, we can use the kinematic equations.
Prepare: Use kinematic equations for constant acceleration. Assume the gannet is in free fall during the dive.
Solve:

$$(v_y)_f^2 = (v_y)_i^2 + 2g\Delta y \Rightarrow \Delta y = \frac{(v_y)_f^2}{2g} = \frac{(32 \text{ m/s})^2}{2(9.8 \text{ m/s}^2)} = 52 \text{ m}$$

Assess: 52 meters seems a reasonable height from which to begin the dive.

P2.49. Strategize: Two objects are in freefall. We must determine where they meet. Since the acceleration is constant during freefall, we can use kinematic equations.
Prepare: We will need to describe the motion of the acrobat and the ball, for which we will use subscripts A and B, respectively. Let us call the vertically upward direction $+y$. To determine when the acrobat catches the ball, we must determine when the two objects have the same vertical positions. We can use Equation 2.12 to describe the change in position of each object. In order for them to meet, we require $\Delta y_A = \Delta y_B + (9.0 \text{ m})$. That is, however much the ball moves, the acrobat must move upward by 15 m more to cover the initial distance between them.

Solve: We apply Equation 2.12 to each object separately:

$$\Delta y_A = \left(v_{Ay}\right)_i \Delta t + \frac{1}{2}a_y\left(\Delta t\right)^2$$

$$\Delta y_B = \left(v_{By}\right)_i \Delta t + \frac{1}{2}a_y\left(\Delta t\right)^2 = \frac{1}{2}a_y\left(\Delta t\right)^2$$

In the second equation we have use the fact that the ball is dropped, not thrown. Note that no subscript is required for the time or the acceleration, since we need the acrobat and ball to be in the same place at the same time, and since both are accelerating only due to gravity. Requiring $\Delta y_A = \Delta y_B + \left(9.0 \text{ m}\right)$, we find

$$\left(v_{Ay}\right)_i \Delta t + \frac{1}{2}a_y\left(\Delta t\right)^2 = \left(9.0 \text{ m}\right) + \frac{1}{2}a_y\left(\Delta t\right)^2$$

Subtracting $\frac{1}{2}a_y\left(\Delta t\right)^2$ from both sides yields

$$\left(v_{Ay}\right)_i \Delta t = \left(9.0 \text{ m}\right) \Rightarrow \Delta t = \left(9.0 \text{ m}\right)/\left(v_{Ay}\right)_i = \left(9.0 \text{ m}\right)/\left(8.0 \text{ m/s}\right) = 1.1 \text{ s}.$$

Assess: Given the length scales of sever meters, a time of 1.1 s for the two objects to meet is reasonable.

P2.51. Strategize: Once the jumper leaves the ground, he or she is in freefall, in which the acceleration is constant. Thus, we can use the kinematic equations.
Prepare: Use the kinematic equation $(v_y)_f^2 = (v_y)_i^2 + 2a_y\Delta y$ where $(v_y)_f^2 = 0$ at the top of the leap.

We assume $a_y = -9.8 \text{ m/s}^2$ and we are given $\Delta y = 1.1 \text{ m}$.
Solve:

$$(v_y)_i^2 = -2a_y\Delta y \quad \Rightarrow \quad (v_y)_i = \sqrt{-2a_y\Delta y} = \sqrt{-2(-9.8 \text{ m/s}^2)(1.1 \text{ m})} = 4.6 \text{ m/s}$$

Assess: This is an achievable take-off speed for good jumpers. The units also work out correctly and the two minus signs under the square root make the radicand positive.

P2.53. Strategize: Once the briefcase is dropped it is in freefall, during which the acceleration is constant. Thus we can use kinematic equations.
Prepare: Since the villain is hanging on to the ladder as the helicopter is ascending, he and the briefcase are moving with the same upward velocity as the helicopter. We can calculate the initial velocity of the briefcase, which is equal to the upward velocity of the helicopter. See the following figure.

Known
$y_i = 130 \text{ m}$ $y_f = 0 \text{ m}$
$t_f = t_i = 6.0 \text{ s}$
$a_y = -g = -9.80 \text{ m/s}^2$

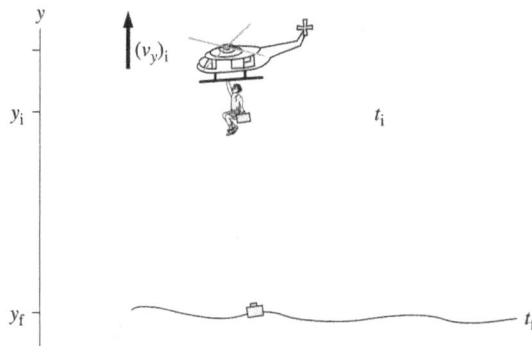

Find
$(v_y)_i$

Solve: We can use Equation 2.12 here. We know the time it takes the briefcase to fall, its acceleration, and the distance it falls. Solving for $(v_y)_i \Delta t$,

$$(v_y)_i \Delta t = (y_f - y_i) - \frac{1}{2}(a_y)\Delta t^2 = -130 \text{ m} - \left[\frac{1}{2}(-9.80 \text{ m/s}^2)(6.0 \text{ s})^2\right] = 46 \text{ m}$$

Dividing by Δt to solve for $(v_y)_i$,

$$(v_x)_i = \frac{46 \text{ m}}{6.0 \text{ s}} = 7.7 \text{ m/s}$$

Assess: Note the placement of negative signs in the calculation. The initial velocity is positive, as expected for a helicopter ascending.

P2.55. Strategize: Since the stones are in freefall after they leave the climber's hands, acceleration will be constant. Thus we can use kinematic equations.
Prepare: There are several steps in this problem, so first draw a picture and, like the examples in the book, list the known quantities and what we need to find.
Call the pool of water the origin and call $t = 0$ s when the first stone is released. We will assume both stones are in free fall after they leave the climber's hand, so $a = -g$. Let a subscript 1 refer to the first stone and a 2 refer to the second.

Known	Find
$(y_1)_i = 50$ m	$(t_2)_f$ or t_f
$(y_2)_i = 50$ m	$(v_2)_i$
$(y_1)_f = 0.0$ m	$(v_1)_f$
$(y_2)_f = 0.0$ m	$(v_2)_f$
$(y_1)_i = -2.0$ m/s	
$(t_2)_f = (t_1)_f$; simply call this t_f	
$(t_2)_i = 1.0$ s	

Solve: (a) Using $(t_1)_i = 0$

$$(y_1)_f = (y_1)_i + (v_1)_i \Delta t + \frac{1}{2}a_y \Delta t^2$$

$$0.0 \text{ m} = 50 \text{ m} + (-2 \text{ m/s})t_f + \frac{1}{2}(-g)t_f^2$$

$$0.0 \text{ m} = 50 \text{ m} - (2 \text{ m/s})t_f - (4.9 \text{ m/s}^2)t_f^2$$

Solving this quadratic equation gives two values for t_f: 3.0 s and -3.4 s, the second of which (being negative) is outside the scope of this problem.
Both stones hit the water at the same time, and it is at $t = 3.0$ s, or 3.0 s after the first stone is released.
(b) For the second stone $\Delta t_2 = t_f - (t_2)_i = 3.0 \text{ s} - 1.0 \text{ s} = 2.0 \text{ s}$. We solve now for $(v_2)_i$.

$$(y_2)_f = (y_2)_i + (v_2)_i \Delta t + \frac{1}{2}a_y \Delta t^2$$

$$0.0 \text{ m} = 50 \text{ m} + (v_2)_i \Delta t_2 + \frac{1}{2}(-g)\Delta t_2^2$$

$$0.0 \text{ m} = 50 \text{ m} + (v_2)_i(2.0 \text{ s}) - (4.9 \text{ m/s}^2)(2.0 \text{ s})^2$$

$$(v_2)_i = \frac{-50 \text{ m} + (4.9 \text{ m/s}^2)(2.0 \text{ s})^2}{2.0 \text{ s}} = -15.2 \text{ m/s}$$

Thus, the second stone is thrown down at a speed of 15 m/s.

(c) Equation 2.11 allows us to compute the final speeds for each stone.

$$(v_y)_f = (v_y)_i + a_y \Delta t$$

For the first stone (which was in the air for 3.0 s):

$$(v_1)_f = -2.0 \text{ m/s} + (-9.8 \text{ m/s}^2)(3.0 \text{ s}) = -31 \text{ m/s}$$

The speed is the magnitude of this velocity, or 31 m/s.
For the second stone (which was in the air for 2.0 s):

$$(v_2)_f = -15.2 \text{ m/s} + (-9.8 \text{ m/s}^2)(2.0 \text{ s}) = -35 \text{ m/s}$$

The speed is the magnitude of this velocity, or 35 m/s.
Assess: The units check out in each of the previous equations. The answers seem reasonable. A stone dropped from rest takes 3.2 s to fall 50 m; this is comparable to the first stone, which was able to fall the 50 m in only 3.0 s because it started with an initial velocity of −2.0 m/s. So we are in the right ballpark. And the second stone would have to be thrown much faster to catch up (because the first stone is accelerating).

P2.57. Strategize: This problem involves two phases, each of which is a constant velocity phase. But in between, the truck speeds up and has non-zero acceleration briefly. Thus, we can apply kinematic equations to either constant-velocity phase of the trip, but not to the trip as a whole.
Prepare: Assume the truck driver is traveling with constant velocity during each segment of his trip.
Solve: Since the driver usually takes 8 hours to travel 440 miles, his usual velocity is

$$v_{\text{usual } x} = \frac{\Delta x}{\Delta t_{\text{usual}}} = \frac{440 \text{ mi}}{8 \text{ h}} = 55 \text{ mph}$$

However, during this trip he was driving slower for the first 120 miles. Usually he would be at the 120 mile point in

$$\Delta t_{\text{usual at 120 mi}} = \frac{\Delta x}{v_{\text{usual at 120 mi } x}} = \frac{120 \text{ mi}}{55 \text{ mph}} = 2.18 \text{ h}$$

He is 15 minutes, or 0.25 hr late. So the time he's taken to get 120 mi is 2.18 hr + 0.25 hr = 2.43 hr. He wants to complete the entire trip in the usual 8 hours, so he only has 8 hr − 2.43 hr = 5.57 hr left to complete 440 mi − 120 mi = 320 mi. So he needs to increase his velocity to

$$v_{\text{to catch up } x} = \frac{\Delta x}{\Delta t_{\text{ to catch up}}} = \frac{320 \text{ mi}}{5.57 \text{ h}} = 57 \text{ mph}$$

where additional significant figures were kept in the intermediate calculations.
Assess: This result makes sense. He is only 15 minutes late.

P2.59. Strategize: The timing between images in Figure P2.59 is constant. We expect to find a a period of constant velocity, followed by a decrease in speed, followed by another period of constant (lower) velocity.
Prepare: We assume that the track, except for the sticky section, is frictionless and aligned along the *x*-axis. Because the motion diagram of Figure P2.59 is made at two frames of film per second, the time interval between consecutive ball positions is 0.5 s.
Solve: (a)

Times (s)	Position
0	−4.0
0.5	−2.0
1.0	0
1.5	1.8
2.0	3.0
2.5	4.0
3.0	5.0
3.5	6.0
4.0	7.0

(b)

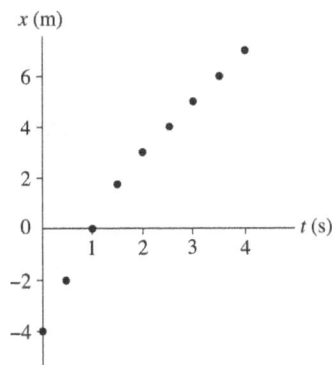

(c) $\Delta x = x$ (at $t = 1$ s) $- x$ (at $t = 0$ s) $= 0$ m $- (-4$ m$) = 4$ m.

(d) $\Delta x = x$ (at $t = 4$ s) $- x$ (at $t = 2$ s) $= 7$ m $- 3$ m $= 4$ m.

(e) From $t = 0$ s to $t = 1$ s, $v_s = \Delta x / \Delta t = 4$ m/s.

(f) From $t = 2$ s to $t = 4$ s, $v_x = \Delta x / \Delta t = 2$ m/s.

(g) The average acceleration is

$$a = \frac{\Delta v}{\Delta t} = \frac{2 \text{ m/s} - 4 \text{ m/s}}{2 \text{ s} - 1 \text{ s}} = -2 \text{ m/s}^2$$

Assess: The sticky section has decreased the ball's speed from 4 m/s, to 2 m/s, which is a reasonable magnitude.

P2.61. Strategize: This is an estimation problem, so a range of answers may be acceptable.

Prepare: We will represent the jetliner's motion to be along the x-axis.

Solve:

(a) Using $a_x = \Delta v / \Delta t$, we have,

$$a_x(t = 0 \text{ to } t = 10 \text{ s}) = \frac{23 \text{ m/s} - 0 \text{ m/s}}{10 \text{ s} - 0 \text{ s}} = 2.3 \text{ m/s}^2 \qquad a_x(t = 20 \text{ s to } t = 30 \text{ s}) = \frac{69 \text{ m/s} - 46 \text{ m/s}}{30 \text{ s} - 20 \text{ s}} = 2.3 \text{ m/s}^2$$

For all time intervals a is 2.3 m/s . In gs this is $(2.3 \text{ m/s })/(9.8 \text{ m/s }) = 0.23g$

(b) Because the jetliner's acceleration is constant, we can use kinematics as follows:

$$(v_x)_f = (v_x)_i + a_x(t_f - t_i) \Rightarrow 80 \text{ m/s} = 0 \text{ m/s} + (2.3 \text{ m/s}^2)(t_f - 0 \text{ s}) \Rightarrow t_f = 34.8 \text{ s}$$

or 35 s to two significant figures.

(c) Using the above values, we calculate the takeoff distance as follows:

$$x_f = x_i + (v_x)_i(t_f - t_i) + \frac{1}{2}a_x(t_f - t_i)^2 = 0 \text{ m} + (0 \text{ m/s})(34.8 \text{ s}) + \frac{1}{2}(2.3 \text{ m/s}^2)(34.8 \text{ s})^2 = 1390 \text{ m}$$

For safety, the runway should be 3×1390 m $= 4.2$ km.

P2.63. Strategize: This is an estimation problem. For all parts, we will read approximate values from the graph.

Prepare: The acceleration is given by $a_x = \dfrac{\Delta v_x}{\Delta t}$, which is the slope of the graph provided. The maximum acceleration is near the beginning of the time shown, where the slope is maximal. The distance traveled is the area under the curve of the graph. It would be difficult to calculate this exactly, but since the curve is roughly linear between the times 0 and 50 ms, we can approximate the area using $\dfrac{1}{2}\left((v_x)_{\text{end}} - (v_x)_{\text{start}}\right)(t_{\text{end}} - t_{\text{start}})$.

Solve: (a) The slope is largest between $t = 0$ and $t = 50$ ms. In this interval, v_x changes from 0 to approximately 0.8 m/s. Thus

$$\left(a_x\right)_{max} = \frac{v_x\left(t = 50 \text{ ms}\right) - v_x\left(t = 0\right)}{50 \text{ ms} - 0} = \frac{\left(0.8 \text{ m/s}\right) - \left(0 \text{ m/s}\right)}{\left(50 \times 10^{-3} \text{ s}\right) - 0} = 16 \text{ m/s}^2$$

Dividing this by 9.8 m/s^2 yields $\left(a_x\right) = 1.6g$.

(b) We can estimate the acceleration at this time by using the velocities at the times just before and just after:

$$a_x\left(t = 150 \text{ ms}\right) = \frac{v_x\left(t = 200 \text{ ms}\right) - v_x\left(t = 100 \text{ ms}\right)}{\left(200 \text{ ms}\right) - \left(100 \text{ ms}\right)} = \frac{\left(1.8 \text{ m/s}\right) - \left(1.3 \text{ m/s}\right)}{\left(100 \times 10^{-3} \text{ s}\right) - 0} = 5 \text{ m/s}^2 \text{ or } 0.5g$$

(c) We estimate the area under the curve to be

$$\Delta x = \frac{1}{2}\left(\left(v_x\right)_{end} - \left(v_x\right)_{start}\right)\left(t_{end} - t_{start}\right) = \frac{1}{2}\left(0.8 \text{ m/s}\right)\left(50 \times 10^{-3} \text{ s}\right) = 0.02 \text{ m or 2 cm.}$$

Assess: Since the large initial acceleration is only applied over a short time, a displacement of 2 cm during the first 50 ms is reasonable.

P2.65. Strategize: Remember that in estimation problems different people may make slightly different estimates. That is OK as long as they end up with reasonable answers that are the same order-of-magnitude. We will assume constant acceleration, such that we can use the kinematic equations.
Prepare: We can use Equation 2.12, and noting that the initial speed is zero, we have

$$x_f = \frac{1}{2}a_x(\Delta t)^2$$

Solve: (a) I guessed about 1.0 cm; this was verified with a ruler and mirror.
(b) We are given a closing time of 0.024 s, so we can compute the acceleration from rearranging the kinematic equations.

$$a_x = \frac{2x_f}{(\Delta t)^2} = \frac{2(1.0 \text{ cm})}{(0.024 \text{ s})^2}\left(\frac{1 \text{ m}}{100 \text{ cm}}\right) = 35 \text{ m/s}^2$$

(c) Since we know the Δt and the a and $v_i = 0.0$ m/s, we can compute the final speed from Equation 2.11:

$$v_f = a\Delta t = (35 \text{ m/s}^2)(0.024 \text{ s}) = 0.84 \text{ m/s}$$

Assess: The uncertainty in our estimates might or might not barely justify two significant figures.
The final speed is reasonable; if we had arrived at an answer 10 times bigger or 10 times smaller we would probably go back and check our work. The lower lid gets smacked at this speed up to 15 times per minute!

P2.67. Strategize: We assume constant acceleration so we can use the kinematic equations.
Prepare: Fleas are amazing jumpers; they can jump several times their body height—something we cannot do. Equation 2.13 relates the three variables we are concerned with in part (a): speed, distance (which we know), and acceleration (which we want).

$$(v_y)_f^2 = (v_y)_i^2 + 2a_y\Delta y$$

In part (b) we use Equation 2.12 because it relates the initial and final velocities and the acceleration (which we know) with the time interval (which we want).

$$(v_y)_f = (v_y)_i + a_y\Delta t$$

Part (c) is about the phase of the jump *after* the flea reaches takeoff speed and leaves the ground. So now it is $(v_y)_i$, that is 1.0 m/s instead of $(v_y)_f$. And the acceleration is not the same as in part (a)—it is now $-g$ (with the positive direction up) since we are ignoring air resistance. We do not know the time it takes the flea to reach maximum height, so we employ Equation 2.13 again because we know everything in that equation except Δy.

Solve: **(a)** Use $(v_y)_i = 0.0$ m/s and rearrange Equation 2.13.

$$a_y = \frac{(v_y)_f^2}{2\Delta y} = \frac{(1.0 \text{ m/s})^2}{2(0.50 \text{ mm})}\left(\frac{1000 \text{ mm}}{1 \text{ m}}\right) = 1000 \text{ m/s}^2$$

(b) Having learned the acceleration from part **(a)** we can now rearrange Equation 2.11 to find the time it takes to reach takeoff speed. Again use $(v_y)_i = 0.0$ m/s.

$$\Delta t = \frac{(v_y)_f}{a_y} = \frac{1.0 \text{ m/s}}{1000 \text{ m/s}^2} = .0010 \text{ s}$$

(c) This time $(v_y)_f = 0.0$ m/s as the flea reaches the top of its trajectory. Rearrange Equation 2.13 to get

$$\Delta y = \frac{-(v_y)_i^2}{2a_y} = \frac{-(1.0 \text{ m/s})^2}{2(-9.8 \text{ m/s}^2)} = 0.051 \text{ m} = 5.1 \text{ cm}$$

Assess: Just over 5 cm is pretty good considering the size of a flea. It is about 10–20 times the size of a typical flea. Check carefully to see that each answer ends up in the appropriate units.
The height of the flea at the top will round to 5.2 cm above the ground if you include the 0.050 cm during the initial acceleration phase before the feet actually leave the ground.

P2.69. Strategize: As soon as the ball leaves the student's hand, it is falling freely and thus kinematic equations hold.
Prepare: A visual overview of the ball's motion that includes a pictorial representation, a motion diagram, and a list of values is shown below. We label the ball's motion along the y-axis. The ball's acceleration is equal to the acceleration due to gravity that always acts vertically downward toward the center of the earth. The initial position of the ball is at the origin where $y = 0$, but the final position is below the origin at $y = -2.0$ m. Recall sign conventions, which tell us that v is positive and a is negative.

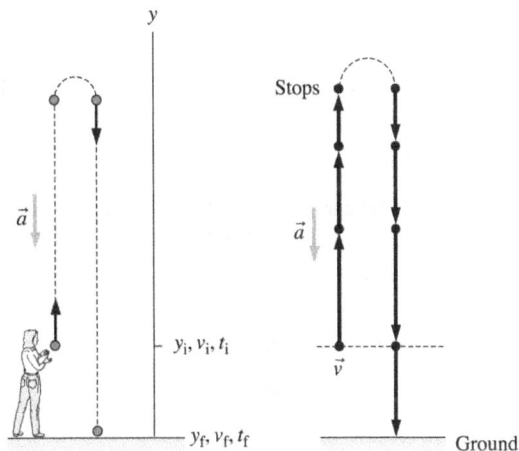

Solve: With all the known information, it is clear that we must use

$$y_f = y_i + v_i \Delta t + \frac{1}{2}a\Delta t^2$$

Substituting the known values

$$-2 \text{ m} = 0 \text{ m} + (15 \text{ m/s})t_f + (1/2)(-9.8 \text{ m/s}^2)t_f^2$$

The solution of this quadratic equation gives $t = 3.2$ s. The other root of this equation yields a negative value for t, which is not physical for this problem.
Assess: A time of 3.2 s is reasonable.

P2.71. Strategize: We treat the diver's motion as one-dimensional (purely vertical). Since the diver falls under the influence of gravity, acceleration is constant and we can use kinematic equations. Let us choose our axes such that $+y$ points vertically upward.

Prepare: We know the initial velocity of the diver, the acceleration due to gravity, and the height above the water. The first part of the question can be solved by applying Equation 2.12: $y_f = y_i + \left(v_y\right)_i \Delta t + \frac{1}{2} a_y \left(\Delta t\right)^2$, and solving for the time. Once the time is known, Equation 2.11 can be used to determine the final speed: $\left(v_y\right)_f = \left(v_y\right)_i + a_y \Delta t$.

Solve: (a) From Equation 2.12, we have a quadratic equation in time. Thus the solutions for the unknown time are given by

$$\Delta t = \left(-\left(v_y\right)_i \pm \sqrt{\left(v_y\right)_i^2 - 4\left(\frac{1}{2} a_y\right)(-\Delta y)} \right) \frac{1}{a_y}$$

$$\Delta t = \left(-(6.3 \text{ m/s}) \pm \sqrt{(6.3 \text{ m/s})^2 - 4\left(\frac{1}{2}(-9.8 \text{ m/s}^2)\right)(-(-3.0 \text{ m}))} \right) \frac{1}{(-9.8 \text{ m/s}^2)}$$

$$\Delta t = -0.37 \text{ s or } 1.7 \text{ s}$$

Because we want to know a duration of time after which the diver will reach the water, we want a positive time, so our answer is 1.7 s.

(b) Using the answer from part (a) (prior to rounding), Equation 2.11 yields

$$\left(v_y\right)_f = \left(v_y\right)_i + a_y \Delta t = (6.3 \text{ m/s}) + \left(-9.8 \text{ m/s}^2\right)(1.66 \text{ s}) = -9.9 \text{ m/s}.$$

We are asked for the speed, not the velocity or any component of the velocity. So we report 9.9 m/s.

Assess: Our answer to the first part fits our intuition that it takes around a second to go from a diving board to the water. The answer to the second part is reasonable for two reasons. Firstly, the component we calculated had the correct sign (since the diver should be moving downward). Secondly, the final speed is greater than the initial speed. This is reasonable since the diver would reach his initial speed as he fell down past the diving board, and would continue to speed up as he approached the water.

P2.73. Strategize: Clearly the acceleration changes in this problem. But during each phase of the motion (speed up, constant velocity, slowing down), let us assume that the acceleration is constant over each interval individually. That way we can apply the kinematic equations to each interval separately.

Prepare: A visual overview of car's motion that includes a pictorial representation, a motion diagram, and a list of values is shown below. We label the car's motion along the x-axis. This is a three-part problem. First the car accelerates, then it moves with a constant speed, and then it decelerates. The total displacement between the stop signs is equal to the sum of the three displacements, that is, $x_3 - x_0 = (x_3 - x_2) + (x_2 - x_1) + (x_1 - x_0)$.

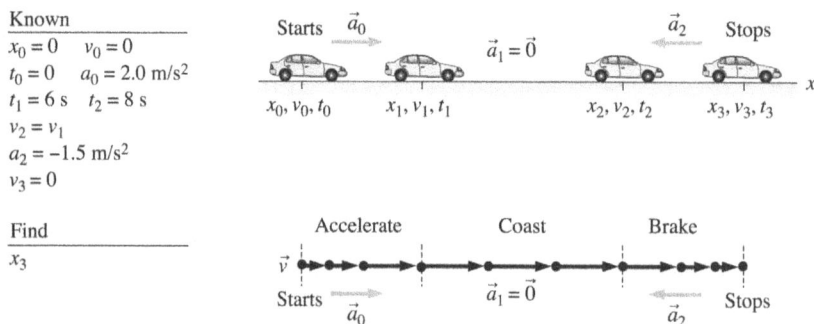

Known		Find
$x_0 = 0$ $v_0 = 0$		x_3
$t_0 = 0$ $a_0 = 2.0 \text{ m/s}^2$		
$t_1 = 6 \text{ s}$ $t_2 = 8 \text{ s}$		
$v_2 = v_1$		
$a_2 = -1.5 \text{ m/s}^2$		
$v_3 = 0$		

Solve: First, the car accelerates:

$$v_1 = v_0 + a_0(t_1 - t_0) = 0 \text{ m/s} + (2.0 \text{ m/s}^2)(6 \text{ s} - 0 \text{ s}) = 12 \text{ m/s}$$

$$x_1 = x_0 + v_0(t_1 - t_0) + \frac{1}{2} a_0(t_1 - t_0)^2 = 0 \text{ m} + \frac{1}{2}(2.0 \text{ m/s}^2)(6 \text{ s} - 0 \text{ s})^2 = 36 \text{ m}$$

Second, the car moves at v :

$$x_2 - x_1 = v_1(t_2 - t_1) + \frac{1}{2}a_1(t_2 - t_1)^2 = (12 \text{ m/s})(8 \text{ s} - 6 \text{ s}) + 0 \text{ m} = 24 \text{ m}$$

Third, the car decelerates:

$$v_3 = v_2 + a_2(t_3 - t_2) \Rightarrow 0 \text{ m/s} = 12 \text{ m/s} + (-1.5 \text{ m/s}^2)(t_3 - t_2) \Rightarrow (t_3 - t_2) = 8 \text{ s}$$

$$x_3 = x_2 + v_2(t_3 - t_2) + \frac{1}{2}a_2(t_3 - t_2)^2 \Rightarrow x_3 - x_2 = (12 \text{ m/s})(8 \text{ s}) + \frac{1}{2}(-1.5 \text{ m/s}^2)(8 \text{ s})^2 = 48 \text{ m}$$

Thus, the total distance between stop signs is

$$x_3 - x_0 = (x_3 - x_2) + (x_2 - x_1) + (x_1 - x_0) = 48 \text{ m} + 24 \text{ m} + 36 \text{ m} = 108 \text{ m}$$

or 110 m to two significant figures.

Assess: A distance of approximately 360 ft in a time of around 16 s with an acceleration/deceleration is reasonable.

P2.75. Strategize: As soon as the rocks are thrown, they fall freely and thus kinematics equations are applicable.
Prepare: A visual overview of the motion of the two rocks, one thrown down by Heather and the other thrown up at the same time by Jerry, that includes a pictorial representation, a motion diagram, and a list of values is shown below. We represent the motion of the rocks along the y-axis with origin at the surface of the water. The initial position for both cases is $y = 50$ m and similarly the final position for both cases is at $y = 0$. Recall sign conventions, which tell us that (v) is positive and (v) is negative.

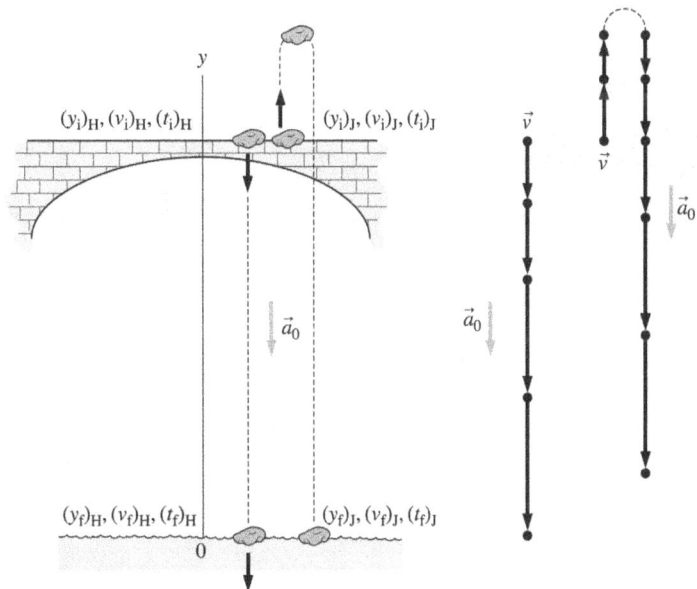

Solve: (a) For Heather,

$$(y_f)_H = (y_i)_H + (v_i)_H[(t_f)_H - (t_i)_H] + \frac{1}{2}a_0[(t_f)_H - (t_i)_H]^2$$

$$\Rightarrow 0 \text{ m} = (50 \text{ m}) + (-20 \text{ m/s})[(t_f)_H - 0 \text{ s}] + \frac{1}{2}(-9.8 \text{ m/s}^2)[(t_f)_H - 0 \text{ s}]^2$$

$$\Rightarrow 4.9 \text{ m/s}^2 \, (t_f)_H^2 + 20 \text{ m/s} \, (t_f)_H - 50 \text{ m} = 0$$

The two mathematical solutions of this equation are −5.83 s and +1.75 s. The first value is not physically acceptable since it represents a rock hitting the water before it was thrown, therefore, $(t) = 1.75$ s.

For Jerry,

$$(y_f)_J = (y_i)_J + (v_i)_J[(t_f)_J - (t_i)_J] + \frac{1}{2}a_0[(t_f)_J - (t_i)_J)]^2$$

$$\Rightarrow 0 \text{ m} = (50 \text{ m}) + (+20 \text{ m/s})[(t_f)_J - 0 \text{ s}] + \frac{1}{2}(-9.8 \text{ m/s}^2)[(t_f)_J - 0 \text{ s}]^2$$

Solving this quadratic equation will yield $(t) = -1.75$ s and $+5.83$ s. Again only the positive root is physically meaningful. The elapsed time between the two splashes is $(t) - (t) = 5.83$ s $- 1.75$ s $= 4.1$ s.
(b) Knowing the times, it is easy to find the impact velocities:

$$(v_f)_H = (v_i)_H + a_0[(t_f)_H - (t_i)_H] = (-20 \text{ m/s}) + (-9.8 \text{ m/s})(1.75 \text{ s} - 0 \text{ s}) = -37 \text{ m/s}$$

$$(v_f)_J = (v_i)_J + a_0[(t_f)_J - (t_i)_J] = (+20 \text{ m/s}) + (-9.8 \text{ m/s}^2)(5.83 \text{ s} - 0 \text{ s}) = -37 \text{ m/s}$$

The two rocks hit the water with equal speeds.
Assess: The two rocks hit the water with equal speeds because Jerry's rock has the same downward speed as Heather's rock when it reaches Heather's starting position during its downward motion.

P2.77. Strategize: When speeding up, we will assume that the acceleration of any creature (horse or human) is constant. Of course, once we are told that the creature reaches its top speed, the acceleration must drop to zero. During a period of constant acceleration, we can apply the kinematic equations.
Prepare: Use the kinematic equations with $(v_x)_i = 0$ m/s in the acceleration phase.
Solve: The man gains speed at a steady rate for the first 1.8 s to reach a top speed of

$$(v_x)_f = (v_x)_i + a_x \Delta t = 0 \text{ m/s} + (6.0 \text{ m/s}^2)(1.8 \text{ s}) = 10.8 \text{ m/s}$$

During this time he will go a distance of

$$\Delta x = \frac{1}{2}a_x(\Delta t)^2 = \frac{1}{2}(6.0 \text{ m/s}^2)(1.8 \text{ s})^2 = 9.72 \text{ m}$$

The man then covers the remaining 100 m $-$ 9.72 m $=$ 90.28 m at constant velocity in a time of

$$\Delta t = \frac{\Delta x}{v_x} = \frac{90.28 \text{ m}}{10.8 \text{ m/s}} = 8.4 \text{ s}$$

The total time for the man is then 1.8 s $+$ 8.4 s $=$ 10.2 s for the 100 m.
We now re-do all the calculations for the horse going 200 m. The horse gains speed at a steady rate for the first 4.8 s to reach a top speed of

$$(v_x)_f = (v_x)_i + a_x \Delta t = 0 \text{ m/s} + (5.0 \text{ m/s}^2)(4.8 \text{ s}) = 24 \text{ m/s}$$

During this time the horse will go a distance of

$$\Delta x = \frac{1}{2}a_x(\Delta t)^2 = \frac{1}{2}(5.0 \text{ m/s}^2)(4.8 \text{ s})^2 = 57.6 \text{ m}$$

The horse then covers the remaining 200 m $-$ 57.6 m $=$ 142.4 m at constant velocity in a time of

$$\Delta t = \frac{\Delta x}{v_x} = \frac{142.2 \text{ m}}{24 \text{ m/s}} = 5.9 \text{ s}$$

The total time for the horse is then 4.8 s $+$ 5.9 s $=$ 10.7 s for the 200 m.
The man wins the race (10.2 s $<$ 10.7 s), but he only went half the distance the horse did.
Assess: We know that 10.2 s is about right for a human sprinter going 100 m. The numbers for the horse also seem reasonable.

P2.79. Strategize: acceleration kinematic equations are applicable because both cars have constant accelerations.
Prepare: A visual overview of the two cars that includes a pictorial representation, a motion diagram, and a list of values is shown below. We label the motion of the two cars along the x-axis. Constant We can easily calculate the times (t) and (t) from the given information.

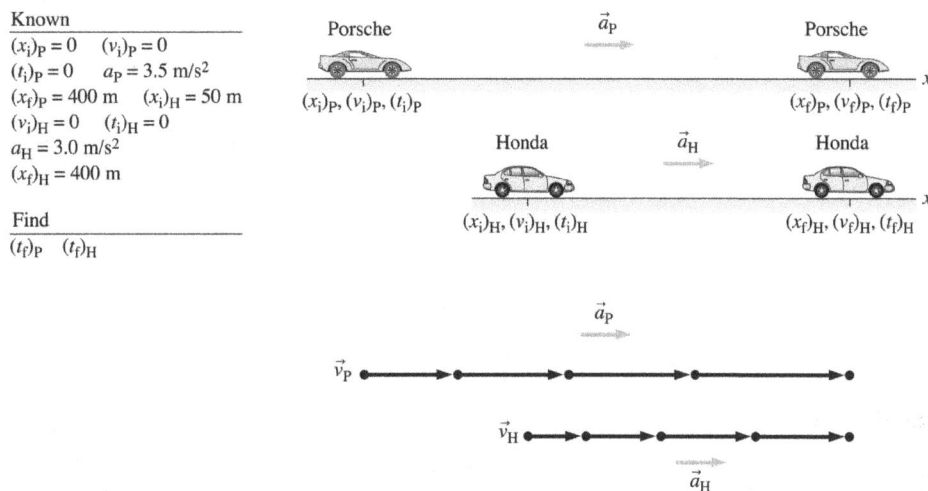

Solve: The Porsche's time to finish the race is determined from the position equation

$$(x_f)_P = (x_i)_P + (v_i)_P((t_f)_P - (t_i)_P) + \frac{1}{2}a_P((t_f)_P - (t_i)_P)^2$$

$$\Rightarrow 400 \text{ m} = 0 \text{ m} + 0 \text{ m} + \frac{1}{2}(3.5 \text{ m/s}^2)((t_f)_P - 0 \text{ s})^2 \Rightarrow (t_f)_P = 15 \text{ s}$$

The Honda's time to finish the race is obtained from Honda's position equation as

$$(x_f)_H = (x_i)_H + (v_i)_H((t_f)_H - (t_i)_H) + \frac{1}{2}a_H((t_f)_H - (t_i)_H)^2$$

$$400 \text{ m} = 100 \text{ m} + 0 \text{ m} + \frac{1}{2}(3.0 \text{ m/s}^2)((t_f)_H - 0 \text{ s})^2 \Rightarrow (t_f)_H = 14 \text{ s}$$

The Honda wins by 1.0 s.
Assess: It seems reasonable that the Honda would win given that it only had to go 300 m. If the Honda's head start had only been 50 m rather than 100 m the race would have been a tie.

P2.81. Strategize: There are two periods of constant acceleration, but the two accelerations are different. We assume constant acceleration of the rocket (and the bolt with it) and then constant acceleration due to gravity during freefall. We can apply kinematic equations to either period individually, but not over both with a single equation.
Prepare: A visual overview of the motion of the rocket and the bolt that includes a pictorial representation, a motion diagram, and a list of values is shown below. We represent the rocket's motion along the y-axis. The initial velocity of the bolt as it falls off the side of the rocket is the same as that of the rocket, that is, (v) = (v) and it is positive since the rocket is moving upward. The bolt continues to move upward with a deceleration equal to g = 9.8 m/s before it comes to rest and begins its downward journey.

Known
$$(y_i)_R = 0 \quad (v_i)_R = 0$$
$$(t_i)_R = 0 \quad (t_f)_R = 4.0 \text{ s}$$
$$(y_i)_B = (y_f)_R \quad (v_i)_B = (v_f)_R$$
$$(t_i)_B = (t_f)_R \quad a_B = -9.8 \text{ m/s}^2$$
$$(y_f)_B = 0 \quad (t_f)_B = 6.05$$

Find
$$a_R$$

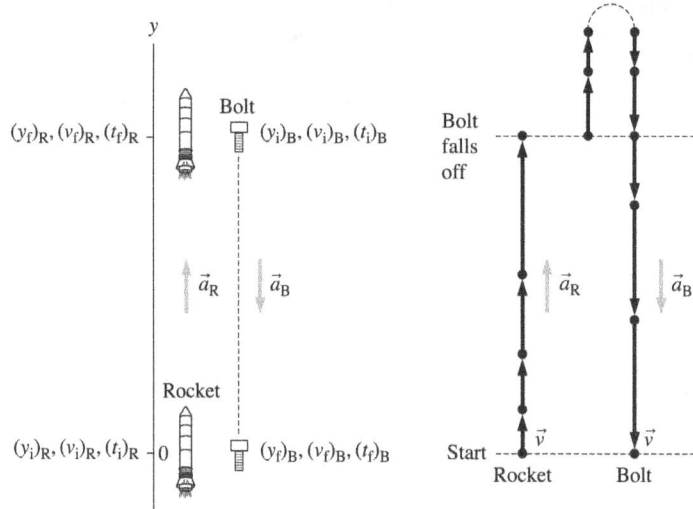

Solve: To find a we look first at the motion of the rocket:

$$(y_f)_R = (y_i)_R + (v_i)_R((t_f)_R - (t_i)_R) + \frac{1}{2}a_R((t_f)_R - (t_i)_R)^2$$

$$= 0 \text{ m} + 0 \text{ m/s} + \frac{1}{2}a_R(4.0 \text{ s} - 0 \text{ s})^2 = 8a_R$$

So we must determine the magnitude of y or y. Let us now look at the bolt's motion:

$$(y_f)_B = (y_i)_B + (v_i)_B((t_f)_B - (t_i)_B) + \frac{1}{2}a_B((t_f)_B - (t_i)_B)^2$$

$$0 = (y_f)_R + (v_f)_R(6.0 \text{ s} - 0 \text{ s}) + \frac{1}{2}(-9.8 \text{ m/s}^2)(6.0 \text{ s} - 0 \text{ s})^2$$

$$\Rightarrow (y_f)_R = 176.4 \text{ m} - (6.0 \text{ s})(v_f)_R$$

Since $(v_f)_R = (v_i)_R + a_R((t_f)_R - (t_i)_R) = 0 \text{ m/s} + 4a_R = 4a_R$ the above equation for (y) yields $(y) = 176.4 - 6.0(4a)$. We know from the first part of the solution that $(y) = 8a$. Therefore, $8a = 176.4 - 24.0a$ and hence $a = 5.5 \text{ m/s}$.
Assess: This seems like a reasonable acceleration for a rocket.

P2.83. Strategize: Azin freefall due to gravity is a constant in either case. On Earth $a_y = -9.8 \text{ m/s}^2$ (constant), and on the moon $a_y = -1.63 \text{ m/s}^2$ (constant). So we can use kinematic equations.

Prepare: We assume that the astronaut's safe landing speed on the moon should be the same as the safe landing speed on the earth.
Solve: The brute force method is to compute the landing speed on the earth with Equation 2.13, and plug that back into the Equation 2.13 for the moon and see what the Δy could be there. This works, but is unnecessarily complicated and gives information (the landing speed) we don't really need to know.
To be more elegant, set up Equation 2.13 for the earth and moon, with both initial velocities zero, but then set the final velocities (squared) equal to each other.

$$(v_{earth})_f^2 = 2(a_{earth})\Delta y_{earth} \quad (v_{moon})_f^2 = 2(a_{moon})\Delta y_{moon}$$
$$2(a_{earth})\Delta y_{earth} = 2(a_{moon})\Delta y_{moon}$$

Dividing both sides by $2(a_{moon})\Delta y_{earth}$ gives

$$\frac{a_{earth}}{a_{moon}} = \frac{\Delta y_{moon}}{\Delta y_{earth}}$$

This result could also be accomplished by dividing the first two equations; the left side of the resulting equation would be 1, and then one arrives at our same result.

Since the acceleration on the earth is six times greater than on the moon, then one can safely jump from a height six times greater on the moon and still have the same landing speed.

So the answer is B.

Assess: Notice that in the elegant method we employed we did not need to find the landing speed (but for curiosity's sake it is 4.4 m/s, which seems reasonable).

VECTORS AND MOTION IN TWO DIMENSIONS

Q3.1. Reason: (a) If one component of the vector is zero, then the other component must not be zero (unless the whole vector is zero). Thus the magnitude of the vector will be the value of the other component. For example, if $A_x = 0$ m and $A_y = 5$ m, then the magnitude of the vector is

$$A = \sqrt{(0 \text{ m})^2 + (5 \text{ m})^2} = 5 \text{ m}$$

(b) A zero magnitude says that the length of the vector is zero, thus each component must be zero.
Assess: It stands to reason that a vector can have a nonzero magnitude with one component zero as long as the other one isn't. It also makes sense that for the magnitude of the vector to be zero *all* the components must be zero.

Q3.3. Reason: Consider two vectors $\vec{A}$ and $\vec{B}$. Their sum can be found using the method of algebraic addition. In Question 3.2 we found that the components of the zero vector are both zero. The components of the resultant of $\vec{A}$ and $\vec{B}$ must then be zero also. So

$$R_x = A_x + B_x = 0$$
$$R_y = A_y + B_y = 0$$

Solving for the components of $\vec{B}$ in terms of $\vec{A}$ gives $B_x = -A_x$ and $B_y = -A_y$. Then the magnitude of $\vec{B}$ is $\sqrt{(B_x)^2 + (B_y)^2} = \sqrt{(-A_x)^2 + (-A_y)^2} = \sqrt{(A_x)^2 + (A_y)^2}$. So then the magnitude of $\vec{B}$ is exactly equal to the magnitude of $\vec{A}$. So, no, two vectors of unequal magnitudes cannot sum to $\vec{0}$.
Assess: For two vectors to add to zero, the vectors must have exactly the same magnitude and point in opposite directions.

Q3.5. Reason: The ones that are constant are v_x, a_x, and a_y. Furthermore, a_x is not only constant, it is zero.

Assess: There are instants when other quantities can be zero, but not throughout the flight. Remember that $a_y = -g$ throughout the flight and that v_x is constant; that is, projectile motion is nothing more than the combination of two simple kinds of motion: constant horizontal velocity and constant vertical acceleration.

Q3.7. Reason: By extending their legs forward, the runners increase their time in the air. As you will learn in chapter 7, the "center of mass" of a projectile follows a parabolic path. By raising their feet so that their feet are closer to their center of mass, the runners increase the time it takes for their feet to hit the ground. By increasing their time of flight, they increase their range. Also, having their feet ahead of them means that their feet will land ahead of where they would have landed otherwise.
Assess: By simply moving their feet, runners can change their time of flight and change the spot where their feet land.

Q3.9. Reason: Ignoring air resistance, the horizontal motion of an object has no effect on its vertical motion. So the ball dropped from a moving car and the ball dropped by a stationary person should have identical vertical motion. Both balls should strike the ground at the same instant.

Assess: Since horizontal and vertical directions are orthogonal, any kinematic variable along the horizontal direction has no effect on a kinematic variable in the vertical direction.

Q3.11. Reason: The lower the angle of the slope, the lower the acceleration down the slope since acceleration along a ramp is given by $g \sin \theta$. Furthermore, the speed at the bottom of a slope will be less if the acceleration is less.

Assess: A lower angle gives the skier a lower velocity and, consequently, better control.

Q3.13. Reason: The time for an object to hit the ground does not depend on its horizontal speed, but only on its height and initial vertical speed. When the pilot goes twice as fast, all that changes is the horizontal speed of the projectile. Therefore the time of flight will be the same in both cases, 2.0 s. However the distance travelled horizontally will be doubled since the horizontal speed is doubled and the time of flight is the same. At the doubled speed, the weight will travel twice as far, or 200 m.

Assess: The above answer assumed no air resistance. Actually the greater speed of the faster projectile will slightly increase the time of flight since a faster object experiences more air resistance.

Q3.15. Reason: The magnitude of the acceleration is $\frac{v^2}{r}$ which is constant since the speed is constant. So the magnitude of the acceleration doesn't change. But the direction does since it always points toward the center of the circle.

Assess: In circular motion, the acceleration is always toward the center of the circle and is called centripetal acceleration.

Q3.17. Reason: To make a tighter turn with a smaller radius, you need to reduce your speed. If you are traveling at the greatest speed which is safe, then you are accelerating at the highest acceleration which is safe. Now if the radius is reduced, this tends to increase the acceleration above safe values. To bring it back down, your speed should be reduced. If we solve the formula for centripetal acceleration, $a = v^2 / r$, for v, we have: $v = \sqrt{ar}$. So v is proportional to the square root of r. For example, if we take a turn with a radius which is four times smaller, we need to cut our speed in half.

Assess: The maximum safe speed is proportional to the square root of the radius of the turn.

Q3.19. Reason: Anna is running away from ball 1. If she observes it approaching her at speed v then the ball must be overcoming Anna's speed and catching up with her at that speed. Ball 2 does not have to be moving at speed v, since Anna is approaching it. Thus ball 1 is moving faster (relative to the ground) than ball 2.

Assess: Anna is reporting speeds relative to herself, whereas we are asked about speeds relative to the ground or to the friends who threw the balls.

Q3.21. Reason: To generate a vector which points to the left, we could add two vectors which point left, one pointing up and the other down. In C, $\vec{Q}$ and $-\vec{P}$ fit this description so their sum points to the left. The various vector combinations are shown.

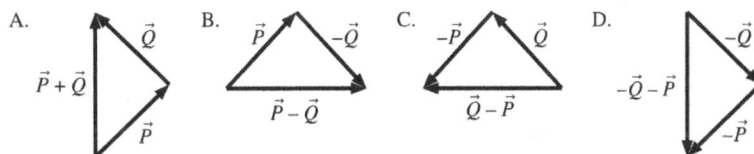

The answer is C.

Assess: If two vectors have equal and opposite components in a certain direction, say the x direction, then when we add the vectors, the equal and opposite components will cancel and leave us with a vector perpendicular to that direction.

Q3.23. Reason: The car is traveling at constant speed, so the only possible cause for accelerations is a change in direction.
(a) At point 1 the car is traveling straight to the right on the diagram, so its velocity is straight to the right. The correct choice is B.
(b) At point 1 the car is traveling at constant speed and not changing direction so its acceleration is zero. The correct choice is E.
(c) The car's velocity at this point on the curve is in the direction of its motion, which is in the direction shown at choice C.
(d) The car is moving on a portion of a circle. The acceleration of an object moving in a circle is always directly toward the center of the circle. The correct choice is D.
(e) The car is moving on a portion of a circle at point 3. The instantaneous velocity vector is directly to the right, which is choice B.
(f) The car is accelerating because it is moving on a portion of a circle. The acceleration is toward the center of the circle, which is in direction A.
Assess: The instantaneous velocity of a particle is always in the direction of its motion at that point in time. For motion in a circle, the direction of the acceleration is always toward the center of the circle.

Q3.25. Reason: The maximum height the ball reaches only depends on the initial velocity it had in the vertical direction. The y-component of the velocity of this ball is

$$(v_y)_i = v_i \sin(\theta) = (23.0 \text{ m/s}) \sin(37.0°) = 13.8 \text{ m/s}$$

In order to reach the same height when being thrown vertically upward the ball's initial velocity must be 13.8 m/s. The correct choice is A.
Assess: Projectile motion is made up of two independent motions: uniform motion at constant velocity in the horizontal direction and free-fall motion in the vertical direction.

Q3.27. Reason: We can apply one kinematic equation to the vertical direction (which we call y) to determine how long the balloon is in the air, then apply a kinematic equation to the horizontal direction (which we call x) to determine how far it can go in that amount of time. Let us use $\Delta y = (v_y)_i \Delta t + \frac{1}{2} a_y (\Delta t)^2$. Noting that the initial velocity is entirely horizontal such that $(v_y)_i = 0$, we find $\Delta t = \sqrt{\dfrac{2\Delta y}{a_y}} = \sqrt{\dfrac{2(-10 \text{ m})}{(-9.8 \text{ m/s}^2)}} = 1.43 \text{ s}$. Now we can use the same kinematic equation again, but for the horizontal direction. This time, we note that there is no horizontal acceleration, such that $\Delta x = (v_x)_i \Delta t + \frac{1}{2} a_x (\Delta t)^2 = (v_x)_i \Delta t = (8.2 \text{ m/s})(1.43 \text{ s}) = 12 \text{ m}$.
The correct answer is C.
Assess: We have separated the motion into horizontal and vertical components and connected them through time.

Q3.29. Reason: Let us try to avoid the calculator as in Question 3.26. Since this question asks for range, we need the horizontal component of the ball's speed. For that you need to know that $\cos 30° = \sqrt{3}/2$ and that $\sqrt{3} \approx 1.73$. The horizontal component is $(v_x)_i = v_i \cos \theta = (20 \text{ m/s})(\sqrt{3}/2) = 17.3 \text{ m/s}$. Since v_x is constant (there is no acceleration in the x direction), the range of the football is given by

$$R = (v_x)_i \Delta t = (17.3 \text{ m/s})(2 \text{ s}) = 34.6 \text{ m} \approx 35 \text{ m}$$

Because we have used the time from the previous question, this answer only has one or two significant figures. The correct choice is C.
Assess: As Questions 3.26 and 3.27 illustrate, by using $g \approx 10 \text{ m/s}^2$, we can estimate the main features of a projectile motion problem such as the time of flight and the range without using a calculator.

Q3.31. Reason: The magnitude of the centripetal acceleration is

$$a_c = \frac{v^2}{r} = \frac{(68 \text{ m/s})^2}{95 \text{ m}} = 48.7 \text{ m/s}^2$$

Divide this by 9.8 m/s^2 to find the answer in units of g: $\frac{48.7 \text{ m/s}^2}{9.8 \text{ m/s}^2} = 5g$ so the correct choice is E.

Assess: People can withstand 5g accelerations, but usually not for a long time.

Problems

P3.1. Strategize: This problem involves vector addition. We add vectors "tip to tail".

Prepare: (a) To find $\vec{A} + \vec{B}$, we place the tail of vector $\vec{B}$ on the tip of vector $\vec{A}$ and then connect vector $\vec{A}$'s tail with vector $\vec{B}$'s tip.

(b) To find $\vec{A} - \vec{B}$, we note that $\vec{A} - \vec{B} = \vec{A} + (-\vec{B})$. We place the tail of vector $-\vec{B}$ on the tip of vector $\vec{A}$ and then connect vector $\vec{A}$'s tail with the tip of vector $-\vec{B}$.

Solve:

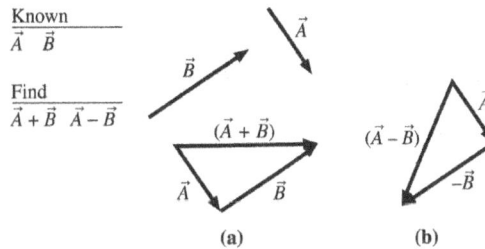

P3.3. Strategize: This problem involves vector addition. We will draw vectors tip-to-tail to add them.

Prepare: In order for the sum of three vectors to be zero, adding them tip to tail must bring us back to the point where we started. We can draw $\vec{A} + \vec{B}$, and then sketch in the vector that brings us back to where $\vec{A}$ began.

Solve: Figure P3.3 (a) shows the copied vectors, (b) shows the correct tip-to-tail addition $\vec{A} + \vec{B}$, and (c) shows the vector $\vec{C}$ required to make $\vec{A} + \vec{B} + \vec{C} = 0$.

Assess: It is particularly clear that this sum is zero, if one thinks of the arrows as displacement vectors. The three displacements clearly would return a person to the starting point.

P3.5. Strategize: This problem involves two orthogonal legs of a triangle. In right triangles, we can use trigonometric functions.

Prepare: The position vector $\vec{d}$ whose magnitude d is 10 m has an x-component of 6 m. It makes an angle θ with the $+x$-axis in the first quadrant. We will use trigonometric relations to find the y-component of the position vector.

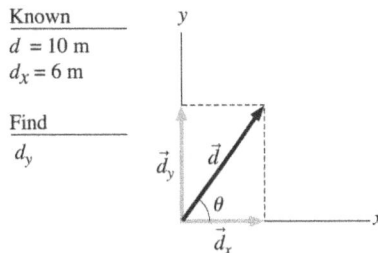

Solve: Using trigonometry, $d_x = d \cos \theta$ or $6 \text{ m} = (10 \text{ m}) \cos \theta$. This gives $\theta = 53.1°$. Thus the y-component of the position vector $\vec{d}$ is $d_y = d \sin \theta = (10 \text{ m}) \sin 53.1° = 8.0 \text{ m}$.

Assess: The y-component is positive since the position vector is in the first quadrant.

P3.7. Strategize: This problem involves a right triangle. In right triangles, we can use trigonometric functions.
Prepare: The figure below shows the components $v_\parallel$ and $v_\perp$, and the angle θ. We will use Tactics Box 3.2 to find the sign attached to the components of a vector.

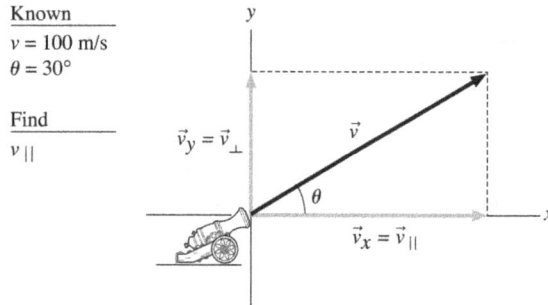

Known
$v = 100$ m/s
$\theta = 30°$

Find
$v_\parallel$

$\vec{v}_y = \vec{v}_\perp$

$\vec{v}$

θ

$\vec{v}_x = \vec{v}_\parallel$

Solve: We have $\vec{v} = \vec{v}_x + \vec{v}_y = \vec{v}_\parallel + \vec{v}_\perp$. Thus, $v_\parallel = v\cos\theta = (100$ m/s$)\cos 30° = 87$ m/s.

Assess: For the small angle of 30°, the obtained value of 87 m/s for the horizontal component is reasonable.

P3.9. Strategize: This problem involves a right triangle. In right triangles, we can use trigonometric functions.
Prepare: We will follow rules given in Tactics Box 3.2.

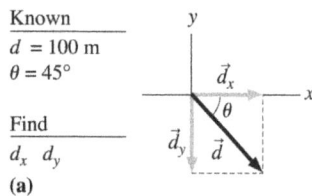

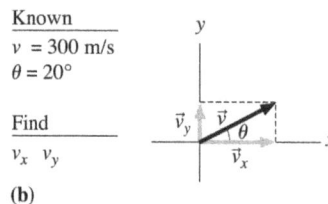

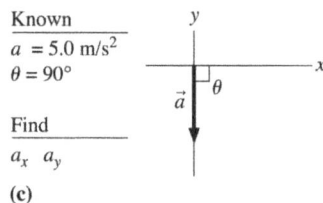

Known
$d = 100$ m
$\theta = 45°$

Find
d_x d_y

$\vec{d}_x$

θ

$\vec{d}_y$ $\vec{d}$

(a)

Known
$v = 300$ m/s
$\theta = 20°$

Find
v_x v_y

$\vec{v}_y$ $\vec{v}$

θ

$\vec{v}_x$

(b)

Known
$a = 5.0$ m/s^2
$\theta = 90°$

Find
a_x a_y

$\vec{a}$ θ

(c)

Solve: (a) Vector $\vec{d}$ points to the right and down, so the components d_x and d_y are positive and negative, respectively:

$$d_x = d\cos\theta = (100\text{ m})\cos 45° = 71\text{ m} \quad d_y = -d\sin\theta = -(100\text{ m})\sin 45° = -71\text{ m}$$

(b) Vector $\vec{v}$ points to the right and up, so the components v_x and v_y are both positive:

$$v_x = v\cos\theta = (300\text{ m/s})\cos 20° = 280\text{ m/s} \quad v_y = v\sin\theta = (300\text{ m/s})\sin 20° = 100\text{ m/s}$$

(c) Vector $\vec{a}$ has the following components:

$$a_x = -a\cos\theta = -(5.0\text{ m/s}^2)\cos 90° = 0.0\text{ m/s}^2 \quad a_y = -a\sin\theta = -(5.0\text{ m/s}^2)\sin 90° = -5.0\text{ m/s}^2$$

Assess: The components have the same units as the vectors. Note the minus signs we have manually inserted according to Tactics Box 3.2.

P3.11. Strategize: This problem involves a right triangle. In right triangles, we can use trigonometric functions and the Pythagorean Theorem.
Prepare: We will draw the vectors to scale as best we can and label the angles from the positive x-axis (positive angles go CCW). We also use Equations 3.7 and 3.8. Make sure your calculator is in degree mode.

Solve: (a)

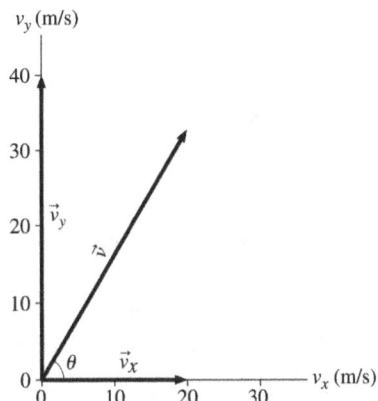

$$v = \sqrt{(v_x)^2 + (v_y)^2} = \sqrt{(20 \text{ m/s})^2 + (40 \text{ m/s})^2} = 45 \text{ m/s}$$

$$\theta = \tan^{-1}\left(\frac{v_y}{v_x}\right) = \tan^{-1}\left(\frac{40 \text{ m/s}}{20 \text{ m/s}}\right) = \tan^{-1}(2) = 63°$$

(b)

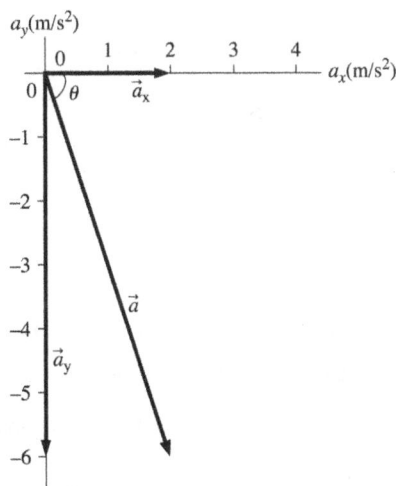

$$a = \sqrt{(a_x)^2 + (a_y)^2} = \sqrt{(2.0 \text{ m/s}^2)^2 + (-6.0 \text{ m/s}^2)^2} = 6.3 \text{ m/s}^2$$

$$\theta = \tan^{-1}\left(\frac{a_y}{a_x}\right) = \tan^{-1}\left(\frac{-6.0 \text{ m/s}^2}{2.0 \text{ m/s}^2}\right) = \tan^{-1}(-3) = -72°$$

Assess: In each case the magnitude is longer than either component, as is required for the hypotenuse of a right triangle. The negative angle in part **(b)** corresponds to a clockwise direction from the positive x-axis.

P3.13. Strategize: This problem is not given in terms of right triangles. We must find orthogonal components and put the problem in terms of a right triangle in order to use trigonometry and the Pythagorean Theorem.

Prepare: Draw a diagram of the situation.

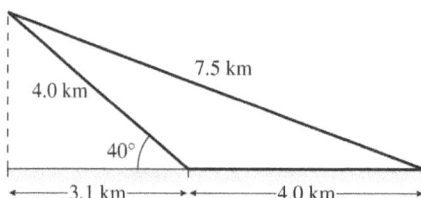

Solve: The x-component of the second leg of the trip is $(4.0 \text{ km})(\cos 40°) = -3.064 \text{ km}$ which is about 3.1 km more west.

The total distance west of their initial position is $4.0 \text{ km} + 3.1 \text{ km} = 7.1 \text{ km}$.

The magnitude of the total displacement can be computed from the Pythagorean theorem:

$$\sqrt{(4.0 \text{ km} + 3.064 \text{ km})^2 + ((4.0 \text{ km}) \sin 40°)^2} = 7.5 \text{ km}$$

Assess: The magnitude of the total displacement can also be computed from the law of cosines:

$$d^2 = (4.0 \text{ km})^2 + (4.0 \text{ km})^2 - 2(4.0 \text{ km})(4.0 \text{ km}) \cos 140° \Rightarrow d = 7.5 \text{ km}$$

P3.15. Strategize: This problem involves a right triangle. In right triangles, we can use trigonometric functions.

Prepare: We'll find v_y and then determine the meters of elevation in 1 min.

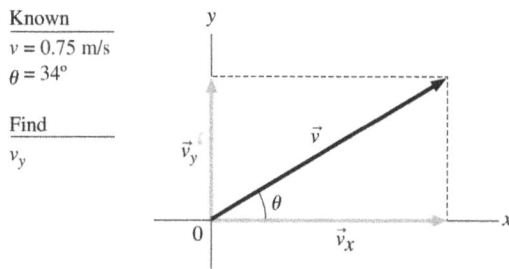

Solve:

$$v_y = v \sin \theta = (0.75 \text{ m/s}) \sin 34° = 0.42 \text{ m/s}$$

$$\Delta y = v_x \Delta t = (0.42 \text{ m/s})(60 \text{ s}) = 25 \text{ m}$$

Assess: This hill is steep, so a 25 m rise in elevation in one minute is reasonable.

P3.17. Strategize: This problem involves motion on a ramp. Since the acceleration is constant (a component of gravity), we can use kinematic equations.

Prepare: A visual overview of the car's motion that includes a pictorial representation, a motion diagram, and a list of values is shown below. We have labeled the x-axis along the incline. Note that the problem "ends" at a turning point, where the car has an instantaneous speed of 0 m/s before rolling back down. The rolling back motion is *not* part of this problem. If we assume the car rolls without friction, then we have motion on a frictionless inclined plane with acceleration $a = -g \sin \theta = -g \sin 5.0° = -0.854 \text{ m/s}^2$.

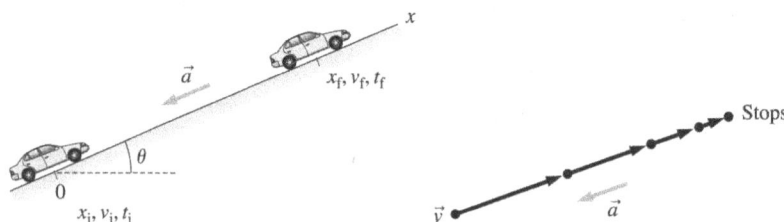

Solve: Constant acceleration kinematics gives

$$v_f^2 = v_i^2 + 2a(x_f - x_i) \Rightarrow 0 = v_i^2 + 2ax_f \Rightarrow x_f = -\frac{v_i^2}{2a} = -\frac{(30 \text{ m/s})^2}{2(-0.854 \text{ m/s}^2)} = 530 \text{ m}$$

Notice how the two negatives canceled to give a positive value for x_f.

Assess: We must include the minus sign because the $\vec{a}$ vector points *down* the slope, which is in the negative x-direction.

P3.19. Strategize: This problem involves motion on a ramp. Since the acceleration is constant (a component of gravity), we can use kinematic equations.
Prepare: Make a sketch with tilted axes with the x-axis parallel to the ramp and the angle of inclination labeled. We must also make a bold assumption that the piano rolls down as if it were an object sliding down with no friction.

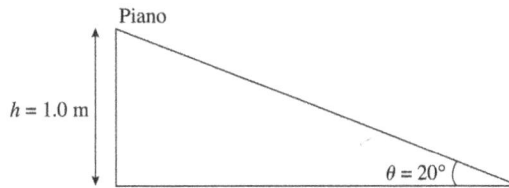

As part of the preparation, compute the length of the ramp in the new tilted x-y coordinates.

$$L = \frac{h}{\sin \theta} = \frac{1.0 \text{ m}}{\sin 20°} = 2.9 \text{ m}$$

The acceleration in the new coordinate system will be $a_x = g \sin \theta = (9.8 \text{ m/s}^2) \sin 20° = 3.4 \text{ m/s}^2$.

Solve: Since this is a case of constant acceleration we can use Equation 2.12 with $x_i = 0.0$ m and $(v_x)_i = 0.0$ m/s.

$$x_f = \frac{1}{2}a_x(\Delta t)^2$$

Solve for Δt, and use $x_f = 2.9$ m and $a_x = 3.4$ m/s^2, which we obtained previously.

$$\Delta t = \sqrt{\frac{2x_f}{a_x}} = \sqrt{\frac{2(2.9 \text{ m})}{3.4 \text{ m/s}^2}} = 1.3 \text{ s}$$

Assess: They may catch it if they have quick reactions, but the piano will be moving 4.5 m/s when it reaches the bottom.

P3.21. Strategize: This problem involves drawing vectors in two dimensions.
Prepare: We can find the positions and velocity and acceleration vectors using a motion diagram.

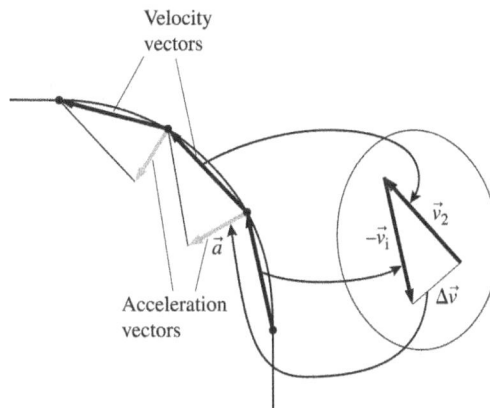

Solve: The figure gives several points along the car's path. The velocity vectors are obtained by connecting successive dots. The acceleration vectors are obtained by subtracting successive velocity vectors. The acceleration vectors point toward the center of the diagram.

Assess: Notice that the acceleration points toward the center of the turn. As you will learn in chapter 4, whenever your car accelerates, you feel like you are being pushed the opposite way. This is why you feel like you are being pushed away from the center of a turn.

P3.23. Strategize: We draw acceleration arrows by finding the difference between adjacent velocity arrows.

Prepare: The first few arrows seem to have constant magnitude and direction. Then the arrows begin to veer downward, so the acceleration vectors must have some downward component.

Solve: (a)

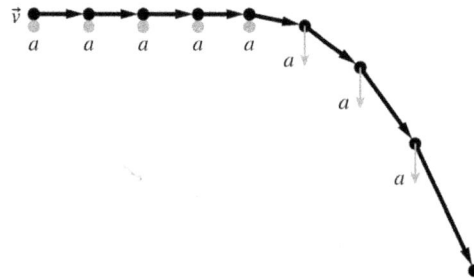

(b) A ball rolls along a table at a constant speed, then rolls off the edge of the table and falls under the influence of gravity.

Assess: The acceleration is constant and equal to zero, then suddenly increases to a downward constant value, as in a ball rolling off a table.

P3.25. Strategize: The acceleration determines the rate of change of the velocity vector.

Prepare: Since the speed is constant, there cannot be any component parallel to the velocity.

Solve: Since the velocity vector is curving to the right, the acceleration vector must point to the right. The answer is C.

Assess: The acceleration vector points in the direction of the change in the velocity vector.

P3.27. Strategize: We will assume the ball is in free fall (*i.e.*, we neglect air resistance). We will use the fact that acceleration is purely in the vertical direction to determine the horizontal velocity as a function of time, and we will use the fact that the acceleration is constant to determine the vertical velocity as a function of time.

Prepare: The trajectory of a projectile is a parabola because it is a combination of constant horizontal velocity $(a_x = 0.0 \text{ m/s}^2)$ combined with constant vertical acceleration $(a_y = -g)$. In this case we see only half of the parabola.

The initial speed given is all in the horizontal direction, that is, $(v_x)_i = 5.0 \text{ m/s}$, which is constant, and $(v_y)_i = 0.0 \text{ m/s}$, which changes linearly in time.

Solve:

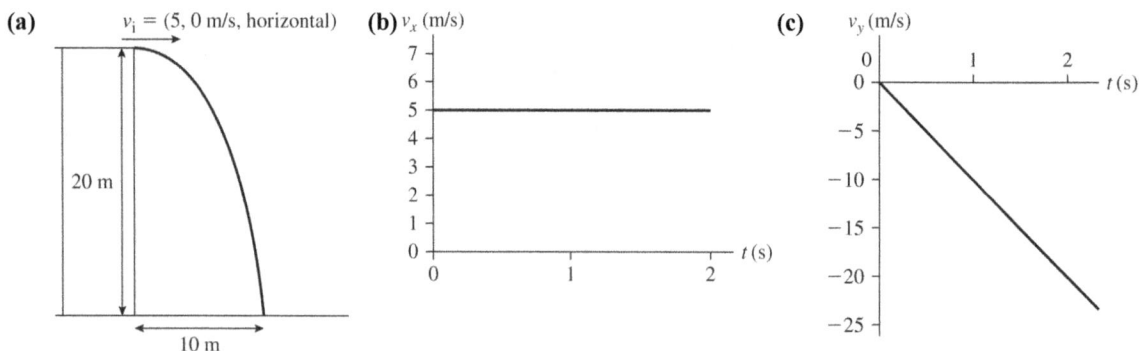

(d) This is a two-step problem. We first use the vertical direction to determine the time it takes, then plug that result into the equation for the horizontal direction.

$$\Delta y = \frac{1}{2} a_y (\Delta t)^2$$

$$\Delta t = \sqrt{\frac{2\Delta y}{a_y}} = \sqrt{\frac{2(-20 \text{ m})}{-9.8 \text{ m/s}^2}} = 2.0 \text{ s}$$

We use the 2.0 s in the equation for the horizontal motion.

$$\Delta x = v_x \Delta t = (5.0 \text{ m/s})(2.0 \text{ s}) = 10 \text{ m}$$

Assess: The answers seem reasonable, and we would get the same answers to two significant figures in a quick mental calculation using $g \approx 10 \text{ m/s}^2$. In fact, I did this before computing the algebra so I would know how to scale the graphs.

P3.29. Strategize: Assume the water is in free fall. We will also treat the water as though it is made up of particles, each of which roughly obeys the projectile motion we expect for solid objects. We ignore

Prepare: We can use the vertical-position equation from Synthesis 3.1 to find the time it takes the water to reach the floor: $y_f = y_i + (v_y)_i \Delta t - \frac{1}{2} g (\Delta t)^2$. The distance the water travels horizontally is governed by the horizontal-position equation from Synthesis 3.1. $x_f = x_i + (v_x)_i \Delta t$.

Solve: Refer to the visual overview shown.

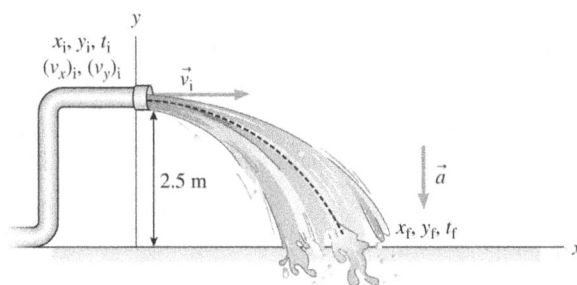

The initial vertical velocity is zero. Take the creek as the origin of coordinates. The ball falls from $y_i = 2.5 \text{ m}$ and lands at $y_f = 0 \text{ m}$. Rearranging the vertical equation to solve for time, we find

$$\Delta t = \sqrt{\frac{-2\Delta y}{g}} = \sqrt{\frac{-2(-2.5 \text{ m})}{(9.8 \text{ m/s}^2)}} = 0.71 \text{ s}.$$

(b) Inserting the result from (a), we find

$$x_f = x_i + (v_x)_i \Delta t = (0.0 \text{ m}) + (1.5 \text{ m/s})(0.714 \text{ s}) = 1.1 \text{ m}.$$

Assess: This is the right order of magnitude, since the water is only in motion for less than one second.

P3.31. Strategize: This problem involves constant acceleration (as in projectile motion). So we can use the kinematic equations in the horizontal and vertical directions.

Prepare: If we consider half the motion from the launch to the peak height, then we can use Equation 2.13 to determine the initial velocity in the vertical direction (which we will call y). Then we can use Equation 2.11 to determine the time that the mountain lion was in the air. We also know that there is no acceleration in the x direction (ignoring air resistance). So once we have that hang time, we can easily determine the initial velocity in the horizontal direction using $(v_x)_i = \Delta x / \Delta t$.

Solve: (a) Considering only half the trip such that the "final" time is at the peak height, Equation 2.13 yields

$(v_y)_f^2 = (v_y)_i^2 + 2a_y\Delta y \Rightarrow (v_y)_i = \sqrt{-2a_y\Delta y} = \sqrt{-2(-9.8 \text{ m/s}^2)(3.0 \text{ m})} = 7.67 \text{ m/s}.$

Now, employing Equation 2.11 to the same time interval, we find

$$(v_y)_f = (v_y)_i + a_y\Delta t \Rightarrow \Delta t = -(v_y)_i / a_y = -(7.67 \text{ m/s})/(-9.8 \text{ m/s}^2) = 0.782 \text{ s}.$$

But note that this is the time for only half the projectile motion of the mountain lion. By symmetry, the full time should be twice that: 1.56 s.

Finally, using $(v_x)_i = \Delta x / \Delta t$ (because there is no acceleration in the horizontal direction), we find

$(v_x)_i = \Delta x / \Delta t = (10 \text{ m})/(1.56 \text{ s}) = 6.4 \text{m/s}.$

Now that we have the horizontal and vertical components of the velocity, we can find its magnitude (the speed) using

the Pythagorean Theorem: $v_i = \sqrt{(v_x)_i^2 + (v_y)_i^2} = \sqrt{(6.4 \text{ m/s})^2 + (7.67 \text{ m/s})^2} = 10 \text{ m/s}.$

(b) Now that we know the horizontal and vertical components of the mountain lion's initial velocity, we can determine the angle from horizontal using simple trigonometry: $\tan(\theta) = opp. / adj. = (v_y)_i / (v_x)_i.$ So

$$\theta = \tan^{-1}\left((v_y)_i / (v_x)_i\right) = \tan^{-1}\left((7.67 \text{ m/s})/(6.40 \text{ m/s})\right) = 50°$$

Assess: Components of velocity equal to a few meters per second are reasonable for a mountain lion. The initial jumping angle fits expectations, since it would need to have a significant initial upward component to stay in the air 1.56 s.

P3.33. Strategize: We can consider vertical and horizontal motion separately to obtain an expression for the distance in terms of the takeoff speed.

Prepare: We are asked to find the take-off speed and horizontal speed of the kangaroo given its initial angle, 20°, and its range. Since the horizontal speed is given by $v_x = v \cos\theta$ and the time of flight is given by $\Delta t = 2v \sin\theta/g$, the range of the kangaroo is given by the product of these: $\Delta x = 2v \sin\theta \cos\theta/g$.

Solve: (a) We can solve the above formula for v and then plug in the range and angle to find the take-off speed:

$$v = g\Delta x/(2 \sin\theta \cos\theta) = (9.8 \text{ m/s}^2)(10 \text{ m})/(2 \sin 20° \cos 20°) = 12.3 \text{ m/s}$$

Its take-off speed is 12 m/s, to two significant figures.

(b) Its horizontal speed is given by $v_x = v\cos\theta = (12.3 \text{ m/s}) \cos 20° = 11.6 \text{ m/s}$ or 12 m/s to two significant figures.

Assess: The reason the horizontal speed and take-off speed appear the same is that 20° is a small angle and the cosine of a small angle is approximately equal to 1.

P3.35. Strategize: This problem involves constant acceleration (as in projectile motion). So we can use the kinematic equations in the horizontal and vertical directions.

Prepare: We can use Equation 2.12 applied to the vertical direction to determine how long the ball is in the air. We can then apply the same equation to the horizontal direction to determine the horizontal distance the ball can travel in that time.

Solve: We have

$$\Delta y = (v_y)_i \Delta t + \frac{1}{2}a_y(\Delta t)^2 \Rightarrow \Delta t = \frac{-(v_y)_i \pm \sqrt{(v_y)_i^2 - 2(-\Delta y)a_y}}{a_y}$$

$$\Delta t = \frac{(12 \text{ m/s})\sin(30°) \pm \sqrt{((12 \text{ m/s})\sin(30°))^2 - 2(6.0\text{m})(-9.8 \text{ m/s}^2)}}{(-9.8 \text{ m/s}^2)} = 0.652\text{s or } -1.88 \text{ s}$$

Here, the positive time is the one we want (it occurs after the ball is thrown). Determining the horizontal distance from the building is now easy: $\Delta x = (v_x)_i \Delta t = (12 \text{ m/s})\cos(30°)(0.652 \text{ s}) = 6.8 \text{ m}.$

Assess: This is a reasonable distance for throwing a ball to a friend.

P3.37. Strategize: We can use the formula for centripetal acceleration, Equation 3.20.
Prepare: We need the radius of the track which is half the diameter: $R = D/2 = (45 \text{ m})/2 = 22.5 \text{ m}$.
Solve:

$$a = \frac{v^2}{r} = \frac{(15 \text{ m/s})^2}{22.5 \text{ m}} = 10 \text{ m/s}^2$$

We now convert this acceleration to units of g:

$$10 \text{ m/s}^2 = 10 \text{ m/s}^2 \frac{g}{9.8 \text{ m/s}^2} = 1.0g$$

Assess: The greyhounds are accelerating at approximately the acceleration of free fall!

P3.39. Strategize: Assuming perfectly circular motion, this is a straightforward application of our equation for centripetal acceleration.
Prepare: We will use $a_c = \dfrac{v^2}{r}$. We are given the distance, and we can determine the speed by noting that $v = 2\pi r / \Delta t$ where Δt is the period of one full revolution.
Solve: As stated we can use $a_c = \dfrac{v^2}{r} = \dfrac{4\pi^2 r}{(\Delta t)^2}$. We first convert the period of one revolution into seconds:

$$27.3 \text{ days}\left(\frac{24 \text{ h}}{1 \text{ day}}\right)\left(\frac{3600 \text{ s}}{1 \text{ h}}\right) = 2.36 \times 10^6 \text{ s}.$$

Then $a_c = \dfrac{4\pi^2 \left(3.84 \times 10^8 \text{ m}\right)}{\left(2.36 \times 10^6 \text{ s}\right)^2} = 2.7 \times 10^{-3} \text{ m/s}^2.$

Assess: This is an extremely small acceleration compared to those we see on Earth.

P3.41. Strategize: The centripetal acceleration, and its dependence on these variables, is described by Equation 3.20.
Prepare: Examine the formula carefully:

$$a = \frac{v^2}{r}$$

Before plugging in numbers, notice that if the speed is held constant (as in part **(a)**), then a and r are inversely proportional to each other: doubling one halves the other. And if r is held constant (as in part **(b)**), then there is a square relationship between a and v: doubling v quadruples a.
Solve: It is convenient to use ratios to solve this problem, because we never have to know any specific values for r or v. We'll use unprimed variables for the original case ($a = 8.0 \text{ m/s}^2$), and primed variables for the new cases.
(a) With the speed held constant, $v' = v$ but $r' = 2r$.

$$\frac{a'}{a} = \frac{\frac{v'^2}{r'}}{\frac{v^2}{r}} = \frac{\frac{v^2}{2r}}{\frac{v^2}{r}} = \frac{1}{2}$$

So $a' = \frac{1}{2}a = \frac{1}{2}(8.0 \text{ m/s}^2) = 4.0 \text{ m/s}^2$.
(b) With the radius held constant, $r'' = r$ but $v'' = 2v$.

$$\frac{a''}{a} = \frac{\frac{v''^2}{r''}}{\frac{v^2}{r}} = \frac{\frac{(2v)^2}{r}}{\frac{v^2}{r}} = 2^2 = 4$$

So $a'' = 4a = 4(8.0 \text{ m/s}^2) = 32 \text{ m/s}^2$.

Assess: Please familiarize yourself with this ratio technique and look for opportunities to use it. The advantage is not needing to know any specific values of r or v—not only not having to know them, but realizing that the result is independent of them.

The daily life lesson is that driving around a curve with a larger radius produces gentler acceleration. Going around a given curve faster, however, requires a much larger acceleration (produced by the friction between the tires and the road), and the relationship is squared. If your tires are bald or the road slippery there won't be enough friction to keep you on the road if you go too fast. And remember that it is a squared relationship, so going around a curve twice as fast requires four times as much friction.

P3.43. Strategize: The magnitude of centripetal acceleration is given in Equation 3.20.
Prepare: The centripetal acceleration is given as 1.5 times the acceleration of gravity, so $a = (1.5)(9.80 \text{ m/s}^2) = 15 \text{ m/s}^2$, We can insert this into Equation 3.20.
Solve:
Using Equation 3.20, the radius of the turn is given by

$$r = \frac{v^2}{a} = \frac{(20 \text{ m/s})^2}{15 \text{ m/s}^2} = 27 \text{ m}.$$

Assess: This seems like a plausible radius for the arc of a bird's flight.

P3.45. Strategize: This problem involves relative motion.
Prepare: For everyday speeds we can use Equation 3.21 to find relative velocities. We will use a subscript A for Anita and a 1 and a 2 for the respective balls; we also use a subscript G for the ground. We will consider all motion in this problem to be along the x-axis (ignore the vertical motion including the fact that the balls also fall under the influence of gravity) and so we drop the x subscript.
It is also worth noting that interchanging the order of the subscripts merely introduces a negative sign. For example, $v_{AG} = 5$ m/s, so $v_{GA} = -5$ m/s.
"According to Anita" means "relative to Anita."
Solve: For ball 1:

$$v_{1A} = v_{1G} + v_{GA} = 10 \text{ m/s} + (-5 \text{ m/s}) = 5 \text{ m/s}$$

For ball 2:

$$v_{2A} = v_{2G} + v_{GA} = -10 \text{ m/s} + (-5 \text{ m/s}) = -15 \text{ m/s}$$

The speed is the magnitude of the velocity, so the speed of ball 2 is 15 m/s.
Assess: You can see that at low speeds velocities simply add or subtract, as the case may be. Mentally put yourself in Anita's place, and you will confirm that she sees ball 1 catching up to her at only 5 m/s while she sees ball 2 speed past her at 15 m/s.

P3.47. Strategize: This problem involves relative motion.
Prepare: We can use the technique of "canceling" subscripts to find relative velocities.
Solve: Anita's friends are standing on the ground, so we can calculate the velocities they threw the balls with by calculating the velocities of the balls relative to the ground. The velocity of ball 1 relative to Anita is $(v_x)_{1A} = +10$ m/s. The velocity of ball 2 relative to Anita is $(v_x)_{2A} = -10$ m/s. Anita's velocity relative to the ground is $(v_x)_{Ag} = +5$ m/s. Then the velocity of ball 1 relative to the ground is

$$(v_x)_{1g} = (v_x)_{1A} + (v_x)_{Ag} = +10 \text{ m/s} + 5 \text{ m/s} = +15 \text{ m/s}$$

The velocity of ball 2 relative to the ground is

$$(v_x)_{2g} = (v_x)_{2A} + (v_x)_{Ag} = -10 \text{ m/s} + 5 \text{ m/s} = -5 \text{ m/s}$$

The speed is the magnitude of the velocity, so the speed of ball 2 is 5 m/s.
Assess: The results make sense. The ball to the left of Anita must be traveling faster than Anita, and the ball to the right must be traveling slower than Anita.

P3.49. Strategize: This is a relative motion problem. We can assume that the boat's ability to push against water is constant, meaning it speed relative to the water is the same on both legs of the journey.

Prepare: Assume motion along the *x*-direction. The velocity of the boat relative to the ground is $(v_x)_{bg}$; the velocity of the boat relative to the water is $(v_x)_{bw}$; and the velocity of the water relative to the ground is $(v_x)_{wg}$. We will use the technique of Equation 3.21: $(\vec{v}_x)_{bg} = (\vec{v}_x)_{bw} + (\vec{v}_x)_{wg}$.

Solve: For travel down the river,

$$(v_x)_{bg} = (v_x)_{bw} + (v_x)_{wg} = \frac{30 \text{ km}}{3.0 \text{ h}} = 10.0 \text{ km/h}$$

For travel up the river,

$$(v_x)_{bg} = -(v_x)_{bw} + (v_x)_{wg} = -\left(\frac{30 \text{ km}}{5.0 \text{ h}}\right) = -6.0 \text{ km/h}$$

Adding these two equations yields $(v_x)_{wg} = 2.0$ km/h. That is, the velocity of the flowing river relative to the earth is 2.0 km/h.

Assess: Note that the speed of the boat relative to the water downstream and upstream are the same.

P3.51. Strategize: This is a basic vector addition (or subtraction) problem. We add vectors using components.

Prepare: The vectors $\vec{A}$, $\vec{B}$, and $\vec{D} = \vec{A} - \vec{B}$ are shown. Because $\vec{A} = \vec{A}_x + \vec{A}_y$ and $\vec{B} = \vec{B}_x + \vec{B}_y$, so the components of the resultant vector are $D_x = A_x - B_x$ and $D_y = A_y - B_y$.

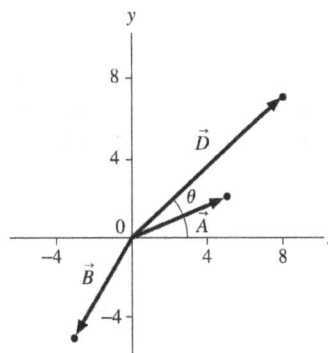

Solve: (a) With $A_x = 5$, $A_y = 2$, $B_x = -3$, and $B_y = -5$, we have $D_x = 8$ and $D_y = 7$.

(b) Vectors $\vec{A}$, $\vec{B}$ and $\vec{D}$ are shown in the above figure.

(c) Since $D_x = 8$ and $D_y = 7$, the magnitude and direction of $\vec{D}$ are

$$D = \sqrt{(8)^2 + (7)^2} = 11 \qquad \theta = \tan^{-1}\left(\frac{D_y}{D_x}\right) = \tan^{-1}\left(\frac{7}{8}\right) = 41°$$

Assess: Since $|D_y| < |D_x|$, the angle θ is less than $45°$, as it should be.

P3.53. Strategize: This is a basic vector addition problem. We add vectors using components.

Prepare: Refer to Figure P3.53 in your textbook. Because $\vec{A} = \vec{A}_x + \vec{A}_y$, $\vec{B} = \vec{B}_x + \vec{B}_y$, and $\vec{C} = \vec{C}_x + \vec{C}_y$, so the components of the resultant vector are $D_x = A_x + B_x + C_x$ and $D_y = A_y + B_y + C_y$. D_x and D_y are given and we will read the components of $\vec{A}$ and $\vec{C}$ off Figure P3.53.

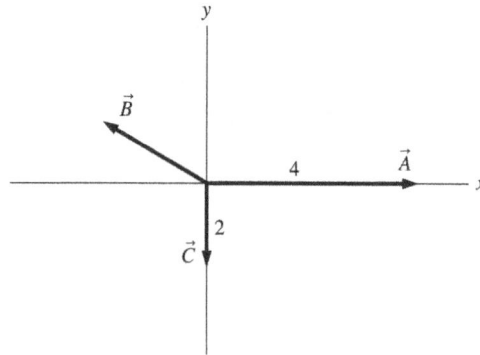

Solve: (a) $A_x = 4$, $C_x = 0$, and $D_x = 2$, so $B_x = A_x - C_x + D_x = -2$. Similarly, $A_y = 0$, $C_y = -2$, and $D_y = 0$, so $B_y = -A_y - C_y + D_y = 2$.

(b) With the components in (a), $B = \sqrt{(-2)^2 + (2)^2} = 2.8$

$$\theta = \tan^{-1}\frac{B_y}{|B_x|} = \tan^{-1}\frac{2}{2} = 45°$$

Since $\vec{B}$ has a negative x-component and a positive y-component, the angle θ made by $\vec{B}$ is with the $-x$-axis and it is above the $-x$-axis. Alternatively, the angle between $\vec{B}$ and the $+x$-axis is 135°.
Assess: Since $|B_y| = |B_x|$, $\theta = 45°$ as is obtained above.

P3.55. Strategize: This problem involves displacements in two dimensions.
Prepare: The diagram below shows the legs of Greg's journey. We simply need to determine the net displacement that results from these. If we find the components of the displacement in the northern and eastern directions, we can find the straight line distance using the Pythagorean Theorem.

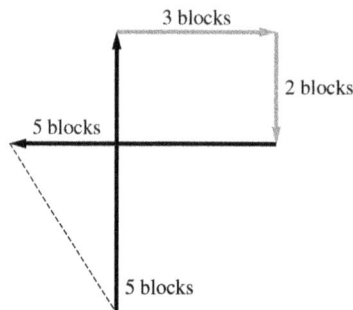

Solve: Clearly the net displacement along the east-west line is 2 blocks west, and that along the north-south line is 3 blocks north. The Pythagorean Theorem tells us $d = \sqrt{(2 \text{ blocks})^2 + (3 \text{ blocks})^2} = 3.61$ blocks. Since each block has a length of 660 ft, the straight line distance is $d = 3.61$ blocks $= 2.4 \times 10^3$ ft.
Assess: An answer of a couple thousand feet is reasonable.

P3.57. Strategize: This problem involves projectile motion, in which acceleration is constant as long as we ignore drag. Thus we can use kinematic equations in the horizontal (x) direction and the vertical (y) direction.
Prepare: We begin by converting to SI units. Equation 2.12 applied to the x direction immediately gives us the x component of the initial velocity. In order to determine the y component of the initial velocity, we may use Equation 2.11 applied to the first half of the projectile motion.
Solve: (a) The given 50-yd distance is equal to 45.7 m. Then

$$\Delta x = (v_x)_i \Delta t + \frac{1}{2}a_x(\Delta t)^2 = (v_x)_i \Delta t \Rightarrow (v_x)_i = \Delta x / \Delta t = (45.7 \text{ m})/(5.0 \text{ s}) = 9.14 \text{ m/s}.$$

In the y direction, let us call the "final" position the peak in the path, such that we consider only the first half of the motion. In that case the final vertical component of the velocity is zero, and we can write

$$\left(v_y\right)_f = \left(v_y\right)_i + a_y \Delta t \Rightarrow \left(v_y\right)_i = -a_y \Delta t = -(-9.8 \text{ m/s}^2)(2.5 \text{ s}) = 24.5 \text{ m/s}.$$

Now that we have the orthogonal components, the Pythagorean Theorem gives us the magnitude of the initial velocity:

$$v_i = \sqrt{\left(v_x\right)_i^2 + \left(v_y\right)_i^2} = \sqrt{(9.14 \text{ m/s})^2 + (24.5 \text{ m/s})^2} = 26 \text{ m/s}$$

(b) The angle between the initial velocity and the ground is described by

$$\tan(\theta) = \frac{\left(v_y\right)_i}{\left(v_x\right)_i} \Rightarrow \theta = \tan^{-1}\left(\frac{\left(v_y\right)_i}{\left(v_x\right)_i}\right) = \tan^{-1}\left(\frac{(24.5 \text{ m/s})}{(9.14 \text{ m/s})}\right) = 70°$$

Assess: The magnitude and angle are reasonable for a human kicking a ball.

P3.59. Strategize: This problem involves motion along an inclined plane. The acceleration can be described using Equation 3.16. Since this acceleration is constant, kinematics can be used.
Prepare: The skier's motion on the horizontal, frictionless snow is not of any interest to us. The skier's speed increases down the incline due to acceleration parallel to the incline, which is equal to $g \sin 10°$. A visual overview of the skier's motion that includes a pictorial representation, a motion representation, and a list of values is shown.

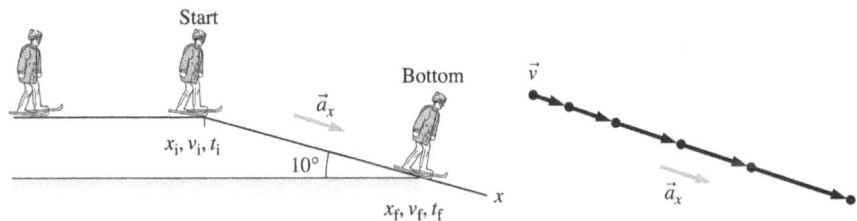

Solve: Using the following constant-acceleration kinematic equations,

$$v_f^2 = v_i^2 + 2a_x(x_f - x_i)$$
$$\Rightarrow (15 \text{ m/s})^2 = (3.0 \text{ m/s})^2 + 2(9.8 \text{ m/s}^2)\sin 10°(x_f - 0 \text{ m}) \Rightarrow x_f = 63 \text{ m}$$
$$v_f = v_i + a_x(t_f - t_i)$$
$$\Rightarrow (15 \text{ m/s}) = (3.0 \text{ m/s}) + (9.8 \text{ m/s}^2)(\sin 10°)t_f \Rightarrow t_f = 7.1 \text{ s}$$

Assess: A time of 7.1 s to cover 63 m is a reasonable value.

P3.61. Strategize: We will apply the velocity and acceleration concepts for projectile motion as shown in Figure 3.29.
Prepare: A visual overview is shown in the following figure.

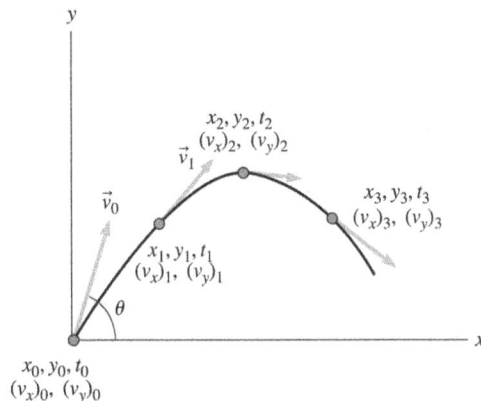

We can use the change in the vertical component of velocity between 1.0 s and 2.0 s to determine the acceleration due to gravity on this planet.

Solve: (a) We know the velocity $\vec{v}_1 = (\vec{v}_x)_1 + (\vec{v}_y)_1$ with $(v_x)_1 = 2.0$ m/s and $(v_y)_1 = 2.0$ m/s at $t = 1$ s. The ball is at its highest point at $t = 2$ s, so $v_y = 0$ m/s. The horizontal velocity is constant in projectile motion, so $v_x = 2.0$ m/s at all times. Thus, $\vec{v}_2 = (\vec{v}_x)_2 + (\vec{v}_y)_2$, with $(v_x)_2 = 2.0$ m/s and $(v_y)_2 = 0$ m/s at $t = 2$ s. We can see that the y-component of velocity *changed* by $\Delta v_y = -2.0$ m/s between $t = 1$ s and $t = 2$ s. Because a_y is constant, v_y changes by -2.0 m/s in *any* 1-s interval. At $t = 3$ s, v_y is 2.0 m/s less than its value of 0 at $t = 2$ s. At $t = 0$ s, v_y must have been 2.0 m/s more than its value of 2.0 m/s at $t = 1$ s. Consequently, at $t = 0$ s,

$$\vec{v}_0 = (\vec{v}_x)_0 + (\vec{v}_y)_0, \text{ with } (v_x)_0 = 2.0 \text{ m/s and } (v_y)_0 = 4.0 \text{ m/s}$$

At $t = 1$ s,

$$\vec{v}_1 = (\vec{v}_x)_1 + (\vec{v}_y)_1, \text{ with } (v_x)_1 = 2.0 \text{ m/s and } (v_y)_1 = 2.0 \text{ m/s}$$

At $t = 2$ s,

$$\vec{v}_2 = (\vec{v}_x)_2 + (\vec{v}_y)_2, \text{ with } (v_x)_2 = 2.0 \text{ m/s and } (v_y)_2 = 0 \text{ m/s}$$

At $t = 3$ s,

$$\vec{v}_3 = (\vec{v}_x)_3 + (\vec{v}_y)_3, \text{ with } (v_x)_3 = 2.0 \text{ m/s and } (v_y)_3 = -2.0 \text{ m/s}$$

(b) Because v_y is changing at the rate -2.0 m/s per s, the y-component of acceleration is $a_y = -2.0$ m/s^2. But $a_y = -g$ for projectile motion, so the value of g on Exidor is $g = 2.0$ m/s^2.

(c) From part (a) the components of $\vec{v}_0$ are $(v_x)_0 = 2.0$ m/s and $(v_y)_0 = 4.0$ m/s. This means

$$\theta = \tan^{-1}\left(\frac{(v_y)_0}{(v_x)_0}\right) = \tan^{-1}\left(\frac{4.0 \text{ m/s}}{2.0 \text{ m/s}}\right) = 63° \text{ above} + x$$

Assess: The y-component of the velocity vector decreases from 2.0 m/s at $t = 1$ s to 0 m/s at $t = 2$ s. This gives an acceleration of -2 m/s^2. All the other values obtained above are also reasonable.

P3.63. Strategize: This problem is somewhat similar to Problem 3.27 with all of the initial velocity in the horizontal direction. We will use the vertical equation for constant acceleration to determine the time of flight and then see how far Captain Brady can go in that time.
Prepare: We will do this two-step problem completely with variables in part (a) and only plug in numbers in part (b). We *could* do part (b) in feet (using $g = 32$ ft/s^2), but to compare with the world record 100 m dash, let's convert to meters. $L = 22$ ft $= 6.71$ m and $h = 20$ ft $= 6.10$ m.
Solve: (a) Given that $(v_y) = 0.0$ ft/s we can use the kinematic equations.

$$(y_f - y_i) = \frac{1}{2}a_y(\Delta t)^2$$

With up as the positive direction, $(y_f - y_i)$ is negative and $a_y = -g$; those signs cancel leaving

$$h = \frac{1}{2}g(\Delta t)^2$$

Solve for Δt.

$$\Delta t = \sqrt{\frac{2h}{g}}$$

Now use that expression for Δt in the equation for constant horizontal velocity.

$$L = \Delta x = v_x \Delta t = v_x \sqrt{\frac{2h}{g}}$$

Finally solve for $v = v_x$ in terms of L and h.

$$v = \frac{L}{\sqrt{\frac{2h}{g}}} = L\sqrt{\frac{g}{2h}}$$

Now plug in the numbers we are given for L and h.

$$v = L\sqrt{\frac{g}{2h}} = (6.71\text{ m})\sqrt{\frac{9.8\text{ m/s}^2}{2(6.10\text{ m})}} = 6.0\text{ m/s}$$

(b) Compare this result ($v = 6.0$ m/s) with the world-class sprinter ($v = 10$ m/s); a fit person could make this leap.

Assess: The results are reasonable, and not obviously wrong. 6.0 m/s ≈ 13 mph, and that would be a fast run, but certainly possible.

By solving the problem first algebraically before plugging in any numbers, we are able to substitute other numbers as well, if we desire, without re-solving the whole problem.

P3.65. Strategize: We will use the initial information (that the marble goes 6.0 m straight up) to find the speed the marble leaves the gun. We also need to know how long it takes something to fall 1.5 m from rest in free fall so we can then use that in the horizontal equation.

Prepare: Assume that there is no air resistance ($a_y = -g$) and that the marble leaves the gun with the same speed (muzzle speed) each time it is fired. We can determine the muzzle speed by applying Equation 2.13 to the vertical case. We can use Equation 2.12 to determine the time to fall 1.5 m in the case of the horizontal launch.

Solve: Equation 2.13 tells us

$$(v_y)_f^2 = (v_y)_i^2 + 2a_y \Delta y$$

where at the top of the trajectory $(v_y)_f = 0.0$ m/s and $\Delta y = 6.0$ m.

$$(v_y)_i^2 = 2g\Delta y \Rightarrow (v_y)_i = \sqrt{2g\Delta y} = 10.8\text{ m/s}$$

We also rearrange Equation 2.12 to find the time for an object to fall 1.5 m from rest: $y_f - y_i = -15$ m now instead of the 6.0 m used previously.

$$\Delta y = \frac{1}{2}a_y(\Delta t)^2$$

$$\Delta t = \sqrt{\frac{2\Delta y}{-g}} = \sqrt{\frac{2(-1.5\text{ m})}{-9.8\text{ m/s}^2}} = 0.553\text{ s}$$

At last we combine this information into the equation for constant horizontal velocity.

$$\Delta x = v_x \Delta t = (10.8\text{ m/s})(0.553\text{ s}) = 6.0\text{ m}$$

Assess: Is it a coincidence that the marble has a horizontal range of 6.0 m when it can reach a height of 6.0 m when fired straight up, or will those numbers always be the same? Well, the 6.0 m horizontal range depends on the height (1.5 m) from which you fire it, so if that were different the range would be different. This leads us to conclude that it *is* a coincidence. You can go back, though, and do the problem algebraically (with no numbers) and find that g cancels and that the horizontal range is 2 times the square root of the product of the vertical height it can reach and the height from which you fire it horizontally.

P3.67. Strategize: This problem involves projectile motion. Ignoring air resistance, the acceleration is constant and we can use kinematic equations.

Prepare: We can apply Equation 2.12 to the horizontal (x) direction to determine the amount of time the paintball is in the air. We can then use that time and apply 2.12 again, this time to the vertical (y) direction, to determine the height above the paintball gun where the paintball strikes the tree.

Solve: First, note that the angle above horizontal at which the gun is fired is
$\theta = \tan^{-1}(\Delta y / \Delta x) = \tan^{-1}((4.0 \text{ m})/(20\text{m})) = 11.3°$. We can determine the time of flight by writing

$$\Delta t = \frac{\Delta x}{(v_x)_i} = \frac{\Delta x}{v_i \cos(\theta)} = \frac{(20 \text{ m})}{(50 \text{ m/s})\cos(11.3°)} = 0.408 \text{ s}$$

Now, for the displacement in the vertical direction, we have

$$\Delta y = (v_y)_i \Delta t + \frac{1}{2}a_y(\Delta t)^2 = (50 \text{ m/s})\sin(11.3°)(0.408 \text{ s}) + \frac{1}{2}(-9.8 \text{ m/s}^2)(0.408 \text{ s})^2 = 3.18 \text{ m}$$

Since the knot was 4.0 m above the gun, the paintball misses by 0.82 m.

Assess: This is consistent with our understanding of parabolic motion. Paintballs should not move in straight lines, but should bend downward from the line-of-sight target.

P3.69. Strategize: This problem involves projectile motion. We will apply the constant-acceleration kinematics equations to the horizontal and vertical motions of the tennis ball as described by Synthesis 3.1.

Prepare: A visual overview is shown as follows. To find whether the ball clears the net, we will determine the vertical fall of the ball as it travels to the net.

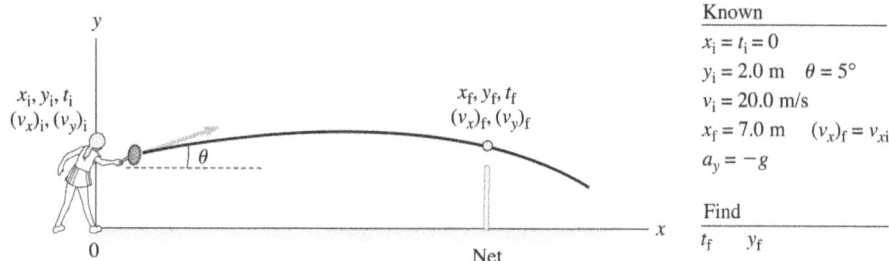

Known
$x_i = t_i = 0$
$y_i = 2.0 \text{ m}$ $\theta = 5°$
$v_i = 20.0 \text{ m/s}$
$x_f = 7.0 \text{ m}$ $(v_x)_f = v_{xi}$
$a_y = -g$

Find
t_f y_f

Solve: The initial velocity is

$$(v_x)_i = v_i \cos 5° = (20 \text{ m/s}) \cos 5° = 19.92 \text{ m/s}$$
$$(v_y)_i = v_i \sin 5° = (20 \text{ m/s}) \sin 5° = 1.743 \text{ m/s}$$

The time it takes for the ball to reach the net is

$$x_f = x_i + (v_x)_i(t_f - t_i) \Rightarrow 7.0 \text{ m} = 0 \text{ m} + (19.92 \text{ m/s})(t_f - 0 \text{ s}) \Rightarrow t_f = 0.351 \text{ s}$$

The vertical position at $t_f = 0.351$ s is

$$y_f = y_i + (v_y)_i(t_f - t_i) + \tfrac{1}{2}a_y(t_f - t_i)^2$$
$$= (2.0 \text{ m}) + (1.743 \text{ m/s})(0.351 \text{ s} - 0 \text{ s}) + \tfrac{1}{2}(-9.8 \text{ m/s}^2)(0.351 \text{ s} - 0 \text{ s})^2 = 2.0 \text{ m}$$

Thus the ball clears the net by 1.0 m.

Assess: The vertical free fall of the ball, with zero initial velocity, in 0.351 s is 0.6 m. The ball will clear by approximately 0.4 m if the ball is thrown horizontally. The initial launch angle of 5° provides some initial vertical velocity and the ball clears by a larger distance. The above result is reasonable.

P3.71. Strategize: This problem involves projectile motion. We will use kinematic equations in the horizontal and vertical directions to determine the components of the final velocity.

Prepare: We can use the equation for vertical motion at constant acceleration to find the time of fall and then use the time to find the final velocity.

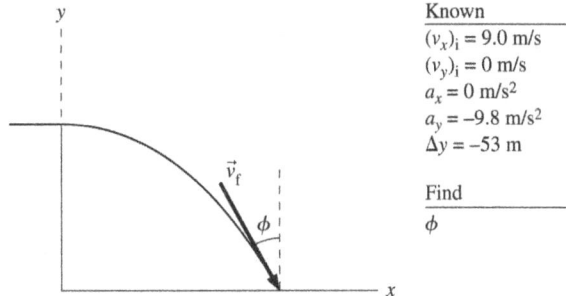

Known
$(v_x)_i = 9.0$ m/s
$(v_y)_i = 0$ m/s
$a_x = 0$ m/s^2
$a_y = -9.8$ m/s^2
$\Delta y = -53$ m

Find
ϕ

Solve: Since the water is launched horizontally, its time of flight and vertical displacement are related by the equation: $\Delta y = -\dfrac{1}{2}g\Delta t^2$. Solving for the time, we have

$$\Delta t = \sqrt{2|\Delta y|/g} = \sqrt{2(53 \text{ m})/9.8 \text{ m/s}^2} = 3.29 \text{ s}$$

The horizontal component of the velocity, v_x, is constant, but the vertical component is given by the equation: $(v_y)_f = (v_y)_i + a_y\Delta t$. At the moment the water strikes the pool, the vertical component is

$$(v_y)_f = 0 \text{ m/s} - (9.8 \text{ m/s}^2)(3.29 \text{ s}) = -32.2 \text{ m/s}$$

At the moment of impact the velocity of the water is: $(9.0 \text{ m/s}, -32.2 \text{ m/s})$. The angle that the water makes with the vertical is given by

$$\phi = \tan^{-1}((9.0 \text{ m/s})/(32.2 \text{ m/s})) = 16°$$

The water is falling at an angle of 16° with the vertical.

Assess: Even though the water is launched at a fairly high speed (9.0 m/s is about 20 mi/hr), it is close to the vertical when it lands because it spends such a long time in the air during which time the absolute value of v_x increases steadily.

P3.73. Strategize: This problem involves projectile motion. Ignoring air resistance the acceleration is constant, such that we can use kinematic equations.

Prepare: We can apply Equation 2.12 to the vertical direction to determine the hang time of the bike. Once we know the time, we can determine the distance from the ramp using $\Delta x = (v_x)_i \Delta t$, since there is no acceleration in the horizontal direction.

Solve: In the vertical direction, we have

$$\Delta y = (v_y)_i \Delta t + \frac{1}{2}a_y(\Delta t)^2 \Rightarrow \Delta t = \frac{-v_i \sin(\theta) \pm \sqrt{(v_i \sin(\theta))^2 + 2a_y\Delta y}}{a_y}$$

$$\Delta t = \frac{-(6.7\,\text{m/s})\sin(40°) \pm \sqrt{((6.7\,\text{m/s})\sin(40°))^2 + 2(-9.8 \text{ m/s}^2)(-1.8 \text{ m})}}{(-9.8 \text{ m/s}^2)} = -0.309 \text{ s or } 1.19\text{s}$$

We want the positive time, as it happens after the bike leaves the ramp. Now, we use this time in our calculation of the horizontal distance covered: $\Delta x = (v_x)_i \Delta t = v_i \cos(\theta)\Delta t = (6.7 \text{ m/s})\cos(40°)(1.19 \text{ s}) = 6.1 \text{ m}$.

Assess: This is a reasonable distance for a bike initially moving at a speed of 6.7 m/s.

P3.75. Strategize: This problem deals with relative motion, since the velocity of the ducks is given relative to the air.

Prepare: A visual overview of the ducks' motion is shown below. The resulting velocity is given by $\vec{v} = \vec{v}_{\text{fly}} + \vec{v}_{\text{wind}}$, where $\vec{v}_{\text{wind}} = 6$ m/s, east) and $\vec{v}_{\text{fly}} = (v_{\text{fly}} \sin \theta, \text{west}) + (v_{\text{fly}} \cos \theta, \text{south})$.

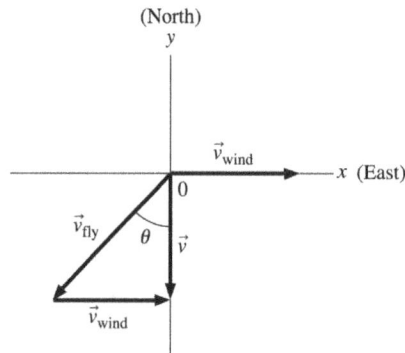

Solve: Substituting the known values we get $\vec{v} = (8 \text{ m/s} \sin \theta, \text{west}) + (8 \text{ m/s} \cos \theta, \text{south}) + (6 \text{ m/s}, \text{east})$. That is, $\vec{v} = (-8 \text{ m/s} \sin \theta, \text{east}) + (8 \text{ m/s} \cos \theta, \text{south}) + (6 \text{ m/s}, \text{east})$. We need to have $v_x = 0$. This means $0 = -8$ m/s $\sin \theta + 6$ m/s, so $\sin \theta = \frac{6}{8}$ and $\theta = 48.6°$. Thus the ducks should head $49°$ west of south (or $41°$ south of west).

P3.77. Strategize: This problem deals with relative motion. We know the speed of the plane relative to the air, and the velocity of the air.

Prepare: A visual overview of the plane's motion is shown in the following figure. The direction the pilot must head the plane can be obtained from $\vec{v}_{\text{pg}} = \vec{v}_{\text{pa}} + \vec{v}_{\text{ag}}$, where $\vec{v}_{\text{pa}} = (v_{\text{pa}} \sin \theta, \text{south}) + (v_{\text{pa}} \cos \theta, \text{east})$, $v_{\text{pa}} = 200$ mph, $\vec{v}_{\text{ag}} = (v_{\text{ag}} \sin 30°, \text{north}) + (v_{\text{ag}} \cos 30°, \text{east})$, and $\vec{v}_{\text{pg}} = (v_{\text{pg}}, \text{east})$.

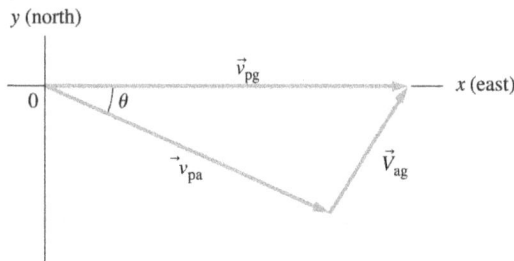

Solve: (a) Writing the equation $\vec{v}_{\text{pg}} = \vec{v}_{\text{pa}} + \vec{v}_{\text{ag}}$ in the form of components

$$(v_{\text{pg}}, \text{east}) = [(200 \text{ mph} \sin \theta, \text{south}) + (200 \text{ mph} \cos \theta, \text{east})] +$$
$$[(50 \text{ mph} \sin 30°, \text{north}) + (50 \text{ mph} \cos 30°, \text{east})]$$

Because $(\vec{v})_{\text{pg}}$ should have no component along north,

$$50 \sin 30° - 200 \sin \theta = 0 \Rightarrow \theta = 7.2°$$

(b) The pilot must head $7.2°$ south of east. Substituting this value of θ in the above velocity equation gives $(v_{\text{pg}}, \text{east}) = (200 \text{ mph} \cos 7.2°, \text{east}) + (50 \text{ mph} \cos 30°, \text{east}) = (240 \text{ mph}, \text{east})$. At a speed of 240 mph, the trip takes $t = 600$ mi/240 mph $= 2.5$ hours.

P3.79. Strategize: This problem deals with relative motion.

Prepare: Assume motion along the x-direction. Let $x_f - x_i$ be the displacement from your gate to the baggage claim. We will use the technique of Equation 3.21: $(\vec{v}_x)_{\text{yg}} = (\vec{v}_x)_{\text{ym}} + (\vec{v}_x)_{\text{mg}}$.

Solve: In the first case, when the moving sidewalk is broken, we can find your velocity

$$v_Y = \frac{(x_f - x_i)}{50 \text{ s}}$$

In the second case, when you stand on the moving sidewalk, the velocity of the sidewalk relative to the ground is

$$v_{sg} = \frac{x_f - x_i}{75 \text{ s}}$$

In the third case, when you walk while riding, we can use the equation

$$(v)_{yg} = (v)_{ys} + (v)_{sg}$$

That is, your velocity relative to the ground, when you are walking on the moving sidewalk, is equal to your velocity relative to the moving sidewalk (which is v_Y) plus the sidewalk's velocity relative to the ground. Thus,

$$\frac{x_1 - x_0}{\Delta t} = \frac{x_1 - x_0}{50 \text{ s}} + \frac{x_1 - x_0}{75 \text{ s}} \Rightarrow \Delta t = 30 \text{ s}$$

Assess: A time smaller than 50 s was expected.

P3.81. Strategize: We will calculate the linear acceleration of the Mustang, and equate this to the expression for centripetal acceleration.
Prepare: We need to convert the radius of the Mini Cooper's turn to meters and convert the final speed of the Mustang to meters per second. The radius of the Mini Cooper's turn is the following:

$$17 \text{ ft} = 17 \text{ ft}\left(\frac{1 \text{ m}}{3.28 \text{ ft}}\right) = 5.18 \text{ m}$$

And the final speed of the Mustang is as follows:

$$60 \frac{\text{mi}}{\text{h}} = 60 \frac{\text{mi}}{\text{h}}\left(\frac{1609 \text{ m}}{1 \text{ mi}}\right)\left(\frac{1 \text{ h}}{3600 \text{ s}}\right) = 26.8 \text{ m/s}$$

The acceleration of the Mustang is given by $a = \Delta v / \Delta t$

$$a = (26.8 \text{ m/s})/(5.6 \text{ s}) = 4.79 \text{ m/s}^2$$

Solve: To match the Mustang's acceleration, the Mini Cooper must have a centripetal acceleration of 4.79 m/s². Given the formula for centripetal acceleration, $a = v^2/r$, we can solve for the necessary radius as follows:

$$v = \sqrt{ar} = \sqrt{(4.79 \text{ m/s}^2)(5.18 \text{ m})} = 4.98 \text{ m/s} = 4.98 \frac{\text{m}}{\text{s}}\left(\frac{1 \text{ mi}}{1609 \text{ m}}\right)\left(\frac{3600 \text{ s}}{1 \text{ h}}\right) = 11 \text{ mph}$$

The Mini Cooper must travel at 5.0 m/s = 11 mph to have the same acceleration as the Mustang.

Assess: Even at a fairly low speed, 11 mph, the acceleration is high. This is because the radius of the turn is so small—17 ft.

P3.83. Strategize: We calculate the centripetal acceleration, and then examine its dependence on speed.
Prepare: We will use Equation 3.20 to relate the acceleration to the speed. But first we need to convert the speed of the car to m/s.

$$40 \frac{\text{mi}}{\text{hr}} = 40 \frac{\text{mi}}{\text{hr}}\left(\frac{1609 \text{m}}{1 \text{ mi}}\right)\left(\frac{1 \text{ hr}}{3600 \text{ s}}\right) = 17.9 \text{ m/s}$$

Solve: (a) Your acceleration is given from the equation $a = v^2/r$

$$\frac{(17.9 \text{ m/s})^2}{110 \text{ m}} = 2.91 \text{ m/s}^2$$

which converts as follows:

$$2.91 \text{ m/s}^2 = (2.91 \text{ m/s}^2)\left(\frac{1g}{9.8 \text{ m/s}^2}\right) = 0.30g$$

The acceleration is 2.9 m/s^2 or $0.30g$.

(b) The formula for centripetal acceleratison, $a = v^2/r$ can be solved for v as follows: $v = \sqrt{ar}$. In this form we see that if the acceleration is doubled, then the velocity is multiplied by $\sqrt{2}$. So we multiply the 40 mph speed limit by $\sqrt{2}$: $(40 \text{ mph})\sqrt{2} = 57 \text{ mph}$. At 57 mph the acceleration would be twice the acceleration at 40 mph.

Assess: As noted in the solution to Problem 3.42, a small change in velocity can produce a large change in centripetal acceleration. Here, with an increase in speed of less than 50%, the acceleration doubles and the friction needed for the turn also doubles.

P3.85. Strategize: Consider how the centripetal acceleration depends on other variables.
Prepare: Equation 3.20 governs circular motion.
Solve: From Equation 3.20 we can see the centripetal acceleration of a car has a quadratic relationship to velocity and an inverse relationship to the radius of the circular motion. If the velocity of the object is decreased, the acceleration decreases. If the radius of the motion is increased the acceleration of the object *decreases*. We can either decrease the velocity or increase the radius. The choices C and D act to increase the velocity, so the correct answer is B.
Assess: Centripetal acceleration decreases with increasing radius.

P3.87. Strategize: This is a projectile motion problem. We know that doubling the distance means doubling the time the slider is in the air.
Prepare: The riders are in free fall during this part of the motion. We can use Synthesis 3.1.
Solve: The riders have no initial velocity in the vertical direction after they leave the slide. Their initial velocity in the horizontal direction will be determined by the first and second sections and is unaffected by changing the height of the slide exit above the water, assuming that the first two sections are unchanged.
The time it takes the riders to hit the water after leaving the slide can be calculated with Equation 2.12.

$$y_f = y_i + (v_y)_i \Delta t - \frac{1}{2}g(\Delta t)^2$$

Taking the origin as the exit of the slide and $(v_y)_i = 0$ m/s this equation becomes

$$\Delta t = \sqrt{-\frac{2y_f}{g}}$$

The initial horizontal velocity of the rider is $(v_x)_i$. The horizontal displacement of the rider once leaving the ramp will be $(v_x)_i \Delta t$. To double the distance the rider travels before hitting the water, we only need to double the time the rider takes to hit the water. From the previous equation, to double the time, we must quadruple the vertical distance. So the answer is $(4)(0.6 \text{ m}) = 2.4 \text{ m}$. The correct choice is C.

Assess: This answer is reasonable, since the distance traveled vertically in free fall has a quadratic relationship to time.

4

FORCES AND NEWTON'S LAWS OF MOTION

Q4.1. Reason: Even if an object is not moving forces can be acting on it. However, the *net* force must be zero. As an example consider a book on a flat table. The forces that act on the book are the weight of the book (a long ranger force) and the normal force exerted by the table (a contact force). There are two forces acting on the book, but it is not moving because the net force on the book is zero.

Assess: The net force, which is the vector sum of the forces acting on an object, governs the acceleration of objects through Equation 4.4.

Q4.3. Reason: No. If you know all of the forces than you know the direction of the acceleration, not the direction of the motion (velocity). For example, a car moving forward could have on it a net force forward if speeding up or backward if slowing down or no net force at all if moving at constant speed.

Assess: Consider carefully what Newton's *second* law says, and what it doesn't say. The net force must *always* be in the direction of the acceleration. This is also the direction of the *change* in velocity, although not necessarily in the direction of the velocity itself.

Q4.5. Reason: What you feel is a contact force between your back and the seat. That could be explained by something pushing you back into the seat, or by the seat pushing against you. The latter is what is actually happening. The car and the seat are accelerating from rest, and the seat has to exert a force on you in order for you to accelerate with the rest of the car.

Assess: There is no force pushing you backward. Your inertia would cause you to remain still while the car accelerates away, except that the seat exerts a forward force on your body.

Q4.7. Reason: The inertia of the ketchup will keep it from moving if it isn't too tightly adhered to the sides of the moving bottle.

Assess: If you hit the bottle downward (while it is upside down) then the ketchup will end up farther from the opening.

Q4.9. Reason: Since there is no source of gravity, you will not be able to feel the weight of the objects. However, Newton's second law is true even in an environment without gravity. Assuming you can exert a reproducible force in throwing both objects, you could throw each and note the acceleration each obtains.

Assess: Mass is independent of the force of gravity and exists even in environments with no sources of gravity.

Q4.11. Reason: If you throw any or all of the items in your toolbelt, you must exert forces on them. They must, therefore, exert forces back on you. If you throw an object, say a hammer, away from the spaceship, then it will exert a force on you toward the spaceship. This will cause a small acceleration of your body toward the spaceship.

Assess: Throwing any item away from the spaceship (with essentially any non-zero speed) should push you toward the spaceship. But the more force you exert on the objects, the more they exert back on you and the faster you reach your ship.

Q4.13. Reason: The force of Josh on Taylor and the force of Taylor on Josh are members of an action/reaction pair, so that the magnitudes of these two forces are the same. However, since Josh is more massive (bigger) than Taylor, his resulting acceleration during the push will be less and hence his final velocity after the push will be less.
Assess: This problem required a correct conceptual understanding of Newton's second and third laws. The third law allows us to conclude that each skater experiences the same force and the second law allows us to understand that the acceleration is inversely proportional to the mass being accelerated.

Q4.15. Reason: As your foot comes into contact with the floor the force of static friction acts in the direction necessary to prevent motion of the foot. The friction prevents your foot from slipping forward, so the direction of the force of the floor on your foot is backward.
Assess: Static friction acts in the direction necessary to prevent motion.

Q4.17. Reason: There are other forces acting on each team. The question is whether or not the tension in the rope is greater than the force of friction between either team's feet and the ground.
Assess: The game is won when the tension becomes greater than other forces holding one of the teams in place.

Q4.19. Reason: The way the tire is twisted indicated the force of the road on the tire is forward. Since this force is likely greater than the backward air resistance force, the net force is also forward; therefore the car is accelerating in the forward direction. This means it must be speeding up.
Assess: If the car were slowing down the net force would point backward (and the indicative wrinkles in the tire would go the other way).

Q4.21. Reason: Since the block glued on to the original block is identical to the original block, the mass of the two together must be twice as large as the mass of the original block. If the force applied is also twice as large, the acceleration will be the same. Explicitly applying Newton's second law to the two blocks glued together gives

$$a_{new} = \frac{2F}{2m} = \frac{F}{m} = a_{old}$$

The correct choice is C.
Assess: In Newton's second law, the acceleration is proportional to the net force and inversely proportional to the mass acccelerated. As a result if you double both the mass and the force, the acceleration will remain the same.

Q4.23. Reason: Drag points opposite to the direction of motion. As the ball is going up, the drag force acts downward. As the ball comes down, the drag force acts upward. The correct choice is D.
Assess: Drag always acts opposite to the direction of motion of an object.

Q4.25. Reason: The direction of the kinetic friction force will be opposite the motion, so the friction points down while the box goes up, and the friction points up while the box slides down.
The answer is D.
Assess: Drawing a free-body diagram (with tilted axes) and applying Newton's second law will support this conclusion.

Q4.27. Reason: To remain stationary there needs to be a zero net force on the scallop. The downward gravitational force is not quite balanced by the upward buoyant force so the thrust force must also be up. For the thrust force on the scallop to be up, it must eject water in the downward direction.
The answer is C.
Assess: Drawing a free-body diagram (with tilted axes) and applying Newton's second law will support this conclusion.

Q4.29. Reason: Newton's third law tells us that the force we exert on the cabinet is always equal in magnitude and opposite in direction to the force the cabinet exerts on us.
The correct answer is C.
Assess: Note that the force we exert on the cabinet is being overcome by some force acting on the cabinet, in order for the cabinet to slow. But that force must be acting on the cabinet to the explain its slowing; any force acting back on us has nothing to do with it. The force overcoming our push is the force of friction between the ground and the cabinet.

Q4.31. Reason: Since block A rides without slipping, it, too, must be accelerating to the right. If it is accelerating to the right there must be a net force to the right, according to Newton's second law. The only object that can exert a force to the right is block B.

This static friction force is to the right to prevent slippage of block A to the left (relative to block B).

The correct choice is B.

Assess: This is one of those cases in which the static friction force can be in the same direction as the motion, to prevent slippage the other way. Verify with a free-body diagram of block A.

Problems

P4.1. Strategize: This problem deals with an object (human head) that has inertia. It will continue with constant velocity or remain at rest until a force acts on it.

Prepare: First note that time progresses to the right in each sequence of pictures. In one case the head is thrown back, and in the other, forward.

Solve: Using the principle of inertia, the head will tend to continue with the same velocity after the collision that it had before.

In the first series of sketches, the head is lagging behind because the car has been quickly accelerated forward (to the right). This is the result of a rear-end collision.

In the second series of sketches, the head is moving forward relative to the car because the car is slowing down and the head's inertia keeps it moving forward at the same velocity (although external forces do eventually stop the head as well). This is the result of a head-on collision.

Assess: Hopefully you haven't experienced either of these in an injurious way, but you have felt similar milder effects as the car simply speeds up or slows down.

It is for this reason that cars are equipped with headrests, to prevent the whiplash shown in the first series of sketches, because rear-end collisions are so common. The laws of physics tell us how wise it is to have the headrests properly positioned for our own height.

Air bags are now employed to prevent injury in the second scenario.

P4.3. Strategize: This problem deals with an object (a human infant) that has inertia. It will continue with constant velocity or remain at rest until a force acts on it.

Prepare: As background, look at question Q4.6 and problem P4.1. Also think about the design and orientation of the seat and how the child rides in the seat. Finally recall Newton's second law.

Solve: As the child rides in the seat, his/her head and back rest against the padded back of the seat. If the car is brought to a rapid stop (as in a head-on collision) the child will continue to move forward at the before-crash speed until he/she hits something. The object hit is the back of the seat (supporting the entire back and head) which is padded and as a result the force increases to the maximum value over a time interval. Granted this time interval may be small but that is considerably better than instantaneous. Also since the head is supported, there will be no whiplash.

Assess: The fact the force acting on the child is spread over a time interval is a critical factor. In later chapters you will learn to call this concept impulse.

P4.5. Strategize: If an object is initially at rest, it will remain at rest if the sum of all forces is zero.

Prepare: Draw the vector sum $\vec{F}_1 + \vec{F}_2$ of the two forces $\vec{F}_1$ and $\vec{F}_2$. Then look for a vector that will "balance" the force vector $\vec{F}_1 + \vec{F}_2$.

Solve: The object will be in equilibrium if $\vec{F}_3$ has the same magnitude as $\vec{F}_1 + \vec{F}_2$ but is in the opposite direction so that the sum of all three forces is zero.

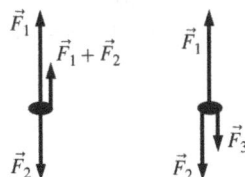

Assess: Adding the new force vector $\vec{F}_3$ with length and direction as shown will cause the object to be at rest.

P4.7. Strategize: This problem involves forces acting on an object. A diagram will be useful.

Prepare: In drawing a free-body diagram, forces will come from objects touching the person, or from gravity. The only object interacting with the person would be the chain supporting the swing. In general, that chain does exert a force on the person (or on the seat, and then the seat exerts the force on the person). This might cause a person to draw the free body diagram shown on the left.

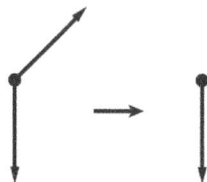

However, note that when the maximum height is reached, you can feel the chains go lax. We suspect therefore that perhaps the diagram on the right is more correct. This is considered more thoroughly below.

Solve: When a person swings, they move in a circular arc. If their speed were constant (or changing very slowly) we could determine the radial force that must be acting on them (along the direction of the chain) using $F_r = ma_r = mv^2 / r$. But at the peak of the swing, the speed is momentarily zero, meaning there would be no radial force at all. This is consistent with our suspicion based on chains going lax at the maximum height. Thus the only force is gravity.

Assess: From the maximum height, the person begins accelerating downward. Only once they have non-zero speed does the tension in the chain once again play a role in curving their path into an arc.

P4.9. Strategize: Gravity can act on objects at a distance (even if the objects are not touching the Earth), but other forces discussed in this chapter require contact with the object.

Prepare: Draw a picture of the situation, identify the system, in this case the baseball player, and draw a closed curve around it. Name and label all relevant contact forces and long-range forces.

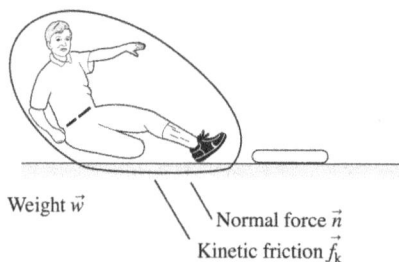

Weight $\vec{w}$ Normal force $\vec{n}$ Kinetic friction $\vec{f}_k$

Solve: There are three forces acting *on* the baseball player due to his interactions with the two agents earth and ground. One of the forces *on* the player is the long-range weight force *by* the earth. Another force is the normal force exerted *by* the ground due to the contact between him and the ground. The third force is the kinetic friction force *by* the ground due to his sliding motion on the ground.

Assess: Note that the kinetic friction force would be *absent* if the baseball player were *not* sliding.

P4.11. Strategize: Gravity can act on objects at a distance (even if the objects are not touching the Earth), but other forces discussed in this chapter require contact with the object.

Prepare: We follow the outline in Tactics Box 4.2. See also Conceptual Example 4.2.

The exact angle of the slope is not critical in this problem; the answers would be very similar for any angle between $0°$ and $90°$.

Solve: The system is the skier.

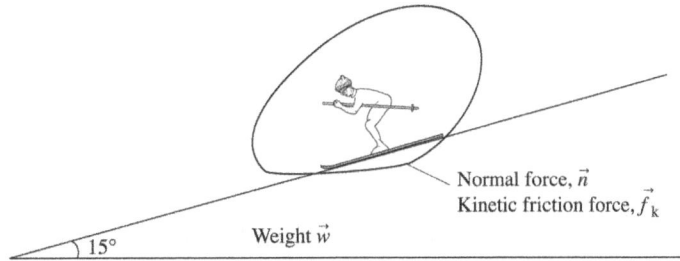

To identify forces, think of objects that are in contact with the object under consideration, as well as any long-range forces that might be acting on it. We are told to not ignore friction, but we will ignore air resistance.

The objects that are in contact with the skier are the snow-covered slope and. . . and that's all (although we will identify two forces exerted by this agent). The long-range force on the skier is the gravitational force of the earth on the skier.

One of the forces, then, is the gravitational force of the earth on the skier. This force points straight toward the center of the earth.

The slope, as we mentioned, exerts two forces on the skier: the normal force (directed perpendicularly to the slope) and the frictional force (directed parallel to the slope, backward from the downhill motion).

Assess: Since there are no other objects (agents) in contact with the skier (we are ignoring the air, remember?) and no other long-range forces we can identify (the gravitational force of the moon or the sun on the skier is also too small to be worth mentioning), then we have probably catalogued them all.

We are not told whether the skier has a constant velocity or is accelerating, and that factor would influence the relative lengths of the three arrows representing the forces. If the motion is constant velocity, then the vector sum of the three arrows must be zero.

P4.13. Strategize: Newton's second law relates force to mass and acceleration. Given a fixed force, different masses will have different accelerations.

Prepare: Refer to Figure P4.13. From force = mass × acceleration or mass = force/acceleration or mass = 1/(acceleration/force), mass is

$$m = \frac{1}{\text{slope of the acceleration-versus-force graph}}$$

A larger slope implies a smaller mass.

Solve: We know $m_2 = 0.20$ kg, and we can find the other masses relative to m_2 by comparing their slopes. Thus

$$\frac{m_1}{m_2} = \frac{1/\text{slope 1}}{1/\text{slope 2}} = \frac{\text{slope 2}}{\text{slope 1}} = \frac{1}{5/2} = \frac{2}{5} = 0.40$$

$$\Rightarrow m_1 = 0.40 \, m_2 = 0.40 \times 0.20 \text{ kg} = 0.080 \text{ kg}$$

Similarly,

$$\frac{m_3}{m_2} = \frac{1/\text{slope 3}}{1/\text{slope 2}} = \frac{\text{slope 2}}{\text{slope 3}} = \frac{1}{2/5} = \frac{5}{2} = 2.50$$

$$\Rightarrow m_3 = 2.50 \, m_2 = 2.50 \times 0.20 \text{ kg} = 0.50 \text{ kg}$$

Assess: From the initial analysis of the slopes, we had expected $m_3 > m_2$ and $m_1 < m_2$. This is consistent with our numerical answers.

P4.15. Strategize: We will use the particle model for the object and use Newton's second law.
Prepare: Assume that the maximum force the road exerts on the car is the same in both cases.
Solve: Use primed quantities for the case with the four new passengers and unprimed quantities for the original case with just the driver. $F' = F$. The original mass was 1200 kg and the new mass is 1600 kg.

$$a' = \frac{F'}{m'} = \frac{F}{m'} = \frac{ma}{m'} = \frac{(1200 \text{ kg})(4 \text{ m/s}^2)}{1600 \text{ kg}} = 3.0 \text{ m/s}^2$$

Assess: We expected the maximum acceleration to be less with the passengers than with the driver only.

P4.17. Strategize: Force and acceleration are related through mass by Newton's second law.
Prepare: The problem may be solved by applying Newton's second law to the present and the new situation.
Solve: (a) We are told that for an unknown force (call it F_o) acting on an unknown mass (call it m_o) the acceleration of the mass is 8.0 m/s². According to Newton's second law

$$F_o = m_o(8.0 \text{ m/s}^2) \quad \text{or} \quad F_o/m_o = 8.0 \text{ m/s}^2$$

For the new situation, the new force is $F_{new} = 2F_o$, the mass is not changed ($m_{new} = m_o$) and we may find the acceleration by

$$F_{new} = m_{new}a_{new}$$

or

$$a_{new} = F_{new}/m_{new} = 2F_o/m_o = 2(F_o/m_o) = 2(8 \text{ m/s}^2) = 16 \text{ m/s}^2$$

(b) For the new situation, the force is unchanged $F_{new} = F_o$, the new mass is half the old mass ($m_{new} = m_o/2$) and we may find the acceleration by

$$F_{new} = m_{new}a_{new}$$

or

$$a_{new} = F_{new}/m_{new} = F_o/2m_o = (F_o/m_o)/2 = (8.0 \text{ m/s}^2)/2 = 4.0 \text{ m/s}^2$$

(c) A similar procedure gives $a = 8.0 \text{ m/s}^2$.
(d) A similar procedure gives $a = 32 \text{ m/s}^2$.
Assess: From the algebraic relationship $a = F/m$ we can see that when (a) the force is doubled, the acceleration is doubled; (b) the mass is doubled, the acceleration is halved; (c) both force and mass are doubled, the acceleration doesn't change; and (d) force is doubled and mass is halved, the acceleration will be four times larger.

P4.19. Strategize: This problem involves forces acting on varying mass to produce different accelerations. We will use Newton's second law.
Prepare: We will assume that the car's ability to produce a force does not depend on the mass. Then we can write Newton's second law for two cases: with a passenger and without. We call the force provided by the car F. Then $F = m_{with}a_{with}$ and $F = m_{without}a_{without}$. We can relate these two and solve for the unknown acceleration.
Solve: Equating the two equations for the force provided by the car, we find

$$F = m_{with}a_{with} = m_{without}a_{without} \Rightarrow a_{with} = \frac{m_{without}}{m_{with}}a_{without} = \left(\frac{1510 \text{ kg}}{1590 \text{ kg}}\right)(0.75)g = (0.71)g.$$

Assess: It is reasonable that an increase in mass should result in a smaller acceleration. But since the increase in mass is small, the decrease in acceleration is also small.

P4.21. Strategize: Force and acceleration are related through mass by Newton's second law.

Prepare: The graph shows acceleration vs. force.

Solve: Newton's second law is $F = ma$. We can read a force and an acceleration from the graph, and hence find the mass. Choosing the force $F = 1$ N gives us $a = 4$ m/s^2. Newton's second law then yields $m = 0.25$ kg.

Assess: Slope of the acceleration-versus-force graph is 4 m/N$\cdot$s^2, and therefore, the inverse of the slope will give the mass.

P4.23. Strategize: Force and acceleration are related through mass by Newton's second law.

Prepare: We can use Newton's second law to find the acceleration of the bear.

Solve: The only forces on the bear are exerted by the girl and boy. A free-body diagram is shown.

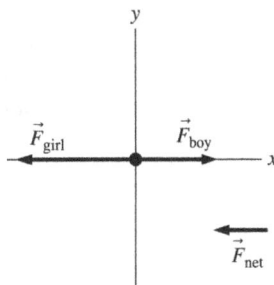

(a) From the free-body diagram shown, the net force in the x-direction is

$$\vec{F}_{Net} = \vec{F}_{Boy} + \vec{F}_{Girl} = 15 \text{ N} - 17 \text{ N} = -2 \text{ N}$$

The net force acting on the bear is 2 N to the left. Since the net force on the bear is not zero, the bear is accelerating. Since at this instant we know nothing about the rate at which the bear's position is changing nothing can be said about the velocity of the bear.

(b) The bear is accelerating, since there is a net force on the bear. From part (a), the net force is 2 N to the left. We can use Newton's second law to find the acceleration of the bear given the mass of the bear.

$$a = \frac{F_{net}}{m} = \frac{-2 \text{ N}}{0.2 \text{ kg}} = -10 \text{ m/s}^2$$

The acceleration is in the same direction as the force, to the left.

Assess: Knowing the mass of an object and the net force acting on it, Newton's second law may be used to determine its acceleration. The acceleration is always in the direction of the net force acting on an object.

P4.25. Strategize: We will apply Newton's second law to relate force to change in velocity.

Prepare: Call the direction of the skater's initial motion the $+x$ direction. We know the skater's mass, initial and final speeds, and the time over which the speeds changed. We treat the acceleration as being constant, such that we can write $\sum \vec{F} = m\vec{a} = m\left(\dfrac{\vec{v}_f - \vec{v}_i}{\Delta t}\right)$.

Solve: The only force in the horizontal direction is the force of kinetic friction between the skates and the ice. Thus, we can write

$$\sum F_x = f_{k,x} = m\left(\frac{(v_x)_f - (v_x)_i}{\Delta t}\right) = (55 \text{ kg})\left(\frac{(2.9 \text{ m/s}) - (3.5 \text{ m/s})}{(5.0 \text{ s})}\right) = -6.6 \text{ N}. \text{ So the magnitude is 6.6 N.}$$

Assess: This is a reasonable magnitude for ice skates.

P4.27. Strategize: We can relate acceleration to force using Newton's second law.

Prepare: Assuming the force described in the graph is the only force acting on the head, we can equate $F_{ave} = ma_{ave}$ and use this to determine the average acceleration required to find the HIC.

Solve: The acceleration is $a_{ave} = F_{ave} / m = (2000 \text{ N}) / (4.5 \text{ kg}) = 444 \text{ m/s}^2$. Inserting this into the expression for the HIC, and reading the duration from the graph, we have

$$\text{HIC} = (a_{avg} / g)^{2.5} \Delta t = \left((444 \text{ m/s}^2) / (9.8 \text{ m/s}^2) \right)^{2.5} (0.080 \text{ s}) = 1.1 \times 10^3 \text{ N}$$

Assess: This collision is likely to cause serious head injury or even death.

P4.29. Strategize: Free-body diagrams show all force vectors acting on a particular object. The object is treated like a single point.

Prepare: The free-body diagram shows two equal and opposite forces such that the net force is zero. The force directed down is labeled as a weight, and the force directed up is labeled as a tension. With zero net force the acceleration is zero. Draw as shown a picture of a real object with two forces to match the given free-body diagram.

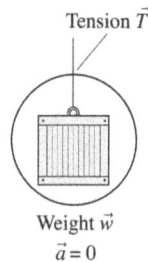

Tension $\vec{T}$

Weight $\vec{w}$

$\vec{a} = 0$

Solve: A possible description is: "An object hangs from a rope and is at rest." Or, "An object hanging from a rope is moving up or down with a constant speed."

Assess: This problem and the following two problems make it clear how important it is to know all forces (and their direction) acting on an object in order to determine the net force acting on the object.

P4.31. Strategize: Free-body diagrams show all force vectors acting on a particular object. The object is treated like a single point.

Prepare: The free-body diagram shows three forces. There is a weight force $\vec{w}$, which is down. There is a normal force labeled $\vec{n}$, which is up. The forces $\vec{w}$ and $\vec{n}$ are shown with vectors of the same length so they are equal in magnitude and the net vertical force is zero. So we have an object on the ground that is not moving vertically. There is also a force $\vec{f}_k$ to the left. This must be a frictional force and we need to decide whether it is static or kinetic friction. The frictional force is the only horizontal force, so the net horizontal force must be $\vec{f}_k$. This means there is a net force to the left producing an acceleration to the left. This all implies motion and therefore the frictional force is kinetic. Draw a picture of a real object with three forces to match the given free-body diagram.

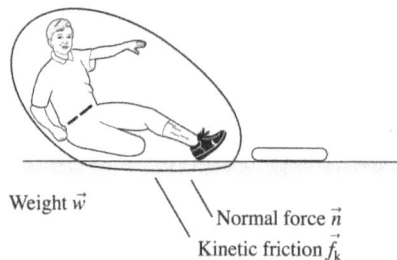

Weight $\vec{w}$

Normal force $\vec{n}$

Kinetic friction $\vec{f}_k$

Solve: A possible description is, "A baseball player is sliding into second base."

Assess: On the free-body diagram, kinetic friction force is the only horizontal force, and it is pointing to the left. This tells us that the baseball player is sliding to the right.

P4.33. Strategize: Free-body diagrams show all force vectors acting on a particular object. The object is treated like a single point.
Prepare: We will follow the procedures in Tactics Box 4.2 and Tactics Box 4.3.
Solve: Your car is the system. See the following diagram.

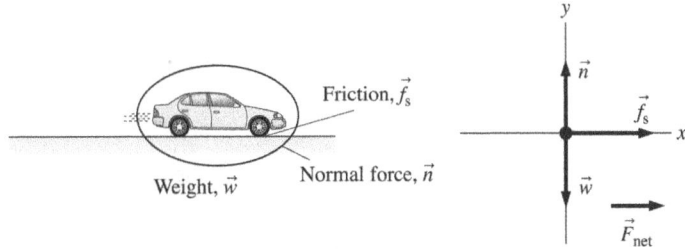

There are contact forces where the car touches the road. One of them is the normal force of the road on the car. The other is the force of static friction between the car's tires and the road, since the car is accelerating from a stop. The only long-range force acting is the weight of the car. Compare to Figure 4.30 and the discussion of propulsion in the text.
Assess: Tactics Box 4.2 and Tactics Box 4.3 give a systematic method for determining all forces on an object and drawing a free-body diagram.

P4.35. Strategize: Free-body diagrams show all force vectors acting on a particular object. The object is treated like a single point.
Prepare: Draw a picture of the situation, identify the system, in this case the physics textbook, and draw a closed curve around it. Name and label all relevant contact forces (the normal force and kinetic friction) and long-range forces (weight).

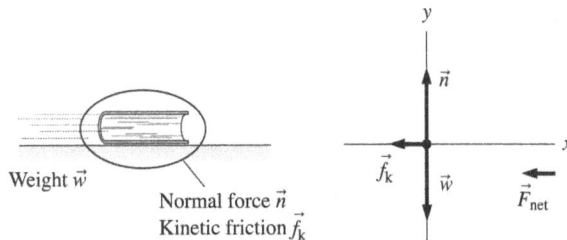

Solve: There are three forces acting *on* the physics textbook due to its interactions with the two agents the earth and the table. One of the forces *on* the book is the long-range weight force *by* the earth. Forces exerted on the book by the table are the normal force and the force of kinetic friction. The normal force exerted *by* the surface of the table is due to the contact between the book and the table. The force of kinetic friction exerted *by* the surface of the table is due to the sliding contact between the book and the table. Since the textbook is slowing down, it has an acceleration and hence net force. The free-body diagram is shown on the right.
Assess: The problem uses the word "sliding." Any real sliding situation involves kinetic friction with the surface the object is sliding over. The force of kinetic friction always opposes the motion of the object.

P4.37. Strategize: Free-body diagrams show all force vectors acting on a particular object. The object is treated like a single point.
Prepare: Follow the steps outlined in Tactics Boxes 4.2 and 4.3. Draw a picture of the situation, identify the system, in this case the car, and draw a closed curve around it. Name and label all relevant contact forces (the normal force and the drag of the air) and long-range forces (weight). Since the road is steep we will assume the car accelerates down the hill; this affects the relative lengths of the arrows we draw.

Solve:

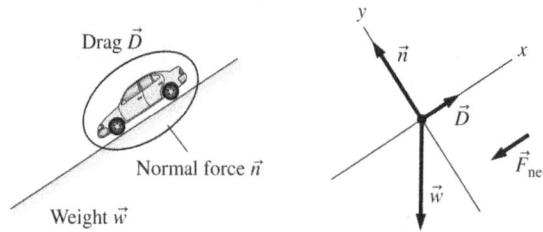

There are three forces acting *on* the car due to its interactions with the two agents earth and the cable. One of the forces *on* the elevator is the long-range weight force *by* the earth. Another force is the normal force of the road on the car. The third is the drag force of the air on the car.

Assess: We ignored friction but not the air resistance and came out with a reasonable answer.

P4.39. Strategize: Free-body diagrams show all force vectors acting on a particular object. The object is treated like a single point.

Prepare: We follow the steps outlined in Tactics Boxes 4.2 and 4.3.

Solve: The system is the box.

The objects in contact with the box are the floor and the rope. The floor exerts an upward normal force and a backwards friction force. The rope exerts a tension force.

The important long-range force is the gravitational force of the earth on the box (i.e., the weight).

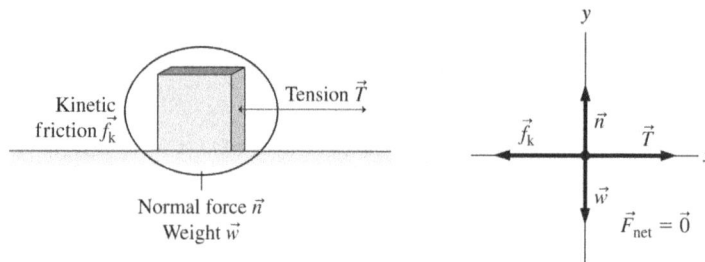

Assess: The net force is zero, as it should be for an object which is moving at constant velocity.

P4.41. Strategize: We will identify action/reaction pairs as described in Newton's third law.

Prepare: When a cannon is fired, gas is ignited and the explosion of hot gas propels the cannonball down the muzzle of the cannon. We will refer to this expanding hot gas as the force of the cannon on the cannonball.

Solve: The force of the cannon on the cannonball and the force of the cannonball back on the cannon make up an action/reaction pair. This can be seen when a cannon lurches backward upon being fired. As the cannon is fired, gravity continues to act on the cannonball. Although there is drag/air resistance on a cannonball, we will ignore that force for this instant at which it is fired.

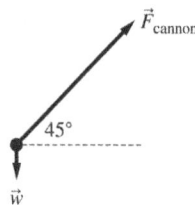

Assess: Note that the force from the cannon on the cannonball is much larger than the cannonball's weight.

P4.43. Strategize: Newton's third law tells us that for every action there is an equal and opposite reaction and that these forces are exerted on different objects.

Prepare: Applying Newton's third law, we can identify, draw and label all the action-reaction pairs. Knowing all the forces acting on skater 2, we can construct a free-body diagram for skater 2.

Solve:

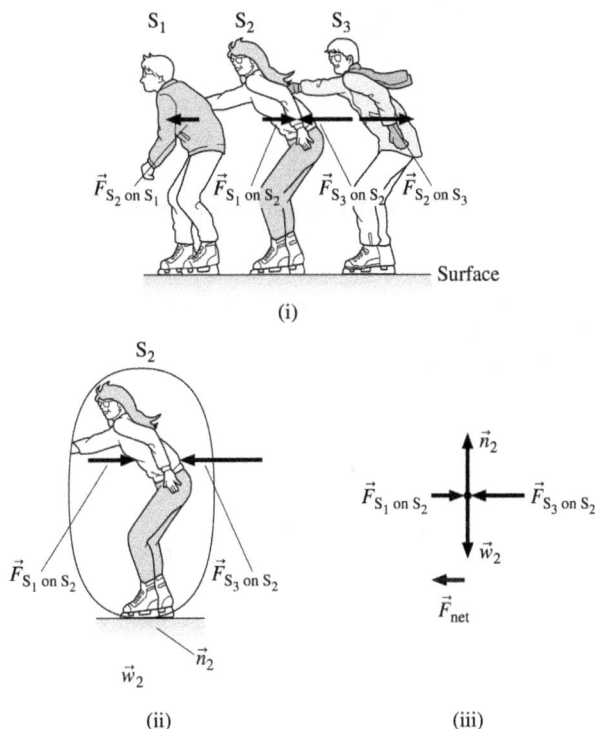

Figure (i) shows all three skaters and the action/reaction forces between them. Note that the lengths of the forces *of each pair* are the same, but the forces of different pairs are different, as explained next. Figure (ii) shows all the forces acting on skater 2. There is the force $\vec{F}_{S_3 \text{ on } S_2}$ due to the skater behind her, and the backward force $\vec{F}_{S_1 \text{ on } S_2}$ due to the skater in front of her. There are also her weight $\vec{w}_2$ and the normal force of the ice $\vec{n}_2$ acting on her.

We are told that skater 3 is pushing on skater 2, so both skaters 1 and 2 will start moving to the left—they are accelerating. Thus the net force on skater 2 (and skater 1) must be to the left. This can only happen if the force $\vec{F}_{S_3 \text{ on } S_2}$ that 3 exerts on 2 is *greater* than the backward force $\vec{F}_{S_1 \text{ on } S_2}$ that skater 1 exerts on 2. These two forces have been drawn so that this is true.

Assess: Newton's third law tells us that the members of action/reaction pairs must have the same magnitude, but we must use Newton's second law to understand the relative magnitudes of forces that are not members of the same action/reaction pair.

P4.45. Strategize: Newton's third law tells us that for every action there is an equal and opposite reaction and that these forces are exerted on different objects.

Prepare: We are told to consider two objects: the road and the car. We can identify all the action-reaction pairs.

Solve: The road exerts an upward normal force on the car, so by the third law the car exerts a downward normal force on the road. The road also exerts a backward kinetic friction force on the car, so the car exerts a kinetic friction force on the road in the opposite direction.

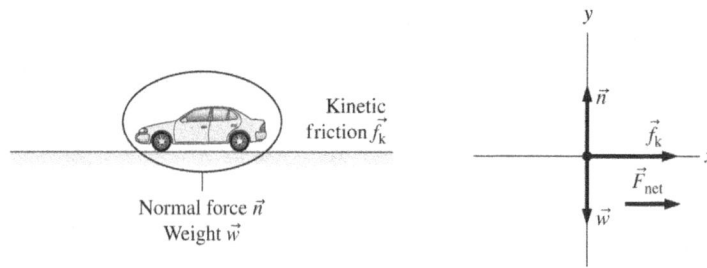

Assess: The road exerts two forces on the car, and the earth exerts a third force on the car; these are the forces that appear in the free-body diagram.

P4.47. Strategize: Motion diagrams show changes in velocity vectors, from which we can determine acceleration. The acceleration and the net force must be in the same direction.
Prepare: Redraw the motion diagram as shown.

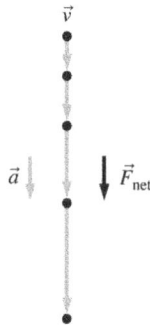

Solve: The previous figure shows velocity as downward, so the object is moving down. The length of the vector increases showing that the speed is increasing (like a dropped ball). Thus, the acceleration is directed down. Since $\vec{F} = m\vec{a}$, the force is in the same direction as the acceleration and must be directed down.

Assess: Since the object is speeding up, the acceleration vector must be parallel to the velocity vector and the net force must be parallel to the acceleration. In order to determine the net force, we had to combine our knowledge of motion diagrams, kinematics, and dynamics.

P4.49. Strategize: Motion diagrams show changes in velocity vectors, from which we can determine acceleration. The acceleration and the net force must be in the same direction.
Prepare: Redraw the motion diagram as shown.

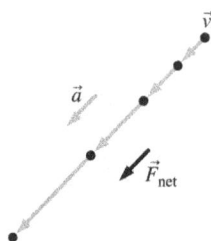

Solve: The velocity vector in the previous figure is shown downward and to the left. So movement is downward and to the left. The velocity vectors get successively longer which means the speed is increasing. Therefore the acceleration is downward and to the left. By Newton's second law $\vec{F} = m\vec{a}$, the net force must be in the same direction as the acceleration. Thus, the net force is downward and to the left.

Assess: Since the object is speeding up, the acceleration vector must be parallel to the velocity vector. This means the acceleration vector must be pointing along the direction of velocity. Therefore the net force must also be downward and to the left.

P4.51. Strategize: Free-body diagrams shown all force vectors acting on a particular object. The object is treated as a point, and the force vectors are drawn on the same scale.
Prepare: Refer to Tactics Box 4.2 and Tactics Box 4.3 for identification of forces and for drawing free-body diagrams. We will draw a correct free-body diagram and compare.
Solve: Your car is the system. See the following diagram.

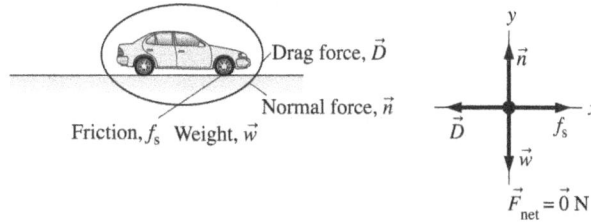

There are contact forces where the car touches the road. One of them is the normal force of the road on the car. The other is the force of static friction between the car's tires and the road since the car is moving.
In addition to this there must be a force in the opposite direction to the car's motion, since the car is moving at constant speed. If only the frictional force acted in the horizontal direction, the car would be accelerating! The diagram omits one of the forces. A possible force that acts in this direction is the force of air drag on the car, which is indicated on the diagram.
The only long-range force acting is the weight of the car.
The diagram also identifies the weight of the car and the normal force on the car as an action/reaction pair. This isn't possible, since both these forces act on the same object, while action/reaction pairs always act on *different* objects. The normal force on an object and its weight are *never* action/reaction pairs.
Assess: In order for an object to be moving at constant velocity, the net force on it must be zero. Action/reaction pairs always act on two different objects.

P4.53. Strategize: The changes in velocity vectors in a motion diagram are related to acceleration, which is determined by the sum of all forces.
Prepare: There are two forces acting *on* the elevator due to its interactions with the two agents the earth and the cable. One of the forces *on* the elevator is the long-range weight force *by* the earth. Another force is the tension force exerted *by* the cable due to the contact between the elevator and the cable. Since the elevator is speeding up as it descends its acceleration is pointing downward. Tension is the only contact force. The downward acceleration implies that $w > T$. Therefore the net force on the elevator must also point downward.
Solve: A force-identification diagram, a motion diagram, and a free-body diagram are shown.

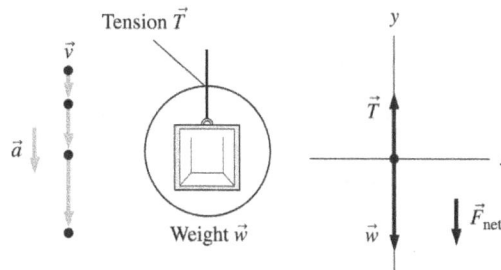

Assess: You now have three important tools in your "Physics Toolbox," motion diagrams, force diagrams, and free-body diagrams. Careful use of these tools will give you an excellent conceptual understanding of a situation.

P4.55. Strategize: The changes in velocity vectors in a motion diagram are related to acceleration, which is determined by the sum of all forces.

Prepare: The normal force is perpendicular to the ground. The thrust force is parallel to the ground and in the direction of acceleration. The drag force is opposite to the direction of motion. There are four forces acting *on* the jet plane due to its interactions with the four agents the earth, the air, the ground, and the hot gases exhausted to the environment. One force on the rocket is the long-range weight force *by* the earth. The second force is the drag force *by* the air. Third is the normal force on the rocket *by* the ground. The fourth is the thrust force exerted on the jet plane *by* the hot gas that is being let out to the environment. Since the jet plane is speeding down the runway, its acceleration is pointing to the right. Therefore, the net force on the jet plane must also point to the right.

Now, draw a picture of the situation, identify the system, in this case the jet plane, and draw a motion diagram. Draw a closed curve around the system, and name and label all relevant contact forces and long-range forces.

Solve: A force-identification diagram, a motion diagram, and a free-body diagram are shown.

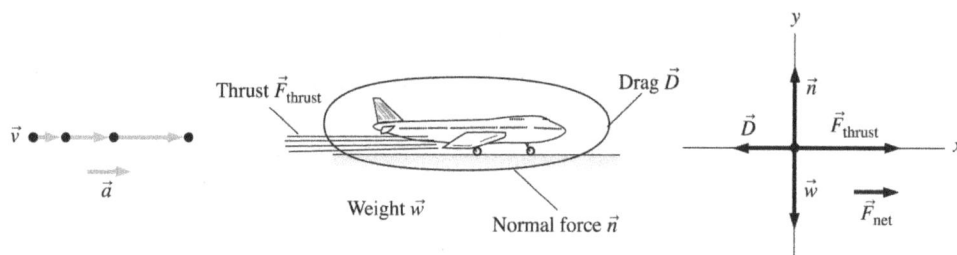

Assess: You now have three important tools in your "Physics Toolbox," motion diagrams, force diagrams, and free-body diagrams. Careful use of these tools will give you an excellent conceptual understanding of a situation.

P4.57. Strategize: The changes in velocity vectors in a motion diagram are related to acceleration, which is determined by the sum of all forces.

Prepare: We will draw a motion diagram with downward velocity vectors that decrease in size as the explorer descends. Our force identification diagram should include the force applied by the rope or chain and gravity. These should also be reflected in our free-body diagram.

Solve:

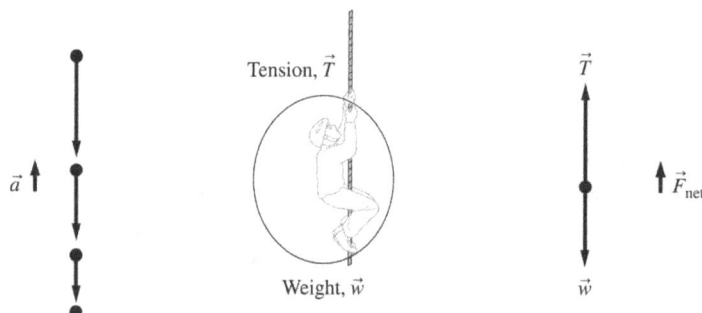

Assess: Note that the upward force from the rope must be greater than the downward weight force in order for the acceleration to be upward.

P4.59. Strategize: The changes in velocity vectors in a motion diagram are related to acceleration, which is determined by the sum of all forces.

Prepare: There are three forces acting on the bale of hay due to its interactions with the two agents: the earth and the bed of the truck. The two contact forces between the bale of hay and the bed of the truck are the normal force and the force of kinetic friction which is dragging the bale of hay forward (even though it is sliding backward). The force at a distance is the force the earth exerts on the bale of hay (the weight). Since the normal force and the weight are equal in magnitude and opposite in direction, there is no net vertical force. Since the force of kinetic friction provides a net horizontal force, the net force acting on the bale of hay and hence the acceleration of the bale of hay is in the direction of the force of kinetic friction. Now, draw a picture of the situation, identify the system, in this case the bale of hay, and draw a motion diagram. Draw a closed curve around the system, and name and label all relevant contact forces and long-range forces.

Solve:

Normal force $\vec{n}$
Weight $\vec{w}$ Kinetic friction $\vec{f}_k$

$\vec{v}$

$\vec{a}$

$\vec{n}$
$\vec{f}_k$
$\vec{w}$
$\vec{F}_{net}$

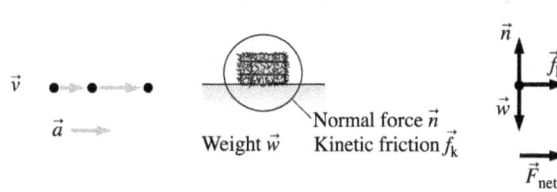

Assess: Since there is a net or unbalanced force acting on the bale of hay, it will experience an acceleration in the direction of this force.

P4.61. Strategize: The changes in velocity vectors in a motion diagram are related to acceleration, which is determined by the sum of all forces.
Prepare: The ball rests on the floor of the barrel because the weight is equal to the normal force. There is a force of the spring to the right, which causes acceleration. Now, draw a picture of the situation, identify the system, in this case the plastic ball, and draw a motion diagram. Draw a closed curve around the system, and name and label all relevant contact forces and long-range forces. Neglect friction
Solve: A force-identification diagram, a motion diagram, and a free-body diagram are shown.

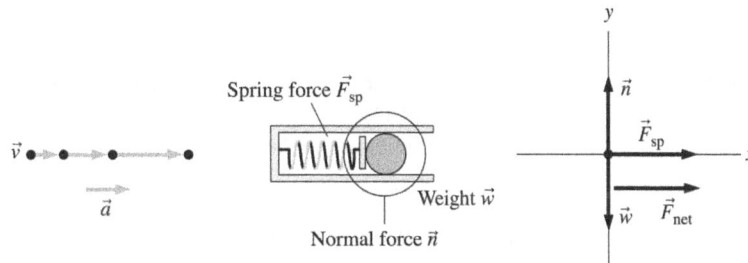

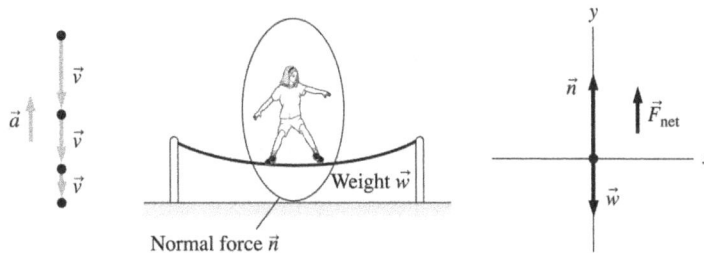

Spring force $\vec{F}_{sp}$
$\vec{v}$
$\vec{a}$
Weight $\vec{w}$
Normal force $\vec{n}$

y
$\vec{n}$
$\vec{F}_{sp}$
x
$\vec{w}$ $\vec{F}_{net}$

Assess: Since the normal force acting on the ball and the weight of the ball are equal in magnitude and opposite in direction, the ball experiences no vertical motion.

P4.63. Strategize: The changes in velocity vectors in a motion diagram are related to acceleration, which is determined by the sum of all forces.
Prepare: The gymnast experiences the long range force of weight. There is also a contact force from the trampoline, which is the normal force of the trampoline on the gymnast. The gymnast is moving downward and the trampoline is decreasing her speed, so the acceleration is upward and there is a net force upward. Thus the normal force must be larger than the weight. The actual behavior of the normal force will be complicated as it involves the stretching of the trampoline and therefore tensions.
Now, draw a picture of the situation, identify the system, in this case the gymnast, and draw a motion diagram. Draw a closed curve around the system, and name and label all relevant contact forces and long-range forces.
Solve: A force-identification diagram, a motion diagram, and a free-body diagram are shown.

$\vec{a}$
$\vec{v}$
$\vec{v}$
$\vec{v}$
Weight $\vec{w}$
Normal force $\vec{n}$

y
$\vec{n}$ $\vec{F}_{net}$
x
$\vec{w}$

Assess: There are only two forces on the gymnast. The weight force is directed downward and the normal force is directed upward. Since the gymnast is slowing down right after making contact with the trampoline, upward normal force must be larger than the downward weight force.

P4.65. Strategize: The changes in velocity vectors in a motion diagram are related to acceleration, which is determined by the sum of all forces.

Prepare: You can see from the motion diagram that the bag accelerates to the left along with the car as the car slows down. According to Newton's second law, $\vec{F} = m\vec{a}$, there must be a force to the *left* acting on the bag. This is friction, but not kinetic friction. The bag is not sliding across the seat. Instead, it is static friction, the force that prevents slipping. Were it not for static friction, the bag would slide off the seat as the car stops. Static friction acts in the direction needed to prevent slipping. In this case, friction must act in the backward (toward the left) direction.

Now, draw a picture of the situation, identify the system, in this case the bag of groceries, and draw a motion diagram. Draw a closed curve around the system, and name and label all relevant contact forces and long-range forces.

Solve: A force-identification diagram, a motion diagram, and a free-body diagram are shown.

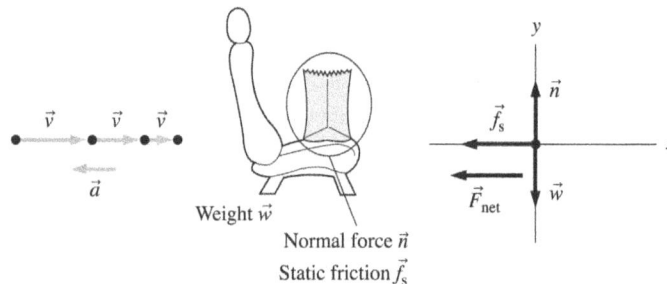

Weight $\vec{w}$
Normal force $\vec{n}$
Static friction $\vec{f}_s$

Assess: Since the normal force acting on the bag of groceries and the weight of the groceries are equal in magnitude and opposite in direction, the bag experiences no vertical motion. The only horizontal force acting on the bag of groceries is static friction, and it provides the net force acting on the bag which results in the acceleration of the bag.

P4.67. Strategize: Acceleration is related to force through Newton's second law.

Prepare: The agent of the force is the object from which the force originates. The force can be found using Newton's second law and the given mass of the greyhound. Finally, the distance travelled in the first 4.0 s can be found using kinematic equations, since the acceleration is constant over that interval.

Solve: (a) The ground is the agent of force. The greyhound exerts a force on the ground and the ground exerts a force back on the greyhound; the latter is the force that propels the greyhound forward.

(b) Assuming the force the ground exerts on the greyhound is the only force in the horizontal direction acting on the greyhound, we can write $\sum F_x = F_{\text{ground greyhound}} = ma_x = (32 \text{ kg})(10 \text{ m/s}^2) = 320 \text{ N}$.

(c) We can use Equation 2.11, and we include the fact that the greyhounds start from rest:

$$\Delta x = \left(v_x\right)_i \Delta t + \frac{1}{2} a_x \left(\Delta t\right)^2 = \frac{1}{2}\left(10 \text{ m/s}^2\right)\left(4 \text{ s}\right)^2 = 80 \text{ m}.$$

Assess: This is a reasonable distance for extremely fast racing dogs.

P4.69. Strategize: We can relate the force to the acceleration through the mass using Newton's second law.

Prepare: Assume the ball undergoes constant acceleration during the pitch so we can use the kinematic equations.

$$\left(v_x\right)_f^2 = \left(v_x\right)_i^2 + 2a_x \Delta x$$

Use coordinates where $+x$ is in the direction the ball is thrown. We are given $\Delta x = 1.0 \text{ m}$, $\left(v_x\right)_f = 47 \text{ m/s}$, and $m = 0.145 \text{ kg}$. Assume $\left(v_x\right)_i = 0.0 \text{ m/s}$.

We'll first solve for a_x and then use Newton's second law to find the average force.

Solve: (a)

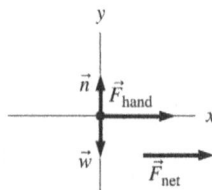

(b) Solve the equation for a_x.

$$a_x = \frac{(v_x)_f^2 - (v_x)_i^2}{2\Delta x} = \frac{(47 \text{ m/s})^2 - (0.0 \text{ m/s})^2}{2(1.0 \text{ m})} = 1100 \text{ m/s}^2$$

$$F_x = ma_x = (0.145 \text{ kg})(1100 \text{ m/s}^2) = 160 \text{ N}$$

(c) Say a typical pitcher weighs 170 lbs.

$$170 \text{ lb}\left(\frac{4.45 \text{ N}}{1 \text{ lb}}\right) \approx 760 \text{ N}$$

Now divide the force from part (b) by this weight to see the fraction.

$$160 \text{ N} \div 760 \text{ N} \approx \frac{1}{5}$$

So the force the pitcher exerted on the ball is about 1/5 his weight.
Assess: The answer to each part seems reasonable. The units also work out.

P4.71. Strategize: We can relate acceleration to force using the mass and Newton's second law. We know that the force the insect exerted on the plate is equal and opposite to the force the plate exerted on the insect, by Newton's third law.
Prepare: We know the maximum force should be exerted with the maximum acceleration is achieved. So we will use the maximum acceleration in our calculation, which appears to be about 87 m/s^2.
Solve: The sum of all forces in the vertical direction (which we will call $+y$) is given by

$$\sum F_y = F_{loc} - mg = ma_y \Rightarrow F_{loc} = ma_y + mg \Rightarrow (F_{loc})_{max} = m(a_y)_{max} + mg$$
$$= (5.0 \times 10^{-4} \text{ kg})((87 \text{ m/s}^2) + (9.8 \text{ m/s}^2)) = 0.084 \text{ N}$$

Assess: The insect had to push with enough force to cancel out gravity just to remain stationary on the plate. The force required to accelerate the insect upward must be added to that weight force to find the total force the insect exerted on the plate.

P4.73. Strategize: This problem involves careful consideration of the sum of all forces acting on an object.
Prepare: We will identify all forces on the beach ball and draw a free-body diagram to consider the magnitude of the net force on the ball.
Solve: The beach ball is our system. See the diagram below.

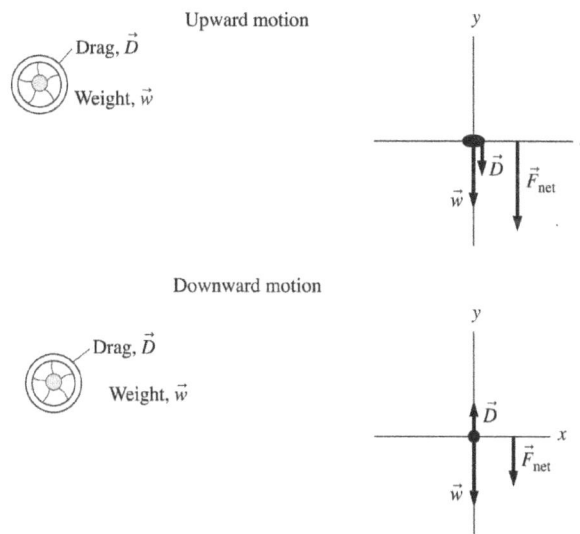

The only contact force is the force of air drag on the ball. The only long-range force is the weight of the ball. Drag always points in the direction opposite to the motion. From the free-body diagram, the net force will always be greater when the beach ball is moving upward compared to downward. There the weight of the ball and the drag force always reinforce each other. The net force is larger on the way up.

Assess: Note that in the downward portion there is a possibility of the net force equaling zero. The ball will not accelerate downward and fall with constant velocity if this happens, in contrast to the case where there is no drag.

P4.75. Strategize: The direction of a tension force is always along the rope that is tense.
Prepare: Tension force is discussed in Section 4.1. We will use Newton's third law to find the force on the rope.
Solve: In the diagram in the problem, the force the left part of the rope exerts on you is mostly in the westerly direction, with a small component to the south. The force that you exert on the left portion of the rope is the reaction force to this force and would be in the exact opposite direction, mostly to the east and a little to the north. The rope is connected directly to the tree. As explained in Section 4.2, the tension force is "transmitted" through the rope by the molecular bonds in the rope. So the force on the tree is directly mostly to the east with a small component to the north. The correct choice is C.
Assess: Tension is "transmitted" along a rope by the molecular bonds in the rope.

P4.77. Strategize: This is a straightforward application of Newton's third law.
Prepare: Consider whether the force the rope exerts on the car and the force the car exerts on the rope form an interaction pair.
Solve: The rope exerts a force on the car and the car exerts a force on the rope. According to Newton's third law, these two forces constitute an action pair and as such they are equal in magnitude and opposite in direction. Hence the force the car exerts on the rope is equal to the force the rope exerts on the car. The correct answer is C.
Assess: Newton's third law informs us that the two forces in an action/reaction-pair are equal in magnitude and opposite in direction.

APPLYING NEWTON'S LAWS

Q5.1. Reason: For an object to be in equilibrium, the net force (i.e., sum of the forces) must be zero. Assume that the two forces mentioned in the question are the only ones acting on the object.

The question boils down to asking if two forces can sum to zero if they aren't in opposite directions. Mental visualization shows that the answer is no, but so does a careful analysis. Set up a coordinate system with the x-axis along one of the forces. If the other force is not along the negative x-axis then there will be a y (or z) component that cannot be canceled by the first one along the x-axis.

Assess: In summary, two forces not in opposite directions cannot sum to zero. Neither can two forces with different magnitudes. However, three can.

Q5.3. Reason: If the blocks remain at rest, then clearly the sum of all forces in the horizontal direction must be zero; otherwise there would be some non-zero acceleration. Thus, in the case of both boxes A and B the force of static friction is exactly cancelling the force applied (along the rope) in the horizontal direction. The frictional force is 30 N to the left on both boxes A and B; they are equal.

Assess: It is easy to get confused by applying $(f_s)_{max} = \mu_s n$. But remember that is the expression for the maximum frictional force before the object starts to slip.

Q5.5. Reason: The reading on the moon will be the moon-weight, or the gravitational force of the moon on the astronaut. This would be about 1/6 of the astronaut's earth-weight or the gravitational force of the earth on the astronaut (while standing on the scales on the earth).

Assess: The astronaut's *mass* does not change by going to the moon.

Q5.7. Reason: The normal force (by definition) is directed perpendicular to the surface.

(a) If the surface that exerts a force on an object is vertical, then the normal force would be horizontal. An example would be holding a picture on a wall by pushing on it horizontally. The wall would exert a normal force horizontally.

(b) In a similar vein, if the surface that exerts a force on an object is horizontal and above the object, then the normal force would be down. One example would be holding a picture on a ceiling by pushing on it. The ceiling would exert a normal force vertically downward. Another example would be the Newton's third law pair force in the case of you sitting on a chair; the chair exerts a normal force upward on you, so you exert a normal force downward on the chair.

Assess: We see that the normal force can be in any direction; it is always perpendicular to the surface pushing on the object in question.

Q5.9. Reason: Increasing the mass does increase the net force on the system, but it also increases the inertia. $a = \dfrac{F_{net}}{m}$. Since both the net force and mass are increased they still cancel, leaving the acceleration the same.

Assess: The m cancels out of every term.

Q5.11. Reason: Before you slip the static friction is keeping you from sliding. But when you slip the friction is kinetic friction, and it is less than the maximum static friction. Before slipping you must have been near the maximum static friction or you wouldn't have slipped, and the net force on you was about zero. So there is less friction on you while sliding than when you weren't sliding, so you will accelerate.

Assess: The static friction is not always the maximum possible amount, but it was close in this case.

Q5.13. Reason: The force of gravity on the trampolinist never changes appreciably. So by "apparent weight" we mean the contact forces pushing up on her in a way similar to how the surface of a bathroom scale pushes up on our feet supporting us against gravity when we weigh ourselves. At point c, there are no such forces, so her apparent weight is definitely minimal at point c. Just before she leaves the trampoline, the trampoline is barely stretched, and is therefore not exerting much force on her. So her apparent weight at point b is not zero, but is small. When she is momentarily at rest, her acceleration is upward (consider the velocity a moment before and a moment after she is at rest). So the upward force of the trampoline must be exceeding the gravitational force acting downward on her. So her apparent weight is even greater than her actual weight at this moment: $\sum F_y = F_t - mg = ma_y \Rightarrow F_t = m(g + a_y)$.

So we find the ranking $w_{app,a} > w_{app,b} > w_{app,c}$.

Assess: It is reasonable that the apparent weight is small when one is in freefall; indeed one feels "weightless" in freefall.

Q5.15. Reason: As the plane's thrust is decreased, the plane will start to decelerate. From Equation 5.12, as the plane's velocity decreases, so does the drag force on the plane. The plane's velocity will decrease until the drag force equals the thrust force. At that point, the plane will stop decelerating as it reaches a new equilibrium. If the drag force continued decreasing, the thrust force would re-accelerate the plane to the point where it would stop accelerating, again reaching the same equilibrium. Therefore the plane will travel with a constant velocity once the new equilibrium is reached.

Assess: Drag force decreases with decrease in velocity.

Q5.17. Reason: See the free-body diagrams below.

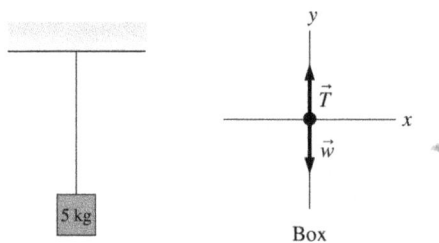

The object is in static equilibrium.

Newton's second law for the 5 kg box gives $T - mg = F_y = ma_y = 0$ N. Then $T = mg = 49$ N

Assess: This makes sense, since the object has a mass of 5 kg.

Q5.19. Reason: Consider the free-body diagrams below.

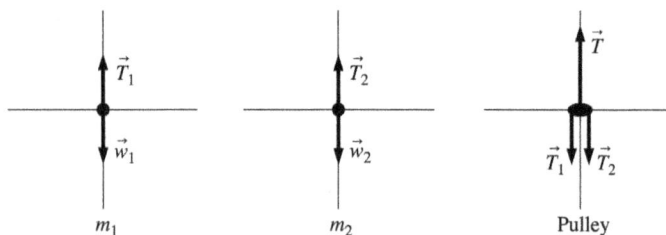

Since each object in the system is in static equilibrium, we can apply Equation 5.1. For m_1, $T_1 = w_1$. For m_2, $T_2 = w_2$. The pulley is massless, so it has no weight. The only forces on the pulley are the three tensions in the diagram, so $T = T_1 + T_2$. The tension in the rope is the sum of the tensions in the ropes directly connected to the objects, which is just the sum of their weights from the equations above. The tension is 98 N.

Assess: This makes sense, since the scale is effectively supporting two 5 kg objects.

Q5.21. Reason: The kinetic friction acts in a direction to oppose the relative motion, so on block 1 the kinetic friction is to the right and on block 2 it is to the left.

Assess: We would expect them to be opposite since they are a Newton third law pair and the forces in a third law pair are always in opposite directions.

Q5.23. Reason: Friction is holding the block in place against the downward pull of gravity. If a massive block is put in position 2, friction will try to keep the block from sliding UP the ramp. Thus, the tension in the rope will have to overcome the component of gravity pulling block 1 down in the incline and the maximum force of static friction. Let us assume the tension just barely overcomes these forces such that the acceleration is momentarily approximately zero. Calling the $+x$ direction up the incline and $+y$ upward and perpendicular to the plane, we start by drawing a free-body diagram.

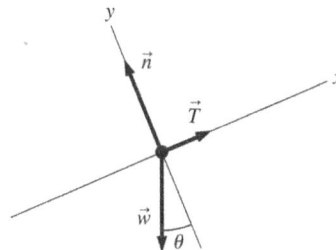

From here, we can write

$$\sum \left(F_{\text{on 1}} \right)_x = T - m_1 g \sin(\theta) - \mu_s n = m_1 a_x \approx 0$$
$$\sum \left(F_{\text{on 1}} \right)_y = -m_1 g \cos(\theta) - n = m_1 a_y = 0$$

Note that $\mu_s = 0.50$ for wood on wood (from Table 5.2).

If the blocks are just barely slipping, then the acceleration of block 2 is also approximately zero, and we can write

$$T - m_2 g \approx 0 \Rightarrow T = m_2 g$$

Inserting this expression for tension and inserting the normal force from the equation for the forces in the y direction into the sum of forces in the x direction, we find

$$m_2 g - m_1 g \sin(\theta) - \mu_s m_1 g \cos(\theta) = 0$$
$$m_2 = m_1 \sin(\theta) + \mu_s m_1 \cos(\theta)$$
$$m_2 = (1.0 \text{ kg}) \sin(20°) + (0.50)(1.0 \text{ kg}) \cos(20°)$$
$$m_2 = 0.81 \text{ kg}$$

The correct answer is B.

Assess: If the tension only needed to overcome a component of gravity, such as if there were no friction, the tension would need to be 0.34. Since there is friction, our answer would need to be somewhat larger than this, which fits with the answer we obtained.

Q5.25. Reason: The ball is in equilibrium. We will use Equation 5.1.
See the free-body diagram below.

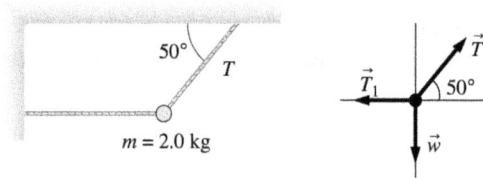

In the vertical direction we have

$$T\sin(50°) - w = T\sin(50°) - mg = 0$$

Solving for T, we obtain

$$T = \frac{mg}{\sin(50°)} = \frac{(2.0 \text{ kg})(9.80 \text{ m/s}^2)}{\sin(50°)} = 26 \text{ N}$$

The correct choice is D.
Assess: Note that we did not need to use the horizontal components of the forces.

Q5.27. Reason: We will use Equation 5.2 since neither the dog nor the floor is in equilibrium.

(a)

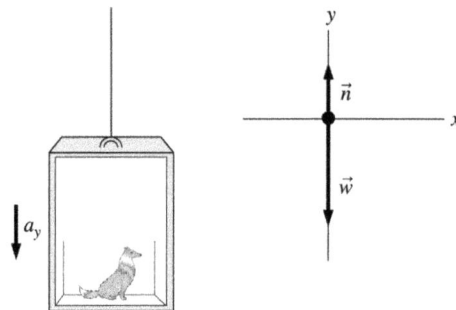

From the free-body diagram above, we have $n - w = ma_y$.
Solving for the normal force,

$$n = w + ma_y = mg + ma_y = (5.0 \text{ kg})(9.80 \text{ m/s}^2) + (5.0 \text{ kg})(-1.20 \text{ m/s}^2) = 43 \text{ N}$$

The correct choice is B.
(b) The normal force on the dog is the force of the floor of the elevator on the dog. The force of the dog on the elevator floor is the reaction force to this. The correct choice is D.
Assess: This result makes sense; the normal force will be less than the weight of the dog, which is 49 N.

Q5.29. Reason: This is still a Newton's second law question; the only twist is that the object is not in equilibrium, i.e., the right side of the second law is not zero.
The forces on Eric are the downward gravitational force of the earth on him w, and the upward normal force of the scale on him n (which we want to know).
We note that $a = -1.7 \text{ m/s}^2$ and $w = mg = (60 \text{ kg})(9.80 \text{ m/s}^2) = 5.88 \text{ N}$.

This is a one-dimensional question in the vertical direction, so the following equations are all in the y-direction.

$$F_{net} = ma$$
$$n - w = ma$$
$$n = ma + w = (60 \text{ kg})(-1.7 \text{ m/s}^2) + 588 \text{ N} = 486 \text{ N} \approx 500 \text{ N}$$

The correct choice is C.

Assess: Because the elevator is accelerating down, we expect the scale to read a bit less than Eric's normal weight. This is the case.

It is important that neither the question nor the answer specify whether the elevator is moving up or down. The elevator can be accelerating down in two ways: It can be moving up and slowing (such as the end of a trip from a low floor to a high floor), or it can be moving down and gaining speed (such as the beginning of a trip from a high floor to a low floor). The answer is the same in both cases.

Q5.31. Reason: We will assume a constant direction so that plus the "constant speed" means no acceleration. The sled is in equilibrium and the net force on it must be zero.

In the horizontal direction there are two forces on the sled: the football player pushing on it, and kinetic friction acting in the opposite direction. These two must have the same magnitude.

Equation 5.8 tells us that $f_k = \mu_k n$, but we don't yet know n.

Independently analyzing the vertical direction reveals that the magnitude of $\vec{n}$ is the same as the magnitude of $w = mg = (60 \text{ kg})(9.80 \text{ m/s}^2) = 590 \text{ N}$.

So the kinetic friction force is $f_k = \mu_k n = (0.30)(590 \text{ N}) = 180 \text{ N}$. And that must also be the magnitude of the football player's pushing force.

The correct answer is C.

Assess: Choices A and B don't seem very strenuous for a football player, but choice D seems like too much. Choice C is in the right range.

Q5.33. Reason: For the Land Rover claim to be true, the vehicle must be able to at least sit on the hill motionless without slipping. So we'll draw a free-body diagram with the vehicle stationary. We use tilted axes with the x-axis running up the slope.

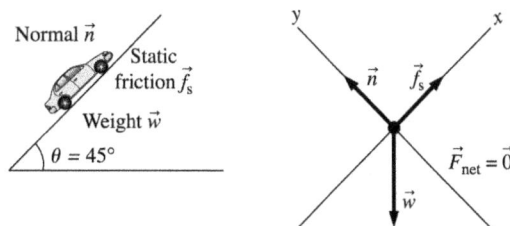

First apply $F_{net} = ma$ in the y-direction.

$$n - w \cos\theta = 0$$

Then apply $F_{net} = ma$ in the x-direction.

$$f_s - w \sin\theta = 0$$

With $f_s = \mu_s n$ we rearrange the pair of equations into

$$\mu_s n = w \sin\theta$$
$$n = w \cos\theta$$

Now the key is to divide the top equation by the bottom one. (This is mathematically legal, because the two sides of the bottom equation are equal to each other, then we are really dividing both sides of the top equation by the same thing.) Remember that $\frac{\sin\theta}{\cos\theta} = \tan\theta$.

$$\mu_s = \tan\theta$$

Insert $\theta = 45°$ and we have $\mu_s = \tan 45° = 1.0$.

The correct choice is D.

Assess: The answer to this question is independent of the mass of the Land Rover! An equivalent way to express this is that w (and n) cancelled out.

Also notice that by solving the equations with a variable θ and only inserting the value of 45° at the end, we are able to solve for the required minimum μ_s for any angle.

Problems

P5.1. Strategize: This problem involves forces in two dimensions. We will separate the forces into components and use Newton's second law.

Prepare: The massless ring is in static equilibrium, so all the forces acting on it must cancel to give a zero net force. The forces acting on the ring are shown on a free-body diagram below.

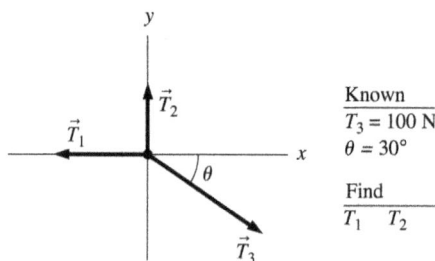

Solve: Written in component form, Newton's first law is

$$(F_{net})_x = \Sigma F_x = T_{1x} + T_{2x} + T_{3x} = 0\text{ N} \quad (F_{net})_y = \Sigma F_y = T_{1y} + T_{2y} + T_{3y} = 0\text{ N}$$

Evaluating the components of the force vectors from the free-body diagram:

$$T_{1x} = -T_1 \quad T_{2x} = 0\text{ N} \quad T_{3x} = T_3 \cos 30°$$

$$T_{1y} = 0\text{ N} \quad T_{2y} = T_2 \quad T_{3y} = -T_3 \sin 30°$$

Using Newton's first law:

$$-T_1 + T_3 \cos 30° = 0\text{ N} \quad T_2 - T_3 \sin 30° = 0\text{ N}$$

Rearranging:

$$T_1 = T_3 \cos 30° = (100\text{ N})(0.8666) = 87\text{ N} \quad T_2 = T_3 \sin 30° = (100\text{ N})(0.5) = 50\text{ N}$$

Assess: Since $\vec{T}_3$ acts closer to the x-axis than to the y-axis, it makes sense that $T_1 > T_2$.

P5.3. Strategize: This problem involves the sum of all forces in two directions. Before release the riders have no acceleration, which tells us the sum of all forces in both the vertical and horizontal directions must be zero. This gives us constraints that we can use to determine the unknown tensions.

Prepare: Let us call the vertically upward direction $+y$ and the direction horizontally to the right $+x$. We will apply Newton's second law to both of these directions. We will draw a free-body diagram, labeling the tension in the right cable as T_r and that in the left cable as T_l.

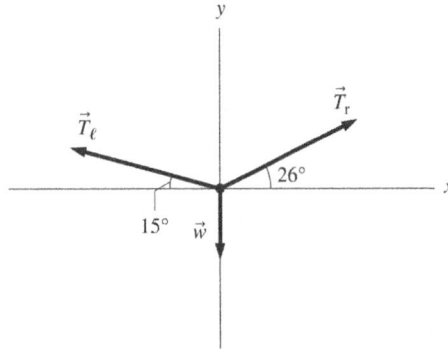

Solve: We have

$$\sum F_x = T_r \cos(25°) - T_1 \cos(15°) = ma_x = 0$$
$$\sum F_y = T_r \sin(25°) + T_1 \sin(15°) - mg = ma_y = 0$$

The first equation tells us $T_r = T_1 \dfrac{\cos(15°)}{\cos(25°)}$, which we then insert into the second equation to obtain

$$T_1\left(\cos(15°)\tan(25°) + \sin(15°)\right) = mg$$

$$T_1 = \frac{mg}{\left(\cos(15°)\tan(25°) + \sin(15°)\right)} = \frac{(270 \text{ kg})(9.8 \text{ m/s}^2)}{\left(\cos(15°)\tan(25°) + \sin(15°)\right)}$$

$$T_1 = 3.6 \times 10^3 \text{ N}$$

Inserting this value (3,683 N, without intermediate rounding) into our expression above for T_r, we find

$$T_r = (3{,}625 \text{ N})\frac{\cos(15°)}{\cos(25°)} = 3.9 \times 10^3 \text{ N}.$$

Assess: We find that the tensions are relatively similar, which is reasonable since both cables are at relatively small angles and similar angles.

P5.5. Strategize: This problem involves forces acting on the patella in two dimensions. We separate the forces into components and use Newton's second law.
Prepare: The femur is in static equilibrium. We can use Equation 5.1.
Solve: See the free-body diagram below.

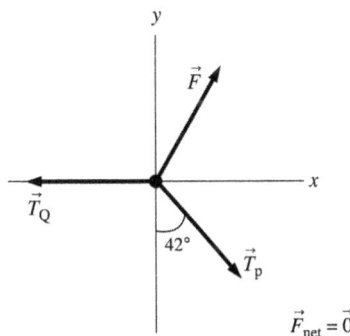

The direction of the force the femur exerts on the patella is indicated roughly on the previous diagram. The sum of the x-components of the forces must be zero. This gives

$$T_Q = T_P \sin(42°) + F_x$$

Solving for F_x,

$$F_x = T_Q - T_P \sin(42°) = 60\text{ N} - (60\text{ N})\sin(42°) = 20\text{ N}$$

The sum of the y-components of the forces must be zero also. This gives

$$F_y = T_P \cos(42°) = (60\text{ N})\cos(42°) = 45\text{ N}$$

The magnitude of the force by the femur on the patella is then

$$F = \sqrt{(F_x)^2 + (F_y)^2} = \sqrt{(20\text{ N})^2 + (45\text{ N})^2} = 49\text{ N}$$

Assess: This result is reasonable in magnitude, considering the magnitude of the forces exerted by the tendons and their directions.

P5.7. Strategize: This problem involves the sum of all forces in two directions. Bethany is at rest, and has no acceleration, which tells us the sum of all forces in both the vertical and horizontal directions must be zero. This gives us constraints that we can use to determine the unknown tensions.

Prepare: Let us call the vertically upward direction $+y$ and the direction horizontally to the right $+x$. We will apply Newton's second law to both of these directions. We will refer to the tension in the rope that makes a $45°$ angle (taken to be the right-most rope) as T_r and that in the left rope as T_l. We will start by drawing a free body diagram.

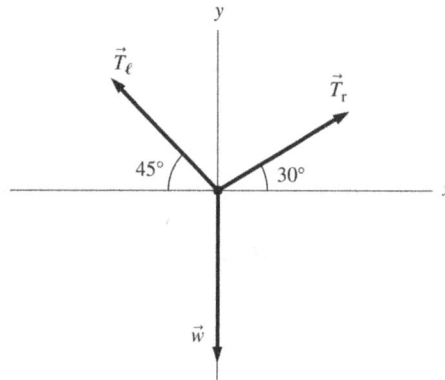

Solve: We have

$$\sum F_x = T_r \cos(45°) - T_l \cos(30°) = ma_x = 0$$
$$\sum F_y = T_r \sin(45°) + T_l \sin(30°) - w = ma_y = 0$$

The first equation tells us $T_r = T_l \dfrac{\cos(30°)}{\cos(45°)}$, which we then insert into the second equation to obtain

$$T_l\left(\cos(30°)\tan(45°) + \sin(30°)\right) = w$$

$$T_l = \frac{w}{\left(\cos(30°)\tan(45°) + \sin(30°)\right)} = \frac{(560\text{ N})}{\left(\cos(30°)\tan(45°) + \sin(30°)\right)}$$

$$T_l = 4.1 \times 10^2 \text{ N}$$

Inserting this value (409.9 N, without intermediate rounding) into our expression above for T_r, we find

$$T_r = (409.9\text{ N})\frac{\cos(30°)}{\cos(45°)} = 5.0 \times 10^2 \text{ N}.$$

Assess: A cursory review of these magnitudes shows that the sum of the magnitudes is larger than Bethany's weight, which must be the case since only one component of each force supports her against gravity.

P5.9. Strategize: We are told the force along the direction of motion and the mass, such that we can determine the acceleration. Given the initial and final speeds, this becomes a simple kinematics problem.

Prepare: The sum of all forces in this case is very simple, with only friction acting to slow the stone. Let us call the direction in which the stone is initially pushed the $+x$ direction. We determine the acceleration and use the kinematic equation $(v_f)_x^2 = (v_i)_x^2 + 2a_x \Delta x$.

Solve: From Newton's second law, we see $a_x = \frac{(F_{net})_x}{m} = \frac{-2.0 \text{ N}}{20 \text{ kg}} = -0.10 \text{ m/s}^2$. Inserting this and given values into the kinematic equation we selected, we find

$$(v_i)_x^2 = (v_f)_x^2 - 2a_x \Delta x = (0)^2 - 2(-0.10 \text{ m/s})(27.9 \text{ m})$$
$$(v_i)_x = \sqrt{-2(-0.10 \text{ m/s})(27.9 \text{ m})} = 2.4 \text{ m/s}$$

Assess: Our answer is of a reasonable order of magnitude and has the correct direction.

P5.11. Strategize: This is a straightforward application of Newton's second law, in two dimensions. There is no need to consider components, since all forces lie along x- or y-axes.

Prepare: The free-body diagram shows five forces acting on an object whose mass is 2.0 kg. We will first find the net force along the x- and the y-axes and then divide these forces by the object's mass to obtain the x- and y-components of the object's acceleration.

Solve: Applying Newton's second law:

$$a_x = \frac{(F_{net})_x}{m} = \frac{4 \text{ N} - 2 \text{ N}}{2 \text{ kg}} = 1.0 \text{ m/s}^2 \quad a_y = \frac{(F_{net})_y}{m} = \frac{3 \text{ N} - 1 \text{ N} - 2 \text{ N}}{2 \text{ kg}} = 0.0 \text{ m/s}^2$$

Assess: The object's acceleration is only along the x-axis.

P5.13. Strategize: We are told a change in speed and a duration, from which we can determine acceleration. Knowing the mass and the acceleration of the ball is enough for us to determine the net force acting on the ball.

Prepare: Let us call the direction of the acceleration $+x$ (although we recognize it may really be at some angle away from horizontal). Let us assume that other forces (such as gravity) on the ball are negligible compared to that of the pitcher's hand. We can check this assumption when we are finished by comparing the force we obtain to the force of gravity. We will use the definition of average acceleration, and Newton's second law.

Solve: We know $(a_x)_{av} = \frac{(v_f)_x - (v_i)_x}{\Delta t} = \frac{(40 \text{ m/s}) - (0 \text{ m/s})}{(50 \times 10^{-3} \text{ s})} = 800 \text{ m/s}^2$. Inserting this into Newton's second law, we have:

$$\sum F_x = ma_x \Rightarrow F_p = ma_x = (0.145 \text{ kg})(800 \text{ m/s}^2) = 1.2 \times 10^2 \text{ N}.$$

Assess: 120 N of force is a very reasonable force for someone to apply to a ball. The skill is in moving the arm fast enough to continue pushing on the ball as it accelerations.

P5.15. Strategize: The expression $w = mg$, is really just a special case of Newton's second law, in which the only force acting is that due to gravity such that $\sum F_y = w = ma_y = mg$. So this should hold on Earth or on the moon. We just need the astronaut's mass.

Prepare: We can find the astronaut's mass on earth and then multiply it with Mars's acceleration due to gravity to find his weight on Mars.

Solve: The mass of the astronaut is

$$m = \frac{w_{earth}}{g_{earth}} = \frac{800 \text{ N}}{9.80 \text{ m/s}^2} = 81.6 \text{ kg}$$

Therefore, the weight of the astronaut on Mars is

$$w_{\text{Mars}} = mg_{\text{Mars}} = (81.6 \text{ kg})(3.76 \text{ m/s}^2) = 310 \text{ N}$$

Assess: The smaller acceleration of gravity on Mars reveals that objects are less strongly attracted to Mars than to the earth, so the smaller weight on Mars makes sense. Also, note that the astronaut's mass stays unchanged.

P5.17. Strategize: We are told the acceleration of the astronauts, and we are given information about forces acting on them. We can determine what the necessary contact force between the floor of the spacecraft and their bodies would be (which is the apparent weight).
Prepare: We will use Newton's second law in the vertical direction (calling vertically upward from the surface of the moon $+y$.
Solve: We have, for the sum of all forces on the astronauts $\sum (F_{\text{on astro}})_y = n - mg = ma_y$. Here g refers to the acceleration due to gravity on the moon. Thus $n = m(a_y + g) = (75 \text{ kg})((3.4 \text{ m/s}^2) + (1.6 \text{ m/s}^2)) = 3.8 \times 10^2$ N. This normal force between the floor and the astronaut is the apparent weight.
Assess: Note that on Earth the weight of a 75 kg astronaut would be about 740 N. It is reasonable that the apparent weight taking off from the moon is still smaller than this since gravity is so weak there and the acceleration of the spacecraft was also small compared to Earth's 9.8 m/s^2 acceleration due to gravity.

P5.19. Strategize: In all cases, we need only apply Newton's second law. We note that the different accelerations will result in different forces on the passenger.
Prepare: The passenger is subject to two vertical forces: the downward pull of gravity and the upward push of the elevator floor. We can use one-dimensional kinematics for the three situations.

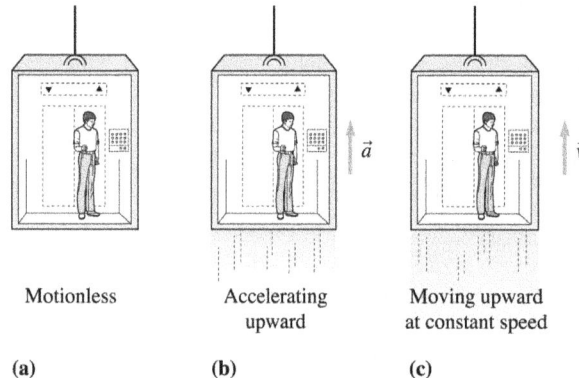

Motionless	Accelerating upward	Moving upward at constant speed
(a)	(b)	(c)

Solve: (a) The apparent weight is

$$w_{\text{app}} = w\left(1 + \frac{a_y}{g}\right) = w\left(1 + \frac{0}{g}\right) = mg = (60 \text{ kg})(9.80 \text{ m/s}^2) = 590 \text{ N}$$

(b) The elevator speeds up from $v_{0y} = 0$ m/s to its cruising speed at $v_y = 10$ m/s. We need its acceleration before we can find the apparent weight:

$$a_y = \frac{\Delta v}{\Delta t} = \frac{10 \text{ m/s} - 0 \text{ m/s}}{4.0 \text{ s}} = 2.5 \text{ m/s}^2$$

The passenger's apparent weight is

$$w_{\text{app}} = w\left(1 + \frac{a_y}{g}\right) = (590 \text{ N})\left(1 + \frac{2.5 \text{ m/s}^2}{9.80 \text{ m/s}^2}\right) = (590 \text{ N})(1.26) = 740 \text{ N}$$

(c) The passenger is no longer accelerating since the elevator has reached its cruising speed. Thus, $w_{app} = w = 590$ N as in part (a).

Assess: The passenger's apparent weight is his normal weight in parts (a) and (c), since there is no acceleration. In part (b), the elevator must not only support his weight but must also accelerate him upward, so it's reasonable that the floor will have to push up harder on him, increasing his apparent weight.

P5.21. Strategize: We are asked for apparent weight, which is the contact force between Zach and the surface supporting him. Thus, we will apply Newton's second law to determine the contact force (which here is a normal force that the floor exerts upward on Zach). We note that since the acceleration is different in parts (a) and (b), we expect the statement of Newton's second law to be different in those cases, meaning we expect to find different apparent weights.

Prepare: We'll assume Zach is a particle moving under the effect of two forces acting in a single vertical line: gravity and the supporting force of the elevator. These forces are shown in Figure 5.9 in a free-body diagram.

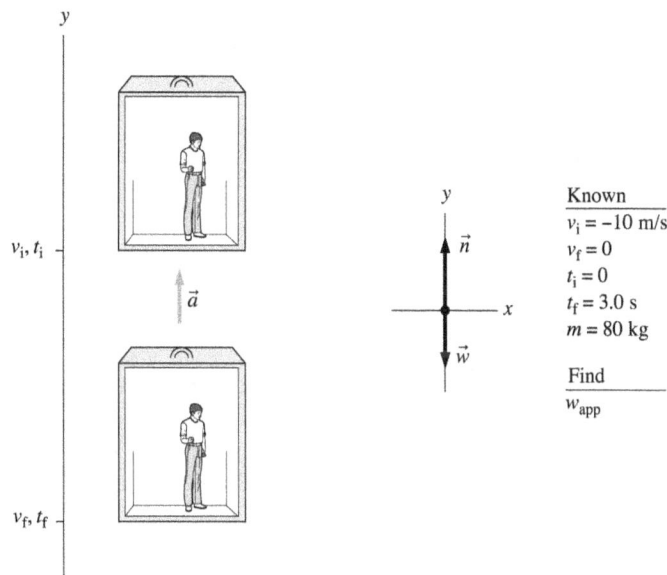

Solve: (a) Before the elevator starts braking, Zach is not accelerating. His apparent weight is

$$w_{app} = w\left(1 + \frac{a}{g}\right) = w\left(1 + \frac{0 \text{ m/s}^2}{g}\right) = mg = (80 \text{ kg})(9.80 \text{ m/s}^2) = 784 \text{ N}$$

or 780 N to two significant figures.

(b) Using the definition of acceleration,

$$a = \frac{\Delta v}{\Delta t} = \frac{v_f - v_i}{t_f - t_i} = \frac{0 - (-10) \text{ m/s}}{3.0 \text{ s}} = 3.33 \text{ m/s}^2$$

$$\Rightarrow w_{app} = w\left(1 + \frac{a}{g}\right) = (80 \text{ kg})(9.80 \text{ m/s}^2)\left(1 + \frac{3.33 \text{ m/s}^2}{9.80 \text{ m/s}^2}\right) = (784 \text{ N})(1 + 0.340) = 1100 \text{ N}$$

Assess: While the elevator is braking, it not only must support Zach's weight but must also push upward on him to decelerate him, so the apparent weight is greater than his normal weight.

P5.23. Strategize: This problem deals with apparent weight, which is the contact force between the passenger and the surface on which he/she is resting. We will use the sum of all forces in the direction of motion (here the y direction) to determine the unknown contact force.

Prepare: The passenger is acted on by only two vertical forces: the downward pull of gravity and the upward force of the elevator floor. Referring to Figure P5.23, the graph has three segments corresponding to different conditions: (1) increasing velocity, meaning an upward acceleration, (2) a period of constant upward velocity, and (3) decreasing velocity, indicating a period of deceleration (negative acceleration). Given the assumptions of our model, we can calculate the acceleration for each segment of the graph.

Solve: The acceleration for the first segment is

$$a_y = \frac{v_f - v_i}{t_f - t_i} = \frac{8 \text{ m/s} - 0 \text{ m/s}}{2 \text{ s} - 0 \text{ s}} = 4 \text{ m/s}^2 \Rightarrow w_{app} = w\left(1 + \frac{a_y}{g}\right) = (mg)\left(1 + \frac{4 \text{ m/s}^2}{9.80 \text{ m/s}^2}\right)$$

$$= (75 \text{ kg})(9.80 \text{ m/s}^2)\left(1 + \frac{4}{9.8}\right) = 1000 \text{ N}$$

For the second segment, $a_y = 0 \text{ m/s}^2$ and the apparent weight is

$$w_{app} = w\left(1 + \frac{0 \text{ m/s}^2}{g}\right) = mg = (75 \text{ kg})(9.80 \text{ m/s}^2) = 740 \text{ N}$$

For the third segment,

$$a_y = \frac{v_3 - v_2}{t_3 - t_2} = \frac{0 \text{ m/s} - 8 \text{ m/s}}{10 \text{ s} - 6 \text{ s}} = -2 \text{ m/s}^2$$

$$\Rightarrow w_{app} = w\left(1 + \frac{-2 \text{ m/s}^2}{9.80 \text{ m/s}^2}\right) = (75 \text{ kg})(9.80 \text{ m/s}^2)(1 - 0.2) = 590 \text{ N}$$

Assess: As expected, the apparent weight is greater than normal when the elevator is accelerating upward and lower than normal when the acceleration is downward. When there is no acceleration the weight is normal. In all three cases the magnitudes are reasonable, given the mass of the passenger and the accelerations of the elevator.

P5.25. Strategize: This is a straightforward application of Newton's second law in two dimensions. The problem will be somewhat simplified if we choose to work with axes that are parallel and perpendicular to the direction of motion.
Prepare: We apply Newton's second law to solve for the value of the normal force.
Solve: (a)

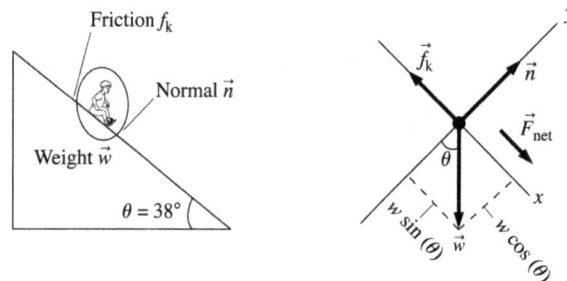

(b) Use tilted axes with the *x*-axis running down the incline. Apply $F_{net} = ma$ in the *y*-direction.

$$n - w \cos\theta = 0 \Rightarrow$$
$$n = mg \cos\theta = (23 \text{ kg})(9.80 \text{ m/s}^2)\cos 38° = 180 \text{ N}$$

Assess: The answer is less than the child's weight of 225 N, as we would expect, since only part of the weight is in the *y*-direction. The value seems to be in the right ballpark. Notice that we solved the problem algebraically before putting numbers in. This not only allows us to solve a similar problem for a different child or incline, but it enables us to check our answer in this case for reasonableness. Take the limit as $\theta \to 0$; the slope approaches zero and *n* tends toward the child's weight as $\cos\theta \to 1$. Then take the limit as $\theta \to 90°$ and the normal force decreases to zero as the incline becomes vertical and there is no normal force on the child.

P5.27. Strategize: This problem requires that we understand the force of kinetic friction and the coefficient of kinetic friction. We can determine the unknown force by using Newton's second law.

Prepare: We assume that the safe is a particle moving only in the *x*-direction. Since it is sliding during the entire problem, the force of kinetic friction opposes the motion by pointing to the left. In the following diagram we give a pictorial representation and a free-body diagram for the safe. The safe is in dynamic equilibrium, since it's not accelerating.

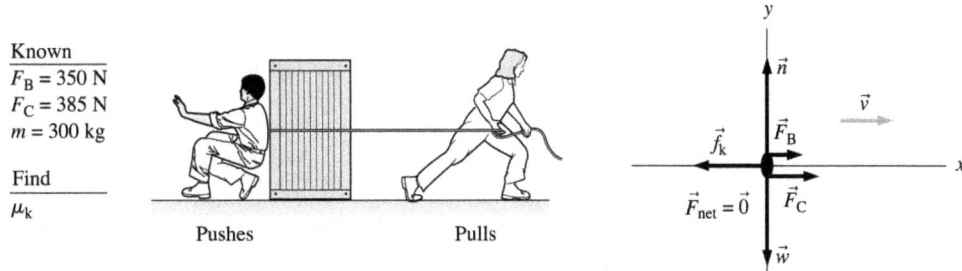

Known
$F_B = 350$ N
$F_C = 385$ N
$m = 300$ kg

Find
μ_k

Pushes   Pulls

Solve: We apply Newton's first law in the vertical and horizontal directions:

$$(F_{net})_x = \Sigma F_x = F_B + F_C - f_k = 0 \text{ N} \Rightarrow f_k = F_B + F_C = 350 \text{ N} + 385 \text{ N} = 735 \text{ N}$$

$$(F_{net})_y = \Sigma F_y = n - w = 0 \text{ N} \Rightarrow n = w = mg = (300 \text{ kg})(9.80 \text{ m/s}^2) = 2940 \text{ N}$$

Then, for kinetic friction

$$f_k = \mu_k n \Rightarrow \mu_k = \frac{f_k}{n} = \frac{735 \text{ N}}{2940 \text{ N}} = 0.25$$

Assess: The value of $\mu_k = 0.25$ is hard to evaluate without knowing the material the floor is made of, but it seems reasonable.

P5.29. Strategize: We can determine the acceleration of the car through a straightforward application of Newton's second law. Assuming the acceleration is constant, we can then use kinematics to determine over what distance the car came to a stop.

Prepare: The car is undergoing skidding, so it is decelerating and the force of kinetic friction acts to the left. We give below an overview of the pictorial representation, a motion diagram, a free-body diagram, and a list of values. We will first apply Newton's second law to find the deceleration and then use kinematics to obtain the length of the skid marks.

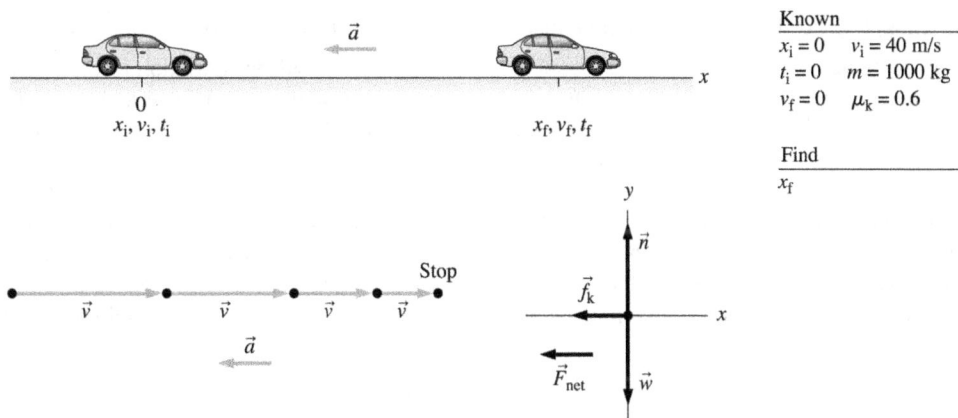

Known
$x_i = 0$ $v_i = 40$ m/s
$t_i = 0$ $m = 1000$ kg
$v_f = 0$ $\mu_k = 0.6$

Find
x_f

Stop

Solve: We begin with Newton's second law. Although the motion is one-dimensional, we need to consider forces in both the *x*- and *y*-directions. However, we know that $a_y = 0$ m/s². We have

$$a_x = \frac{(F_{net})_x}{m} = \frac{-f_k}{m} \qquad a_y = 0 \text{ m/s}^2 = \frac{(F_{net})_y}{m} = \frac{n-w}{m} = \frac{n-mg}{m}$$

We used $(f_k)_x = -f_k$ because the free-body diagram tells us that $\vec{f_k}$ points to the left. The force of kinetic friction relates $\vec{f_k}$ to $\vec{n}$ with the equation $f_k = \mu_k n$. The y-equation is solved to give $n = mg$. Thus, the kinetic friction force is $f_k = \mu_k mg$.

Substituting this into the x-equation yields

$$a_x = \frac{-\mu_k mg}{m} = -\mu_k g = -(0.6)(9.80 \text{ m/s}^2) = -5.88 \text{ m/s}^2$$

The acceleration is negative because the acceleration vector points to the left as the car slows. Now we have a constant-acceleration kinematics problem. Δt isn't known, so use

$$v_f^2 = 0 \text{ m}^2/\text{s}^2 = v_i^2 + 2a_x \Delta x \Rightarrow \Delta x = -\frac{(40 \text{ m/s})^2}{2(-5.88 \text{ m/s}^2)} = 140 \text{ m}$$

Assess: The skid marks are 140 m long. This is ≈ 430 feet, reasonable for a car traveling at ≈ 80 mph. It is worth noting that an algebraic solution led to the m canceling out.

P5.31. Strategize: This problem requires us to understand that the force of kinetic friction is always given by $f_k = \mu_k n$, whereas the force of static friction can be anything up to a maximum value of $(f_s)_{max} = \mu_s n$. If the crate does not slip across the surface of the conveyor belt, then we will use static friction; if it does slip we will use kinetic friction. We will write down Newton's second law to help us find the required accelerations.

Prepare: We show below the free-body diagrams of the crate when the conveyer belt runs at constant speed (part (a)) and the belt is speeding up (part (b)).

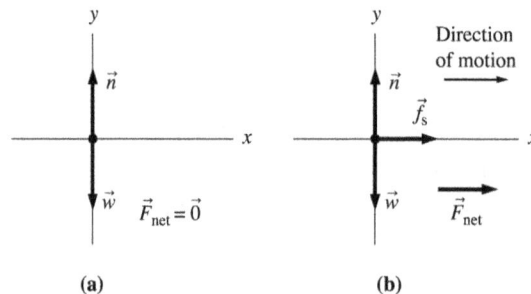

(a)   (b)

Solve: (a) When the belt runs at constant speed, the crate has an acceleration $\vec{a} = \vec{0}$ m/s^2 and is in dynamic equilibrium. Thus $\vec{F}_{net} = \vec{0}$. It is tempting to think that the belt exerts a friction force on the crate. But if it did, there would be a *net* force because there are no other possible horizontal forces to balance a friction force. Because there is no net force, there cannot be a friction force. The only forces are the upward normal force and the crate's weight. (A friction force would have been needed to get the crate moving initially, but no horizontal force is needed to keep it moving once it is moving with the same constant speed as the belt.)

(b) If the belt accelerates gently, the crate speeds up without slipping on the belt. Because it is accelerating, the crate must have a net horizontal force. So *now* there is a friction force, and the force points in the direction of the crate's motion. Is it static friction or kinetic friction? Although the crate is moving, there is *no* motion of the crate relative to the belt. Thus, it is a *static* friction force that accelerates the crate so that it moves without slipping on the belt.

(c) The static friction force has a maximum possible value $(f_s)_{max} = \mu_s n$. The maximum possible acceleration of the crate is

$$a_{max} = \frac{(f_s)_{max}}{m} = \frac{\mu_s n}{m}$$

If the belt accelerates more rapidly than this, the crate will not be able to keep up and will slip. It is clear from the free-body diagram that $n = w = mg$. Thus,

$$a_{max} = \mu_s g = (0.50)(9.80 \text{ m/s}^2) = 4.9 \text{ m/s}^2$$

(d) The acceleration of the crate will be $a = \mu_k g = (0.30)(9.80 \text{ m/s}^2) = 2.9 \text{ m/s}^2$.

P5.33. Strategize: If we assume that rolling friction is the only force acting on the locomotive in the horizontal direction, we can write down Newton's second law and determine the acceleration.
Prepare: Assume the locomotive is on level ground and the acceleration is constant.

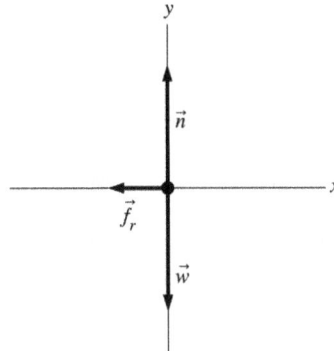

Solve: Since the locomotive does not accelerate in the vertical direction, the free-body diagram shows that $n = w = mg$. The friction force is $f_r = \mu_r mg$.

$$F_{net} = f_r = \mu_r mg \Rightarrow a = \frac{F_{net}}{m} = \mu_r g$$

The definition of acceleration $a = \Delta v / \Delta t$ gives

$$\Delta t = \frac{\Delta v}{a} = \frac{\Delta v}{\mu_r g} = \frac{10 \text{ m/s}}{(0.002)(9.80 \text{ m/s}^2)} = 510.2 \text{ s} \approx 500 \text{ s}$$

We now use the kinematic equation to find how far the locomotive will move during this time.

$$\Delta x = \frac{1}{2} a (\Delta t)^2 = \frac{1}{2}(0.002)(9.80 \text{ m/s}^2)(510.2 \text{ s})^2 = 2550 \text{ m} \approx 3000 \text{ m}$$

Assess: We are impressed, but not surprised, by the long time it would take the locomotive to coast to a stop without brakes. And it covers almost 3 km in that time. The mass was irrelevant in this problem.

P5.35. Strategize: We are not given information about the buoyant force or weight of the submersible, but we are given information about the horizontal direction. We will write the sum of all forces in the horizontal direction.
Prepare: Let us call the direction of horizontal motion of the submersible the $+x$ direction. We can determine the drag force in that direction and relate that to the horizontal component of the tension in the cable. Since the velocity is constant, the acceleration in the horizontal direction is zero. In calculating drag, note that the density of seawater is really somewhat greater than that of freshwater, approximately $1,025 \text{ kg/m}^3$. But because this number is variable (depending on salinity) and because it is only slightly different from the density of freshwater, we will use $1,000 \text{ kg/m}^3$. The forces acting on the submersible are shown in the free-body diagram.

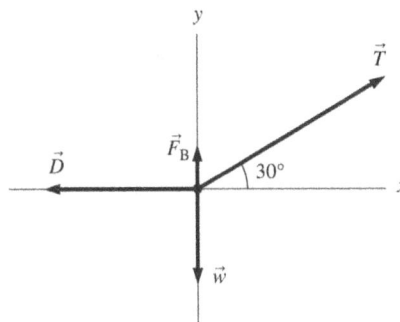

Solve: Newton's second law tells us

$$\sum F_x = T\cos(\theta) - D = ma_x = 0$$

$$T = \frac{D}{\cos(\theta)} = \frac{1}{\cos(\theta)}\frac{1}{2}C_D\rho Av^2 = \frac{1}{\cos(30°)}\frac{1}{2}(1.2)(1000 \text{ kg/m}^3)(1.3 \text{ m}^2)(5.1 \text{ m/s})^2$$

$$T = 2.4 \times 10^4 \text{ N}$$

Assess: This is a tremendous force, which explains why a hefty steel cable must be used.

P5.37. Strategize: It is likely that the Reynold's number is very small in this case. One can guess this based on Example 5.16, or by doing an order of magnitude estimate. The speed of the particle could not possibly be greater than 1 m/s, and is likely much less. Using that speed, we find a Reynold's number of 0.16, so our suspicion that it is low was correct. Thus, we will use Stokes'Law to determine the drag force.

Prepare: The particles will begin to settle and will quickly reach their terminal velocity. We calculate their terminal speed by equating the force of gravity to $D = 6\pi\eta rv$, and solving for the speed. Then we will use the speed to determine the settling time, using $\Delta y \approx (v_{\text{term}})_y \Delta t$. We look up the viscosity of air from Table 5.3, and use half the diameter of the particles as the characteristic radius.

Solve: The sum of all forces in the vertical direction yields

$$\sum F_y = D - mg = 6\pi\eta rv - mg = ma_y = 0$$

$$v = \frac{mg}{6\pi\eta r} = \frac{(1.4\times10^{-14} \text{ kg})(9.8 \text{ m/s}^2)}{6\pi(1.8\times10^{-5} \text{ Pa}\cdot\text{s})(1.25\times10^{-6} \text{ m})} = 3.23\times10^{-4} \text{ m/s}$$

Finally $\Delta t \approx \Delta y / (v_{\text{term}})_y = (0.15 \text{ m})/(3.23\times10^{-4} \text{ m/s}) = 4.6\times10^2 \text{ s}$ or about 7.7 min.

Assess: It is reasonable that it takes several minutes for such fine grains of pollution to settle.

P5.39. Strategize: The skydiver will have a fairly high Reynold's number, so we will use the quadratic drag model, from Equation 5.11.

Prepare: We will write out the sum of all forces in the vertical (y) direction, for a general case. Then we will specify that the velocity is ½ the terminal velocity, and see what we obtain.

Solve: In general, the sum of all forces in the vertical direction is $\sum F_y = D - mg = ma_y$. If we write down the expression for the terminal case, there is a trivial modification describing the drag at ½ the terminal speed:

$$\frac{1}{2}C_D\rho Av_{\text{term}}^2 - mg = 0 \Rightarrow \frac{1}{2}C_D\rho A\left(\frac{v_{\text{term}}}{2}\right)^2 = \frac{mg}{4}$$

Finally, writing the sum of all forces in the case where the speed is ½ the eventual terminal speed, we find

$$\sum F_y = D - mg = ma_y \Rightarrow \frac{mg}{4} - mg = ma_y \Rightarrow a_y = -\frac{3}{4}g = -7.4 \text{ m/s}^2.$$

We are asked for the magnitude, which is 7.4 m/s².

Assess: This answer has the right sign and is slightly smaller than the acceleration due to gravity, which makes sense.

P5.41. Strategize: We will write down Newton's second law for the car and truck separately, and we will use Newton's third law to describe interactions between the car and truck.

Prepare: The car and the truck will be denoted by the symbols C and T, respectively. The ground will be denoted by the symbol G. A visual overview shows a pictorial representation, a list of known and unknown values, and a free-body diagram for both the car and the truck. Since the car and the truck move together in the positive x-direction, they have the same acceleration.

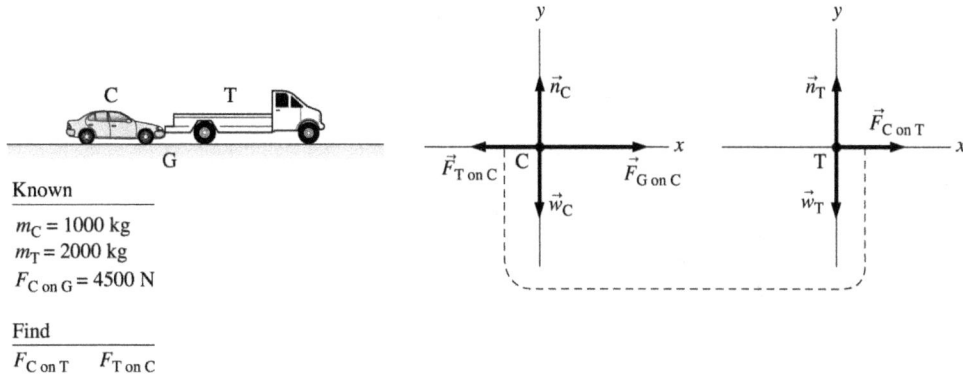

Known

$m_C = 1000$ kg
$m_T = 2000$ kg
$F_{C\ on\ G} = 4500$ N

Find

$F_{C\ on\ T}$ $F_{T\ on\ C}$

Solve: (a) The x-component of Newton's second law for the car is

$$\Sigma(F_{on\ C})_x = F_{G\ on\ C} - F_{T\ on\ C} = m_C a_C$$

The x-component of Newton's second law for the truck is

$$\Sigma(F_{on\ T})_x = F_{C\ on\ T} = m_T a_T$$

Using $a_C = a_T = a$ and $F_{T\ on\ C} = F_{C\ on\ T}$, we get

$$(F_{C\ on\ G} - F_{C\ on\ T})\left(\frac{1}{m_C}\right) = a \quad (F_{C\ on\ T})\left(\frac{1}{m_T}\right) = a$$

Combining these two equations,

$$(F_{C\ on\ G} - F_{C\ on\ T})\left(\frac{1}{m_C}\right) = (F_{C\ on\ T})\left(\frac{1}{m_T}\right) \Rightarrow F_{C\ on\ T}\left(\frac{1}{m_C} + \frac{1}{m_T}\right) = (F_{C\ on\ G})\left(\frac{1}{m_C}\right)$$

$$\Rightarrow F_{C\ on\ T} = (F_{C\ on\ G})\left(\frac{m_T}{m_C + m_T}\right) = (4500\ \text{N})\left(\frac{2000\ \text{kg}}{1000\ \text{kg} + 2000\ \text{kg}}\right) = 3000\ \text{N}$$

(b) Due to Newton's third law, $F_{T\ on\ C} = 3000$ N.

P5.43. Strategize: This problem involves heavy use of Newton's third law. We will write down Newton's second law for each object, and relate them by requiring that Newton's third law hold. For example: $(F_{1\ on\ 2})_x = -(F_{2\ on\ 1})_x$.

Prepare: The blocks are denoted as 1, 2, and 3. The surface is frictionless and along with the earth it is a part of the environment. The three blocks are our three systems of interest. The force applied on block 1 is $F_{A\ on\ 1} = 12$ N. The acceleration for all the blocks is the same and is denoted by a. A visual overview shows a pictorial representation, a list of known and unknown values, and a free-body diagram for the three blocks.

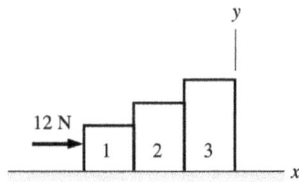

Known
$$m_1 = 1.0 \text{ kg}$$
$$m_2 = 2.0 \text{ kg}$$
$$m_3 = 3.0 \text{ kg}$$
$$F_{A \text{ on } 1} = 12 \text{ N}$$

Find
$$F_{2 \text{ on } 3}$$
$$F_{2 \text{ on } 1}$$

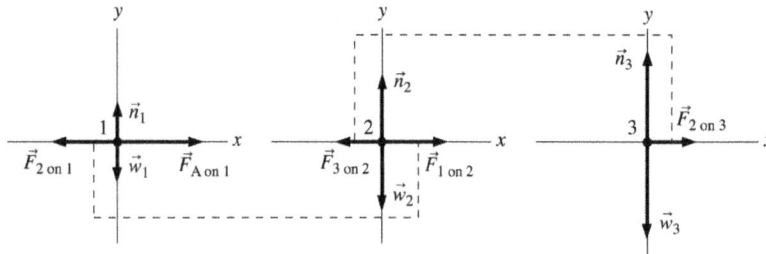

Solve: Newton's second law for the three blocks along the x-direction is

$$\Sigma(F_{\text{on } 1})_x = F_{A \text{ on } 1} - F_{2 \text{ on } 1} = m_1 a \quad \Sigma(F_{\text{on } 2})_x = F_{1 \text{ on } 2} - F_{3 \text{ on } 2} = m_2 a \quad \Sigma(F_{\text{on } 3})_x = F_{2 \text{ on } 3} = m_3 a$$

Adding these three equations and using Newton's third law $(F_{2 \text{ on } 1} = F_{1 \text{ on } 2}$ and $F_{3 \text{ on } 2} = F_{2 \text{ on } 3})$, we get

$$F_{A \text{ on } 1} = (m_1 + m_2 + m_3)a \Rightarrow (12 \text{ N}) = (1.0 \text{ kg} + 2.0 \text{ kg} + 3.0 \text{ kg})a \Rightarrow a = 2.0 \text{ m/s}^2$$

(a) Using this value of a, the force equation on block 3 gives

$$F_{2 \text{ on } 3} = m_3 a = (3.0 \text{ kg})(2.0 \text{ m/s}^2) = 6.0 \text{ N}$$

(b) Substituting into the force equation on block 1,

$$12 \text{ N} - F_{2 \text{ on } 1} = 12 \text{ N} - (1.0 \text{ kg})(2.0 \text{ m/s}^2) \Rightarrow F_{2 \text{ on } 1} = 10 \text{ N}$$

Assess: Because all three blocks are pushed forward by a force of 12 N, the value of 10 N for the force that the 2.0 kg block exerts on the 1.0 kg block is reasonable.

P5.45. Strategize: This is a straightforward application of Newton's second law.
Prepare: If we take our "object" to be the painter and the chair combined, the rope is connected to this object at two points. The tension acting upward at each of those connections is equal. Let us call the vertically upward direction $+y$. We will write the sum of all forces in the vertical direction and solve for the tension. We will use m to refer to the total 80 kg mass of the object.

Solve: We have $\sum F_y = 2T - mg = ma_y = 0 \Rightarrow T = \dfrac{mg}{2} = \dfrac{(80 \text{ kg})(9.8 \text{ m/s}^2)}{2} = 3.9 \times 10^2 \text{ N}.$

Assess: This may seem strange that the tension is less than the weight of the object. But the fact that tension acts at each end of a taut rope is one reason why we use pulley systems.

P5.47. Strategize: This problem involves Newton's second law. In particular, we can group blocks together into one composite mass to determine what tension is required in the rope adjacent to / attached to that group of blocks.
Prepare: Since each block has the same acceleration as all the others they must each experience the same net force. Each block will have one more newton pulling forward than the force pulling back on it from the blocks behind.
Solve:
(a) 1 N
(b) 50 N
Assess: Since 100 N accelerates 100 blocks then n Newtons accelerates n blocks.

P5.49. Strategize: This problem involves Newton's second law in two dimensions. Because the piano is to descend at a steady speed, it is in dynamic equilibrium.

Prepare: The following shows a free-body diagram of the piano and a list of values.

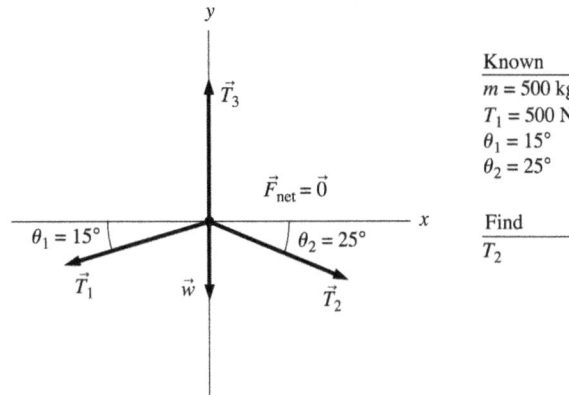

Known
$m = 500$ kg
$T_1 = 500$ N
$\theta_1 = 15°$
$\theta_2 = 25°$

Find
T_2

Solve: **(a)** Based on the free-body diagram, Newton's second law is

$$(F_{net})_x = 0 \text{ N} = T_{1x} + T_{2x} = T_2 \cos \theta - T_1 \cos \theta_1$$
$$(F_{net})_y = 0 \text{ N} = T_{1y} + T_{2y} + T_{3y} + w_y = T_3 - T_1 \sin \theta_1 - T_2 \sin \theta_2 - mg$$

Notice how the force components all appear in the second law with *plus* signs because we are *adding* vector forces. The negative signs appear only when we *evaluate* the various components. These are two simultaneous equations in the two unknowns T_2 and T_3. From the x-equation we find

$$T_2 = \frac{T_1 \cos \theta_1}{\cos \theta_2} = \frac{(500 \text{ N})\cos 15°}{\cos 25°} = 530 \text{ N}$$

(b) Now we can use the y-equation to find

$$T_3 = T_1 \sin \theta_1 + T_2 \sin \theta_2 + mg = 5300 \text{ N}$$

P5.51. Strategize: To find the net force at a given time, we need the acceleration at that time. Then we can simply apply Newton's second law.

Prepare: Because the times where we are asked to find the net force fall on distinct slopes of the velocity-versus-time graph, we can use the constant slopes of the three segments of the graph to calculate the three accelerations.

Solve: For t between 0 s and 3 s,

$$a_x = \frac{\Delta v_x}{\Delta t} = \frac{12 \text{ m/s} - 0 \text{ s}}{3 \text{ s}} = 4 \text{ m/s}^2$$

For t between 3 s and 6 s, $\Delta v_x = 0$ m/s, so $a_x = 0$ m/s^2. For t between 6 s and 8 s,

$$a_x = \frac{\Delta v_x}{\Delta t} = \frac{0 \text{ m/s} - 12 \text{ m/s}}{2 \text{ s}} = -6 \text{ m/s}^2$$

From Newton's second law, at $t = 1$ s we have

$$F_{net} = ma_x = (2.0 \text{ kg})(4 \text{ m/s}^2) = 8 \text{ N}$$

At $t = 4$ s, $a_x = 0$ m/s^2, so $F_{net} = 0$ N.

At $t = 7$ s,

$$F_{net} = ma_x = (2.0 \text{ kg})(-6.0 \text{ m/s}^2) = -12 \text{ N}$$

Assess: The magnitudes of the forces look reasonable, given the small mass of the object. The positive and negative signs are appropriate for an object first speeding up, then slowing down.

P5.53. Strategize: To solve this problem we can apply Newton's second law, and solve for tension.
Prepare: The box is acted on by two forces: the tension in the rope and the pull of gravity. Both the forces act along the same vertical line which is taken to be the y-axis. The free-body diagram for the box is shown below.

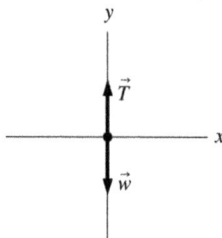

Solve: (a) Since the box is at rest, $a_y = 0$ m/s² and the net force on it must be zero:

$$F_{net} = T - w = 0 \text{ N} \Rightarrow T = w = mg = (50 \text{ kg})(9.80 \text{ m/s}^2) = 490 \text{ N}$$

(b) The velocity of the box is irrelevant, since only a *change* in velocity requires a nonzero net force. Since $a_y = 5.0$ m/s²,

$$F_{net} = T - w = ma_y = (50 \text{ kg})(5.0 \text{ m/s}^2) = 250 \text{ N} \Rightarrow T = 250 \text{ N} + w = 250 \text{ N} + 490 \text{ N} = 740 \text{ N}$$

Assess: For part (a) the zero acceleration immediately implies that the box's weight must be exactly balanced by the upward tension in the rope. For part (b) the tension not only has to support the box's weight but must also accelerate it upward, hence, T must be greater than w.

P5.55. Strategize: In both cases we can simply apply Newton's second law and solve for the unknown tension.
Prepare: The box is acted on by two forces: the tension in the rope and the pull of gravity. Both the forces act along the same vertical line which is taken to be the y-axis. The following shows the free-body diagram for the box.

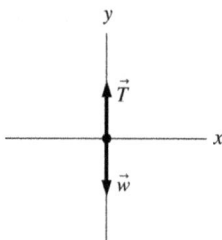

Solve: (a) Since the box is rising at a constant speed, $a_y = 0$ m/s² and the net force on it must be zero:

$$F_{net} = T - w = 0 \text{ N} \Rightarrow T = w = mg = (50 \text{ kg})(9.80 \text{ m/s}^2) = 490 \text{ N}$$

(b) Since the box is slowing down, $a_y = -5.0$ m/s² and we have

$$F_{net} = T - w = ma_y = (50 \text{ kg})(-5.0 \text{ m/s}^2) = -250 \text{ N}$$
$$\Rightarrow T = -250 \text{ N} + w = -250 \text{ N} + 490 \text{ N} = 240 \text{ N}$$

Assess: For part (a) the zero acceleration immediately implies that the box's weight must be exactly balanced by the upward tension in the rope. For part (b), when the box accelerates downward, the rope need not support the entire weight, hence, T is less than w.

P5.57. Strategize: This problem deals with apparent weight, which is the contact force between a surface and the people resting on it. We can use kinematics to determine the acceleration. From there we can use Newton's second law to relate that acceleration to the forces involved, and specifically to the apparent weight.

Prepare: Use the kinematic equations twice. The first time find out the velocity of the rider at the end of the 2.0 s and then use that as the initial velocity during the second part to compute the acceleration.

Solve: During the free fall phase the initial velocity is zero, so

$$v_f = a\Delta t = (-9.80 \text{ m/s}^2)(2.0 \text{ s}) = -19.6 \text{ m/s}$$

then, as the rider slows,

$$a = \frac{\Delta v}{\Delta t} = \frac{0 \text{ m/s} - (-19.6 \text{ m/s})}{0.50 \text{ s}} = 39.2 \text{ m/s}^2$$

This is an upward acceleration. The apparent weight is

$$w_{app} = m(g + a_y) = (65 \text{ kg})(9.80 \text{ m/s}^2 + 39.2 \text{ m/s}^2) = 3200 \text{ N}$$

This is 5.0x the rider's actual weight.

Assess: We would say the rider experiences 5g's, which is a significant acceleration, but it lasts for only a short while.

P5.59. Strategize: We can use the stopping distances to determine the acceleration in each case. Then we can relate the acceleration to the forces involved using Newton's second law.

Prepare: We can assume the person is moving in a straight line under the influence of the combined decelerating forces of the air bag and seat belt or, in the absence of restraints, the dashboard or windshield. The following is an overview of the situation in a pictorial representation and the occupant's free-body diagram is shown below. Note that the occupant is brought to rest over a distance of 1 m in the former case, but only over 5 mm in the latter.

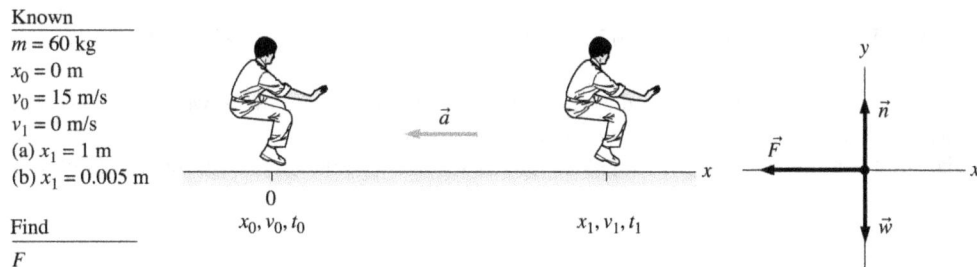

Known
$m = 60$ kg
$x_0 = 0$ m
$v_0 = 15$ m/s
$v_1 = 0$ m/s
(a) $x_1 = 1$ m
(b) $x_1 = 0.005$ m

Find
F

Solve: (a) In order to use Newton's second law for the passenger, we'll need the acceleration. Since we don't have the stopping time,

$$v_f^2 = v_i^2 + 2a(x_f - x_i) \Rightarrow a = \frac{v_f^2 = v_i^2}{2(x_f - x_i)} = \frac{0 \text{ m}^2/\text{s}^2 - (15 \text{ m/s}^2)}{2(1 \text{ m} - 0 \text{ m})} = -112.5 \text{ m/s}^2$$

$$\Rightarrow F_{net} = F = ma = (60 \text{ kg})(-112.5 \text{ m/s}^2) = -6750 \text{ N}$$

The net force is 6800 N to the left.

(b) Using the same approach as in part (a),

$$F = ma = m \frac{v_f^2 = v_i^2}{2(x_f - x_i)} = (60 \text{ kg})\frac{0 \text{ m}^2/\text{s}^2 - (15 \text{ m/s}^2)}{2(0.005 \text{ m})} = -1,350,000 \text{ N}$$

The net force is 1.4×10^6 N to the left.

(c) The passenger's weight is $mg = (60 \text{ kg})(9.80 \text{ m/s}^2) = 590$ N. The force in part (a) is 11 times the passenger's weight. The force in part (b) is 2300 times the passenger's weight.

Assess: An acceleration of 11g is well within the capability of the human body to withstand. A force of 2300 times the passenger's weight, on the other hand, would surely be catastrophic.

P5.61. Strategize: This problem is a straightforward application of Newton's second law. For the first part, we will solve for an unknown acceleration, and in the second part we will solve for an unknown mass.

Prepare: The rocket is moving along the y-axis under the influence of two forces: the rocket's thrust and the force of gravity. Its free-body diagram is shown, which, along with Newton's second law (and using the rocket's initial mass), will help us find its initial acceleration. At 5000 m the acceleration has increased because the rocket mass has decreased. Utilizing the free-body diagram, Newton's second law, and the increased acceleration, we can determine the decreased mass of the rocket.

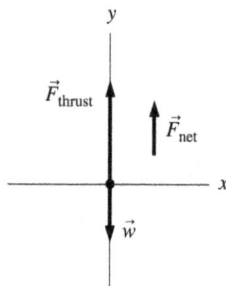

Solve: (a) The y-component of Newton's second law is

$$a_y = a = \frac{(F_{net})_y}{m} = \frac{F_{thrust} - mg}{m} = \frac{300,000 \text{ N}}{20,000 \text{ kg}} - 9.80 \text{ m/s}^2 = 5.2 \text{ m/s}^2$$

(b) Solving the equation of part (a) for m gives

$$m_{5000 \text{ m}} = \frac{F_{thrust}}{a_{5000 \text{ m}} + g} = \frac{300,000 \text{ N}}{6.00 \text{ m/s}^2 + 9.80 \text{ m/s}^2} = 18,990 \text{ kg}$$

The mass of fuel burned is $m_{fuel} = m_{initial} - m_{5000} = 1010 \text{ kg}$ or 1000 kg to two significant figures.

P5.63. Strategize: This is a straightforward application of Newton's second law in two dimensions. In particular, we will be concerned with the components of forces perpendicular to the slide.

Prepare: The child is not accelerating in the y-direction, so we can use Equation 5.1 for the forces perpendicular to the incline.

Solve:

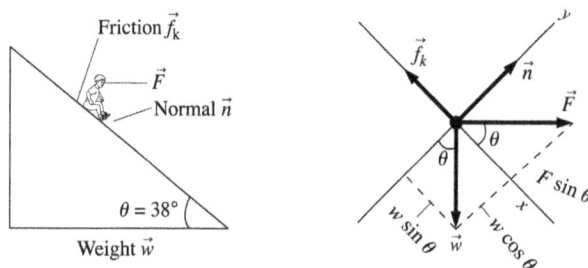

There are three forces with components in the y-direction, the normal force, the weight of the child, and the force of the rope. Equation 5.1 gives

$$n + F\sin(\theta) - w\cos(\theta) = 0$$

Solving for the normal force, we have

$$n = w\cos(\theta) - F\sin(\theta) = mg\cos(\theta) - F\sin(\theta) = (23 \text{ kg})(9.80 \text{ m/s}^2)\cos(38°) - (30 \text{ N})\sin(38°) = 160 \text{ N}$$

Assess: This is less than the child's weight, as expected. Note that the force from the rope acts to decrease the normal force on the child since it tends to pull the child away from the incline.

P5.65. Strategize: This problem involves two stages of motion: one without friction and one with friction. We must split up the problem and apply Newton's second law to each stage.

Prepare: The length of the hill is $\Delta x = h/\sin\theta$. The acceleration is $g\sin\theta$.

Solve: First use the kinematic equation, with $v_i = 0$ m/s at the top of the hill, to determine the speed at the bottom of the hill.

$$(v_f)_1^2 = (v_i)_1^2 + 2a\Delta x \Rightarrow (v_f)_1^2 = 2(g\sin\theta)(h/\sin\theta) = 2gh$$

Now apply the same kinematic equation to the horizontal patch of snow, only this time we want Δx. To connect the two parts $(v_f)_1 = (v_i)_2$. The final speed is zero: $(v_f)_2 = 0$.

$$(v_f)_2^2 = (v_i)_2^2 + 2a\Delta x = (v_f)_1^2 + 2a\Delta x = 2gh + 2a\Delta x = 0$$

The friction force is the net force, so $a = -f_k / m$. Note $f_k = \mu_k n = \mu_k mg$. Solve for Δx.

$$\Delta x = \frac{-2gh}{2a} = \frac{-gh}{-f_k/m} = \frac{gh}{\mu_k mg/m} = \frac{h}{\mu_k} = \frac{3.0 \text{ m}}{0.05} = 60 \text{ m}$$

Assess: It seems reasonable to glide 60 m with such a low coefficient of friction. It is interesting that we did not need to know the angle of the (frictionless) slope; this will become clear in the chapter on energy. The answer is also independent of Josh's mass.

P5.67. Strategize: We assume that the bicyclist's speed will continue to increase until the force of drag is equal to the component of gravity pulling him down the hill. We can equate these two forces and determine the speed.

Prepare: The Reynold's number for this situation is large, so we can use the quadratic expression for drag: $D = \frac{1}{2}C_D A\rho v^2$. Let us call the direction of motion (down the hill) the $+x$ direction. We recognize that the component of gravity acting along this direction is $mg\sin(\theta)$.

Solve: From Newton's second law, we have

$$\sum F_x = mg\sin(\theta) - \frac{1}{2}C_D A\rho v^2 = ma_x = 0$$

$$v = \sqrt{\frac{2mg\sin(\theta)}{C_D A\rho}} = \sqrt{\frac{2(70 \text{ kg})(9.8 \text{ m/s}^2)\sin(3.5°)}{(0.88)(0.32 \text{ m}^2)(1.22 \text{ kg/m}^3)}} = 16 \text{ m/s}$$

Assess: This is a reasonable terminal speed given the small slope and the fact that the rider is just coasting.

P5.69. Strategize: The Reynold's number for this situation is high, so we will use the quadratic expression for the drag. The second part is a straightforward kinematics problem.

Prepare: Let us call the direction of the pitch the $+x$ direction. We can determine the drag using $D = \frac{1}{2}C_D A\rho v^2$, and the direction is clearly $-x$. If that is the only force on the ball in the horizontal direction, then clearly $D_x = -\frac{1}{2}C_D A\rho v^2 = ma_x \Rightarrow a_x = -\frac{1}{2m}C_D A\rho v^2$. Table 5.4 from the chapter gives us the drag coefficient for a baseball as $C_D = 0.35$. For part (b), we can use $(v_f)_x^2 = (v_i)_x^2 + 2a_x\Delta x$ to determine the final speed.

Solve: (a) Inserting the given values, we have

$$a_x = -\frac{1}{2m}C_D A\rho v^2 = -\frac{1}{2(0.145 \text{ kg})}(0.35)\pi(0.037 \text{ m})^2(1.22 \text{ kg/m}^3)(40.2 \text{ m/s})^2 = -10.2 \text{ m/s}^2$$

Here we have reported to three significant digits for use in part (b). To the correct precision, the acceleration due to the drag force is 10 m/s^2 opposite the direction of the pitch.

(b) Inserting the answer to part (a) and given quantities, we obtain

$$(v_f)_x^2 = (v_i)_x^2 + 2a_x\Delta x \Rightarrow (v_f)_x = \sqrt{(40.2 \text{ m/s})^2 + 2(-10.2 \text{ N})(18.4 \text{ m})} = 35 \text{ m/s}$$

Assess: We note that as the speed is reduced, so is the drag. So the baseball is not actually slowed this much during a normal pitch. Still, our answer is a reasonable order of magnitude.

P5.71. Strategize: In either case, the grip force will result in an equal and opposite normal force that the steel exerts on the hand. We can determine the maximum force of static friction in either case.
Prepare: Let us call the vertically upward direction $+y$. The heavy rod will be supported by static friction. Since we are asked to maximize the weight, we will max out the static friction, meaning $f_s = (f_s)_{max} = \mu_s n$. Here the normal force is the same as the grip.

Solve: (a) The sum of all forces in the vertical direction yields $\sum F_y = \mu_s n - mg = ma_y = 0 \Rightarrow m = \dfrac{\mu_s F_{grip}}{g}$.

Inserting the given values, we have

$$m_{dry} = \frac{(0.27)(400 \text{ N})}{(9.8 \text{ m/s}^2)} = 11 \text{ kg}.$$

(b) Inserting the static coefficient of friction for wet hands, we have $m_{dry} = \dfrac{(1.4)(400 \text{ N})}{(9.8 \text{ m/s}^2)} = 57 \text{ kg}.$

Assess: The weights of these rods are 24 lbs and 126 lbs, respectively. Heavy steel objects that are made to be held are often textured to allow a better grip. But if the bar were smooth, it is reasonable that nothing heavier than 126 lbs could be kept from slipping straight down by a 400 N grip.

P5.73. Strategize: We will apply Newton's second law to the book.
Prepare: The book is in static equilibrium so Equation 5.1 can be applied. The maximum static frictional force the person can exert will determine the heaviest book he can hold.
Solve: Consider the free-body diagram below. The force of the fingers on the book is the reaction force to the normal force of the book on the fingers, so is exactly equal and opposite the normal force on the fingers.

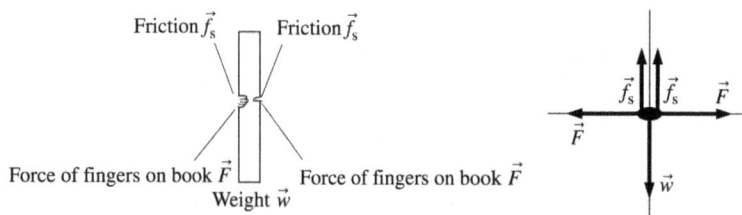

The maximal static friction force will be equal to $f_{s\,max} = \mu_s n = (0.80)(6.0 \text{ N}) = 4.8 \text{ N}$. The frictional force is exerted on both sides of the book. Considering the forces in the y-direction, the weight supported by the maximal frictional force is

$$w = f_{s\,max} + f_{s\,max} = 2f_{s\,max} = 9.6 \text{ N}$$

We now find the mass of a 9.6 N book.

$$m = \frac{w}{g} = \frac{9.6 \text{ N}}{9.80 \text{ m/s}^2} = 0.98 \text{ kg}$$

Assess: Note that the force on both sides of the book are exactly equal also because the book is in equilibrium.

P5.75. Strategize: We will apply Newton's second law to the block. We must be especially careful with the direction of friction.

Prepare: We show below the free-body diagram of the 1 kg block. The block is initially at rest, so initially the friction force is static friction. If the 12 N pushing force is too strong, the box will begin to move up the wall. If it is too weak, the box will begin to slide down the wall. And if the pushing force is within the proper range, the box will remain stuck in place.

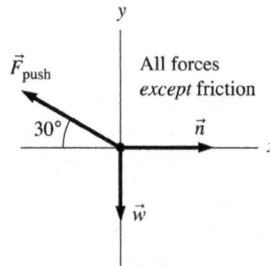

Solve: First, let's evaluate the sum of all the forces *except* friction:

$$\Sigma F_x = n - F_{push} \cos 30° = 0 \text{ N} \Rightarrow n = F_{push} \cos 30°$$

$$\Sigma F_y = F_{push} \sin 30° - w = F_{push} \sin 30° - mg = (12 \text{ N}) \sin 30° - (1 \text{ kg})(9.80 \text{ m/s}^2) = -3.8 \text{ N}$$

In the first equation we have utilized the fact that any motion is parallel to the wall, so $a_x = 0 \text{ m/s}^2$.

The two forces in the second y-equation add up to -3.8 N. This means the static friction force will be able to prevent the box from moving if $f_s = +3.8$ N. Using the x-equation we get

$$f_{s\,max} = \mu_s n = \mu_s F_{push} \cos 30° = 5.2 \text{ N}$$

where we used $\mu_s = 0.5$ for wood on wood. The static friction force $\vec{f}_s$ needed to keep the box from moving is *less* than $f_{s\,max}$. Thus the box will stay at rest.

P5.77. Strategize: We will write out Newton's second law for each block to determine the unknown tensions.

Prepare: With no friction, the only forces along the incline are the tension T and a component of the weight, $mg \sin \theta$. Since the blocks aren't accelerating then $T = mg \sin \theta$.

Solve: For tension 1:

$$T_1 = (5 \text{ kg})(9.80 \text{ m/s}^2) \sin 20° = 16.76 \text{ N}$$

or 17 N to two significant figures.

For tension 2 the mass is the sum of both blocks:

$$T_2 = (5 \text{ kg} + 3 \text{ kg})(9.80 \text{ m/s}^2) \sin 20° = 26.81 \text{ N}$$

or 27 N to two significant figures.

Assess: We would expect $T_2 > T_1$.

P5.79. Strategize: We want to pull as hard as possible, which means we want the top block to accelerate as much as possible before it slips. This means we need the force of friction to be as large as possible. With this information, we will use Newton's second law.

Prepare: Call the force we seek F, and the mass of one block m. The maximum force without slippage is when the friction force between the blocks is maximum: $f_s = \mu_s n = \mu_s mg$.

Solve: For the two-block system F is the net force.

$$F = (2m)a$$

Considering only the top block, f_s is the net force, so $f_s = ma.$

$$f_s = \mu_s n = \mu_s mg = ma$$

Now insert our latest expression for ma into the equation for the two-block system.

$$F = 2ma = 2(\mu_s mg) = 2(\mu_s mg) = 2(0.35)(9.80\,\text{m/s}^2)(2.0\,\text{kg}) = 14\,\text{N}$$

Assess: 14 N seems reasonable.

P5.81. Strategize: We must consider how the ropes act on different objects and where the tension in a rope must be uniform and when it can change. We will rely heavily on Newton's second law.
Prepare: Let us call the vertically upward direction $+y$. We know that the tension everywhere in our (massless) ropes must be the same, but when two ropes are joined, or where they attach to objects, the tension can be different on one side and the other. We will use this to help us write out Newton's second law for the knot where ropes 1 and 2 are attached, and for the pulley that supports rope 1 (the lower pulley). We will call the weight of the climber w_c.
Solve: (a) We start by writing the sum of all forces on the knot and on the pulley in the vertical direction:

$$\sum \left(F_{\text{knot}}\right)_y = T_2 + T_1 - w_c = 0$$
$$\sum \left(F_{\text{pulley 1}}\right)_y = T_2 - 2T_1 = 0 \Rightarrow T_2 = 2T_1$$

Inserting the bottom equation into the top, we find

$$3T_1 = w_c \Rightarrow T_1 = w_c / 3 = (660\,\text{N}) / 3 = 220\,\text{N}.$$

(b) From the answer to (a) and the sum of all forces in the vertical direction on pulley 1, it follows immediately that $T_2 = 2T_1 = 440\,\text{N}.$
Assess: Note that the two ropes are joined together in lifting the injured climber, and indeed the sum of the two tensions is equal to the weight of the climber.

P5.83. Strategize: We will apply Newton's second law to each block separately, and relate the two expressions through the tension in the rope.
Prepare: Call the 10 kg block m_2 and the 5.0 kg block m_1. Assume the pulley is massless and frictionless.
Solve: On block 2 use tilted axes.

$$\Sigma F_x = T - m_2 g \sin\theta = m_2 a_2$$

Block 1 is also accelerating.

$$\Sigma F_y = T - m_1 g = m_1 a_1$$

The acceleration constraint is $(a_2)_x = -(a_1)_y = a.$ Solve for T in the second equation and insert in the first. $T = m_1(g - a).$

$$m_1(g - a) - m_2 g \sin\theta = m_2 a$$
$$m_1 g - m_2 g \sin\theta = m_2 a + m_1 a$$
$$a = \frac{g(m_1 - m_2 \sin\theta)}{m_1 + m_2} = \frac{(9.80\,\text{m/s}^2)(1.0\,\text{kg} - (2.0\,\text{kg})\sin 40°)}{1.0\,\text{kg} + 2.0\,\text{kg}} = -0.93\,\text{m/s}^2$$

Or 0.93 m/s², down the ramp.
Assess: The answer depends on θ; for a shallow angle the block accelerates up the ramp, for a steep angle the block accelerates down the ramp. This is expected behavior.

P5.85. Strategize: This is a straightforward application of Newton's second law in one dimension.
Prepare: We can use Newton's second law to calculate the force needed to bring the stone up to speed.
Solve: Ignoring friction, the only force on the stone is the force the curler applies. Using Newton's second law, we can calculate the force given the acceleration and mass of the stone. The acceleration of the stone is given by the definition of acceleration.

$$a_x = \frac{\Delta v_x}{\Delta t} = \frac{3.0 \text{ m/s}}{2.0 \text{ s}} = 1.5 \text{ m/s}^2$$

The force on the stone is then

$$F_x = ma_x = (20 \text{ kg})(1.5 \text{ m/s}^2) = 30 \text{ N}$$

The correct choice is C.
Assess: We will check that ignoring friction is a reasonable assumption in Problem 5.87.

P5.87. Strategize: We will assume the acceleration is constant, such that we can use kinematic equations. We can then apply Newton's second law.
Prepare: Let us call the direction of the initial velocity of the stone the $+x$ direction.
Solve: Using Equation 2.13 with $v_f = 0$ m/s, $v_i = 3$ m/s, and $\Delta x = 40$ m, we have

$$a_x = -\frac{(v_i)^2}{2\Delta x} = \frac{(3 \text{ m/s})^2}{2(40 \text{ m})} = 0.1125 \text{ m/s}^2$$

where additional significant figures have been kept in this intermediate result.
The force on the stone is then

$$F_x = ma_x = (20 \text{ kg})(0.1125 \text{ m/s}^2) = 2 \text{ N}$$

The correct choice is B.
Assess: Compare to Problem 5.85. The force of friction is much less than the force the curler applies, so ignoring friction in Problem 5.85 is reasonable.

CIRCULAR MOTION, ORBITS, AND GRAVITY

Q6.1. Reason: Acceleration is a change in *velocity*. Since velocity is a vector, it can change by changing direction, even while the magnitude (speed) remains constant. The cyclist's acceleration is *not* zero in uniform circular motion. She has a centripetal (center-seeking) acceleration.

Assess: In everyday usage, acceleration usually means only a change in speed (specifically a speeding up), hence the confusion. But in physics we must use words very carefully to communicate clearly. Everyday usage is fine outside the physics context, but while doing physics we must use the precise physics definitions of the words.

Q6.3. Reason: A turn does not need to be perfectly circular, but for any small segment of a smooth turn, we can draw a circle with the appropriate radius and center point, such that the small segment of the turn is approximately circular. Thus, in order for an airplane to turn, there must be some component of some force (or forces) pointing toward the center of this circle. We call this net inward force the centripetal force. In this case, the centripetal force is coming from the lift on the wings. So the airplane must turn on its side at least slightly so that this lift is not purely vertical, but has some component toward the center of the circle approximating the plane's path.

Assess: If you have ever seen a plane turn, or been inside a plane as it turns, you know that airplanes turn quite a lot on their sides as they bank.

Q6.5. Reason: The discussion in the section on maximum walking speed leads to the equation $v_{\max} = \sqrt{gr}$ where r is the length of the leg. For a leg as short as a chickadee's this produces a walking speed that is simply too slow to be practical, so they hop or fly.

Assess: The longer the leg the greater the maximum walking speed, and the formula produces reasonable walking speeds for pheasants.

Q6.7. Reason: At the lowest point, the acceleration is upward. Thus, the tension must be greater than the weight for the net force to be upward. The tension in the string not only offsets the weight of the ball, but additionally provides the centripetal force to keep it moving in a circle.

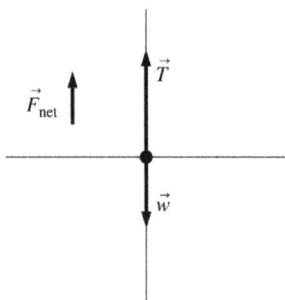

Assess: The string must have a higher strength rating than the weight of the ball in order for the ball to swing in a vertical circle.

Of course, at the top of the circle the weight itself points centripetally, so the tension in the string can be less than at the bottom.

Q6.9. Reason: (a) The moon's orbit around the earth is fairly circular, and it is the gravitational force of the earth on the moon that provides the centripetal force to keep the moon in its circular motion.

(b) The riders in the Gravitron carnival ride (Section 6.3) have a centripetal acceleration caused by the normal force of the walls on them.

Another example would be the biological sample in a centrifuge. The test tube walls exert a normal force on the sample toward the center of the circle.

Assess: The point is that centripetal forces are not a new *kind* of force; it is just the name we give to the force (or sum of forces) that points toward the center of the circle and keeps the object from flying off in a straight line.

Q6.11. Reason: The car is traveling along a circle and so it must have centripetal acceleration which points downward. From Newton's second law, if an object is accelerating downward, the total force on the object must be downward. The answer is C because only there is the downward force (the weight of the car) greater than the upward force (the normal force on the car) so that the total force is downward.

Assess: It makes sense that the normal force on the car would be less than the weight of the car because, from experience, you know that you feel lighter going over a hill in your car and normal force tells you how heavy you feel. In the same way, the normal force on the car will be less than its weight.

Q6.13. Reason: When a pickup truck turns suddenly there *isn't* a force that pushes the riders toward the outside of the curve, throwing them out. Instead, the riders' inertia tends to keep them moving in the same straight-line motion while the truck turns beneath them. The crux of the danger in a pickup truck is that the walls are so low that they don't provide much centripetal force to hold the riders in the truck as it turns. If you must ride in the back of a pickup truck, sit down low so the walls will be able to exert a centripetal force on you and keep you moving with the truck (around the corner).

Assess: This reasoning carried farther says that riding on the back of a flat-bed truck is that much more dangerous. However, in a cab the door (and seat belt) can provide the centripetal force needed to keep a rider moving around the turn.

Q6.15. Reason: The radius of the loop decreases as the carts enter and exit the loop. The centripetal acceleration is smaller for larger radius loops and larger for smaller radius loops. This means the centripetal acceleration increases from a minimum at the entry to the loop to a maximum at the top of the loop and then decreases as the cars exit the loop. This prevents a sudden change of acceleration, which can be painful. This also limits the largest accelerations to the top of the loop, so that riders only experience the maximum acceleration for a portion of the trip.

Assess: This is reasonable. If the cars entered a small radius loop directly, the centripetal acceleration would increase suddenly.

Q6.17. Reason: When we walk on the ground we push off with one foot while pivoting on the other; the weight force brings us back down from the push-off for the next step. In an orbiting station, which is in free fall along with the astronaut, after one foot pushes off there isn't a force to bring the astronaut back to the "floor" for the next step; the first push-off sends the astronaut across the cabin.

Assess: If the spacecraft is designed to rotate to provide an artificial gravity then one can walk fairly normally around on the inside; "up" would be toward the center of the circular motion, "down" would be "out"; but that probably isn't the origin of the phrase "down and out."

Q6.19. Reason: An object's weight is defined to be the gravitational force of the earth on the object. And the gravitational force of the earth on an object decreases with distance (as $1/r^2$), where we measure r from center to center. At the top of a mountain the climber's center is farther from the center of the earth, and so the gravitational force (i.e., the weight) is less, even though the climber's mass hasn't changed.

Assess: This is not just a change in *apparent* weight (what the scales read); this is a change in the real weight (the gravitational force).

Doubling the height of the mountain would decrease the weight by a factor of 4—but only if you take the height of the mountain to be r (from the center of the earth), *not* the height above sea level.

Q6.21. Reason: Originally, the ball is going around once every second. When the ball is sped up so that it goes around once in only half a second, it is moving twice as fast. Consequently its acceleration, which is given by $a = \omega^2 r$ will be four times as great. From Newton's second law, force is directly proportional to acceleration, so if we multiply the acceleration by 4, we must multiply the tension by 4. Thus the tension in the string will be four times as great, or 24 N. The answer is D.

Assess: This accords with our experience that when we swing an object around a circle, as the speed increases, the tension in the string increases.

Q6.23. Reason: The static friction is directed centripetally and is the net force. The radius of the turn is 95 m.

$$F_{net} = ma = m\frac{v^2}{r} = (610 \text{ kg})\frac{(68 \text{ m/s})^2}{95 \text{ m}} = 30,000 \text{ N}$$

The correct choice is E.

Assess: This large friction force is only possible if the wings help push the car into the track.

Q6.25. Reason: The force that holds the car to the road as it accelerates through a turn is the force of friction (acting inward toward the center of the circular path). Thus, we have from Newton's second law $f_s = mv^2/r$. Since we are asked about the maximum speeds, we must be maxing out the force of static (rolling) friction. Thus $\mu_s n = mv^2/r$ and $\mu_s mg = mv^2/r \Rightarrow v = \sqrt{\mu_s gr}$. If we write this for the snowy day (subscript "c" for "cold") and on the sunny day (subscript "h" for "hot"), we can combine the expressions noting that gravity and the radius of curvature are constant:

$$\frac{v_h}{v_c} = \frac{\sqrt{\mu_{s,h} gr}}{\sqrt{\mu_{s,c} gr}} \Rightarrow v_h = v_c \sqrt{\frac{\mu_{s,h}}{\mu_{s,c}}} = (20 \text{ mph})\sqrt{\frac{(1.0)}{(0.5)}} = 28 \text{ mph}.$$

So the correct answer is C.

Assess: It makes sense that the maximum safe speed at which you can take a curve should increase if the road is dry and free of snow and ice.

Q6.27. Reason: The speed of a satellite in low orbit is $v = \sqrt{gr}$. Use ratios to find v_{Jup}/v_{Earth}.

$$\frac{v_{Jup}}{v_{Earth}} = \frac{\sqrt{g_{Jup} r_{Jup}}}{\sqrt{g_{Earth} r_{Earth}}} = \frac{\sqrt{(2.5 g_{Earth})(11 r_{Earth})}}{\sqrt{g_{Earth} r_{Earth}}} = \sqrt{(2.5)(11)} = 5.2$$

The speed of a satellite in low Jupiter orbit is 5.2 times the speed of a satellite in low Earth orbit, so the correct choice is A.

Assess: Both factors made the speed greater around Jupiter.

Q6.29. Reason: Equation 6.18 gives

$$g_{planet} = \frac{GM_{planet}}{R_{planet}^2}$$

If the mass stays the same while the radius doubles, then the new g will be 1/4 of the old one. Since $g \approx 10 \text{ m/s}^2$ now, then one quarter of that is 2.5 m/s^2.

The correct choice is A.

Assess: Especially note that in part (b) the magnitude of the force of the floor on you is not the same as the magnitude of the earth's gravitational force on you, as it would have been if you hadn't been pushing on the ceiling.

Q6.31. Reason: We need to use Equation 6.22 (also known as Kepler's Third Law) because it relates the orbital period T to the orbital radius r. We are given that $r_2 = 4r_1$.

Write Equation 6.22 for each planet (write planet 2 first) and then divide the two equations:

$$T_2^2 = \left(\frac{4\pi^2}{GM}\right)r_2^3$$

$$T_1^2 = \left(\frac{4\pi^2}{GM}\right)r_1^3$$

$$\frac{T_2^2}{T_1^2} = \frac{r_2^3}{r_1^3}$$

$$\frac{T_2^2}{T_1^2} = \frac{(4r_1)^3}{r_1^3}$$

Multiply both sides by T_1^2 and cancel r_1^3:

$$T_2^2 = T_1^2(4)^3$$

Take square roots:

$$T_2 = T_1\sqrt{(4)^3} = T_1\sqrt{64} = 8T_1$$

The correct choice is D.

Assess: When the orbital radius quadruples, the period increases by a factor of eight because planet 2 has not only farther to go, but also moves slower. It is instructive to test this relationship with real data. According to Example 6.15, communication satellites have an orbital radius of $4.22 \times 10^7\,\text{m}$ and we know from the table inside the back cover of the book that the moon's orbital radius is 3.84×10^8 m. Combining these, we have $r_{moon} \approx 9r_{satellite}$, so using the math above with the new number, $T_{moon} \approx \sqrt{9^3}\,T_{satellite} = \left(\sqrt{9}\right)^3 T_{satellite} = 27T_{satellite} = 27\,\text{d}$. From Question 6.28, we know that this is the length of one month.

Problems

P6.1. Strategize: This problem requires that we relate the speed of an object in uniform circular motion to the period and radius of that motion.
Prepare: Find the speed of an object in uniform circular motion. We are given $r = 2.5$ m (half of the diameter). A preliminary calculation will give ω.

$$\omega = 2\pi\ \text{rad}/4.0\ \text{s} = 1.57\ \text{rad/s}$$

Solve:

$$v = \omega r = (1.57\ \text{rad/s})(2.5\ \text{m}) = 3.9\ \text{m/s}$$

Assess: A speed of 3.9 m/s seems reasonable for a merry-go-round turning this fast.

P6.3. Strategize: This problem is essentially one of unit conversion. It also requires us to understanding the meaning of frequency and period.
Prepare: Assume uniform circular motion.
Solve: (a) Converting revolutions per minute to revolutions per second

$$\left(33\frac{1}{3}\frac{\text{revolutions}}{\text{minute}}\right)\left(\frac{1\ \text{minute}}{60\ \text{s}}\right) = 0.56\ \text{rev/s}$$

(b) Using the equation from the text

$$T = \frac{1}{f} = \frac{1}{0.56 \text{ rev/s}} = 1.8 \text{ s}$$

Assess: This seems reasonable, if you're old enough to remember LPs. They are making a comeback now.

P6.5. Strategize: This problem requires that we relate several concepts important to circular motion. In particular, we will use the relationship between frequency and period, as well as the relationship between centripetal acceleration and speed.

Prepare: We know $f = 1/T$, and we know that speed is given simply by distance over time (in this case $v = 2\pi r / T$). Finally we use $a_c = v^2 / r$.

Solve: (a) $f = 1/T = 1/(0.43 \text{ s}) = 2.3$ Hz.

(b) The speed is given by $v = \dfrac{2\pi r}{T} = \dfrac{2\pi(2.1 \text{ m})}{(0.43 \text{ s})} = 30.7$ m/s. We note this value to three significant digits for use in part (c), but we round our final answer here to 31 m/s.

(c) $a_c = v^2 / r = \dfrac{(30.7 \text{ m/s})^2}{(2.1 \text{ m})} = 4.5 \times 10^2 \text{ m/s}^2$.

Assess: The first two are reasonable kinematic numbers; the acceleration is quite impressive and requires great strength.

P6.7. Strategize: We treat Earth as a particle orbiting around the sun in roughly uniform circular motion.

Prepare: We need to use the relationship between speed, radius and period: $v = \dfrac{2\pi r}{T}$, as well as the equation describing centripetal acceleration in uniform circular motion: $a_c = v^2 / r$.

Solve: (a) The magnitude of the earth's velocity is displacement divided by time:

$$v = \frac{2\pi r}{T} = \frac{2\pi(1.50 \times 10^{11} \text{ m})}{365 \text{ days} \times \frac{24 \text{ hr}}{1 \text{ day}} \times \frac{3600 \text{ s}}{1 \text{ hr}}} = 3.0 \times 10^4 \text{ m/s}$$

(b) The centripetal acceleration is

$$a_r = \frac{v^2}{r} = \frac{(3.0 \times 10^4 \text{ m/s})^2}{1.5 \times 10^{11} \text{ m}} = 6.0 \times 10^{-3} \text{ m/s}^2$$

Assess: A tangential velocity of 3.0×10^4 m/s or 30 km/s is large, but needed for the earth to go through a displacement of $2\pi(1.5 \times 10^{11} \text{ m}) \approx 9.4 \times 10^8$ km in 1 year.

P6.9. Strategize: We will apply the known relationship between speed along a circular path and centripetal acceleration.

Prepare: We know $a_c = v^2 / r$, and we are given all the required information.

Solve: (a) $a_c = v^2 / r = (4.2 \text{ m/s})^2 / (0.35 \text{ m}) = 50.4 \text{ m/s}^2 = 5.1g$.

(b) $(5.1g)/(9g) = 57\%$.

Assess: This is an impressive acceleration, as we expected.

P6.11 Strategize: Since the astronaut will move in a circular track at a fixed frequency, the astronaut's speed will be constant. We can use results from uniform circular motion.

Prepare: We will use Equation 6.5 to express the centripetal acceleration in terms of the radius of the circular path and the frequency. We can then rearrange to solve for the frequency.

Solve: Equation 6.5 reads

$$a_r = \left(2\pi f\right)^2 r \Rightarrow f = \sqrt{\frac{a_r}{4\pi^2 r}}$$

Inserting the given values, we have

$$f = \sqrt{\frac{1.4\left(9.8 \text{ m/s}^2\right)}{4\pi^2\left(1.1 \text{ m}\right)}} = 0.56 \text{ Hz}$$

Assess: This is a reasonable frequency, since the astronaut would complete one revolution approximately every 2 seconds.

P6.13. Strategize: The objects are in uniform circular motion. Since the circular paths are horizontal and no forces are shown acting on the particles other than tension, we will assume that all centripetal force is provided by tension.
Prepare: The equation in the text tells us the tension:

$$T = m\frac{v^2}{r}$$

Because all four are moving at the same speed, we need only consider the effect of m and r on T. A small r and a large m would make for a large T, as in case 3.
Solve: $T_3 > T_1 = T_4 > T_2$
Assess: Case 4 is the same as case 1 because both the mass and radius are doubled.

P6.15. Strategize: We use Newton's second law applied to a horizontal circular path to determine the required coefficient of friction.
Prepare: Let us call the direction inward toward the center of the circular path the $+r$ direction. Let us assume there are no other forces acting on the cheetah in this direction.
Solve: Newton's second law reads $\sum F_r = f_s = ma_c$. Assuming the minimum possible coefficient of friction, we can write $\mu_s mg = ma_c \Rightarrow \mu_s = a_c / g = \left(18 \text{ m/s}^2\right)/\left(9.8 \text{ m/s}^2\right) = 1.8$.

Assess: This is far above coefficients of friction that are commonly seen. But if the padding on a cheetah's feet are exceptionally good at gripping, it might be possible. It is more likely that the cheetah exerted forces on some objects (stone, indentations in the dirt) that were not perfectly horizontal surfaces.

P6.17. Strategize: We are using the particle model for the car in uniform circular motion on a flat circular track.
Prepare: There must be friction between the tires and the road for the car to move in a circle. A pictorial representation of the car, its free-body diagram, and a list of values are shown below.

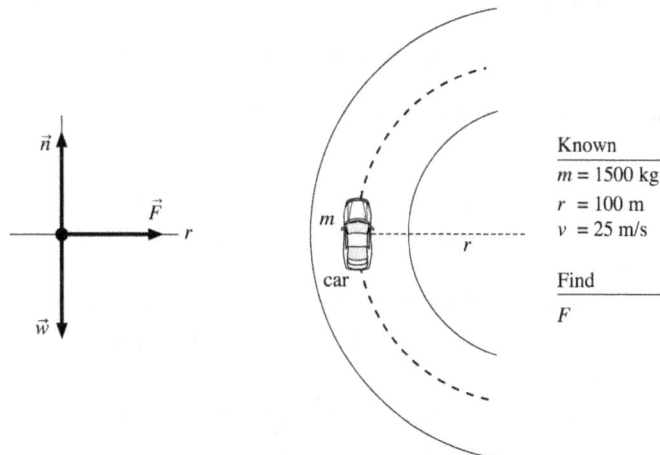

Solve: The equation in the text gives the centripetal acceleration

$$a = \frac{v^2}{r} = \frac{(25 \text{ m/s})^2}{100 \text{ m}} = 6.25 \text{ m/s}^2$$

The acceleration points to the center of the circle, so the net force is

$$\vec{F} = m\vec{a} = (1500 \text{ kg})(6.25 \text{ m/s}^2, \text{ toward center}) = (9400 \text{ N}, \text{ toward center})$$

This force is provided by static friction:

$$f_s = F_r = 9400 \text{ N}$$

P6.19. Strategize: We treat the ball as a particle moving in uniform circular motion. We can relate the know speed and radius of the circular path to the required centripetal acceleration, and force.
Prepare: We can calculate the ball's centripetal acceleration and the centripetal force.
Solve: Refer to the following figure.

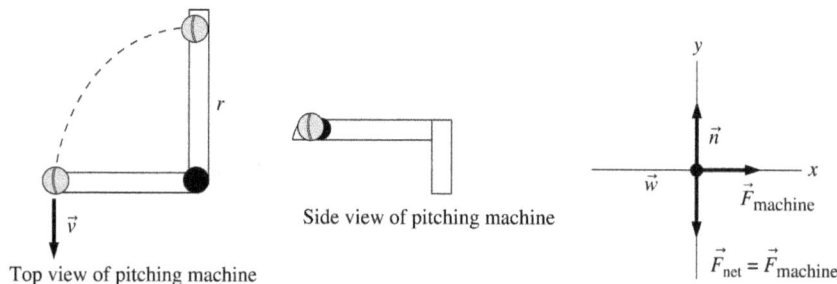

Top view of pitching machine   Side view of pitching machine

(a) Converting the velocity of the ball to meters per second, we have

$$v = (85 \text{ mph})\left(\frac{0.447 \text{ m/s}}{1 \text{ mph}}\right) = 38 \text{ m/s}$$

The centripetal acceleration of the ball is then

$$a = \frac{v^2}{r} = \frac{(38 \text{ m/s}^2)}{0.85 \text{ m}} = 1.7 \times 10^3 \text{ m/s}^2$$

(b) From the free-body diagram in the figure above, the net force on the ball is in the centripetal direction and so is equal to the centripetal force on the ball.

$$F_{net} = ma = (0.144 \text{ kg})(1700 \text{ m/s}^2) = 240 \text{ N}$$

Assess: The centripetal acceleration is large. The centripetal force needed during the launch of the ball is about 54 pounds.

P6.21. Strategize: We know that the force of static friction is the force holding the steel toolbox to the bed of the truck. If the centripetal force required to keep the toolbox moving in the circular path exceeds the maximum force of static friction, the toolbox will slide.
Prepare: We know the coefficient of static friction between two steel surfaces from Table 5.2: $\mu_s = 0.80$. We will call the direction inward toward the center of the circular path the $+r$ direction, and we will write the sum of all forces on the toolbox in this direction.
Solve: $\sum F_r = f_s = ma_c \Rightarrow \mu_s mg = mv_{max}^2 / r \Rightarrow v_{max} = \sqrt{\mu_s gr} = \sqrt{(0.80)(9.8 \text{ m/s}^2)(20 \text{ m})} = 13 \text{ m/s}.$
Assess: This is equivalent to about 28 mph, which is a reasonable maximum speed for such a tight turn.

P6.23. Strategize: We treat the gibbon as a point-like particle undergoing uniform circular motion. In reality, the gibbon speeds up/ slows down a little during the swing. But at the instant described, we will assume the speed is changing relatively little.
Prepare: At the bottom there are two forces on the gibbon, the upward tension force in the arm (modeled as a massless rod) and the downward force of gravity.
Solve: At the bottom of the swing the tension force in the rod must be greater than the weight in order to provide and upward centripetal acceleration.

$$F_{net} = T - mg = ma = m\frac{v^2}{r} \Rightarrow T = m\frac{v^2}{r} + mg = m\left(\frac{v^2}{r} + g\right) = (9.0 \text{ kg})\left(\frac{(3.5 \text{ m/s}^2)}{0.60 \text{ m}} + 9.80 \text{ m/s}^2\right) = 270 \text{ N}$$

The branch must be able to provide this much support without breaking.
Assess: The branch must be able to support about three times the weight of the gibbon.

P6.25. Strategize: We will write out the sum of all forces in the direction radially inward toward the center of the athlete's circular path. We require that this sum yield the necessary centripetal force.
Prepare: At the top of the circular arc, gravity acts straight down toward the center of the circle and the pole exerts some upward force on the athlete. Let us call the direction toward the center of the circle the $+r$ direction.
Solve: Writing the sum of all forces in the radial direction, we have $\sum F_r = mg - F_{pole} = mv^2 / r$ or equivalently

$$F_{pole} = m\left(g - v^2 / r\right) = (55 \text{ kg})\left((9.8 \text{ m/s}^2) - (2.5 \text{ m/s})^2 / (5.1 \text{ m})\right) = 4.7\times10^2 \text{ N}.$$

Assess: Note that this is somewhat less than the athlete's weight. This is reasonable, since there must be some net downward force to change the athlete's direction along the circular arc.

P6.27. Strategize: Model the roller coaster car as a particle undergoing uniform circular motion along a loop. Newton's second law will help us relate forces to the acceleration.
Prepare: A pictorial representation of the car, its free-body diagram, and a list of values are shown. Note that the normal force $\vec{n}$ of the seat pushing on the passenger is the passenger's apparent weight, and in this problem the apparent weight is equal to the true weight: $w_{app} = n = mg.$

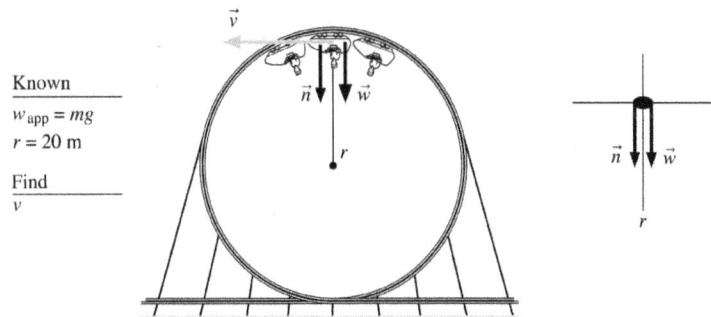

Solve: We have

$$\Sigma F = n + w = \frac{mv^2}{r} = mg + mg \Rightarrow v = \sqrt{2rg} = \sqrt{2(20 \text{ m})(9.80 \text{ m/s}^2)} = 20 \text{ m/s}$$

Assess: A speed of 20 m/s or 44 mph on a roller coaster ride is reasonable. The mass cancels out of the calculation.

P6.29. Strategize: This question asks about apparent weight, so we are interested in the contact force between the driver and the seat. We will use the sum of all forces on the driver in the radial direction to determine this.
Prepare: If the top of the hill can be modeled as an arc in a circle, the center of the circle would be beneath the road. This means that downward forces are radially inward. We will write out the sum of all forces in this radially inward $(+r)$ direction. Before we can insert numbers we must convert to SI units:

$$\left(\frac{75 \text{ mi}}{1 \text{ h}}\right)\left(\frac{1610 \text{ m}}{1 \text{ mi}}\right)\left(\frac{1 \text{ h}}{3600 \text{ s}}\right) = 33.5 \text{ m/s}.$$

Solve: Noting all forces acting on the driver, we have $\sum F_r = mg - n = mv^2r \Rightarrow n = m(g - v^2/r)$. We want the

fraction of this apparent weight to the usual weight: $\dfrac{n}{mg} = \left(1 - \dfrac{v^2}{gr}\right) = \left(1 - \dfrac{(33.5 \text{ m/s})^2}{(9.8 \text{ m/s}^2)(525 \text{ m})}\right) = 0.78$, or 78%.

Assess: Anyone who has ever driven quickly over the top of a hill knows that there is a moment when they feel lighter. This reduction in apparent weight of 22% is reasonable.

P6.31. Strategize: We will use the particle model for the test tube, which is in uniform circular motion. We can relate the frequency and radius to the centripetal acceleration.
Prepare: The radius to the end of the tube from the axis of rotation is 10 cm or 0.1 m. We will use kinematic equations and work with SI units.
Solve: (a) The acceleration is

$$a = r\omega^2 = (0.1 \text{ m})\left(4000\frac{\text{rev}}{\text{min}} \times \frac{1 \text{ min}}{60 \text{ s}} \times \frac{2\pi \text{ rad}}{1 \text{ rev}}\right)^2 = 1.8 \times 10^4 \text{ m/s}^2$$

(b) An object falling 1 meter has a speed calculated as follows:

$$v_f^2 = v_i^2 + 2a_y(y_f - y_i) = 0 \text{ m} + 2(-9.8 \text{ m/s}^2)(-1.0 \text{ m}) \Rightarrow v_1 = 4.43 \text{ m/s}$$

When this object is stopped in 1×10^{-3} s upon hitting the floor,

$$v_f = v_i + a_y(t_f - t_i) \Rightarrow 0 \text{ m/s} = -4.43 \text{ m/s} + a_y(1 \times 10^{-3} \text{ s}) \Rightarrow a_y = 4.4 \times 10^3 \text{ m/s}^2$$

This result is one-fourth of the above radial acceleration in part (a).
Assess: The radial acceleration of the centrifuge is large, but it is also true that falling objects are subjected to large accelerations when they are stopped by hard surfaces.

P6.33. Strategize: Treat the spacecraft as a point-like particle undergoing uniform circular motion.
Prepare: Assume the radius of the satellite's orbit is about the same as the radius of Mars itself.

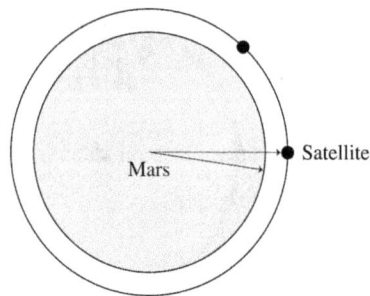

As a preliminary calculation, compute the angular velocity of the satellite:

$$\omega = \frac{2\pi}{T} = \frac{2\pi \text{ rad}}{110 \text{ min}}\left(\frac{1 \text{ min}}{60 \text{ s}}\right) = 9.52 \times 10^{-4} \text{ rad/s}$$

Solve: Since $T = \dfrac{2\pi}{\omega}$ and $a = \omega^2 r \Rightarrow \omega = \sqrt{\dfrac{a}{r}}$, then

$$T = \frac{2\pi}{\omega} = \frac{2\pi}{\sqrt{\dfrac{a}{r}}} = \frac{2\pi}{\sqrt{\dfrac{3.8 \text{ m/s}^2}{3.37 \times 10^6 \text{ m}}}} = 5900 \text{ s}$$

This answer is equal to about 99 min.
Assess: This is between the orbital period for a satellite in low earth orbit and one in low moon orbit, which sounds right.

P6.35. Strategize: We will use Newton's Law of Universal Gravitation to write an expression for the gravitational attraction between each planet and the star.

Prepare: Call the mass of the star M. Write Newton's law of gravitation for each planet.

$$F_1 = \frac{GMm_1}{r_1^2}$$

$$F_2 = \frac{GMm_2}{r_2^2} = \frac{GM(2m_1)}{(2r_1)^2}$$

Solve: Divide the two equations to get the ratio desired.

$$\frac{F_2}{F_1} = \frac{\frac{GM(2m_1)}{(2r_1)^2}}{\frac{GMm_1}{r_1^2}} = \frac{1}{2}$$

Assess: The answer is expected. Even with twice the mass, because the radius in the denominator is squared, we expect the force on planet 2 to be less than the force on planet 1.

P6.37. Strategize: We will use Newton's Law of Universal Gravitation to write an expression for the free-fall acceleration on you due to each object, and then compare them.

Prepare: Model the sun (s) and the earth (e) as spherical masses. Due to the large difference between your size and mass and that of either the sun or the earth, a human body can be treated as a particle.

Solve: $F_{\text{s on you}} = \dfrac{GM_e m_y}{r_{s-e}^2}$ and $F_{\text{e on you}} = \dfrac{GM_e m_y}{r_e^2}$

Dividing these two equations gives

$$\frac{F_{\text{s on y}}}{F_{\text{e on y}}} = \left(\frac{M_s}{M_e}\right)\left(\frac{r_e}{r_{s-e}}\right)^2 = \left(\frac{1.99\times10^{30}\ \text{kg}}{5.98\times10^{24}\ \text{kg}}\right)\left(\frac{6.37\times10^6\ \text{m}}{1.5\times10^{11}\ \text{m}}\right)^2 = 6.0\times10^{-4}$$

Assess: The result shows the smallness of the sun's gravitational force on you compared to that of the earth.

P6.39. Strategize: We will use Newton's Law of Universal Gravitation and known physical characteristics of Jupiter to determine the freefall acceleration on this new planet.

Prepare: Look up the data for Jupiter. $M_{\text{Jupiter}} = 1.90\times10^{27}$ kg, $R_{\text{Jupiter}} = 6.99\times10^7$ m.

Solve: From the equation in the text,

$$g = \frac{GM}{R^2} = \frac{G(0.43M_{\text{Jupiter}})}{(1.7R_{\text{Jupiter}})^2} = \frac{(6.67\times10^{-11}\ \text{N}\cdot\text{m}^2/\text{kg}^2)\big((0.43)(1.90\times10^{27}\ \text{kg})\big)}{((1.7)(6.99\times10^7\ \text{m}))^2} = 3.9\ \text{m/s}^2$$

Assess: This is in the range of g for other planets.

P6.41. Strategize: We will use Newton's Law of Universal Gravitation to determine the attractive forces between each pair of astronomical objects.

Prepare: Model the sun (s), the earth (e), and the moon (m) as spherical masses.

Solve: (a) $F_{\text{s on e}} = \dfrac{Gm_s m_e}{r_{s-e}^2} = \dfrac{(6.67\times10^{-11}\ \text{N}\cdot\text{m}^2/\text{kg}^2)(1.99\times10^{30}\ \text{kg})(5.98\times10^{24}\ \text{kg})}{(1.50\times10^{11}\ \text{m})^2} = 3.53\times10^{22}\ \text{N}$

(b) $F_{\text{m on e}} = \dfrac{GM_m M_e}{r_{m-e}^2} = \dfrac{(6.67\times10^{-11}\ \text{N}\cdot\text{m}^2/\text{kg}^2)(7.36\times10^{22}\ \text{kg})(5.98\times10^{24}\ \text{kg})}{(3.84\times10^8\ \text{m})^2} = 1.99\times10^{20}\ \text{N}$

(c) The moon's force on the earth as a percent of the sun's force on the earth is

$$\left(\frac{1.99\times10^{20}\ \text{N}}{3.53\times10^{22}\ \text{N}}\right)\times100 = 0.564\%$$

P6.43. Strategize: We will use Newton's Law of Universal Gravitation and we will look up the known physical properties of Mars and Jupiter to determine the acceleration due to gravity on those planets.

Prepare: Model Mars (m) and Jupiter (J) as spherical masses.

Solve: (a) $g_{\text{Mars surface}} = \dfrac{(6.67 \times 10^{-11} \text{ N} \cdot \text{m}^2/\text{kg}^2)(6.42 \times 10^{23} \text{ kg})}{(3.37 \times 10^6 \text{ m})^2} = 3.77 \text{ m/s}^2$

(b) $g_{\text{Jupiter surface}} = \dfrac{GM_J}{R_J^2} = \dfrac{(6.67 \times 10^{-11} \text{ N} \cdot \text{m}^2/\text{kg}^2)(1.90 \times 10^{27} \text{ kg})}{(6.99 \times 10^7 \text{ m})^2} = 25.9 \text{ m/s}^2$

P6.45. Strategize: We know how to relate orbital periods to radii of orbits and the masses at the center of the orbit. We will do this for each moon and compare expressions.

Prepare: For each moon, we will consider Equation 6.22: $T^2 = \left(\dfrac{4\pi^2}{GM}\right) r^3$, and then we will compare the periods.

Solve: For Deimos and Phobos separately, we have $T_D^2 = \left(\dfrac{4\pi^2}{GM}\right) r_D^3$, $T_P^2 = \left(\dfrac{4\pi^2}{GM}\right) r_P^3$. The mass here refers to the mass at the center of the orbit (Mars) which is the same for both moons. Dividing one expression by the other, we obtain

$$\left(\frac{T_D}{T_P}\right)^2 = \frac{r_D^3}{r_P^3} \Rightarrow \frac{T_D}{T_P} = \left(\frac{r_D}{r_P}\right)^{3/2} = \left(\frac{(23,500 \text{ km})}{(9,380 \text{ km})}\right)^{3/2} = 4.0.$$

Assess: This means Deimos requires 4 times as long to complete one orbit as Phobos.

P6.47. Strategize: We will relate the orbital period to other orbital information and then convert from seconds to years.

Prepare: We will start with Equation 6.22: $T^2 = \left(\dfrac{4\pi^2}{GM}\right) r^3$.

Solve: Inserting known values, we have

$$T = \sqrt{\left(\frac{4\pi^2}{GM}\right) r^3} = \sqrt{\left(\frac{4\pi^2}{(6.67 \times 10^{-11} \text{ N} \cdot \text{m}^2 / \text{kg}^2)(1.99 \times 10^{30} \text{ kg})}\right)(6.4 \times 10^{12} \text{ m})^3} = 8.83 \times 10^9 \text{ s}$$

Converting, we have $(8.83 \times 10^9 \text{ s})\left(\dfrac{1 \text{ h}}{3600 \text{ s}}\right)\left(\dfrac{1 \text{ day}}{24 \text{ h}}\right)\left(\dfrac{1 \text{ y}}{365.25 \text{ days}}\right) = 280 \text{ y}.$

Assess: The radius is much larger than Earth's orbital radius, so it makes sense that the period would be much longer.

P6.49. Strategize: Let us model the earth (e) as a spherical mass and the space station (s) as a point particle. Then the space station is undergoing uniform circular motion.

Prepare: The shuttle with mass m_s and velocity v_s orbits the earth in a circle of radius r_s. We will denote the earth's mass by M_e. As a reminder, the gravitational force between the earth and the shuttle provides the necessary centripetal acceleration for circular motion.

Solve: Newton's second law is

$$\frac{GM_e m_s}{r_s^2} = \frac{m_s v_s^2}{r_s} \Rightarrow v_s^2 = \frac{GM_e}{r_s} \Rightarrow v_s = \sqrt{\frac{GM_e}{r_s}}$$

Because $r_s = R_e + 250 \text{ miles} = 6.37 \times 10^6 \text{ m} + 4.023 \times 10^5 \text{ m} = 6.77 \times 10^6 \text{ m}$,

$$v_s = \sqrt{\frac{(6.67 \times 10^{-11} \text{ N} \cdot \text{m}^2/\text{kg}^2)(5.98 \times 10^{24} \text{ kg})}{(6.77 \times 10^6 \text{ m})}} = 7675 \text{ m/s} \approx 7700 \text{ m/s}$$

$$T_s = \frac{2\pi r_s}{v_s} = \frac{2\pi(6.77 \times 10^6 \text{ m})}{7.675 \times 10^3 \text{ m/s}} = 5542 \text{ s} = 92 \text{ min}$$

Assess: An orbital period of 92.4 minutes is reasonable for a 250 mile high orbit. As comparison, the orbital period is 1440 minutes for a geostationary orbit at a distance of approximately 25,000 miles.

P6.51. Strategize: We can model the planet as a point-particle and assume it is moving in uniform circular motion around its star.
Prepare: From Equation 6.22 for circular orbits we solve for r.
Solve: Ratios are a good way to solve this problem.

$$T^2 = \frac{4\pi^2}{GM}r^3 \Rightarrow r^3 = \frac{GMT^2}{4\pi^2}$$

Compare with data from our solar system.

$$\frac{r_2^3}{r_1^3} = \frac{M_2 T_2^2}{M_1 T_1^2} \Rightarrow r_2 = r_1 \sqrt[3]{\frac{M_2}{M_1}\left(\frac{T_2}{T_1}\right)^2} = (1.0 \text{ au})\sqrt[3]{\frac{1.1}{1}\left(\frac{2.7 \text{ d}}{365 \text{ d}}\right)^2} = 0.039 \text{ au}$$

Assess: This is a very small orbital radius because the period so short. We have no planets like this in our solar system. This answer can also be obtained without ratios. Preliminary calculations give $2.7 \text{ d} = 2.33 \times 10^5 \text{ s}$.

$$r = \sqrt[3]{\frac{T^2 GM}{4\pi^2}} = \sqrt[3]{\frac{(2.33 \times 10^5 \text{s})^2 (6.67 \times 10^{-11} \text{ N} \cdot \text{m}^2/\text{kg}^2)(1.1)(1.99 \times 10^{30} \text{ kg})}{4\pi^2}} = 5.86 \times 10^8 \text{ m} = 0.039 \text{ au}$$

P6.53 Strategize: We can model the planet as a point-particle and assume it is moving in uniform circular motion around its star.
Prepare: From the equation for circular orbits we solve for T. Preliminary calculations give $0.0058 \text{ au} = 8.70 \times 10^8 \text{ m}$ and $0.13 M_{\text{sun}} = 2.59 \times 10^{29} \text{ kg}$.
Solve: The speed is

$$T = 2\pi\sqrt{\frac{r^3}{GM}} = 2\pi\sqrt{\frac{(8.70 \times 10^8 \text{ m})^3}{(6.67 \times 10^{-11} \text{ N} \cdot \text{m}^2/\text{kg}^2)(2.59 \times 10^{29} \text{ kg})}} = 11 \text{ h}$$

Assess: This is an extremely short year. This problem can also be solved using ratios.

P6.55. Strategize: If we consider segments of road to be arcs of circular paths, we can use what we know about uniform circular motion and centripetal acceleration to solve this.
Prepare: Since the speed is constant the acceleration tangent to the path at each point is zero.
Solve: Since $a = v^2/r$ and v is constant, we see that the radius of curvature of the road at point A is about three times larger than the radius of curvature at point C, so the car's centripetal acceleration at point C is three times larger than at point A.

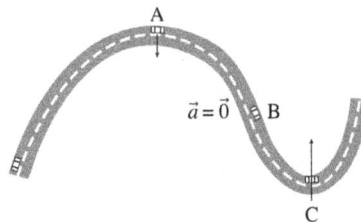

At point B there is no curvature, so there is no centripetal acceleration.
Assess: When you drive on windy roads you know that the tighter the curve the more acceleration you feel, and it is often wise to *not* keep your speed constant. Slowing down for tight curves keeps the centripetal acceleration manageable (it must be produced by the centripetal force of friction of the road on the tires).

P6.57. Strategize: The sum of all forces in the radial direction must provide the necessary centripetal force. We will write down Newton's second law and solve for the unknown normal force.

Prepare: As the car drives over the hill that is approximated by a circular arc, the center of that circle is below the road. Thus we call down toward the center of the circle ($+r$). We draw a free body diagram to help us with Newton's second law.

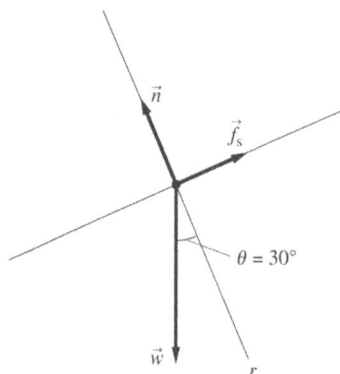

Solve: We could write Newton's second law for both the radial and tangential directions. But in this case, we need only do so for the radial direction:

$$\sum F_r = mg\cos(\theta) - n = \frac{mv^2}{r} \Rightarrow n = m\left(g\cos(\theta) - \frac{v^2}{r}\right) = (1400 \text{ kg})\left((9.8 \text{ m/s}^2)\cos(30°) - \frac{(27 \text{ m/s})^2}{(430 \text{ m})}\right) = 9.5\times10^3 \text{ N}.$$

Assess: This is slightly less than the weight of the car. This makes sense, because the weight must be larger than the normal force in order for the net force to point downward, toward the center of the circle.

P6.59. Strategize: We calculate the magnitude of the vector sum of forces the ground exerts on a runner's feet in two cases. First, we consider a runner moving in a straight line; second we consider a runner going through a curve, in which both friction and the normal force play significant roles.

Prepare: In the case of the runner moving in a straight line, there is no horizontal force and the only force the track exerts on the runner is the normal force. In the case of the runner rounding a curve, the only force acting radially inward toward the center of the circular path will be friction, and there will still be a normal force in the vertical direction. We calculate the magnitude of the vector sum of these forces and compare.

Solve: For the straight path, the magnitude of all forces the track exerts on the feet is $F_{\text{straight}} = n = mg$. For the curved path, the friction must provide the centripetal acceleration: $\sum F_r = f_s = mv^2/r$. We also have $n = mg$ orthogonal to that, such that the magnitude of the sum of these vectors is (using the Pythagorean Theorem) $F_{\text{curved}} = m\sqrt{g^2 + v^4/r^2}$. Comparing the two forces, we have

$$\frac{F_{\text{curved}}}{F_{\text{straight}}} = \sqrt{1 + v^4/g^2r^2} = \sqrt{1 + \frac{(10 \text{ m/s})^4}{(9.8 \text{ m/s}^2)^2(20 \text{ m})^2}} = 1.12$$

This means that the force during the run around the curve is larger by about 12%.

Assess: This is at least a partial explanation of why runners slow, responding to the natural increase in forces on their feet (not to mention some rotational effects in their ankles, which requires material to be discussed in later chapters).

P6.61. Strategize: The horizontal force that can provide the necessary centripetal force to change the mouse's velocity is friction. We will use Newton's second law and our understanding of uniform circular motion to find the necessary force of friction.

Prepare: We note that the situation described is that in which the mouse is running as fast as possible without slipping, such that we can write $f_s = (f_s)_{max} = \mu_s n$. We will assume everything is on level ground, such that the only horizontal force is friction.

Solve: Newton's second law in the radial direction tells us $\sum F_r = f_s = mv^2 / r$ and using the fact that the mouse is on the verge of slipping this becomes

$$\mu_s n = mv^2 / r \Rightarrow \mu_s mg = mv^2 / r \Rightarrow \mu_s = v^2 / (gr) = (1.29 \text{ m/s})^2 / \left((9.8 \text{ m/s}^2)(0.15 \text{ m})\right) = 1.1.$$

Assess: This is a very high coefficient of static friction, higher even than for most rubber on concrete. But if the mouse's feet have some adhesive properties, it could be reasonable. This seems likely, since mice are known to run up walls that have only moderate roughness to them.

P6.63. Strategize: If the motorcycles are moving very fast, then there must be a very large normal force acting inward toward the center of the circular path, acting as the centripetal force. If the normal force is very large, then the force of friction might be large enough to suppose the weight of the motorcycle and rider.

Prepare: We will refer to the mass of the motorcycle and rider together as m. We make a free-body diagram to help us in writing out Newton's second law for the radial and vertically upward directions.

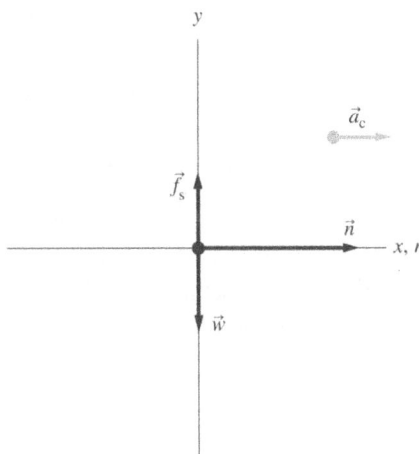

We note that since we are asked for the minimum speed of the motorcycle, this corresponds to being just on the verge of slipping such that $f_s = (f_s)_{max} = \mu_s n$.

Solve: In the radial and vertical directions, respectively, we have

$$\sum F_r = n = mv^2 / r$$
$$\sum F_y = f_s - mg = ma_y = 0 \Rightarrow \mu_s n = mg$$

Combining the two equations by solving one for the normal vector and inserting that into the other equation, we find

$$\frac{\mu_s mv^2}{r} = mg \Rightarrow v = \sqrt{gr / \mu_s} = \sqrt{(9.8 \text{ m/s}^2)(5.0 \text{ m}) / (0.90)} = 7.4 \text{ m/s}$$

Assess: This is about 17 mph, which is certainly possible during such a stunt. But we don't recommend trying it!

P6.65. Strategize: Treat the coin as a particle which is undergoing circular motion.
Prepare: A visual overview of the coin's circular motion is shown below in the following pictorial representation, free-body diagram, and list of values. The force of static friction between the coin and the turntable, as long as the coin does not slide, causes the centripetal acceleration needed for circular motion. The force of static friction is $f_s = \mu_s n = \mu_s mg$. This force is equivalent to the maximum centripetal force that can be applied without sliding. Work with SI units.

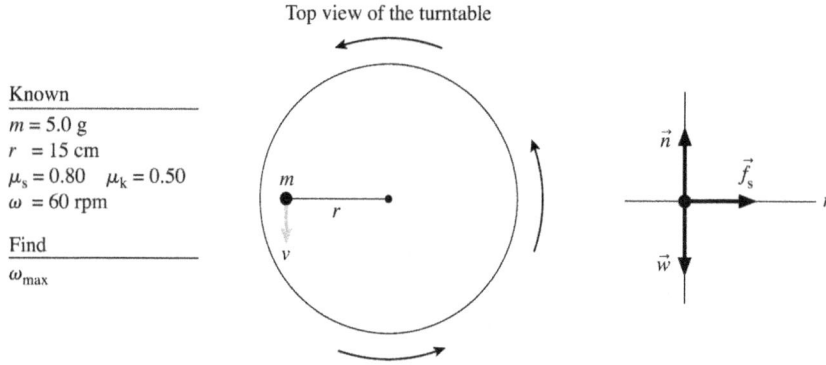

Top view of the turntable

Known

$m = 5.0$ g
$r = 15$ cm
$\mu_s = 0.80$ $\mu_k = 0.50$
$\omega = 60$ rpm

Find

ω_{max}

Solve: That is,

$$\mu_s mg = m\frac{v^2}{r} = m(r\omega_{max}^2) \Rightarrow \omega_{max} = \sqrt{\frac{\mu_s g}{r}} = \sqrt{\frac{(0.80)(9.8 \text{ m/s}^2)}{0.15 \text{ m}}} = 7.23 \text{ rad/s}$$

$$= 7.23 \frac{\text{rad}}{\text{s}} \times \frac{1 \text{ rev}}{2\pi \text{ rad}} \times \frac{60 \text{ s}}{1 \text{ min}} = 69 \text{ rpm}$$

So, the coin will stay still on the turntable.

Assess: A rotational speed of approximately 1 rev per second for the coin to stay stationary seems reasonable.

P6.67. Strategize: We will treat the puck as a point-like particle undergoing uniform circular motion. We will use Newton's second law to relate the forces on the puck (and hanging block) to the puck's motion.

Prepare: Since the hanging block is at rest, the total force on it is zero. The two forces are the tension in the string, T, and the weight of the puck, $-mg$. Since the revolving puck is moving at constant speed in a circle, the total force on the puck is the centripetal force. We must write the equations and solve them.

Solve: The total force on the block is $T - mg$. From Newton's second law, the total force is zero so we write:

$$T = mg = (1.20 \text{ kg})(9.80 \text{ m/s}^2) = 11.8 \text{ N}$$

The centripetal acceleration of the puck is caused by the tension in the string, so $mv^2 / r = T$. We solve this to obtain:

$$v = \sqrt{Tr / m} = \sqrt{(11.8 \text{ N})(0.50 \text{ m})/(0.20 \text{ kg})} = 5.4 \text{ m/s}$$

The puck must rotate at a speed of 5.4 m/s.

Assess: It is remarkable that a block can be supported by a puck moving horizontally. But both the puck and the block are able to pull on the string—the block pulls downward on one end and the puck pulls outward on the other end. The relatively small mass of the puck is compensated by its high speed of 5.4 m/s.

P6.69. Strategize: Treat the car as a particle which is undergoing circular motion. We will use Newton's second law to relate forces to the motion of the car.

Prepare: The car is in circular motion with the center of the circle below the car. A visual overview of the car's circular motion is shown below in the following pictorial representation, free-body diagram, and list of values.

Known

$r = 50$ m

Find

v_{max}

Solve: Newton's second law at the top of the hill is

$$F_{net} = \sum F_y = w - n = mg - n = ma = \frac{mv^2}{r} \Rightarrow v^2 = r\left(g - \frac{n}{m}\right)$$

This result shows that maximum speed is reached when $n = 0$ and the car is beginning to lose contact with the road. Then,

$$v_{max} = \sqrt{rg} = \sqrt{(50 \text{ m})(9.80 \text{ m/s}^2)} = 22 \text{ m/s}$$

Assess: A speed of 22 m/s is equivalent to 50 mph, which seems like a reasonable value.

P6.71. Strategize: Model Earth (e) as a spherical mass. We will use Newton's Law of Universal Gravitation to relate the distance between the observatory and the center of Earth to the freefall acceleration there.
Prepare: We will take the free-fall acceleration to be 9.83 m/s^2 and $R_e = 6.37 \times 10^6$ m. A pictorial representation of the situation is shown.

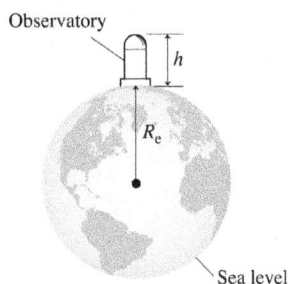

Solve: $g_{observatory} = \frac{GM_e}{(R_e + h)^2} = \frac{GM_e}{R_e^2(1 + \frac{h}{R_e})^2} = \frac{g_{earth}}{(1 + \frac{h}{R_e})^2} = (9.83 - 0.0075) \text{ m/s}^2$

Here $g_{earth} = GM_e/R_e^2$ is the free-fall acceleration. Solving for h,

$$h = \left(\sqrt{\frac{9.83}{9.8225}} - 1\right)R_e = 2400 \text{ m}$$

Assess: This altitude is relative to the sea level and is at reasonable altitude.

P6.73. Strategize: Model the planet Z as a spherical mass.

Prepare: We will use Newton's Law of Universal Gravitation: $F_{G,z} = mg_{Z \text{ surface}} = \frac{GM_Z m}{R_Z^2} \Rightarrow g_{Z \text{ surface}} = \frac{GM_Z}{R_Z^2}$.

Solve: **(a)** $g_{Z \text{ surface}} = \frac{GM_Z}{R_Z^2} \Rightarrow 8.0 \text{ m/s}^2 = \frac{(6.67 \times 10^{-11} \text{ N} \cdot \text{m}^2/\text{kg}^2)M_Z}{(5.0 \times 10^6 \text{ m})^2} \Rightarrow M_Z = 3.0 \times 10^{24} \text{ kg}$

(b) Let h be the height above the north pole. Thus,

$$g_{above \text{ N pole}} = \frac{GM_Z}{(R_Z + h)^2} = \frac{GM_Z}{R_Z^2\left(1 + \frac{h}{R_Z}\right)^2} = \frac{g_{Z \text{ surface}}}{\left(1 + \frac{h}{R_Z}\right)^2} = \frac{8.0 \text{ m/s}^2}{\left(1 + \frac{10.0 \times 10^6 \text{ m}}{5.0 \times 10^6 \text{ m}}\right)^2} = 0.89 \text{ m/s}^2$$

P6.75. Strategize: We will use Newton's Law of Universal Gravitation to determine the forces acting on the small mass due to each of the two spheres.
Prepare: We place the origin of the coordinate system on the 20 kg sphere (m_1). The sphere (m) with a mass of 10 kg is 20 cm away on the x-axis, as shown below. The point at which the net gravitational force is zero must lie between the masses m_1 and m_2. This is because on such a point, the gravitational forces due to m_1 and m_2 are in opposite directions. As the gravitational force is directly proportional to the two masses and inversely proportional to the square of distance between them, the mass m must be closer to the 10-kg mass. The small mass m, if placed either to the left of m_1 or to the right of m_2, will experience gravitational forces from m_1 and m_2 pointing in the same direction, thus always leading to a nonzero force.

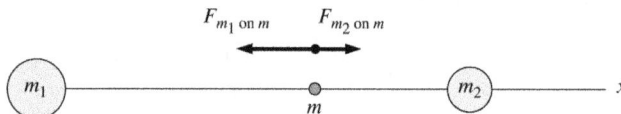

Solve:

$$F_{m_1 \text{ on } m} = F_{m_2 \text{ on } m} \Rightarrow G\frac{m_1 m}{x^2} = G\frac{m_2 m}{(0.20 - x)^2} \Rightarrow \frac{20}{x^2} = \frac{10}{(0.20 - x)^2} \Rightarrow 10x^2 - 8x + 0.8 = 0$$

The value $x = 68.3$ cm is unphysical in the current situation, since this point is not between m_1 and m_2. Thus, the point $(x, y) = (11.7 \text{ cm}, 0 \text{ cm}) \approx (12 \text{ cm}, 0 \text{ cm})$ is where a small mass is to be placed for a zero gravitational force.

P6.77. Strategize: Model Mars (m) and Phobos as spherical masses.

Prepare: Equation 6.22 relates orbital period and radius of orbit to mass: $T^2 = \left(\dfrac{4\pi^2}{GM_m}\right)r^3$

Solve: Solving Equation 6.22 for the mass, we can use Phobos' orbit to find the mass of Mars:

$$M_m = \frac{4\pi^2 r^3}{GT^2} = \frac{4\pi^2 (9.4 \times 10^6 \text{ m})^3}{(6.67 \times 10^{-11} \text{ N} \cdot \text{m}^2/\text{kg}^2)(2.7540 \times 10^4 \text{ s})^2} = 6.5 \times 10^{23} \text{ kg}$$

Assess: The mass of Mars is 6.42×10^{23} kg. The slight difference is likely due to Phobos' orbit being somewhat noncircular.

P6.79. Strategize: As the astronaut walks, the hip moves in a circular arc. We will use Newton's Law of Universal Gravitation to determine what force of gravity is present to provide the astronaut with the necessary centripetal acceleration.

Prepare: According to the discussion in Section 6.2 the maximum walking speed is $v_{max} = \sqrt{gr}$. The astronaut's leg is about 0.70 m long whether on earth or on Europa, but g will be different.

$$g_{Europa} = \frac{GM_{Europa}}{(R_{Europa})^2} = \frac{(6.67 \times 10^{-11} \text{ N} \cdot \text{m}^2/\text{kg}^2)(4.8 \times 10^{22} \text{ kg})}{(3.1 \times 10^6 \text{ m})^2} = 0.333 \text{ m/s}^2$$

Solve:

$$v_{max} = \sqrt{gr} = \sqrt{(0.333 \text{ m/s}^2)(0.70 \text{ m})} = 0.48 \text{ m/s}$$

Assess: The answer is about 1 mph or about 1/6 of the speed the astronaut could walk on the earth. This is reasonable on a small celestial body. Astronauts may adopt a hopping gait like some did on the moon.
Carefully analyze the units in the preliminary calculation to see that g ends up in m/s^2 or N/kg.

P6.81. Strategize: The spacecraft is undergoing uniform circular motion.
Prepare: Equation 6.13 which gives the orbital speed in terms of the free-fall acceleration and orbital radius can be used. The radius is half the diameter, $r_{Moon} = 1.75 \times 10^6$ m.
Solve: Applying the equation for orbital speed,

$$v_{orbit} = \sqrt{rg} = \sqrt{(1.75 \times 10^6 \text{ m})(1.6 \text{ m/s}^2)} = 1700 \text{ m/s}$$

The correct choice is C.

Assess: Even though the free-fall acceleration on the moon is much less than the free-fall acceleration on earth, the moon's orbital speed is still very high. At $F = \dfrac{\tau}{r \sin \theta} = \dfrac{20 \text{ N} \cdot \text{m}}{(0.15 \text{ m})(\sin 90°)} = 133 \text{ N} \approx 130 \text{ N}$, it is still faster than an airplane.

P6.83. Strategize: The spacecraft is undergoing uniform circular motion, and we wish to preserve that. This means maintaining a constant radial distance from the center of the Moon, and maintaining a constant speed.
Prepare: The centripetal acceleration will be constant if the velocity and radius of the orbit remain the same.
Solve: The gravitational force is stronger on the spacecraft when it is orbiting the near side of the moon. The net centripetal force must remain the same so the spacecraft should compensate for the increased gravitational force towards the center of the moon by firing its rockets so that they exert a force away from the center of the moon. The correct choice is A.
Assess: Another way to keep the radius of the orbit the same is to fire the rockets in the direction of motion of the spacecraft. However, if the spacecraft were fired in the direction of motion the velocity of the spacecraft would increase.

ROTATIONAL MOTION

Q7.1. Reason: Looking down from above the player runs around the bases in a counterclockwise direction, hence the angular velocity is positive.

Assess: Note that looking from below (from under the grass) the motion would be clockwise and the angular velocity would be negative. We assumed the bird's eye view because it is standard to do so, and it is difficult to view the game from below the ground. This is akin to setting up a coordinate system with the positive x-axis pointing left and the positive y-axis pointing down; the real-life physics wouldn't change any, and the calculations of measurable quantities would produce the same results, and it might even be occasionally convenient. But unless there is a clear reason to do otherwise, the usual conventions should be your first thought.

Q7.3. Reason: By convention, clockwise rotations are negative and counterclockwise rotations are positive. As a result, an angular acceleration that decreases/increases a negative angular velocity is positive/negative. In like manner, an angular acceleration that decreases/increases a positive angular velocity is negative/positive. Knowing this we can establish the situation for each figure. Figure (a) the pulley is rotating clockwise ($\omega = -$), however since the large mass is on the left it is decelerating ($\alpha = +$).

Figure (b) the pulley is rotating counterclockwise ($\omega = +$) and since the large mass is on the left it is accelerating ($\alpha = +$).

Figure (c) the pulley is rotating clockwise ($\omega = -$) and since the large mass is on the right it is accelerating ($\alpha = -$).

Figure (d) the pulley is rotating counterclockwise ($\omega = +$), however since the large mass is on the right it is decelerating ($\alpha = -$).

Assess: It is important to know the sign convention for all physical quantities that are vectors. This is especially important when working with rotational motion.

Q7.5. Reason: Torque is the product of the force and the moment arm. Thus, we can achieve a significant torque with a relatively small force if we exert that force far from the axis of rotation (as on a steering wheel). In order to achieve the same torque with a very small moment are (as in opening a jar), more force is required.

Assess: To maintain a fixed amount of torque, if the moment arm decreases, the force must increase.

Q7.7. Reason: The question properly identified where the torques are computed about (the hinge). Torques that tend to make the door rotate counterclockwise in the diagram are positive by convention (general agreement) and torques that tend to make the door rotate clockwise are negative.

(a) +
(b) −
(c) +
(d) −
(e) 0

Assess: Looking at the diagram we see that $\vec{F}_a$ and $\vec{F}_c$ are parallel and are both creating a negative or counterclockwise torque. But since $\vec{F}_c$ is farther from the hinge, its torque will be greater. A similar argument can be made for $\vec{F}_b$ and $\vec{F}_d$. Note that $\vec{F}_e$ causes no torque since it has no moment arm.

Q7.9. Reason: The reason for large-diameter steering wheels in trucks is that more torque is needed to turn the wheels due to the greater mass of the truck. Making the steering wheel larger means that more torque is exerted on the steering shaft for the same force from the driver's hands.
Assess: Most light cars employ a rack-and-pinion steering system, while larger SUVs and trucks often employ a recirculating-ball steering system; however both systems can be assisted by pressurized hydraulic fluid (power steering), so steering, even in trucks, can be much easier than in the old days.

Q7.11. Reason: Any object free to rotate about a pivot will come to rest with its center of gravity directly below the pivot. When the center of gravity is below the pivot there are no torques to rotate it. That means the center of gravity is somewhere on the blue and, for pivot 2, somewhere on the red line. Only the intersection of the two lines satisfies both requirements.
Assess: A clever way to find the center of gravity of an irregularly shaped object.

Q7.13. Reason: Since $\alpha = \tau / I$, and the torque is a fixed value, the sphere with the smaller moment of inertia will have the larger angular acceleration. One can look up expressions for the moments of inertia of solid $\left(\frac{2}{5}mR^2\right)$ and hollow $\left(\frac{2}{3}mR^2\right)$ spheres and choose the sphere that has the smaller one (the solid sphere). But one can also use a qualitative understanding of the moment of inertia. An object with all the mass far from the center will be hard to spin, whereas an object with some mass near the axis will be easier to spin, meaning the solid sphere will reach the ground first.
Assess: If an object has all mass far from the center, then an angular acceleration requires the mass to have a large tangential acceleration, and this is difficult. If an object has a lot of mass near the axis of rotation, that part of the mass will not have a very large tangential acceleration, and this is easier.

Q7.15. Reason: It is easier to stay on the log if it is more difficult to get it rolling. It is true that a larger diameter, means that forces from your feet can cause a larger torque, according to $\tau = Fr_\perp$. But the angular acceleration of the log depends on both torque and the moment of inertia: $\alpha = \tau / I$, or for the case of a solid cylinder: $\alpha = 2FR\sin(\theta)/\left(mR^2\right)$. A log with a larger diameter will have a larger moment of inertia, and (if the log is solid) a much larger mass. Since the moment of inertia increases quadratically with diameter, as does the mass, and the torque only increases linearly, the angular acceleration for a large log will be smaller than for a small log.
Assess: Torque depends linearly on the level arm. The moment of inertia depends quadratically on the distance from a small region of mass to the axis, and the total moment of inertia depends on the total mass.

Q7.17 Reason: A review of Section 7.7 in the text will prepare you to answer this question. Let's separate the rolling motion into translation and rotation. Figure (a) below shows the translational velocity of the points of interest. Figure (b) below shows the velocity due to rotation at all points of interest. Finally, Figure (c) below combines the velocity vectors due to both translation and rotation.

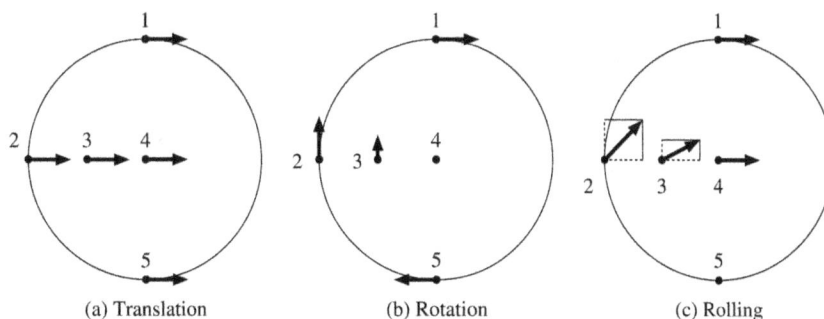

(a) Translation   (b) Rotation   (c) Rolling

Looking at the magnitude of the resulting velocity vectors we can write the following:

$$v_1 > v_2 > v_3 > v_4 > v_5 = 0$$

Assess: We know that the magnitudes of $\vec{v}_2$ and $\vec{v}_3$ are greater than the magnitude of $\vec{v}_4$ because their horizontal component is equal to the magnitude of $\vec{v}_4$.

Q7.19. Reason: Assuming all the forces have the same magnitude, the largest torque will be exerted on the nut by choice C. Choice A has a lesser torque than C because the moment arm of the force is smaller than in choice C. This can be seen from Equation 7.10. Choice D has a lesser torque than choice C because the component of the force perpendicular to the radial line is smaller. This can be seen from Equation 7.9. In choice B there is no torque exerted on the nut since the radial line and the force are parallel. The torque in this case is zero from Equation 7.11.
Assess: The torque created by a given force can be increased by increasing the moment arm of a force or by increasing the component of the force perpendicular to the radial line.

Q7.21. Reason: Since the center of gravity of piece 2 is to the right of the center of gravity of piece 1, the horizontal position of the center of gravity of the two pieces should be between the center of gravity of the two pieces. The same argument applies to the vertical position of the center of gravity of the pieces. The only point that is located between the two centers of gravity is point D.
Assess: Our solution to the problem is based on the fact that we can replace piece 2 with a single mass point (with the same mass as piece 2) located at the center of gravity of piece 2 and we can replace piece 1 with a single mass point (with the same mass as piece 1) located at the center of gravity of piece 1. We now need to find the center of gravity of these two mass points and our knowledge of the physics involved makes us aware that it must be somewhere on the line connecting them. Only point D satisfies this condition.

Q7.23. Reason: The vertical and horizontal components of the center of gravity can be found independently. Since both centers of gravity given (for the upper body and lower body, separately) have the same vertical position, the total center of gravity must also have that vertical position. Thus point A is the position of the total center of gravity.
Assess: Note that the horizontal position would depend on the relative weight of the torso and legs. But since both have the same vertical position, we do not need to know relative weights to know the vertical position of the center of gravity for the whole body.

Q7.25. Reason: Assume the rods are both uniform. Then the downward gravitational force acts as if concentrated at the center of the rod. The moment of inertia for a rod swinging from an axis through one end is $I = \frac{1}{3}ML^2$. Use ratios so lots of factors cancel out.

$$\frac{\alpha_2}{\alpha_1} = \frac{\frac{\tau_2}{I_2}}{\frac{\tau_1}{I_1}} = \frac{\frac{r_2 F_2}{I_2}}{\frac{r_1 F_1}{I_1}} = \frac{\frac{(L)(2Mg)}{(\frac{1}{3}(2M)(2L)^2)}}{\frac{(L/2)(Mg)}{(\frac{1}{3}ML^2)}} = \frac{\frac{2}{8}}{\frac{1}{2}} = \frac{1}{2}$$

Because the ratio of the angular acceleration for rod 2 to rod 1 is less than 1 then the angular acceleration of rod 1 must be greater. The correct choice is A.
Assess: Note that the increase of the total mass of the disk needed to be included in the calculation.

Q7.27. Reason: The skater turns one-and-a-half revolutions in 0.5 s. One-and-a-half revolutions is 3π radians. Her angular velocity is

$$\omega = \frac{\Delta\theta}{\Delta t} = \frac{3\pi \text{ rad}}{0.5 \text{ s}} = 20 \text{ rad/s}$$

The correct choice is D.
Assess: This result is reasonable. She makes three revolutions in one second, which is 6π radians per second.

Q7.29. Reason: In Question 7.27 we found that the angular velocity of the skater is 20 rad/s. Estimating the length of her arm to be about 0.75 m from the center of rotation, we can calculate the speed of her hand.

$$v = \omega r = (20 \text{ rad/s})(0.75 \text{ m}) = 15 \text{ m/s}$$

The correct choice is D.
Assess: This is actually a high velocity, over 30 mph.

Problems

P7.1. Strategize: This problem deals with angular measurements in radians. Recall that 2π rad = 360°, and that angle measurements start at the $+x$ axis in standard Cartesian coordinates.
Prepare: The position of the minute hand is determined by the number after the colon. There are 60 minutes in an hour so the number of minutes after the hour, when divided by 60, gives the fraction of a circle which has been covered by the minute hand. Also, the minute hand starts at $\pi/2$ rad and travels clockwise, thus decreasing the angle. If we get a negative angle, we can make it positive by adding 2π rad.

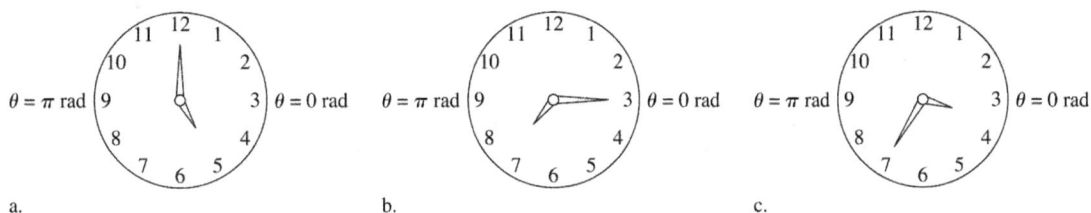

a.   b.   c.

Solve: (a) The angle is calculated as described above. Since the number after the colon is 0, we subtract nothing from $\pi/2$ rad, so $\theta = \pi/2$.
(b) We subtract 15/60 of 2π rad from the starting angle, so we have:

$$\theta = \frac{\pi}{2} - \left(\frac{15}{60}\right)(2\pi) = 0$$

(c) As before, the angle is given by:

$$\theta = \frac{\pi}{2} - \left(\frac{35}{60}\right)(2\pi) = -\frac{2}{3}\pi$$

Since this angle is negative, we can add 2π rad to obtain: $\theta = -2\pi/3 + 2\pi = 4\pi/3$.
Assess: The first two parts make sense from our experience with clocks. In part (a), the minute hand is straight up. In part (b), it points to the right.

P7.3. Strategize: This problem deals with angular speed in radians per second. Recall that 2π rad = 360°.
Prepare: To compute the angular speed ω we use the equation in the text and convert to rad/s. The minute hand takes an hour to complete one revolution.
Solve:

$$\omega = \frac{\Delta\theta}{\Delta t} = \frac{1.0 \text{ rev}}{60 \text{ min}}\left(\frac{2\pi \text{ rad}}{1 \text{ rev}}\right)\left(\frac{1 \text{ min}}{60 \text{ s}}\right) = 0.0017 \text{ rad/s} = 1.7 \times 10^{-3} \text{ rad/s}$$

Assess: This answer applies not just to the tip, but the whole minute hand. The answer is small, but the minute hand moves quite slowly.
The second hand moves 60 times faster, or 0.10 rad/s. This too seems reasonable.

P7.5. Strategize: This problem deals with distances on the circumference of a circle, and how they relate to an angular separation.

Prepare: The airplane is to be treated as a particle in uniform circular motion on the equator around the center of the earth. We show the following pictorial representation of the problem and a list of values. To convert radians into degrees, we note that 2π rad $= 360°$.

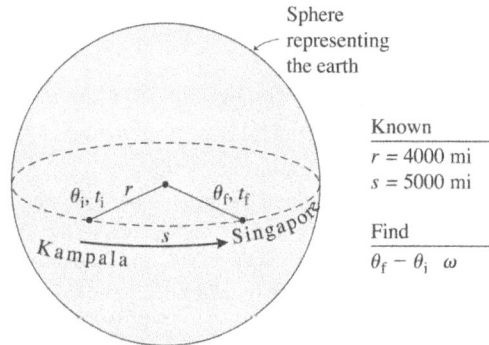

Solve: (a) The angle you turn through is

$$\theta_f - \theta_i = \frac{s}{r} = \frac{5000 \text{ mi}}{4000 \text{ mi}} = 1.25 \text{ rad} = 1.25 \text{ rad} \times \frac{180°}{\pi \text{ rad}} = 71.6°$$

So, the angle is 1.3 rad or $72°$.

(b) The plane's angular speed is

$$\omega = \frac{\theta_f - \theta_i}{t_f - t_i} = \frac{1.25 \text{ rad}}{9 \text{ h}} = 0.139 \text{ rad/h} = 0.139 \frac{\text{rad}}{\text{h}} \times \frac{1 \text{ h}}{3600 \text{ s}} = 3.9 \times 10^{-5} \text{ rad/s}$$

Assess: An angular displacement of approximately one-fifth of a complete rotation is reasonable because the separation between Kampala and Singapore is approximately one-fifth of the earth's circumference.

P7.7. Strategize: This problem involves angular displacement at a constant speed. We note that 1 rev $= 2\pi$ rad $= 360°$.

Prepare: We'll use the equation in the text to compute the angular displacement. We are given $\theta_i = 0.45$ rad and that $\Delta t = 8.0 \text{ s} - 0 \text{ s} = 8.0 \text{ s}$.

We'll do a preliminary calculation to convert $\omega = 78$ rpm into rad/s:

$$78 \text{ rpm} = 78 \frac{\text{rev}}{\text{min}} \left(\frac{2\pi \text{ rad}}{1 \text{ rev}} \right) \left(\frac{1 \text{ min}}{60 \text{ s}} \right) = 8.17 \text{ rad/s}$$

Solve: Solve the equation for θ_f:

$$\theta_f = \theta_i + \omega \, \Delta t = 0.45 \text{ rad} + (8.17 \text{ rad/s})(8.0 \text{ s}) = 65.8 \text{ rad} = 10.474 \times 2\pi \text{ rad}$$
$$= 10 \times 2\pi \text{ rad} + 0.474 \times 2\pi \text{ rad} = 10 \times 2\pi \text{ rad} + 2.98 \text{ rad}$$

So the speck completed almost ten and a half revolutions. An observer would say the angular position is 3.0 rad (to two significant figures) at $t = 8.0$ s.

Assess: Ask your grandparents if they remember the old records that turned at 78 rpm. They turned quite fast and so the music didn't last long before it was time to turn the record over.

Singles came on smaller records that turned at 45 rpm, and later "long play" (LP) records turned at 33 rpm.

CDs don't have a constant angular velocity, instead they are designed to have constant linear velocity, so the motor has to change speeds. For the old vinyl records the recording had to take into account the changing linear velocity because they had constant angular velocity.

P7.9. Strategize: This problem involves angular velocity and displacement. The angular acceleration is not constant, since the angular speed changes at the 10 s mark. This means we cannot use angular kinematic equations, unless we split it up into time intervals before and after the 10 s mark.

Prepare: To find angular displacement we simply subtract the initial value of the angle from the final value of the angle: $\Delta\theta = \theta_F - \theta_I$.

Solve: (a) At $t = 5$ s, the angle is 50 rad and at $t = 15$ s, the angle is 62.5 rad. Thus

$$\Delta\theta = 62.5 \text{ rad} - 50 \text{ rad} = 12.5 \text{ rad}$$

(b) The angular velocity of the wheel at 15 s is the slope of the θ vs. t graph at 15 s. We can find this slope by comparing the angle at 10 s and the angle at 20 s:

$$\omega = \frac{\theta_F - \theta_I}{t_F - t_I} = \frac{50 \text{ rad} - 75 \text{ rad}}{20 \text{ s} - 10 \text{ s}} = -2.5 \text{ rad/s}$$

(c) Angular velocity is related to translational velocity by $v = \omega R$ such that $v_{max} = \omega_{max}R = (75 \text{ rad/s})(0.15 \text{ m}) = 11 \text{ m/s}$.

Assess: The angular velocity is negative at $t = 15$ s because the angle is decreasing, that is, the wheel is rotating clockwise.

P7.11. Strategize: The smooth rotation assures constant angular velocity.

Prepare: This is essentially a conversion problem, in which we will use that 1 rev = 2π rad, and 1 min = 60 s. We will use $v = \omega r$ to relate angular and translational variables.

Solve: (a) The second hand has a rotational speed of one revolution per minute, or 1 rpm. We simply need to convert this to SI units.

$$1 \frac{\text{rev}}{\text{min}} = \left(1 \frac{\text{rev}}{\text{min}}\right)\left(\frac{2\pi \text{ rad}}{1 \text{ rev}}\right)\left(\frac{1 \text{ min}}{60 \text{ s}}\right) = 0.105 \frac{\text{rad}}{\text{s}}$$

(b) The tip of the hand has speed

$$v = \omega R = (0.105 \text{ rad/s})(0.0100 \text{ m}) = 0.00105 \text{ m/s}$$

Assess: This is about a millimeter per second, which is about right.

P7.13. Strategize: This problem involves rotational motion with a constant angular acceleration. This means we can use angular kinematic equations.

Prepare: We'll assume a constant acceleration during the one revolution. We'll use the second and third equations for circular motion in Synthesis 7.1., the third to find α and then the second to find ω_f.

Known
$r = \frac{1}{2}D = .90$ m
$\Delta t = 1.0$ s
$\Delta\theta = 1.0 \text{ rev} = 2\pi$ rad
Find
$v_f = r\omega_f$

Solve: Using Synthesis 7.1 allows us to solve for α. That $\omega_0 = 0$ makes it easier.

$$\Delta\theta = \frac{1}{2}\alpha(\Delta t)^2$$

$$\alpha = \frac{2\Delta\theta}{(\Delta t)^2} = \frac{2(2\pi \text{ rad})}{(1.0 \text{ s})^2} = 12.6 \text{ rad/s}^2$$

Synthesis 7.1 gives ω_f:

$$\omega_f = \omega_0 + \alpha\Delta t = 0 \text{ rad/s} + (12.6 \text{ rad/s}^2)(1.0 \text{ s}) = 12.6 \text{ rad/s}$$

Finally, we compute $v_f = r\omega_f = (.90 \text{ m})(12.6 \text{ rad/s}) = 11 \text{ m/s}$.

Assess: This speed seems reasonable, about 1/4 of a baseball fast pitch. The hammer throw is similar to the discus, but the weight is on a wire so the radius of the circular motion is a bit longer than the arm and the release speed is a bit larger, hence the distance it goes before landing is a few meters more.

P7.15. Strategize: This problem involves rotational motion with a constant angular acceleration. This means we can use angular kinematic equations.
Prepare: The motion has two parts: the disk accelerates for some time until it reaches a constant angular velocity and then rotates at this constant angular velocity for the remainder of the time. We can use the equations in Synthesis 7.1 to calculate the number of turns the disk has made.
Solve: One additional significant figure has been kept in each of the intermediate results. The disk starts from rest, so $\omega_i = 0$ rad/s. The disk's final angular velocity converted to rad/s is

$$\omega_f = 7200 \text{ rpm} = \left(\frac{7200 \text{ rev}}{\text{min}}\right)\left(\frac{2\pi \text{ rad}}{\text{rev}}\right)\left(\frac{1 \text{ min}}{60 \text{ s}}\right) = 754 \text{ rad/s}$$

We can find the time the disk takes to accelerate to this angular velocity using Synthesis 7.1.

$$\Delta t_{\text{accelerating}} = \frac{\omega_f - \omega_i}{\alpha} = \frac{754 \text{ rad/s} - 0 \text{ rad/s}}{190 \text{ rad/s}^2} = 3.97 \text{ s}$$

Using Synthesis 7.1., the total angular displacement of the disk during this time is

$$\Delta\theta_{\text{accelerating}} = \frac{1}{2}\alpha(\Delta t_{\text{accelerating}})^2 = \frac{1}{2}(190 \text{ rad/s}^2)(3.97 \text{ s})^2 = 1500 \text{ rad}$$

After it has reached its final angular velocity ω_f, the disk spins with that angular velocity for the remainder of the time. The angular displacement during this time is given by Synthesis 7.1.

$$\Delta\theta_{\text{constant } \omega} = \omega_f\Delta t_{\text{constant } \omega} = (754 \text{ rad/s})(10.0 \text{ s} - 3.97 \text{ s}) = 4550 \text{ rad}$$

The total angular displacement of the disk is $\Delta\theta = 1500 \text{ rad} + 4550 \text{ rad} = 6050 \text{ rad}$.
Converting this result to revolutions, we have $\Delta\theta = 960$ revolutions.
Assess: It's important to realize that there are two different parts to this motion.

P7.17. Strategize: This problem involves rotational motion with a constant angular acceleration. This means we can use angular kinematic equations.
Prepare: We assume constant angular acceleration; then we can use Synthesis 7.1.

Known
$\Delta t = 2.0 \text{ s}$
$\omega_0 = 0$
$\omega_f = 3000 \text{ rpm}$
Find
α
$\Delta\theta$

Convert $\Delta\omega$ to rad/s.

$$\Delta\omega = \omega_f - \omega_0 = 3000 \frac{\text{rev}}{\text{min}}\left(\frac{2\pi \text{ rad}}{1 \text{ rev}}\right)\left(\frac{1 \text{ min}}{60 \text{ s}}\right) = 314 \text{ rad/s}$$

Solve: (a)

$$\alpha = \frac{\Delta\omega}{\Delta t} = \frac{314 \text{ rad/s}}{2.0 \text{ s}} \approx 160 \text{ rad/s}^2$$

(b) We'll use Synthesis 7.1.2, using $\omega_0 = 0$.

$$\Delta\theta = \omega_0 \Delta t + \frac{1}{2}\alpha(\Delta t)^2 = \frac{1}{2}(157 \text{ rad/s}^2)(2.0 \text{ s})^2 = 314 \text{ rad}$$

Finally, convert to revolutions:

$$314 \text{ rad} = 314 \text{ rad}\left(\frac{1 \text{ rev}}{2\pi \text{ rad}}\right) = 50 \text{ rev}$$

Assess: 50 rev seems like a lot in 2 s, but it is reasonable with the large angular acceleration and the final angular velocity of 3000 rpm.

P7.19. Strategize: This problem involves the relationship between force, lever arm, and torque.
Prepare: The magnitude of the torque in each case is $\tau = rF$ because $\sin\phi = 1$.
Solve: $\quad \tau_1 = rF \qquad \tau_2 = r2F \qquad \tau_3 = 2rF \qquad \tau_4 = 2r2F$
Examining the above we see that $\tau_1 < \tau_2 = \tau_3 < \tau_4$. Since for each case $\tau = rF$ (because $\sin\phi = 1$), in order to determine the torque we have just kept track of each force (F), the magnitude of the position vector (r) which locates the point of application of the force, and finally the product (rF).
Assess: As expected, both the force and the lever arm contribute to the torque. Larger forces and larger lever arms make larger torques. Case 4 has both the largest force and the largest lever arm, hence the largest torque.

P7.21. Strategize: This problem involves the sum of torques on a rigid object.
Prepare: Torque by a force is defined as $\tau = Fr\sin\phi$ [Equation 7.11], where ϕ is measured counterclockwise from the $\vec{r}$ vector to the $\vec{F}$ vector. The radial line passing through the axis of rotation is shown below by the broken line. We see that the 20 N force makes an angle of $+90°$ relative to the radius vector r_2, but the 30 N force makes an angle of $-90°$ relative to r_1.

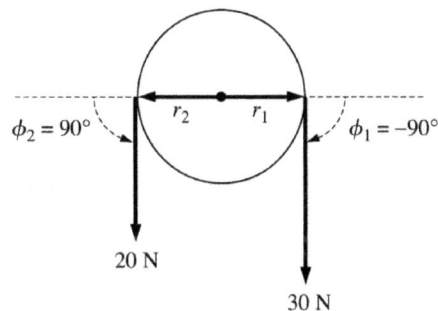

Solve: The net torque on the pulley about the axle is the torque due to the 30 N force plus the torque due to the 20 N force:

$$(30 \text{ N})r_1 \sin\phi_1 + (20 \text{ N})r_2 \sin\phi_2 = (30 \text{ N})(0.02 \text{ m})\sin(-90°) + (20 \text{ N})(0.02 \text{ m})\sin(90°)$$
$$= (-0.60 \text{ N}\cdot\text{m}) + (0.40 \text{ N}\cdot\text{m}) = -0.20 \text{ N}\cdot\text{m}$$

Assess: A negative torque will cause a clockwise rotation of the pulley.

P7.23. Strategize: This problem involves the sum of all torques equaling zero, in which case the rigid object does not experience any angular acceleration.
Prepare: We can use Equations 7.11 and 7.9 to calculate the torques due to each force and then set the net torque to zero to solve for F_2.

Solve: Using Equation 7.10 to calculate the torque due to F_1 with $\phi_1 = 45°$,

$$\tau_1 = +r_1 F_1 \sin \phi_1 = r_1(20.0 \text{ N}) \sin 45°$$

At this point, we can't calculate the numerical value of this torque, since r_1 is not given. Since F_2 is perpendicular to the radial line the torque due to F_2 is

$$\tau_2 = -r_2(F_2) = -r_2 F_2 = -\left(\frac{r_1}{2}\right) F_2$$

A negative sign has been inserted since F_2 tends to turn the rod clockwise.

Setting the net torque to zero gives the equation

$$\tau_1 + \tau_2 = r_1(20.0 \text{ N}) \sin 45° - \left(\frac{r_1}{2}\right) F_2 = 0$$

Solving for F_2 gives

$$F_2 = 2(20.0 \text{ N}) \sin 45° = 28.3 \text{ N}$$

Assess: This answer makes sense. F_1 is twice as far from the pivot but its contribution to the torque is diminished by almost a factor of two due to the angle it makes with the radial line. The force F_2 should be a larger than the force F_1.

P7.25. Strategize: This problem involves addition of torques. Torque can be caused by any force, although the careful attention must be paid to the radial vector and the angle between the radial vector and the force. Some forces may produce no torque.

Prepare: Assume the bar is weightless. We can calculate the torques due to both forces using Equation 7.9.

Solve: Refer to the diagram below.

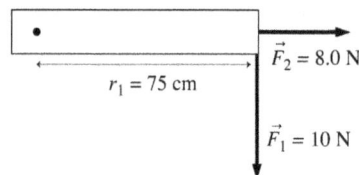

For F_1

$$\tau_1 = r_1(F_1)_\perp = (0.75 \text{ m})(10 \text{ N}) = 7.5 \text{ N} \cdot \text{m}$$

This torque is positive since F_1 tends to turn the bar counterclockwise.

For F_2

$$\tau_2 = r_2(F_2)_\perp = (0.75 \text{ m})(0 \text{ N}) = 0 \text{ N} \cdot \text{m}$$

The net torque about the pivot is $\tau_1 + \tau_2 = 7.5 \text{ N} \cdot \text{m}$.

Assess: The torque due to F_2 is zero because it has no component perpendicular to the radial line.

P7.27. Strategize: This problem involves addition of torques. Torque can be caused by any force, although the careful attention must be paid to the radial vector and the angle between the radial vector and the force. Some forces may produce no torque.

Prepare: Knowing that torque may be determined by $\tau = rF_\perp$, that counterclockwise torque is positive and clockwise torque negative we can determine the net torque acting on the bar.

Solve:

$$\tau_{\text{clockwise}} = -(0.25 \text{ m})(8.0 \text{ N}) = -2.0 \text{ N} \cdot \text{m}$$
$$\tau_{\text{counterclockwise}} = (0.75 \text{ m})(10 \text{ N}) = 7.5 \text{ N} \cdot \text{m}$$
$$\tau_{\text{net}} = \tau_{\text{counterclockwise}} + \tau_{\text{clockwise}} = 7.5 \text{ N} \cdot \text{m} - 2.0 \text{ N} \cdot \text{m} = 5.5 \text{ N} \cdot \text{m}$$

Since the net torque is $+5.5$ N·m, the bar will rotate in the counterclockwise direction around the dot.

Assess: The counterclockwise torque had both the larger r and the larger F, so the net torque was also counterclockwise. The numbers also seem reasonable, and the units work out.

P7.29. Strategize: This problem involves addition of torques. Torque can be caused by any force, although the careful attention must be paid to the radial vector and the angle between the radial vector and the force.
Prepare: We can use Equation 7.9 to calculate the net torque on the bar. Recall that clockwise torques are negative and counterclockwise torques are positive.
Solve: Refer to the diagram below.

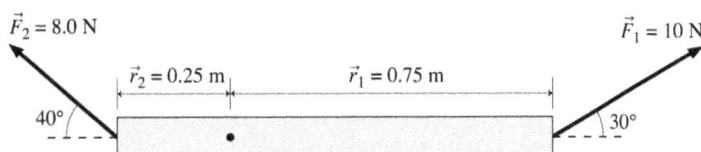

Calculate the torque associated with F_1

$$\tau_1 = r_1(F_1)_\perp = r_1 F_1 \sin(\phi_1) = (0.75 \text{ m})(10 \text{ N})\sin(30°) = 3.75 \text{ N} \cdot \text{m}$$

An additional significant figure has been kept in the intermediate result above.
Calculate the torque associated with F_2

$$\tau_2 = -r_2(F_2)_\perp = -r_2 F_2 \sin(\phi_2) = -(0.25 \text{ m})(8.0 \text{ N})\sin(40°) = -1.29 \text{ N} \cdot \text{m}$$

The net torque is

$$\tau_1 + \tau_2 = 3.75 \text{ N} \cdot \text{m} - 1.29 \text{ N} \cdot \text{m} = +2.46 \text{ N} \cdot \text{m}$$

Since all information is given to two significant figures, the result should be reported to two significant figures as

$$\tau_{\text{net}} = 2.5 \text{ N} \cdot \text{m}$$

Assess: This result makes sense. Force F_1 is at a much larger distance from the axis than F_2, while both forces are close in magnitude and in the angle they make with the radial line. The net torque should act to turn the bar counterclockwise.

P7.31. Strategize: This problem involves torque due to gravity. In such a problem, the force of gravity can be treated as though it acts at a single point: the center of gravity.
Prepare: First let's divide the object into two parts. Let's call part #1 the part to the left of the point of interest and part #2 the part to the right of the point of interest. Next using our sense of center of gravity, we know the center of mass of part #1 is at 12.5 m and the center of mass of part #2 is at +37.5 cm. We also know the mass of part #1 is one fourth the total mass of the object and the mass of part #2 is three fourths the total mass of the object. Finally, we can determine the gravitational torque of each part using any of the three expressions for torque as shown below:
Equation 7.3 $\tau = rF_\perp$ is straightforward to use because the forces are perpendicular to the position vectors, which locate the point of application of the force.
Equation 7.4 $\tau = r_\perp F$ is straightforward to use because the position vectors that locate the point of application of the forces are also the moment arms for the forces.
Equation 7.4 $\tau = rF \sin \phi$ is straightforward to use because the angles are either $90°$ or $270°$.

Solve: Using Equation 7.4 we obtain the following

$$\tau_{net} = \tau_1 + \tau_2 = m_1 g r_1 \sin(90°) + m_2 g r_2 \sin(-90°)$$

$$= (0.5 \text{ kg})(9.8 \text{ m/s}^2)(-0.125 \text{ m})(1) + (0.75 \text{ kg})(9.8 \text{ m/s}^2)(0.375 \text{ m})(-1) = -2.1 \text{ N} \cdot \text{m}$$

Assess: According to this answer, if released the object should rotate in a clockwise direction. Looking at the figure this is exactly what we would expect to happen. It might be easier to consider the whole rod as if acting at its center of mass and then the distance would be just 25 cm and there would be only one torque calculation.

P7.33. Strategize: This problem involves addition of torques, and the torques are caused by the force of gravity. In such problems, the force of gravity can be treated as though it acts at a single point for each object: the object's center of gravity.
Prepare: Model the arm as a uniform rigid rod. Its mass acts at the center of mass. The force in each case is the weight, $w = mg$.

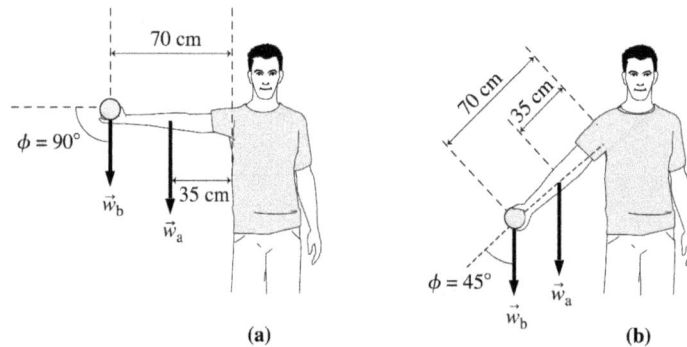

(a)   (b)

Solve: (a) The torque is due both to the weight of the ball and the weight of the arm.

$$\tau = \tau_{ball} + \tau_{arm} = r_b(m_b g)\sin 90° + r_a(m_a g)\sin 90°$$
$$= (0.70 \text{ m})(3.0 \text{ kg})(9.80 \text{ m/s}^2) + (0.35 \text{ m})(4.0 \text{ kg})(9.80 \text{ m/s}^2) = 34 \text{ N} \cdot \text{m}$$

(b) The torque is reduced because the moment arms are reduced. Both forces act at $\phi = 45°$ from the radial line, so

$$\tau = \tau_{ball} + \tau_{arm} = r_b(m_b g)\sin 45° + r_a(m_a g)\sin 45°$$
$$= (0.70 \text{ m})(3.0 \text{ kg})(9.80 \text{ m/s}^2)(0.707) + (0.35 \text{ m})(4.0 \text{ kg})(9.80 \text{ m/s}^2)(0.707) = 24 \text{ N} \cdot \text{m}$$

Assess: This problem could also have been done by first finding the center of mass of the arm-ball system and computing the torque due to the combined weight acting at that point. The final answers would be the same.

P7.35. Strategize: This problem involves the center of gravity of a collection of three uniformly dense particles. We will treat all the mass as though it is concentrated at the center of the spheres shown.
Prepare: The procedure in Tactics Box 7.1 can be used to calculate the center of gravity of the coins.
Solve: Refer to the figure below.

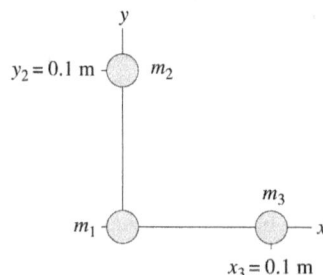

The coordinates for the three masses are

$$x_1 = 0 \text{ m} \qquad\qquad y_1 = 0 \text{ m}$$
$$x_2 = 0 \text{ m} \qquad\qquad y_2 = 0.100 \text{ m}$$
$$x_3 = 0.100 \text{ m} \qquad y_3 = 0 \text{ m}$$

All the coins have the same mass since they are identical. The x- and y-coordinates of the center of gravity of the coins may be determined by

$$x_{cg} = \frac{x_1 m_1 + x_2 m_2 + x_3 m_3}{m_1 + m_2 + m_3} = \frac{x_1 m + x_2 m + x_3 m}{m + m + m} = \frac{x_3 m}{3m} = \frac{1}{3} x_3 = \frac{1}{3}(0.100 \text{ m}) = 0.0333 \text{ m}$$

$$y_{cg} = \frac{y_1 m_1 + y_2 m_2 + y_3 m_3}{m_1 + m_2 + m_3} = \frac{y_1 m + y_2 m + y_3 m}{m + m + m} = \frac{y_2 m}{3m} = \frac{1}{3} y_2 = \frac{1}{3}(0.100 \text{ m}) = 0.0333 \text{ m}$$

Assess: Most of the mass lies near the x- and y-axes, so this answer makes sense.

P7.37. Strategize: This problem involves the center of gravity of objects with uniform mass distribution.
Prepare: Because the three nickels are identical and because they have uniform mass distribution, we expect their geometric center to be the same as their center of gravity. But we will calculate this using $y_{cg} = \frac{1}{m_{tot}} \sum_i m_i y_i$, where we are calling the vertical direction (above the table) the $+y$ direction.

Solve: $y_{cg} = \frac{1}{3m} \sum_i m y_i = \frac{1}{3}\left(\frac{1.95\times10^{-3} \text{ m}}{2} + \frac{3\left(1.95\times10^{-3} \text{ m}\right)}{2} + \frac{5\left(1.95\times10^{-3} \text{ m}\right)}{2} \right) = 2.93 \text{ mm}$

Assess: As our intuition suggested, this is exactly half-way up a stack of total height $3\left(1.95\times10^{-3} \text{ m}\right) = 5.85 \text{ mm}$.

P7.39. Strategize: This problem involves calculating the moment of inertia of a rigid object, in this case a thin spherical shell.
Prepare: We look up the equation for the moment of inertia of a thin spherical shell in Table 7.1. The radius is half the diameter.
Solve:

$$I = \tfrac{2}{3} M R^2 = \tfrac{2}{3}(0.0027 \text{ kg})(0.020 \text{ m})^2 = 7.2\times10^{-7} \text{ kg}\cdot\text{m}^2$$

Assess: The answer is small, but then again, it isn't hard to start a table tennis ball rotating or stop it from doing so. By the way, this calculation can be done in one's head without a calculator by writing the data in scientific notation and mentally keeping track of the significant figures:

$$I = \frac{2}{3}(2.7\times10^{-3} \text{ kg})(2.0\times10^{-2} \text{ m})^2 = \frac{2}{3}(27\times10^{-4} \text{ kg})(2\times10^{-2} \text{ m})^2 = \frac{27}{3}\cdot 2\cdot 2^2 \times10^{-4}\times10^{-4} \text{ kg}\cdot\text{m}^2$$
$$= 9\cdot 8\times10^{-8} \text{ kg}\cdot\text{m}^2 = 72\times10^{-8} \text{ kg}\cdot\text{m}^2 = 7.2\times10^{-7} \text{ kg}\cdot\text{m}^2$$

P7.41. Strategize: This problem involves the moments of inertia for two different objects. Consider how the mass is distributed differently in the two objects.
Prepare: We can obtain expressions for the moment of inertia for a cylinder and a sphere from the table in the text. Given that the masses and the moments of inertia are the same, we can obtain a relationship between their radii and then determine the radius of the sphere.
Solve: Knowing that the moment of inertia of the cylinder is equal to the moment of inertia of the sphere we may write

$$I_{cylinder} = I_{sphere}$$

Inserting expressions for the moment of inertia from Table 7.1.

$$M_c R_c^2 / 2 = 2 M_s R_s^2 / 5$$

Cancel the masses (recall they are equal) and solve for R_s

$$R_s = (5/4)^{1/2} R_c = 4.5 \text{ cm}$$

Assess: Knowing that the masses are equal and that the coefficient for the cylinder is 0.50 and for the sphere 0.40, we should expect the radius of the sphere to be greater than the radius of the cylinder in order for them to have the same moment of inertia.

P7.43. Strategize: This problem involves the moment of inertia of a rigid object that can be modeled approximately as a ring/hoop.
Prepare: Treating the bicycle rim as a hoop, we can use the expression given in Table 7.1 for the moment of inertia of a hoop. Manipulate this expression to obtain the mass.
Solve: The mass of the rim is determined by

$$m = I/R^2 = 0.19 \text{ kg} \cdot \text{m}^2/(0.65 \text{ m}/2)^2 = 1.8 \text{ kg}$$

Assess: Note that the units reduce to kg as expected. This amount seems a little heavy, but since we are not told what type of bicycle it is not an unreasonable amount.

P7.45. Strategize: Torque can cause an angular acceleration. The moment of inertia describes how difficult it is to cause an angular acceleration for a particular rigid body. This problem deals with the relationship between torque, moment of inertia, and angular acceleration.
Prepare: We can calculate the angular acceleration using Equation 7.21.
Solve: Using Equation 7.21 we find

$$\tau = I\alpha = (14.0 \times 10^{-5} \text{ kg} \cdot \text{m}^2)(150 \text{ rad/s}^2) = 6.0 \times 10^{-3} \text{ N} \cdot \text{m}$$

Assess: This result is reasonable. The moment of inertia of the grinding wheel is small.

P7.47. Strategize: This problem explores the relationship between torque, angular acceleration, and the moment of inertia.
Prepare: We are given the moment of inertia I and the angular acceleration α of the object. Since $\tau = I\alpha$ is the rotational analog of Newton's second law $F = ma$, we can use this relation to find the net torque on the object.
Solve: $\tau = (2.0 \text{ kg} \cdot \text{m}^2)(4.0 \text{ rad/s}^2) = 8.0 \text{ kg} \cdot \text{m}^2/\text{s}^2 = 8.0 \text{ N} \cdot \text{m}$
Assess: The units are correct and the relatively small torque magnitude is consistent with the size of the moment of inertia and the angular acceleration.

P7.49. Strategize: This problem involves torque causing an angular acceleration to a rigid body for which we can calculate the moment of inertia.
Prepare: A circular plastic disk rotating on an axle through its center is a rigid body. Assume the axis is perpendicular to the disk. Since $\tau = I\alpha$ is the rotational analog of Newton's second law $F = ma$, we can use this relation to find the net torque on the object. To determine the torque (τ) needed to take the plastic disk from $\omega_i = 0$ rad/s to $\omega_f = 1800$ rpm $= (1800)(2\pi)/60$ rad/s $= 60\pi$ rad/s in $t_f - t_i = 4.0$ s, we need to determine the angular acceleration (α) and the disk's moment of inertia (I) about the axle in its center. The radius of the disk is $R = 10.0$ cm.
Solve: We have

$$I = \tfrac{1}{2}MR^2 = \tfrac{1}{2}(0.200 \text{ kg})(0.10 \text{ m})^2 = 1.0 \times 10^{-3} \text{ kg} \cdot \text{m}^2$$

$$\omega_f = \omega_i + \alpha(t_f - t_i) \Rightarrow \alpha = \frac{\omega_f - \omega_i}{t_f - t_i} = \frac{60\pi \text{ rad/s} - 0 \text{ rad/s}}{4.0 \text{ s}} = 15\pi \text{ rad/s}^2$$

Thus, $\tau = I\alpha = (1.0 \times 10^{-3} \text{ kg} \cdot \text{m}^2)(15\pi \text{ rad/s}^2) = 0.047 \text{ N} \cdot \text{m}$.

Assess: The solution to this problem required a knowledge of torque, moment of inertia, rotational dynamics and rotational kinematics. You should consider it an accomplishment to have mastered these concepts and then combined them to solve a problem.

P7.51. Strategize: This problem involves a net torque producing an angular acceleration for a rigid body. We have enough information to determine the moment of inertia of the rigid body (the pulley).

Prepare: We'll apply the rotational version of Newton's second law. We'll write the net torque as $\Sigma \tau$ to emphasize that we are summing the two given torques; the 12 N force is producing a positive (counterclockwise) torque, while the 10 N force is producing a negative (clockwise) torque. For each torque $R = 0.30$ m.

We will assume that the rope comes off tangent to the pulley on each side, so that $\phi = 90°$ and $\sin \phi = 1$.

Looking up the formula for I of a cylinder in Table 7.1, and using $M = 0.80$ kg, gives

$$I = \tfrac{1}{2}MR^2 = \tfrac{1}{2}(0.80 \text{ kg})(0.30 \text{ m})^2 = 0.036 \text{ kg} \cdot \text{m}^2$$

Solve:

$$\alpha = \frac{\Sigma \tau}{I} = \frac{(0.30 \text{ m})(12 \text{ N}) - (0.30 \text{ m})(10 \text{ N})}{0.036 \text{ kg} \cdot \text{m}^2} = \frac{(0.30 \text{ m})(12 \text{ N} - 10 \text{ N})}{0.036 \text{ kg} \cdot \text{m}^2} = 17 \text{ rad/s}^2$$

Assess: This result answers the question. The proper units cancel to give α in rad/s^2.

Notice that the specific angles the ropes make with the vertical do not matter, as long as they are exerting torques in opposite directions and coming off of the pulley tangentially.

P7.53. Strategize: This problem involves torque from a known force (friction) causing an angular acceleration. We will assume the angular acceleration is constant, such that we can use angular kinematic equations.

Prepare: Equation 7.21 can be used to calculate the frictional torque once the angular acceleration is known. Using Synthesis 7.1 we can calculate the angular acceleration.

Solve: The wheel stops spinning in 12 seconds from an initial angular velocity of 0.72 revolutions per second. The initial angular velocity in radians per second is

$$\omega_i = 0.72 \text{ rev/s} = (0.72 \text{ rev/s})\left(\frac{2\pi \text{ rad}}{\text{rev}}\right) = 4.5 \text{ rad/s}$$

The angular acceleration of the wheel is

$$\alpha = \frac{\Delta \omega}{\Delta t} = \frac{-4.5 \text{ rad/s}}{12 \text{ s}} = -0.38 \text{ rad/s}^2$$

The only force acting on the wheel is friction. Using the results above in Equation 7.21 the *magnitude* of the frictional torque is

$$\tau = I\alpha = (0.30 \text{ kg} \cdot \text{m}^2)(0.38 \text{ rad/s}^2) = 0.11 \text{ N} \cdot \text{m}$$

Assess: This is a relatively large amount of friction. Assuming the torque of the wheel bearings acts at a radius of around a centimeter the frictional force exerted by the bearings is over 10 N. This shows you just one of the problems you will have if you do not properly maintain your bicycle.

P7.55. Strategize: This problem involves a torque from a known force causing an angular acceleration of a rigid body. Because the applied force (and hence torque) is constant, we know the angular acceleration should be constant. Thus we can use angular kinematic equations.

Prepare: We can use the equations in Synthesis 7.1., but it is also clear that we will need to do a preliminary calculation to get α. The string will come off tangentially so that $\phi = 90°$. We will also interpret "to get the top spinning" as $\omega_0 = 0$. The radius is half the diameter.

Known
$r = 0.025$ m
$F = 0.30$ N
$\phi = 90°$
$I = 3.0 \times 10^{-5}$ kg·m²
$\Delta\theta = 5.0$ rev=31.4 rad
$\omega_o = 0$

Find
α (Preliminary)
Δt

$$\alpha = \frac{\tau_{\text{net}}}{I} = \frac{rF \sin \phi}{I} = \frac{(0.025 \text{ m})(0.30 \text{ N})(1)}{3.0 \times 10^{-5} \text{ kg} \cdot \text{m}^2} = 250 \text{ rad/s}^2$$

Solve: Solve the angular displacement equation for circular motion Synthesis 7.1 (with $\omega_0 = 0$) for Δt:

$$\Delta \theta = \tfrac{1}{2}\alpha(\Delta t)^2$$

$$\Delta t = \sqrt{\frac{2\Delta\theta}{\alpha}} = \sqrt{\frac{2(31.4 \text{ rad})}{250 \text{ rad/s}^2}} = 0.50 \text{ s}$$

Assess: As you may know from personal experience, it doesn't take long for a toy top to complete five revolutions, and this bears that out.

P7.57. Strategize: This problem involves rolling without slipping. This means that the segment of tire that makes contact with the road is momentarily at rest relative to the road.
Prepare: We can use the rolling constraint to find the speed of the dot and the angular speed of the tires.
Solve: (a) Since the tires are rolling without slipping the angular velocity of the tires is given by Equation 7.24.

$$\omega = \frac{v_{\text{cm}}}{R} = \frac{5.6 \text{ m/s}}{0.40 \text{ m}} = 14 \text{ rad/s}$$

(b) The blue dot is undergoing translation and rotation. At the top of the tire, it has a translational velocity equal to the speed of the bike, and an additional velocity equal to ωR due to the rotation of the tire. See Figure 7.41. The speed of the dot at this point is

$$v = v_{\text{cm}} + \omega R = 2v_{\text{cm}} = 2(5.6 \text{ m/s}) = 11 \text{ m/s}$$

(c) See the diagram below.

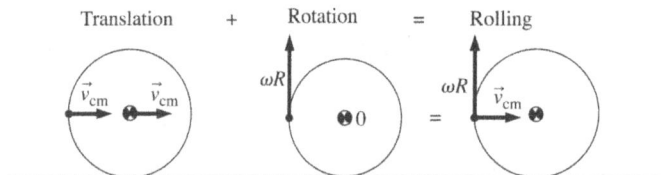

The dot has a translational velocity equal to the velocity of the center of mass of the tire in the horizontal direction. The tangential velocity of the dot is in the vertical direction and has a magnitude equal to ωR. The velocity of the dot is equal to the sum of these two vectors. Since the tire is rolling without slipping, $v_{\text{cm}} = \omega R$.
See the vector diagram below.

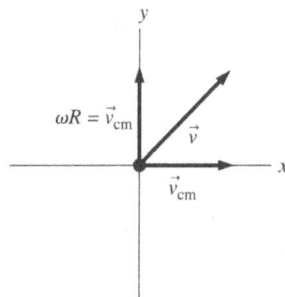

The speed of the dot is equal to

$$v = \sqrt{(v_{\text{cm}})^2 + (v_{\text{cm}})^2} = 7.9 \text{ m/s}$$

Assess: Note that the speed of the dot during the downward motion for part (c) is equal to the speed of the dot during the upward motion shown in the diagram.

P7.59. Strategize: This problem involves the center of gravity, and also involves rolling without slipping. In particular, it involves the fact that the center of gravity of a rolling wheel does not move (relative to the ground) at the same speed as a point on the top of the wheel or the bottom of the wheel.

Prepare: Both the slab and the rollers move to the right, but not at the same speed. The initial distance between the center of gravity of the slab and the right-most roller is 1.0 m.

Solve: Because the speed of the top of a wheel is twice the speed of the center of mass, the slab moves at twice the translational speed of the rollers. In the same time the slab moves 2.0 m the rollers on the right will move 1.0 m. At that point the center of gravity will have caught up with the right-most roller. So the slab moves 2.0 m.

Assess: This matches experience.

P7.61. Strategize: This problem involves a graphical understanding of angular variables, specifically relating angular speed to angular position.

Prepare: The particle is moving in a circle and a motion of 2_π radians corresponds to one full rotation. Angular velocity at any given time is defined as $\omega = \Delta\theta / \Delta t$, or the slope of the angular position-versus-time graph.

Solve: (a) From $t = 0$ s to $t = 1$ s the particle rotates clockwise from the angular position $+4\pi$ rad to -2π rad. Therefore, $\Delta\theta = -2\pi - (+4\pi) = -6\pi$ rad in one sec, or $\omega = -6\pi$ rad/s. From $t = 1$ s to $t = 2$ s, $\omega = 0$ rad/s. From $t = 2$ s to $t = 4$ s the particle rotates counterclockwise from the angular position -2π rad to 0 rad. Thus $\Delta\theta = 0 - (-2\pi) = 2\pi$ rad rad and $\omega = +\pi$ rad/s.

(b)

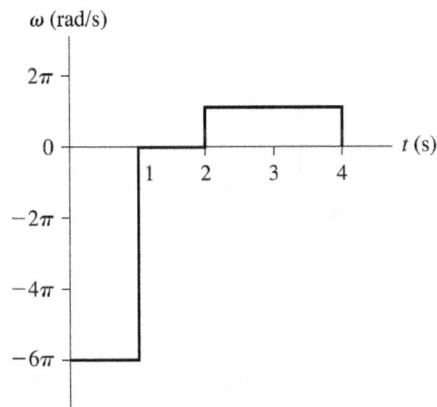

Assess: Since we take positive angular displacements as counterclockwise and negative angular displacements clockwise, we know the particle is traveling around the circle in a clockwise direction. Knowing that the slope of the angular position-versus-time plot is the angular velocity, we can establish a plot of angular velocity-versus-time.

P7.63 Strategize: This problem deals with the relationship between angular variables and linear variables, which are connected through the radius of the rolling wheel.

Prepare: Knowing the kinematic equations and the fact that the distance the car travels is some number of revolutions (or circumferences) of the tire, we can solve this problem.

Solve: The acceleration of the car may be obtained from the expression:

$$v = v_o + at \text{ which gives } a = \frac{v - v_o}{t} = \frac{v}{t} \text{ since } v_o = 0 \text{ m/s}$$

The distance traveled by the car during the time it is accelerating may be determined by:

$$2a\Delta x = v^2 - v_o^2 \text{ which gives } \Delta x = \frac{v^2 - v_o^2}{2a} = \frac{v^2}{2a} = \frac{v^2}{2(v/t)} = \frac{vt}{2}$$

Finally, the number of times the tires rotate (i.e. the number of circumferences of the tires) may be determined by:

$$\Delta x = N(2\pi r) = N\pi d \text{ which gives } N = \frac{\Delta x}{\pi d} = \frac{(vt/2)}{\pi d} = \frac{vt}{2\pi d} = \frac{(20 \text{ m/s})(10 \text{ s})}{2[\pi(0.58 \text{ m})/\text{rotation}]} = 55 \text{ rotations}$$

Assess: As with many kinematics problems, we can check our work by approaching the problem in a different manner. For example, since the acceleration is constant, the distance the car travels is just the average velocity times the time of travel. This may be expressed as follows:

$$\Delta x = v_{ave}t = \left(\frac{v + v_o}{2}\right)t = \frac{vt}{2}$$ which is the same as the expression obtained above for the distance traveled.

Note also that the final units are in rotations and that 55 is a reasonable number of rotations for a tire in 10 seconds.

P7.65. Strategize: This problem involves determining torques caused by various forces with different lever arms.
Prepare: The disk is a rotating rigid body and it rotates on an axle through its center. We will use Equation 7.11 to find the net torque.

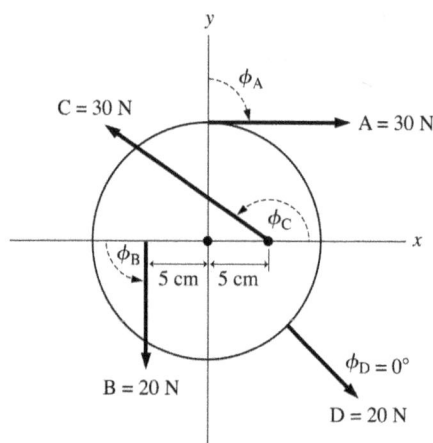

Solve: The net torque on the axle is

$$\tau = F_A r_A \sin\phi_A + F_B r_B \sin\phi_B + F_C r_C \sin\phi_C + F_D r_D \sin\phi_D$$

$$= (30 \text{ N})(0.10 \text{ m}) \sin(-90°) + (20 \text{ N})(0.05 \text{ m}) \sin 90° + (30 \text{ N})(0.05 \text{ m}) \sin 135° + (20 \text{ N})(0.10 \text{ m}) \sin 0°$$

$$= -3 \text{ N} \cdot \text{m} + 1 \text{ N} \cdot \text{m} + 1.0607 \text{ N} \cdot \text{m} = -0.94 \text{ N} \cdot \text{m}$$

Assess: A negative net torque means a clockwise acceleration of the disk.

P7.67. Strategize: This problem involves the center of gravity of a compound object (a person).
Prepare: Equation 7.15 tells us the center of gravity of a compound object. If we take all the rest of the body other than the arms as one object (call it the trunk, even though it includes head and legs) then we can write

$$y_{cg} = \frac{y_{trunk} m_{trunk} + 2y_{arm} m_{arm}}{M}$$

Where $M = m_{trunk} + m_{arm} = 70 \text{ kg}$ (the mass of the whole body).

The language "by how much does he raise his center of gravity" makes us think of writing Δy_{cg}.

Since we have modeled the arm as a uniform cylinder 0.75 m long, its own center of gravity is at its geometric center, 0.375 m from the pivot point at the shoulder. So raising the arm from hanging down to straight up would change the height of the center of gravity of the arm by twice the distance from the pivot to the center of gravity: $(\Delta y_{cg})_{arm} = 2(0.375 \text{ m}) = 0.75 \text{ m}$.

Solve:

$$\Delta(y_{cg})_{body} = (y_{cg})_{\text{with arms up}} - (y_{cg})_{\text{with arms down}} = \frac{(y_{cg})_{trunk} m_{trunk} + 2(y_{cg})_{arm, up} m_{arm}}{M} - \frac{(y_{cg})_{trunk} m_{trunk} + 2(y_{cg})_{arm, down} m_{arm}}{M}$$

$$= \frac{2m_{arm}}{M}((y_{cg})_{arm, up} - (y_{cg})_{arm, down}) = \frac{2m_{arm}}{M}(\Delta y_{cg})_{arm} = \frac{2(3.5\text{ kg})}{70\text{ kg}}(0.75\text{ m}) = 0.075\text{ m} = 7.5\text{ cm}$$

Assess: 7.5 cm seems like a reasonable amount, not a lot, but not too little. The trunk term subtracted out, which is both expected and good because we didn't know $(y_{cg})_{trunk}$.

P7.69. Strategize: This problem involves the moment of inertia of a composite object (an ice skater). Since the skater has mass concentrated farther from the axis with her arms outstretched, we expect her to have a larger moment of inertia in that position.
Prepare: We can use Equation 7.20, and Table 7.1.
Solve: Refer to the figure below.

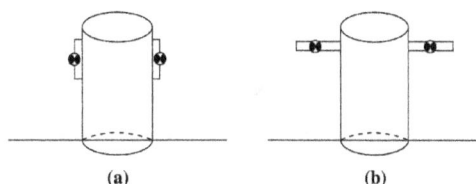

(a)   (b)

(a) The moment of inertia of the skater will be the moment of inertia of her body plus the moment of inertia of her arms. The center of mass of each arm is at her side, 20 cm from the axis of rotation. The mass of each arm is half of one eighth of the mass of her body, which is 4 kg. The mass of her body is $64\text{ kg} - 8\text{ kg} = 56\text{ kg}$. With her arms at her sides, her total moment of inertia is

$$I = I_{body} + I_{arm} + I_{arm} = \tfrac{1}{2}M_{body}(R_{body})^2 + M_{arm}(R_{arm})^2 + M_{arm}(R_{arm})^2$$
$$= \tfrac{1}{2}(56\text{ kg})(0.20\text{ m})^2 + (4\text{ kg})(0.20\text{ m})^2 + (4\text{ kg})(0.20\text{ m})^2 = 1.4\text{ kg}\cdot\text{m}^2$$

(b) With her arms outstretched, the center of mass of her arms is now 50 cm from the axis of rotation. Her moment of inertia is now

$$I = I_{body} + I_{arm} + I_{arm} = \tfrac{1}{2}M_{body}(R_{body})^2 + M_{arm}(R_{arm})^2 + M_{arm}(R_{arm})^2$$
$$= \tfrac{1}{2}(56\text{ kg})(0.20\text{ m})^2 + (4\text{ kg})(0.50\text{ m})^2 + (4\text{ kg})(0.50\text{ m})^2 = 3.1\text{ kg}\cdot\text{m}^2$$

Her moment of inertia has increased.
Assess: Her moment of inertia with arms outstretched is almost twice as large as with them at her side. This is reasonable, since their distance from the axis of rotation is much larger.

P7.71. Strategize: This problem requires a knowledge of translational ($F_{net} = ma$) and rotational ($\tau_{net} = I\alpha$) dynamics.

Prepare: Notice that the counterclockwise torque is greater than the clockwise torque, hence the system will rotate counterclockwise. Let's agree to call any force that tends to accelerate the system positive and any force that tends to decelerate the system negative. Also let's agree to call the small disk M_1 and the large disk M_2.
Solve:

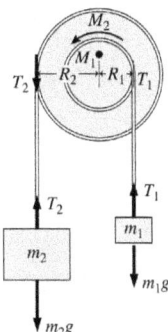

Write Newton's second law equation for m_2 and m_1 as follows:

$$m_2 g - T_2 = m_2 a_2 = m_2 R_2 \alpha \quad \text{or} \quad T_2 = m_2 g - m_2 R_2 \alpha$$

and

$$T_1 - m_1 g = m_1 a_1 = m_1 R_1 \alpha \quad \text{or} \quad T_1 = m_1 g + m_1 R_1 \alpha$$

The net torque acting on the system may be determined by

$$\tau = R_2 T_2 - R_1 T_1 = R_2 (m_2 g - m_2 \alpha R_2) - R_1 (m_1 g + m_1 \alpha R_1)$$

The moment of inertia of the system is

$$I = I_1 + I_2 = (M_1 R_1^2 / 2) + (M_2 R_2^2 / 2)$$

Knowing

$$\tau = I\alpha$$

we may combine the above to get

$$R_2 (m_2 g - m_2 \alpha R_2) - R_1 (m_1 g + m_1 \alpha R_1) = [(M_1 R_1^2 / 2) + (M_2 R_2^2 / 2)]\alpha$$

which may be solved for α to obtain

$$\alpha = \frac{(m_2 R_2 - m_1 R_1)g}{R_2^2 (m_2 + M_2/2) + R_1^2 (m_1 + M_1/2)} = 3.5 \text{ rad/s}^2$$

Assess: This angular acceleration amounts to speeding up about a half revolution per second every second. That is not an unreasonable amount.

P7.73. Strategize: This problem involves a known torque producing an angular acceleration in a massive rigid object. Since the torque is constant, the acceleration will be also and we can use angular kinematic equations.
Prepare: The flywheel is a rigid body rotating about its central axis as shown below. The initial angular speed will be taken as $\omega_i = 0$ and the final angular speed is $\omega_{max} = \omega_f = 1200$ rpm, which must be converted to SI units of rad/s. Noting that 1 revolution corresponds to 2π radians, $\omega_f = 1200$ rpm $(2\pi/60)$ rad/s $= 40\pi$ rad/s.

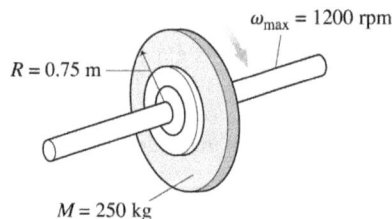

Solve: The radius of the flywheel is $R = 0.75$ m and its mass is $M = 250$ kg. The moment of inertia about the axis of rotation is that of a disk:

$$I = \tfrac{1}{2} MR^2 = \tfrac{1}{2}(250 \text{ kg})(0.75 \text{ m})^2 = 70.31 \text{ kg} \cdot \text{m}^2$$

The angular acceleration is calculated as follows:

$$\tau_{net} = I\alpha \Rightarrow \alpha = \tau_{net}/I = (50 \text{ N} \cdot \text{m})/(70.31 \text{ kg} \cdot \text{m}^2) = 0.711 \text{ rad/s}^2$$

Using the kinematics equation for angular velocity gives

$$\omega_f = \omega_i + \alpha(t_f - t_i) = 1200 \text{ rpm} = 40\pi \text{ rad/s} = 0 \text{ rad/s} + 0.711 \text{ rad/s}^2 (t_f - 0 \text{ s}) \Rightarrow t_f = 180 \text{ s}$$

Assess: To solve this problem, you used your knowledge of moment of inertia, rotational dynamics, and rotational kinematics. As you move farther into physics, you will find that you need mastery of multiple concepts in order to solve the problems. You should enjoy this process.

P7.75. Strategize: This problem involves a compound object rotating about its center of gravity. We can determine the torque required to cause a certain angular acceleration. If we assume this angular acceleration is going to be constant, then we can use angular kinematic equations.

Prepare: Assume that the size of the balls is small compared to 1 m. We have placed the origin of the coordinate system on the 1.0 kg ball. Since $\tau = I_{about\ cg}\alpha$, we need the moment of inertia and the angular acceleration to be able to calculate the required torque.

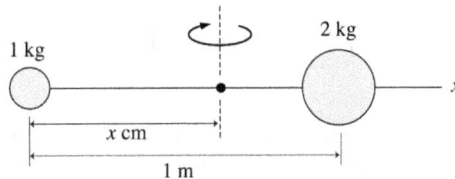

Solve: The center of gravity and the moment of inertia are

$$x_{cm} = \frac{(1.0\ \text{kg})(0\ \text{m}) + (2.0\ \text{kg})(1.0\ \text{m})}{(1.0\ \text{kg} + 2.0\ \text{kg})} = 0.667\ \text{m} \quad \text{and} \quad y_{cm} = 0\ \text{m}$$

$$I_{about\ cm} = \sum m_i r_i^2 = (1.0\ \text{kg})(0.667\ \text{m})^2 + (2.0\ \text{kg})(0.333\ \text{m})^2 = 0.667\ \text{kg} \cdot \text{m}^2$$

We have $\omega_f = 0\ \text{rad/s}$, $t_f - t_i = 5.0\ \text{s}$, and $\omega_i = 20\ \text{rpm} = 20(2\pi\ \text{rad}/60\ \text{s}) = \frac{2}{3}\pi\ \text{rad/s}$, so $\omega_f = \omega_i + \alpha(t_f - t_i)$ becomes

$$0\ \text{rad/s} = \left(\frac{2\pi}{3}\ \text{rad/s}\right) + \alpha(5.0\ \text{s}) \Rightarrow \alpha = -\frac{2\pi}{15}\ \text{rad/s}^2$$

Having found I and α, we can now find the torque τ that will bring the balls to a halt in 5.0 s:

$$\tau = I_{about\ cm}\alpha = \left(\frac{2}{3}\ \text{kg} \cdot \text{m}^2\right)\left(-\frac{2\pi}{15}\ \text{rad/s}^2\right) = -\frac{4\pi}{45}\ \text{N} \cdot \text{m} = -0.28\ \text{N} \cdot \text{m}$$

Assess: The minus sign with the torque indicates that the torque acts clockwise.

P7.77. Strategize: This problem involves a torque from an unknown force causing an angular acceleration of a rigid object. We have sufficient information to determine the moment of inertia of the rigid object (the grindstone). And since the force is constant, the angular acceleration will be constant, such that we can use angular kinematic equations.

Prepare: This is an excellent review problem. In order to solve this problem you will need a working knowledge of rotational kinematics ($\omega_f = \omega_o + \alpha t$), moment of inertia of a cylinder ($I = MR^2/2$), rotational dynamics ($\tau = I\alpha$), torque ($\tau = Rf_k \sin \phi$), and kinetic friction ($f_k = \mu_k N$).

Solve: First determine an expression for the angular acceleration

$$\alpha = (\omega_f - \omega_o)/\Delta t$$

Next obtain an expression for the moment of inertias of the grindstone

$$I = MR^2/2$$

Then obtain an expression for the torque acting on the grindstone

$$\tau = I\alpha = \left(\frac{MR^2}{2}\right)\left(\frac{\omega_f - w_o}{\Delta t}\right)$$

Write a second expression for the torque in terms of the force of friction and then the normal force

$$\tau = f_k R = \mu_k NR$$

Finally, equate the last two expressions for the torque and solve for N (the force with which the man presses the knife against the grindstone).

$$N = \frac{MR(\omega_f - \omega_o)}{2\mu_k \Delta t} = \frac{(28\,\text{kg})(0.15\,\text{m})(180\,\text{rev/min} - 200\ \text{rev/min})(2\pi\,\text{rad/rev})(\text{min}/60\,\text{s})}{2(0.2)(10\,\text{s})} = 2.2\,\text{N}$$

Assess: This is a reasonable force (i.e. one that the man could easily exert and yet not grind the knife to a sliver in a matter of minutes).

P7.79. Strategize: This problem involves a rigid object moving through a known angle in a known amount of time. We are told we can assume that angular acceleration is constant, so we can use angular kinematic equations.
Prepare: The sacs are constrained by the stamens.

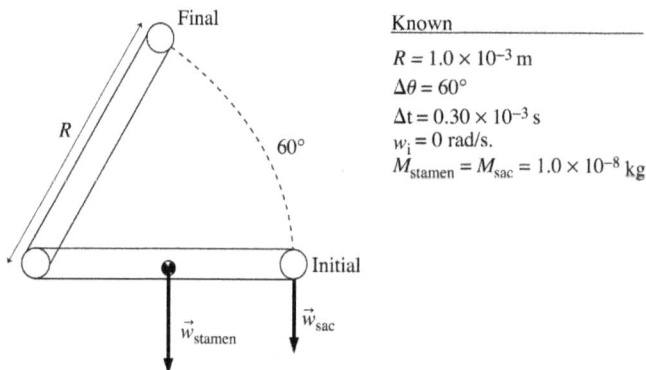

Known
$R = 1.0 \times 10^{-3}\,\text{m}$
$\Delta\theta = 60°$
$\Delta t = 0.30 \times 10^{-3}\,\text{s}$
$w_i = 0\ \text{rad/s}.$
$M_{stamen} = M_{sac} = 1.0 \times 10^{-8}\,\text{kg}$

Solve: Using Synthesis 7.1 with the result from the previous problem gives

$$\omega_f = \alpha\Delta t = 7.0\times10^3\ \text{rad/s}$$

With the angular speed we can now compute the speed.

$$v = \omega r = (7.0\times10^3\ \text{rad/s})(0.001\ \text{m}) = 7.0\ \text{m/s}$$

The correct choice is B.
Assess: These results seem reasonable.

P7.81. Strategize: This problem involves an understanding of the meaning of a center of gravity, and how it moves under the influence of external forces.
Prepare: The center of gravity must follow a parabolic path, so the answer isn't C.
While it is possible to bend over and get one's center of gravity outside the body (search the Web for the Fosbury flop technique in the high jump), in this case the center of gravity of the dancer stays in the body, so the answer isn't D.
Solve: The correct answer is B, with the reasoning being that, as shown in the diagram, the head can stay quite level by bringing up the arms and legs and then lowering them again. This allows the center of gravity to follow the parabola while the head stays about level.
Assess: The whole point of movement is to raise and lower the center of gravity, and this is done by raising and lowering the arms and legs.

P7.83. Strategize: This problem involves torque due to gravity, and torque on a compound object (a dancer).

Prepare: While the dancer is in the air, the gravitational force on her acts as if her mass were concentrated at her center of gravity.

Solve: The correct answer is B; there is no lever arm to create a torque because the gravitational force is directly through the center of gravity. Even if the arms and legs move, the center of gravity moves and the gravitational force is still through the center of gravity. So if the axis of rotation is through her center of gravity then there can be no gravitational force.

Assess: Of course, computing the gravitational torque about some axis other than through the center of gravity could produce a torque, but we were told to use an axis through the center of gravity.

EQUILIBRIUM AND ELASTICITY

Q8.1. Reason: See the figure below for a few possible forces.

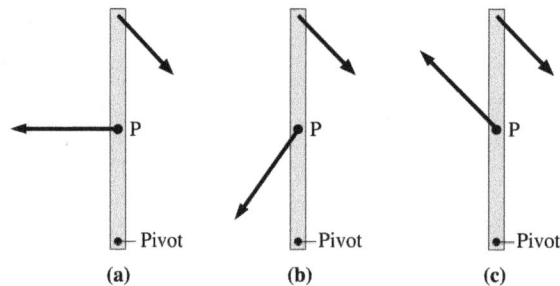

(a)   (b)   (c)

Calculate torques about the pivot point. The moment arm for forces exerted at point P is smaller than that for the force at the end of the rod, so the component of the force at point P perpendicular to the radial line must be larger than that of the force at the tip of the rod. As long as the force has the appropriate magnitude component perpendicular to the radial line, the net torque will be zero.

Assess: Note that the pivot provides an additional force that keeps the rod in static equilibrium in the horizontal and vertical direction.

Q8.3. Reason: Without the pole the line of action of the tension force is close to the pivot point, so the perpendicular distance from the pivot to the line is smaller. With the pole and pulley the perpendicular distance from the pivot to the line of action of the tension force is greater, so even with the same tension the torque is greater.

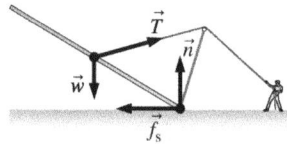

Assess: Imagine actually lifting a mast this way, and the answer will make intuitive sense.

Q8.5. Reason: In order to be stable, the centers of gravity of the people must lie over their base of support. The center of gravity of the person on the left must lie over his foot. For the person on the right, the center of gravity must be somewhere over her feet. Since she is leaning sideways, the center of gravity is shifted to the left. See the diagram below.

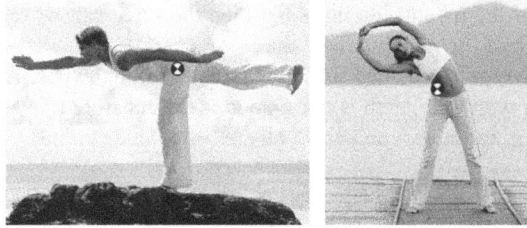

Assess: This result makes sense.

Q8.7. Reason: The position of your center of gravity depends on the orientation of all your body parts. Leaning as described in the question would put your center of gravity very far from the wall and no longer above your feet. In calculating the torque due to gravity, we can treat the force of gravity as though it acts at the center of mass; thus there is a torque due to gravity serving to rotate your body. If the center of gravity is above your feet, you can compensate for a small gravitational torque by using your feet. Specifically you can exert a little more or less force with your toes to produce a torque opposing that from gravity.

Assess: If the torque from gravity can be counteracted by one from your feet, you will not fall. But with your center of gravity no longer above your feet, any force from your feet would cause a torque in the same direction as the torque due to gravity.

Q8.9. Strategize: This spring has a restoring force that depends linearly on the stretch of the spring. This means it obeys Hooke's law. We can use this to determine the spring constant.

Prepare: We see that the equilibrium length of the spring (where there is no restoring force) is 10 cm. Hooke's law tells us that $(F_{sp})_y = -k\Delta y$. The "−" sign means that if the spring is stretched upward, the restoring force will be downward. In the plot, we are given the tension required to counteract the spring's restoring force, meaning $T = -(F_{sp})_y = k\Delta y$. Thus, we can determine the spring constant from the ratio of the tension to the extension of the spring (which is the slope in graph provided).

Solve: We could solve for the spring constant in terms of the tension or the spring's restoring force, since the two only differ by a sign. Choosing the latter, the spring constant is

$$k = -\frac{(F_{sp})_y}{\Delta y} = -\frac{(-10\ \text{N})}{(0.10\ \text{m})} = 100\ \text{N/m}$$

Assess: Note that the force used in Hooke's Law is the force that the spring exerts. In this case, the tension exerts an upward force on the spring. The force exerted by the spring is the reaction force to this and is downward. This is the reason for the double negative sign in the equation above.

Q8.11. Reason: Since both halves of the spring are made of the same material and constructed the same way, the spring constant of each half will be twice the spring constant of the original long spring since it will stretch only half as much under the same tension.

Assess: Hooke's Law does not depend on the length of a spring.

Q8.13. Reason: In order for the wheel barrow to maintain one orientation, the sum of all torques around the axis of rotation (which would be the wheel) must be zero. If the load is placed very near the wheel, its weight force will produce a small torque because of the small lever arm. Since you are exerting a force far from the axis, at the handles, you should only have to exert a small force to create sufficient torque.

Assess: Note that if the load is always closer to the wheel than the handles, the force you need to apply is always less than the weight of the load.

Q8.15. Reason: The muscles that close the jaw have a fixed maximum force with which they can contract, and they are attached to the jaw at a fixed position , a fixed distance from the axis around which the jaw rotates. This means the torque being applied to the jaw by muscles is fixed. Thus the torque required to stop the jaw from moving, caused by something between the upper and lower jaw is also fixed. If that torque is caused by a force that has a larger lever arm (near the front of the mouth, like incisors) then the force can be relatively small. If, however, the force acts near the axis at the back of the mouth (where the molars are) the force would have a very small lever are; the force would need to be much larger to provided sufficient torque to stop the jaw from rotating.

Assess: Biting on a hard material with your teeth is not a good idea. But if you have ever tried to use your teeth to break something like packaging, for example, you know you are better able to pinch or crush it with your molars than with your incisors.

Q8.17. Reason: The force needed to bend a "beam," whether it's a nail or a steel wool fiber, depends on the thickness-to-length ratio. The diameter (thickness) of a steel wool fiber is *much* less, relative to its length, than that of a steel nail. Thus it takes only a very small force to bend and flex the thin fibers of steel wool, but a very large force to bend a steel nail.

Assess: Fiberglass is also flexible while a thicker glass rod is not, for the same reasons.

The extreme case is carbon nanotubes that are *so* thin that they bend easily, but if made into a solid bulk substance as thick as nails would be more resistant to bending.

Q8.19. Reason: Use equilibrium calculations $\Sigma \tau = 0$ around the suspended end. The weight of the rod acts at its center ($L/2$ away from the right end) and the normal force acts at L from the right end. For $\Sigma \tau = 0$ the normal force must be half the force at twice the distance, so the answer is A.

Assess: For $\Sigma F = 0$ the tension in the suspension string must also be 7.0 N. This also makes sense when computing the torques around the center of the rod.

Q8.21. Reason: The fact that the board is "very light" means we will neglect its mass (which we weren't given anyway). We know that the student weighs 165 lbs because the downward force of gravity on the student must equal the upward sum of the two scale readings for the student to be in equilibrium.

Also required for equilibrium is $\Sigma \tau = 0$ and we are free to choose the axis around which we compute the torques anywhere we want. It would be most convenient to select a point above one of the scales so that the upward normal force due to that scale will not produce a torque. Furthermore, since we want to know the distance from the right hand scale, choose it as the pivot.

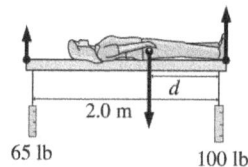

$$\Sigma \tau = d(165\,\text{lb}) - (2.0\,\text{m})(65\,\text{lb}) = 0\,\text{lb}\cdot\text{m}$$

where the counterclockwise torque is positive and the clockwise torque is negative:

$$d = \frac{(2.0\,\text{m})(65\,\text{lb})}{165\,\text{lb}} = 0.79\,\text{m} \approx 0.8\,\text{m}$$

So the correct choice is B.

Assess: Not only *could* we have chosen the pivot point at the left scale and produced the same answer (using $L - d$ as the lever arm), but we *should* do so as a check.

$$\Sigma \tau = (2.0\,\text{m})(100\,\text{lb}) - (L - d)(165\,\text{lb}) = 0\,\text{lb}\cdot\text{m}$$

$$L - d = \frac{(2.0\,\text{m})(100\,\text{lb})}{165\,\text{lb}}$$

$$d = 2.0\,\text{m} - \frac{(2.0\,\text{m})(100\,\text{lb})}{165\,\text{lb}} = 0.79\,\text{m} \approx 0.8\,\text{m}$$

Q8.23. Reason: If the gymnast is in static equilibrium, then the gymnast's weight must be completely canceled out by the upward restoring force: $-mg + F_{restor.} = 0$. We also know from Q8.22 that $\dfrac{|F_{restor.}|}{|\Delta y|} = 3200$ N/m. So we require

$$\frac{mg}{\Delta y} = 3200 \text{ N/m} \quad \text{or} \quad \Delta y = \frac{mg}{(3200 \text{ N/m})} = \frac{(65 \text{ kg})(9.8 \text{ m/s}^2)}{(3200 \text{ N/m})} = 0.20 \text{ m}. \text{ The correct answer is D.}$$

Assess: This is a reasonable stretch of a trampoline with a person standing on it.

Q8.25. Reason: Since the hanging mass is in equilibrium, initially we have $-k\Delta y_i = mg$, or $\Delta y_i = -mg/k$. Later, with two ropes attached, nothing has changed about the stretchiness of the ropes, but there are now two of them. So we can write $-2k\Delta y_f = mg$, or $\Delta y_f = -mg/2k$. Comparing this with the initial case, we immediately see $\Delta y_f = \Delta y_i / 2 = (20 \text{ cm})/2 = 10 \text{ cm}$. The correct answer is B.

Assess: It makes perfect sense that two ropes would be stretched by half as much as a single rope.

Q8.27. Reason: Because $\Delta L = FL/AY$ if L is doubled then A must be doubled too to keep ΔL the same. To double A requires an increase in the diameter by a factor of $\sqrt{2}$. So the answer is B.

Assess: Doubling the diameter (an incorrect answer) would actually make the twice-as-long wire stretch less.

Problems

P8.1. Strategize: This problem involves static equilibrium. The sum of all torques and the sum of all forces on the board must be zero.

Prepare: Because the board is "very light" we will assume that it is massless and does not contribute to the scale reading, nor does it contribute any torques. The sum of the two scale readings must equal the woman's weight: $w = mg = (64 \text{ kg})(9.80 \text{ m/s}^2) = 627 \text{ N} \approx 630 \text{ N}$.

Solve: Compute the torques around the point the board rests on the left scale. The woman's weight creates a clockwise (negative) torque; and the normal force n_{right} of the right scale creates a counterclockwise (positive) torque.

$$\Sigma \tau = (2.0 \text{ m})(n_{right}) - (1.5 \text{ m})(627 \text{ N}) = 0 \text{ N} \cdot \text{m}$$

The right scale reads n_{right}:

$$n_{right} = \frac{(1.5 \text{ m})(627 \text{ N})}{2.0 \text{ m}} = 470 \text{ N}$$

By simple subtraction the left scale reads

$$n_{left} = 627 \text{ N} - 470 \text{ N} = 160 \text{ N}$$

Assess: The answer is reasonable. Since the woman is three times farther from the left scale than the right one, it (the left one) reads three times less. And the two scale readings sum to the woman's weight, as required.

Not only *could* we have chosen the pivot point at the right scale and produced the same answer, but we *should* do so as a check.

$$\Sigma \tau = (0.5 \text{ m})(627 \text{ N}) - (2.0 \text{ m})(n_{left}) = 0 \text{ N} \cdot \text{m}$$
$$n_{left} = \frac{(0.5 \text{ m})(627 \text{ N})}{2.0 \text{ m}} = 160 \text{ N}$$

And so

$$n_{right} = 627 \text{ N} - 157 \text{ N} = 470 \text{ N}$$

It is true that $\Sigma \tau = 0$ around *any* point (for equilibrium), but we picked the two we did (the second as a check) because then the resulting torque equations each had only one unknown in them.

P8.3. Strategize: If we consider the point where the table is just about to tip, then the table is in static equilibrium, so the sum of all torques and the sum of all forces on the table must be zero.

Prepare: Compute the torques around the bottom of the right leg of the table. The horizontal distance from there to the center of gravity of the table is $\frac{2.10 \text{ m}}{2} = -.55 \text{ m} = 0.50 \text{ m}$. Note also that if the table is just about to tip, the normal force the floor exerts on the left leg is zero.

Solve: Call the horizontal distance from the bottom of the right leg to the center of gravity of the man x.

$$\Sigma\tau = (56 \text{ kg})(9.8 \text{ m/s}^2)(0.50 \text{ m}) - (70 \text{ kg})(9.8 \text{ m/s}^2)x = 0 \Rightarrow x = 0.40 \text{ m}$$

The distance from the right edge of the table is now $0.55 \text{ m} - 0.40 \text{ m} = 0.15 \text{ m} = 15 \text{ cm}$.

Assess: It seems likely that the table would tip if the man were closer than 15 cm to the edge.

P8.5. Strategize: Since the pole is remaining at rest, the sum of all torques and the sum of all forces on the pole must be zero.

Prepare: Assume the pole is uniform in diameter and density. We will use the equilibrium equation $\Sigma\tau = 0$. The weight of the pole is $w = mg = (25 \text{ kg})(9.80 \text{ m/s}^2) = 245 \text{ N}$.

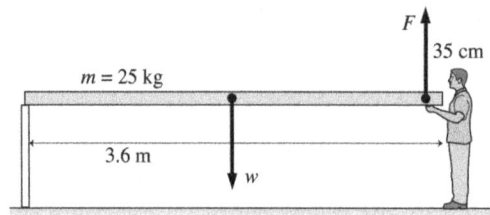

Solve: Compute the torques around the left end, where the pole rests on the fence. The weight of the pole (acting at its center of gravity) will produce a clockwise (negative) torque, and force F near the right end of the pole will produce a counterclockwise (positive) torque.

$$\Sigma\tau = (3.6 \text{ m} - 0.35 \text{ m})F - (1.8 \text{ m})(245 \text{ N})$$
$$= (3.25 \text{ m})F - (1.8 \text{ m})(245 \text{ N})$$
$$= 0 \text{ N} \cdot \text{m}$$

Now solve for F:

$$F = \frac{(1.8 \text{ m})(245 \text{ N})}{3.25 \text{ m}} = 140 \text{ N}$$

Assess: This really *is* a rest because you only have to exert a force of just over half the pole's weight, instead of the whole weight when you were carrying it. The fence is helping hold up the pole.

Think carefully about the figure and imagine moving your hands toward the fence. The upward force you would have to exert to keep $\Sigma\tau = 0$ would increase, and when you support the pole at its center of gravity the torque equation says your force is equal to the weight of the pole. At that point the fence is no longer helping (and you aren't resting) as it exerts no upward force. If you tried moving even farther toward the fence, past the center of gravity, there would be no way to keep the pole in equilibrium and it would rotate, fall, and hit the ground.

P8.7. Strategize: The rod is in static equilibrium, so the sum of all torques on the rod must be zero.

Prepare: Consider the right end of the rod from which the sign hangs. Assume the length of the rod is L. Simply apply the $\Sigma\tau = 0$ equilibrium condition. The two forces are the downward weight force and the tension in the cable. The angle between the cable and the rod is $30°$, so the perpendicular distance from the left end of the rod to the cable is $L\sin 30°$.

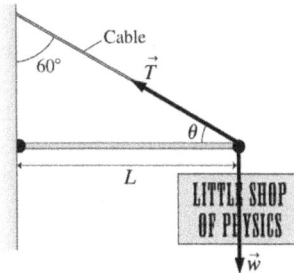

Solve: Compute the sum of the torques around the left end of the rod.

$$\Sigma \tau = T(L\sin\theta) - mgL = 0 \Rightarrow T = \frac{mgL}{L\sin\theta} = \frac{mg}{\sin\theta} = \frac{(8.0 \text{ kg})(9.80 \text{ m/s}^2)}{\sin 30°} = 160 \text{ N}$$

Assess: Limiting cases assure the correctness of the answer; if the angle were smaller the tension would be greater.

P8.9. Strategize: The object balanced on the pivot is a rigid body. Since the object is balanced on the pivot, it is in both translational equilibrium and rotational equilibrium.

Prepare: There are three forces acting on the object: the weight $\vec{w}_1$ acting through the center of gravity of the long rod, the weight $\vec{w}_2$ acting through the center of gravity of the short rod, and the normal force $\vec{P}$ on the object applied by the pivot.

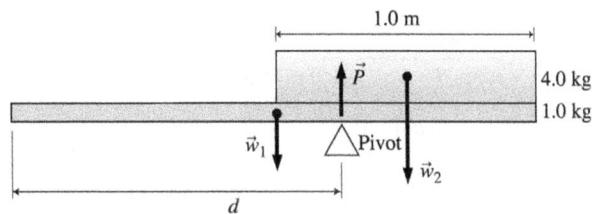

Solve: The translational equilibrium equation $(F_{\text{net}})_y = 0 \text{ N}$ is

$$-w_1 - w_2 + P = 0 \text{ N} \Rightarrow P = w_1 + w_2 = (1.0 \text{ kg})(9.8 \text{ m/s}^2) + (4.0 \text{ kg})(9.8 \text{ m/s}^2) = 49.0 \text{ N}$$

Measuring torques about the left end, the equation for rotational equilibrium $\tau_{\text{net}} = 0$ is

$$Pd - w_1(1.0 \text{ m}) - w_2(1.5 \text{ m}) = 0 \text{ N} \cdot \text{m} \Rightarrow (49.0 \text{ N})d - (1.0 \text{ kg})(9.8 \text{ m/s}^2)(1.0 \text{ m}) - (4.0 \text{ kg})(9.8 \text{ m/s}^2)(1.5 \text{ m})$$
$$= 0 \text{ N} \Rightarrow d = 1.4 \text{ m}$$

Thus, the pivot is 1.4 m from the left end.

Assess: From geometry, the distance d is expected to be more than 1.0 m but less than 1.5 m. A value of 1.4 m is reasonable.

P8.11. Strategize: The board is in static equilibrium, so the sum of all torques and the sum of all forces on the board must be zero.

Prepare: In this problem we are given the mass of the board, but because of the symmetry of the distances (the fulcrum, or right support, is right in the center) we don't need it. We'll compute the torques around the right support, which is under the center of gravity of the board, so the weight of the board won't produce a torque.

The weight of the diver is $w = mg = (60 \text{ kg})(9.80 \text{ m/s}^2) = 590 \text{ N}$.

Solve: Because of the symmetry of the situation (the two lever arms are equal in length), we can examine the torque equation in our heads and realize that the force exerted by the hinge must have the same magnitude as the weight of the diver. Therefore, the force the hinge exerts on the board is 590 N.

Assess: We computed the torques around the right support because of the symmetry, but also because had we computed them around the hinge on the left, we would be eliminating from the equation the very torque we need to answer the problem.

Since both the weight of the diver and the force of the hinge act in the downward direction, it is also clear that the right support must exert a force of twice the diver's weight in the upward direction.

P8.13. Strategize: The beam is in static equilibrium, so the sum of all torques on the beam is zero.
Prepare: We can apply Equation 8.1 using the given mass and distances.
Solve: Refer to the diagram below.

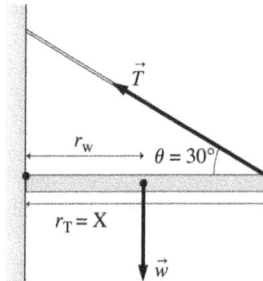

The net torque on the beam is zero. Calculating torques about the pivot, the torque equation in Equation 8.1 gives

$$r_T T_\perp - (r_w)_\perp w = r_T T \sin\theta - r_w w = 0$$

Solving for the tension,

$$T = \frac{r_w w}{r_T \sin\theta} = \frac{(0.5\text{ m})(10\text{ kg})(9.80\text{ m/s}^2)}{(1.0\text{ m})\sin 30°} = 98\text{ N}$$

Assess: Note that using different forms of the definition of torque led to simpler calculations.

P8.15. Strategize: The critical angle is when the center of gravity is just over the edge of the cabinet.
Prepare: Refer to the diagram below, in which the relevant distances have be labeled.

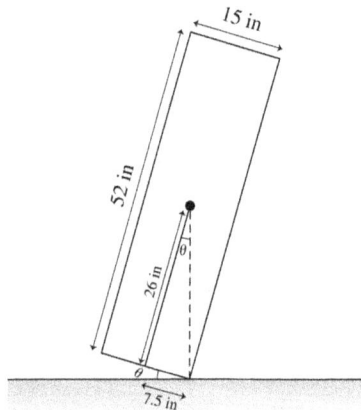

Solve: Solve the right triangle for the desired angle.

$$\theta_c = \tan^{-1}\left(\frac{7.5\text{ in}}{26\text{ in}}\right) = 16°$$

Assess: The file cabinet is tall, so the critical angle is small. Loading the lower drawers more would lower the center of gravity and increase the critical angle.

P8.17. Strategize: The critical angle is when the center of gravity is just over the tire.
Prepare: Refer to the diagram below, in which the relevant angles and distances have been labeled.

Solve: From the diagram we see that the Static Stability Factor (SSF) is equal to $\tan(\theta_c)$.

$$\theta_c = \tan^{-1}(1.2) = 50°$$

Assess: One can see that increasing the track width increases the stability of the SUV. More information is at http://www.suu.edu/faculty/penny/RolloverPaper/RolloverPaper.pdf.

P8.19. Strategize: In order not to tip, the vehicle's center of gravity must be over its base of support.
Prepare: Refer to the following figure.

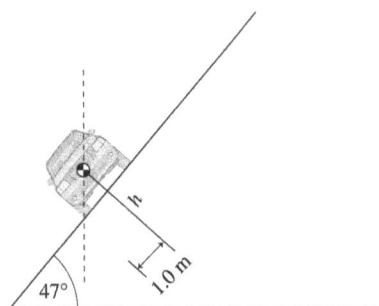

The figure shows the vehicle just as it is about to tip. The center of gravity is at the edge of the base of support.
Solve: Solving for the height,

$$h = \frac{1.0 \text{ m}}{\tan(47°)} = 0.93 \text{ m}$$

Assess: Note that as the angle decreases the height required decreases.

P8.21. Strategize: This problem involves Hooke's law, in which compression in one direction produces a restoring force in the opposite direction.
Prepare: Hooke's law is given in Equation 8.3, $(F_{sp})_y = -k\Delta y$. We'll just compute the magnitude (ignoring the minus sign).
Solve:

$$(F_{sp})_y = k\Delta y = (160 \text{ N/m})(0.0060 \text{ m}) = 0.96 \text{ N}$$

Assess: This force pushes the jaw open.

P8.23. Strategize: Because the DNA is elastic, it will obey Hooke's law for small displacements.
Prepare: We can use Equation 8.3: $F_x = -k\Delta x$ to determine the spring constant.
Solve: The force the DNA exerts is equal and opposite to the force the tweezers exert. See the figure below.

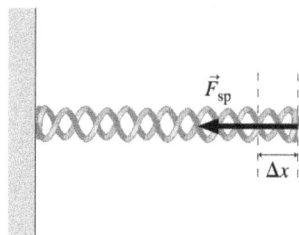

Applying Equation 8.3,

$$k = -\frac{F_{sp}}{\Delta x} = -\frac{-1.5 \times 10^{-9}\ \text{N}}{5.0 \times 10^{-9}\ \text{m}} = 0.30\ \text{N/m}$$

Assess: Note that Equation 8.3 refers to the force the spring exerts, not the force applied to the spring.

P8.25. Strategize: This problem involves an object held in static equilibrium by forces including the elastic restoring force of a spring. We will assume an ideal spring that obeys Hooke's law.
Prepare: A visual overview below shows the details, including a free-body diagram, of the problem.

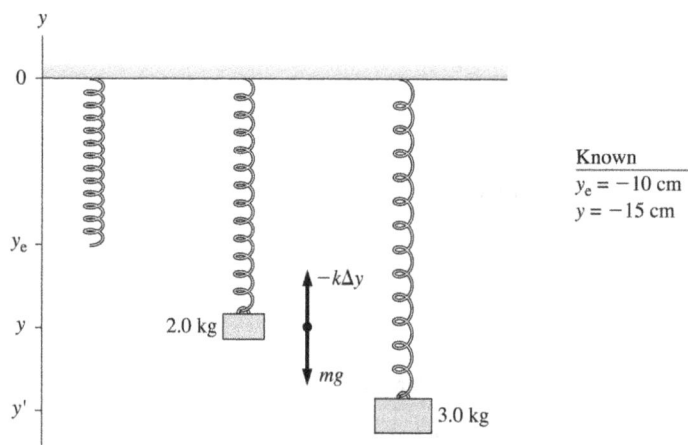

Solve: (a) The spring force on the 2.0 kg mass is $F_{sp} = -k\Delta y$. Notice that Δy is negative, so F_{sp} is positive. This force is equal to mg, because the 2.0 kg mass is at rest. We have $-k\Delta y = mg$. Solving for k:

$$k = -(mg/\Delta y) = -(2.0\ \text{kg})\,(9.80\ \text{m/s}^2)/(-0.15\ \text{m} - (-0.10\ \text{m})) = 392\ \text{N/m} = 390\ \text{N/m}$$

(b) Again using $-k\Delta y = mg$:

$$\Delta y = -mg/k = -(3.0\ \text{kg})(9.80\ \text{m/s}^2)/(392\ \text{N/m}) = -0.075\ \text{m}$$

$$y' - y_e = -0.075\ \text{m} \Rightarrow y' = y_e - 0.075\ \text{m} = -0.10\ \text{m} - 0.075\ \text{m} = -0.175\ \text{m} = -18\ \text{cm}$$

The length of the spring is 18 cm when a mass of 3.0 kg is attached to the spring.
Assess: The *position* of the end of the spring is negative because it is below the origin, but length must be a positive number. We expected the length to be a little more than 15 cm.

P8.27. Strategize: This problem the elastic restoring force of a spring. We will assume an ideal spring that obeys Hooke's law.

Prepare: A visual overview below shows the details, including a free-body diagram, of the problem. We will assume an ideal spring that obeys Hooke's law.

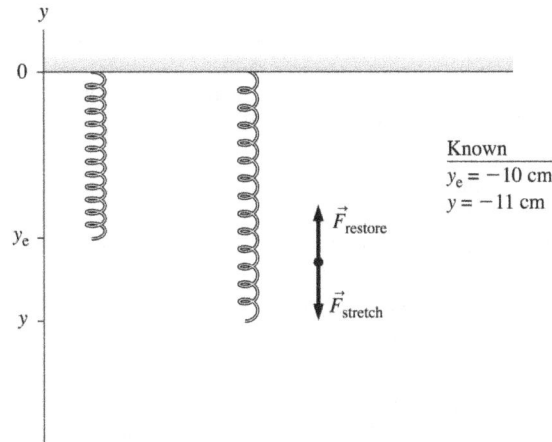

Solve: (a) The spring force or the restoring force is $F_{sp} = -k\Delta y$. For $\Delta y = -1.0$ cm and the force in Newtons,

$$F_{sp} = F = -k\Delta y \Rightarrow k = -F/\Delta y = -F/(-0.01\,\text{m}) = 100F \text{ N/m}$$

Notice that Δy is negative, so F_{sp} is positive.

We can now calculate the new length for a restoring force of $3F$:

$$F_{sp} = 3F = -k\Delta y = (-100\,F)\Delta y \Rightarrow \Delta y = -0.03\,\text{m}$$

From $\Delta y = y - y_e = -0.03$ m, or $y = -0.03$ m $+ y_e$, or $y = -0.03$ m $+ (-0.10$ m$) = -0.13$ m, the length of the spring is 0.13 m.

(b) The new compressed length for a restoring force of $2F$ can be calculated as:

$$F_{sp} = 2F = -k\Delta y = (-100\,F)\Delta y \Rightarrow \Delta y = -0.02\,\text{m}$$

Using $\Delta y = y_e - y = -0.02$ m, or $y = 0.02$ m $+ y_e$, or $y = 0.02$ m $+ (-0.10$ m$) = -0.08$ m, the length of the compressed spring is 0.08 m.

Assess: The stretch Δx is proportional to the applied force, as both parts of this problem demonstrate. Of course, this bet is off if the spring is stretched or compressed far enough to take it out of the linear region.

P8.29. Strategize: This problem involves the elastic restoring force from a spring, We will assume that the spring is ideal and obeys Hooke's law.

Prepare: According to Hooke's law, the spring force acting on a mass (m) attached to the end of a spring is given as $F_{sp} = k\Delta y$, where Δy is the change in length of the spring. If the mass m is at rest, then F_{sp} is also equal to the weight $w = mg$.

Solve: We have $F_{sp} = k\Delta y = mg$. We want a 0.10 kg mass to give $\Delta y = 0.010$ m. This means

$$k = mg/\Delta y = (0.10\,\text{kg})\,(9.80\,\text{m/s}^2)/(0.010\,\text{m}) = 98 \text{ N/m}$$

Assess: If you double the mass and hence the weight, the displacement of the end of the spring will double as well.

P8.31. Strategize: This problem involves the dependence e of the elastic restoring force on an object's properties like its length and cross-sectional area. The notions of stress and strain may be useful.

Prepare: Rearrange Equation 8.5 to see that the stretch is proportional to the length (for part (a)) and inversely proportional to the area (for part (b)).

$$\Delta L = \frac{LF}{AY}$$

Solve: (a) In this part everything on the right side of the equation stays constant except the length L. Since the length of the second wire is twice the length of the first wire, then the second wire will stretch twice as much by the same force. So the answer is 2 mm.

(b) In this part everything on the right side of the equation stays constant except the cross-sectional area A. The cross-sectional area of the third wire is four times the area of the first wire, since $A = \pi r^2 = \pi(D/2)^2$ and the diameter of the third wire is twice the diameter of the first wire, so the third wire will stretch one-quarter as much by the same force. The answer is 0.25 mm.

Assess: This problem is worth mentally reviewing to make sure the explanation given makes sense, and to tuck the results away as tidbits of practical knowledge.

P8.33. Strategize: This problem involves an object (the hanging mass) creating a tensile stress in the wire.

Prepare: The force (F) pulling on the wire, which is simply the weight (mg) of the hanging mass, produces tensile stress given by F/A, where A is the cross-sectional area of the wire. We will use Equation 8.5 to find the hanging mass.

Solve: From the definition of Young's modulus, we have

$$Y = \frac{mg/A}{\Delta L/L} \Rightarrow m = \frac{(\pi r^2)Y\Delta L}{gL} = \frac{\pi(2.50\times10^{-4}\text{ m})^2(20\times10^{10}\text{ N/m}^2)(1.0\times10^{-3}\text{ m})}{(9.80\text{ m/s}^2)(2.0\text{ m})} = 2.0\text{ kg}$$

Assess: A 1.0 mm stretch of a 2.0 m wire by 2.0 kg hanging mass is reasonable.

P8.35. Strategize: This problem involves stress and strain on a steel cable. They are related through Young's modulus.

Prepare: Equation 8.5 relates the quantities in question; solve it for ΔL, which is what we want to know.

Look up Young's modulus for steel in Table 8.1: $Y_{\text{steel}} = 20\times10^{10}\text{ N/m}^2$.

Assume a circular cross section: $A = \pi r^2 = \pi(D/2)^2 = \pi(0.0125\text{ m})^2 = 4.91\times10^{-4}\text{ m}^2$.

We are also given that $L = 800$ m. Compute $F = w = mg = (1000\text{ kg})(9.80\text{ m/s}^2) = 9800\text{ N}$.

Solve:

$$\Delta L = \frac{LF}{AY} = \frac{(800\text{ m})(9800\text{ N})}{(4.91\times10^{-4}\text{ m}^2)(10\times10^{10}\text{ N/m}^2)} = 0.16\text{ m} = 16\text{ cm}$$

Assess: 16 cm is quite a stretch, but 1000 kg (times g) is quite a bit of weight, and 800 m is quite a long cable, so the answer is reasonable. The design of the shaft would have to take this 16 cm stretch into account. Also check to see that the units work out.

P8.37. Strategize: This problem involves stress and strain on wooden legs of a bar stool. They are related through Young's modulus.

Prepare: Equation 8.5 relates the quantities in question; the fractional decrease in length will be $\Delta L/L$, so rearrange the equation so $\Delta L/L$ is isolated.

Look up Young's modulus for Douglas fir in Table 8.1: $Y_{\text{Douglas fir}} = 1\times10^{10}\text{ N/m}^2$.

The total cross section will be three times the area of one leg:

$$A_{\text{tot}} = 3(\pi r^2) = 3\left(\pi\left(\frac{D}{2}\right)^2\right) = 3\pi(0.010\text{ m})^2 = 9.42\times10^{-4}\text{ m}^2$$

Compute $F = w = mg = (75\text{ kg})(9.80\text{ m/s}^2) = 735\text{ N}$.

Solve:

$$\frac{\Delta L}{L} = \frac{F}{AY} = \frac{735 \text{ N}}{(9.42\times10^{-4} \text{ m}^2)(1\times10^{10} \text{ N/m}^2)} = 7.8\times10^{-5}$$

This is a 0.0078% change in length.

Assess: We were not given the original length of the stool legs, but regardless of the original length, they decrease in length by only a small percentage—0.0078%—because F isn't large but A is.

P8.39. Strategize: This problem involves stress and strain on a concrete column. They are related through Young's modulus.

Prepare: The load supported by a concrete column creates compressive stress in the concrete column. The weight of the load produces tensile stress given by F/A, where A is the cross-sectional area of the concrete column and F equals the weight of the load.

Solve: From the definition of Young's modulus,

$$Y = \frac{F/A}{\Delta L/L} \Rightarrow \Delta L = \left(\frac{F}{A}\right)\left(\frac{L}{Y}\right) = \left(\frac{200\,000 \text{ kg}\times9.8 \text{ m/s}^2}{\pi(0.25 \text{ m})^2}\right)\left(\frac{3.0 \text{ m}}{3\times10^{10} \text{ N/m}^2}\right) = 1.0 \text{ mm}$$

Assess: A compression of 1.0 mm of the concrete column by a load of approximately 200 tons is reasonable.

P8.41. Strategize: This problem involves the tensile strength of a fiber optic cable, and the strain resulting from this stress.

Prepare: The stress is F/A. Solve for F. The maximum stress is the tensile strength; for glass Table 8.2 says the strength is 60×10^6 N/m². Table 8.1 says for glass $Y = 7\times10^{10}$ N/m².

Solve: (a)

$$F = (\text{tensile strength})(\text{area}) = (60\times10^6 \text{ N/m}^2)(\pi(4.5\times10^{-6} \text{ m})^2) = 3.8\times10^{-3} \text{ N}$$

So the tension required to break the glass fiber is only 3.8 mN.

(b) Use the force from part (a) to compute the stretch.

$$\Delta L = \frac{LF}{AY} = \frac{(10 \text{ m})(3.8\times10^{-3} \text{ N})}{(\pi(4.5\times10^{-6} \text{ m})^2)(7\times10^{10} \text{ N/m}^2)} = 8.6\times10^{-3} \text{ m}$$

The data in Table 8.2 was given to only one significant figure, so we report this as 9 mm.

Assess: A stretch of just under 1 cm sounds reasonable for a 10-meter fiber.

P8.43. Strategize: This problem involves stress and strain on a human Achilles tendon. They are related through Young's modulus.

Prepare: Table 8.3 says for tendon $Y = 0.15\times10^{10}$ N/m².

Solve: (a)

$$\Delta L = \frac{LF}{AY} = \frac{(0.15 \text{ m})(8(70 \text{ kg})(9.8 \text{ m/s}^2))}{(110 \text{ mm}^2)(\frac{1 \text{ m}}{1000 \text{ mm}})^2(0.15\times10^{10} \text{ N/m}^2)} = 5.0 \text{ mm}.$$

The tendon will stretch 5.0 mm.

(b) The fraction of the tendon's length is

$$\frac{0.50 \text{ cm}}{15 \text{ cm}} = 0.033 = 3.3\%.$$

Assess: A 3.3% stretch sounds reasonable.

P8.45. Strategize: This is an estimation problem, so a range of answers may be acceptable. Let us assume that during the crushing motion, the angular acceleration is very small, such that we can consider this to be very close to static equilibrium.

Prepare: Treating the upper handle as though it is in static equilibrium, we can write $\sum \tau = 0$. We will treat the axis of rotation as the hinge in the garlic press. There are two torques: one from the force exerted by the person's hand, and one from the contact force of the garlic upward on the top handle. We estimate the lever arms for these two forces to be about 8.5 cm, and 1.3 cm respectively. In this estimate we have considered that the forces appear to act approximately perpendicular to the top arm of the press.

Solve: Inserting values into the sum of all torques, we obtain:

$$\sum \tau = -F_{hand} r_{hand} + F_{garlic} r_{garlic} = 0 \Rightarrow F_{garlic} = F_{hand} \frac{r_{hand}}{r_{garlic}} = (12 \text{ N}) \frac{(8.5 \text{ cm})}{(1.3 \text{ cm})} = 78 \text{ N}$$

This is an estimation problem, so a range of answers is acceptable.

Asses: We expect an answer considerably larger than 12 N; otherwise there would be no point in using the press.

P8.47. Strategize: This problem involves a human forearm in both translational equilibrium and rotational equilibrium. Thus the sum of all forces and the sum of all torques on the forearm must both be zero.

Prepare: Compute the torques around the pivot at the elbow as shown. All the forces are in the vertical direction.

Solve: Call the force in the triceps muscle T.

$$\sum \tau = (90 \text{ N})(30 \text{ cm}) - T(2.4 \text{ cm}) = 0 \text{ N} \cdot \text{m} \Rightarrow T = \frac{(90 \text{ N})(30 \text{ cm})}{2.4 \text{ cm}} = 1100 \text{ N}$$

Assess: This is a pretty big force, but not too hard for your triceps.

P8.49. Strategize: This problem involves a human arm in both translational equilibrium and rotational equilibrium. Thus the sum of all forces and the sum of all torques on the arm must be zero.

Prepare: The Compute the torques around the pivot at the shoulder as shown.

Solve: (a) Call the force in the deltoid muscle T.

$$\sum \tau = T(17 \text{ cm})(\sin 15°) - (4.0 \text{ kg})(9.8 \text{ m/s}^2)(38 \text{ cm}) = 0 \text{ N} \cdot \text{m}$$

$$T = \frac{(4.0 \text{ kg})(9.8 \text{ m/s}^2)(38 \text{ cm})}{(17 \text{ cm})(\sin 15°)} = 340 \text{ N}$$

(b) Divide both sides of the previous equation by mg.

$$\frac{T}{mg} = \frac{(38 \text{ cm})}{(17 \text{ cm})(\sin 15°)} = 8.6$$

Thus, the deltoid muscle must be able to exert a force 8.6 times greater than the weight of the arm just to let you keep it outstretched at rest.

Assess: Holding something in your hand would greatly increase the clockwise torque, so the force in the deltoid muscle would need to be even larger.

P8.51. Strategize: To stay in place, the beam must be in both translational equilibrium ($\vec{F}_{net} = \vec{0} \text{ N}$) and rotational equilibrium ($\tau_{net} = 0 \text{ N} \cdot \text{m}$).

Prepare: The beam is a rigid body of length 3.0 m and the student is a particle. $\vec{F}_1$ and $\vec{F}_2$ are the normal forces on the beam due to the supports, $\vec{w}_{beam}$ is the weight of the beam acting at the center of gravity, and $\vec{w}_{student}$ is the student's weight. The student is 1 m away from support 2.

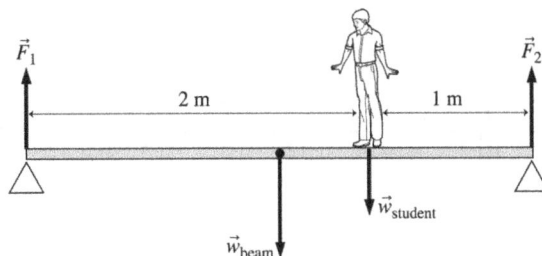

Solve: The condition on the forces is

$$\Sigma F_y = -w_{beam} - w_{student} + F_1 + F_2 = 0 \text{ N} \Rightarrow F_1 + F_2 = w_{beam} + w_{student} = (100 \text{ kg} + 80 \text{ kg})(9.80 \text{ m/s}^2) = 1764 \text{ N}$$

Taking the torques about the left end of the beam, the second condition is

$$-w_{beam}(1.5 \text{ m}) - w_{student}(2.0 \text{ m}) + F_2(3.0 \text{ m}) = 0 \text{ N} \cdot \text{m}$$

$$-(100 \text{ kg})(9.80 \text{ m/s}^2)(1.5 \text{ m}) - (80 \text{ kg})(9.80 \text{ m/s}^2)(2.0 \text{ m}) + F_2(3.0 \text{ m}) = 0 \text{ N} \cdot \text{m} \Rightarrow F_2 = 1013 \text{ N}$$

From $F_1 + F_2 = 1764$ N, we get $F_1 = 1764 \text{ N} - 1013 \text{ N} = 750 \text{ N}$.

Assess: To establish rotational equilibrium, the choice for the pivot is arbitrary. We can take torques about any point of interest.

P8.53. Strategize: Since the pole is motionless, it is in equilibrium, so $\tau_{net} = 0$ and $F_{net} = 0$.

Prepare: Assume that the flagpole has a uniform diameter so that its center of mass is at the center, $7.5 \text{ m}/2 = 3.75 \text{ m}$ from either end.
The magnitude of the counterclockwise torque due to the tension in the rope must equal the magnitude of the clockwise torque due to the weight of the flagpole.
Use Equation 7.11 for the torques: $\tau = rF \sin\phi$.

Solve: The clockwise torque is due to the weight of the flagpole. Draw a quick right triangle to see that the angle between r and F (that is, w) is 60°.

$$\tau_{clockwise} = rF \sin\phi = rw \sin\phi = r(mg)\sin\phi = (3.75 \text{ m})(28 \text{ kg})(9.80 \text{ m/s}^2)\sin 60° = 891 \text{ N} \cdot \text{m}$$

In the counterclockwise direction we want to know the force (tension in the rope), so solve the torque equation for F and put in the previous result for the torque. This time r is the distance from the pivot at the bottom of the pole to where the rope is attached at the top of the pole, or 7.5 m.

$$F_{rope} = \frac{\tau}{r \sin\phi} = \frac{891 \text{ N} \cdot \text{m}}{(7.5 \text{ m})(\sin 20°)} = 350 \text{ N}$$

Assess: The man must exert 350 N of force because the angle 20° is so small. One way to increase the angle (and the sine of the angle) is to use a longer rope and stand farther back. This will slightly decrease the needed force. By doing this r is not changed; it is still the length of the pole if the rope is attached to the top of the pole.

P8.55. Strategize: For the beam not to fall over, it must be both in translational equilibrium ($\vec{F}_{net} = \vec{0}$ N) and rotational equilibrium ($\vec{\tau}_{net} = 0$ N·m).

Prepare: Model the beam as a rigid body. The boy walks along the beam a distance x, measured from the left end of the beam. There are four forces acting on the beam. $\vec{F}_1$ and $\vec{F}_2$ are from the two supports, $\vec{w}_b$ is the weight of the beam, and $\vec{w}_B$ is the weight of the boy.

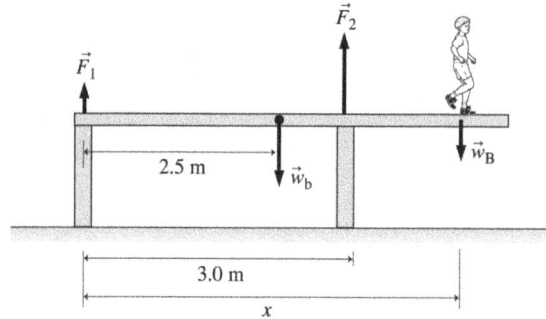

Solve: We pick our pivot point on the left end through the first support. The equation for rotational equilibrium is

$$-w_b(2.5 \text{ m}) + F_2(3.0 \text{ m}) - w_B x = 0 \text{ N} \cdot \text{m}$$

$$-(40 \text{ kg})(9.80 \text{ m/s}^2)(2.5 \text{ m}) + F_2(3.0 \text{ m}) - (20 \text{ kg})(9.80 \text{ m/s}^2)x = 0 \text{ N} \cdot \text{m}$$

The equation for translation equilibrium is

$$\Sigma F_y = 0 \text{ N} = F_1 + F_2 - w_b - w_B \Rightarrow F_1 + F_2 = w_b + w_B = (40 \text{ kg} + 20 \text{ kg})(9.80 \text{ m/s}^2) = 588 \text{ N}$$

Just when the boy is at the point where the beam tips, $F_1 = 0$ N. Thus $F_2 = 588$ N. With this value of F_2, we can simplify the torque equation to

$$-(40 \text{ kg})(9.80 \text{ m/s}^2)(2.5 \text{ m}) + (588 \text{ N})(3.0 \text{ m}) - (20 \text{ kg})(9.80 \text{ m/s}^2)x = 0 \text{ N} \cdot \text{m} \Rightarrow x = 4.0 \text{ m}$$

Thus, the distance from the right end is $5.0 \text{ m} - 4.0 \text{ m} = 1.0 \text{ m}$.

P8.57. Strategize: This problem involves an object being held in equilibrium by the sum of a spring restoring force. Assume that the spring is ideal and it obeys Hooke's law.
Prepare: We model the 5.0 kg mass as a particle. We will use the subscript s for the scale and sp for the spring. With the y-axis representing vertical positions, pictorial representations and free-body diagrams are shown for parts (a) through (c).

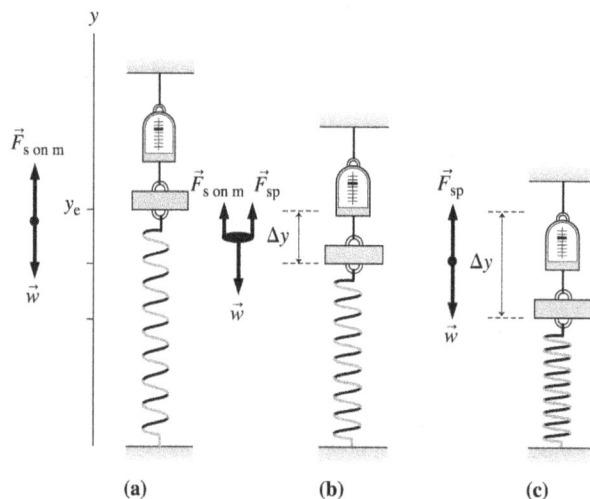

Solve: **(a)** The scale reads the upward force $F_{s \text{ on } m}$ that it applies to the mass. Newton's second law gives

$$\Sigma(F_{\text{on } m})_y = F_{s \text{ on } m} - w = 0 \Rightarrow F_{s \text{ on } m} = w = mg = (5.0 \text{ kg})(9.80 \text{ m/s}) = 49 \text{ N}$$

(b) In this case, the force is

$$\Sigma(F_{\text{on m}})_y = F_{\text{s on m}} + F_{\text{sp}} - w = 0 \Rightarrow 20 \text{ N} + k\Delta y - mg = 0 \Rightarrow k = (mg - 20 \text{ N})/\Delta y$$
$$= (49 \text{ N} - 20 \text{ N})/0.02 \text{ m} = 1450 \text{ N/m} \approx 1500 \text{ N/m}$$

(c) In this case, the force is

$$\Sigma(F_{\text{on m}})_y = F_{\text{sp}} - w = 0 \Rightarrow k\Delta y - mg = 0 \Rightarrow \Delta y = mg/k = (49 \text{ N})/(1450 \text{ N/m}) = 0.0338 \text{ m} = 3.4 \text{ cm}$$

P8.59. Strategize: When the block is in its equilibrium position, $\Sigma \vec{F} = \vec{0}$.
Prepare: The force exerted to the right by the left spring must be balanced by a force exerted to the left by the right spring. The force exerted by the left spring is

$$(F_{\text{sp}})_1 = (10 \text{ N/m})(0.020 \text{ m}) = 0.20 \text{ N}.$$

In part (b) the left spring will exert a force to the left (because it will be stretched to the right); the right force will also exert a force to the left since it will be quite compressed. Remember, the block is no longer in equilibrium, and so we do not expect $\Sigma \vec{F} = \vec{0}$.
Solve: (a) The force exerted by the right spring must be 0.20 N. This will be produced if

$$\Delta x_2 = \frac{F_{\text{sp}}}{k_2} = \frac{0.20 \text{ N}}{20 \text{ N/m}} = 0.010 \text{ m} = 1.0 \text{ cm}$$

(b) The spring on the left will be stretched $15 \text{ cm} - 2.0 \text{ cm} = 13 \text{ cm}$ beyond its unstretched length. The force will be toward the left.

$$(F_{\text{sp}})_1 = k_1 \Delta x_1 = (10 \text{ N/m})(0.13 \text{ m}) = 1.3 \text{ N}$$

The spring on the right will be compressed $15 \text{ cm} + 1.0 \text{ cm} = 16 \text{ cm}$ from its uncompressed length. This force will also be toward the left.

$$(F_{\text{sp}})_2 = k_2 \Delta x_2 = (20 \text{ N/m})(0.16 \text{ m}) = 3.2 \text{ N}$$

Both forces are to the left so we simply add them up.

$$F_{\text{net}} = (F_{\text{sp}})_1 + (F_{\text{sp}})_2 = 1.3 \text{ N} + 3.2 \text{ N} = 4.5 \text{ N}$$

to the left.
Assess: In part (b) the right spring exerted a greater force because its k is greater and its length is farther from its unstretched/uncompressed length.

P8.61. Strategize: This problem combines several concepts. The plank is in static equilibrium, such that that sum of all forces and the sum of all torques on the plank must be zero. One of the forces that produces a torque on the plank must be calculated by assuming the tensile strength of the rope is reached.
Prepare: Assume the plank is massless. Compute the torques around the pivot at the left end of the plank. The cross section area of the rope is $A = \pi\left(\dfrac{7.0 \text{ mm}}{2}\right)^2 = 3.848 \times 10^{-5} \text{ m}^2$.

Solve: (a) The maximum tension the rope can support is $T = (6.0 \times 10^7 \text{ N/m}^2)(3.848 \times 10^{-5} \text{ m}^2) = 2300 \text{ N}$.
(b) Call the distance from the pivot to the machinery at maximum distance x.

$$\Sigma \tau = T(3.5 \text{ m}) - (800 \text{ kg})(9.8 \text{ m/s}^2)x = 0$$

$$x = \frac{(2300 \text{ N})(3.5 \text{ m})}{(800 \text{ kg})(9.8 \text{ m/s}^2)} = 1.0 \text{ m}$$

Assess: 1.0 m isn't very far along the plank, but the machinery is heavy. We could have moved the machinery farther out if the rope had been thicker.

P8.63. Strategize: This problem involves relating stress and strain through the Young's modulus.

Prepare: Model the disk as a short wide rod. We are asked for the strain—the fractional change in length of the disk. We can solve for $\Delta L/L$ from Equation 8.6.

Assume that the disk is circular so that the area is $A = \pi R^2 = \pi (D/2)^2 = \pi (0.020 \text{ m})^2 = 0.00126 \text{ m}^2$.

The force is half the weight of the person: $F = \frac{1}{2} mg = \frac{1}{2} (65 \text{ kg})(9.80 \text{ m/s}^2) = 319 \text{ N}$.

Young's modulus for cartilage is not given in the chapter, but is in the problem: $Y = 1.0 \times 10^6 \text{ N/m}^2$.

Solve: Solve Equation 8.6 for $\Delta L/L$.

$$\frac{\Delta L}{L} = \frac{F}{YA} = \frac{319 \text{ N}}{(1.0 \times 10^{10} \text{ N/m}^2)(0.00126 \text{ m}^2)} = 0.000025 = 0.0025\%$$

Assess: This means the disk compresses by only a tiny amount. This seems reasonable. Notice that the actual thickness of the disk, given as 0.50 cm, is not needed in the calculation of the fractional compression.

P8.65. Strategize: This problem involves scaling and the concept of tensile strength.

Prepare: We'll use the data from Example 8.10: $m_{original} = 70 \text{ kg}$ and $A_{original} = 4.8 \times 10^{-4} \text{ m}^2$.

The femur is not solid cortical bone material; we model it as a tube with an inner diameter and an outer diameter. Look up Young's modulus for cortical bone in Table 8.1.

Solve: (a) Both the inner and outer diameters are increased by a factor of 10; however, the cross-sectional area of the bone material does not increase by a factor of 10. Instead, because $A = \pi R^2$, the outer cross-sectional area and the inner cross-sectional area (the "hollow" of the tube) both increase by a factor of 100. But this means that the cross-sectional area of the bone material (the difference of the outer and inner areas) also increases by a factor of 100. So the new area is $A_{new} = 100(4.8 \times 10^{-4} \text{ m}^2) = 4.8 \times 10^{-2} \text{ m}^2$.

(b) Since volume is a three-dimensional concept, if we increase each linear dimension by a factor of 10 then the volume increases by a factor of $10^3 = 1000$. We assume the density of the man is the same as before, so his mass increases by the same factor as the volume: $m_{new} = 1000(70 \text{ kg}) = 70\,000 \text{ kg}$.

(c) We follow the strategy of Example 8.10. The force compressing the femur is the man's weight, $F = mg = (70\,000 \text{ kg})(9.80 \text{ m/s}^2) = 690\,000 \text{ N}$. The resulting stress on the femur is

$$\frac{F}{A} = \frac{690\,000 \text{ N}}{4.8 \times 10^{-2} \text{ m}^2} = 1.4 \times 10^7 \text{ N/m}^2$$

A stress of $1.4 \times 10^7 \text{ N/m}^2$ is 14% of the tensile strength of cortical bone given in Table 8.4.

Assess: This scaling problem illustrates clearly why animals of different sizes have different proportions. Because the volume scales with the cube of the linear dimensions and the area scales with the square of the linear dimensions then the force in F/A grows more quickly than the cross sectional area does.

P8.67. Strategize: If the woman is in static equilibrium, the sum of all torques and the sum of all forces on her must be zero.

Prepare: Compute torques around the knees to find the normal force on the hands. Then use $\Sigma F = 0$ to find the normal force on the knees.

Solve:

(a)

$$\Sigma \tau = (580 \text{ N})(54 \text{ cm}) - n(76 \text{ cm}) = 0 \Rightarrow n = 412 \text{ N}$$

Divide this by two to find the normal force on each hand: $412 \text{ N}/2 = 206 \text{ N} \approx 210 \text{ N}$

(b) The sum of the upward normal force on the hands and the upward normal force on the knees must equal the weight. Therefore the normal force on both knees is $580 \text{ N} - 412 \text{ N} = 168 \text{ N}$. Divide this by two to get the force on each knee: $168 \text{ N}/2 = 84 \text{ N}$.

Assess: It makes sense that the force on the hands is greater than the force on the knees since the center of gravity is closer to the arms than the knees.

P8.69. Strategize: This problem involves Young's modulus which is the ratio of stress/strain.

Prepare: The Young's modulus is the slope of the stress vs. strain plot we are given. We will use endpoints of the line to minimize error.

Solve: $Y = \dfrac{\text{stress}}{\text{strain}} = \dfrac{\left(5.0 \times 10^5 \text{ N/m}^2\right) - (0)}{(0.60) - (0)} = 8.3 \times 10^5 \text{ N/m}^2$. The correct answer is B.

Assess: This Young's modulus is several orders of magnitude smaller than the smallest one listed in Table 8.1, which fits with the description of the ligament as "much more elastic" than other ligaments.

P8.71. Strategize: This problem involves the definitions of stress and strain, and the relationship between them through Young's modulus.

Prepare: Inserting definitions for stress and strain, we can write $F / A = Y \Delta L / L$. We can use this to relate changes in cross-sectional area A to other variables.

Solve: Note that Y is a material constant, and we presume that the strain is held fixed in this experiment, such that the entire right hand side of $F / A = Y \Delta L / L$ is constant. That means the ratio F / A cannot change. Since $A = \pi (D / 2)^2$, doubling the diameter quadruples the area. The force must therefore also increase by a factor of 4. The correct answer is C.

Assess: It is reasonable that increasing the area by a factor of 4 means you must increase the force by a factor of 4 to maintain the same stress.

FORCE AND MOTION

PptI.1. Reason: The orbital speed depends on the mass of the central body. If there were no dark matter then the stars would orbit around the Milky Way slower so the period would be longer. The answer is A.
Assess: If the sun's mass were smaller then the earth would take more than 365 days to orbit.

PptI.3. Reason: If the orbital speeds are the same, then we simply compare the centripetal accelerations.

$$\frac{a_S}{a_e} = \frac{\dfrac{v_S^2}{R_S}}{\dfrac{v_e^2}{R_e}} = \frac{\dfrac{v_e^2}{10R_e}}{\dfrac{v_e^2}{R_e}} = \frac{1}{10}$$

The answer is A.
Assess: The acceleration is greater with the dark matter than without.

PptI.5. Reason: Solve $\vec{a} = \Delta\vec{v}/\Delta t$ for Δt.

$$\Delta t = \frac{\Delta\vec{v}}{\vec{a}} = \frac{20\text{ m/s}}{6.0\text{ m/s}^2} = 3.2\text{ s}$$

The correct answer is C.
Assess: It would take a greyhound less time to reach top speed, but 3.2 s seems reasonable for a horse.

PptI.7. Reason: Solve $a_c = v^2/r$ for r.

$$r = \frac{v^2}{a_c} = \frac{(15\text{ m/s})^2}{7.1\text{ m/s}^2} = 31.7\text{ m}$$

The correct answer is B.
Assess: From the photograph it appears 32 m is in the ballpark for the radius of the turn.

PptI.9. Reason: Because $\vec{F}_{net} = m\vec{a}$ the net force on the dog must be in the same direction as the dog's acceleration. In uniform circular motion the acceleration is toward the center of the circle, so the answer is D.
Assess: The force is provided by friction of the ground acting on the dog's paws.

PptI.11. Reason: The acceleration is constant so we can use kinematics. Solve $v_f^2 = v_i^2 + 2a\Delta x$ for Δx. The equation is simple because $v_f = 0$.

$$\Delta x = \frac{-v_i^2}{2a} = \frac{-(0.25 \text{ mm/s})^2}{2(-1.812 \text{ m/s}^2)} = 1.72 \times 10^{-8} \text{ m} \approx 0.02 \ \mu\text{m}$$

The answer is A.
Assess: The paramecium comes to rest in a distance much less than its own length.

PptI.13. Reason: At terminal speed the acceleration is zero, so the net force is also zero. The answer is B.
Assess: At terminal speed the magnitude of the drag force is the same as the magnitude of the gravitational force.

PptI.15. Reason: The mass of the falcon is $(8.0 \text{ N})/g = 0.816$ kg.

$$F_{\text{net}} = F_{\text{lift}} - mg = ma \quad \Rightarrow \quad F_{\text{lift}} = m(a + g) = (0.816 \text{ kg})(15 \text{ m/s}^2 + 9.8 \text{ m/s}^2) = 20 \text{ N}$$

The correct answer is D.
Assess: The gravitational force cannot be neglected in this problem.

PptI.17. Reason: F and d are directly proportional, so if the force (due to the weight of the person) is only half, then the deflection will also be only half. The answer is B.
Assess: This makes intuitive sense since the variables are proportional.

PptI.19. Reason: With all other variables held constant, the deflection is proportional to L^3.

$$d = F\left[\frac{4}{Ywt^3}\right]L^3$$

If the length is decreased to half, then the deflection will only be $(1/2)^3 = 1/8$ as much.

$$d' = \frac{1}{8}(4.0 \text{ cm}) = 0.50 \text{ cm}$$

The correct choice is A.
Assess: This makes sense since the variables are proportional.

PptI.21. Reason:
(a) For uniform circular motion there must be a net force toward the center of the circle.

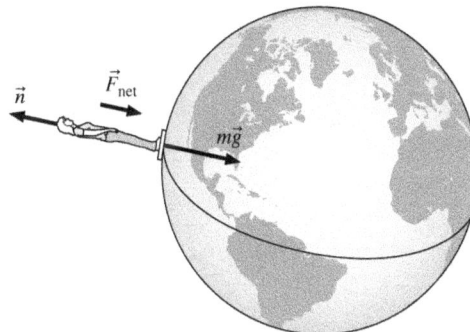

(b) The new reading is n, and the amount the reading is reduced is $mg - n$. The radius of the earth is 6.37×10^6 m. The mass of the person is $(800 \text{ N})/(9.8 \text{ m/s}^2) = 81.6$ kg. $1\text{d} = 86400$ s

$$mg - n = \Sigma F = ma = m\left(\frac{v^2}{r}\right) = m\frac{\left(\frac{2\pi r}{\Delta t}\right)^2}{r} = m\left(\frac{2\pi}{\Delta t}\right)^2 r =$$

$$(81.6 \text{ kg})\left(\frac{2\pi}{86400 \text{ s}}\right)^2 (6.37 \times 10^6 \text{ m}) = 2.8 \text{ N}$$

Assess: The reading is only 2.8 N less than 800 N.

MOMENTUM

Q9.1. Reason: Momentum depends on mass and velocity: $\vec{p} = m\vec{v}$. Increasing either the mass of the bowling ball or the velocity with which you roll it would increase the momentum.

Assess: We know from experience that at a given speed, a heavier object is more likely to knock over a target than a lightweight object. And we know that a fast-moving object is more likely to knock over a target than a slow-moving one.

Q9.3. Reason: (a) The acceleration of the puck with the smaller mass will be four times greater than the acceleration of the puck with the larger mass. The puck with the larger mass takes longer to travel the distance. The time it takes an object to move a given distance from rest can be calculated from the kinematic equation

$$\Delta x = \tfrac{1}{2} a_x (\Delta t)^2$$

Solving for the time,

$$\Delta t = \sqrt{\frac{2\Delta x}{a_x}}$$

Since the acceleration of the smaller puck is four times that of the larger puck, the time it takes the smaller puck to travel the distance is half the time it takes the larger puck.

(b) The force on each puck is the same. Since the smaller puck takes half the time to travel the distance, the impulse of the smaller puck is half the impulse of the larger puck. From the impulse-momentum theorem, the change in momentum of the smaller puck is half the change in momentum of the larger puck. The larger puck has the greater momentum after completing the distance.

Assess: The final speed of the larger puck is also twice the final speed of the smaller puck.

Q9.5. Reason: When you reach the ground, you have considerable downward momentum. Some impulse is required to change your momentum and bring you to rest, according to $\Delta \vec{p} = \vec{F} \Delta t$ (treating the force as constant).

If your knees bend, your torso is stopped over a greater distance, and over a greater time. Since $\vec{F} = \dfrac{\Delta \vec{p}}{\Delta t}$, a longer duration means the average force is smaller.

Assess: This fits our experience. Landing straight-legged can be painful, and can injure you.

Q9.7. Reason: The sum of the momenta of the three pieces must be the zero vector. Since the first piece is traveling east, its momentum will have the form $(p_1, 0)$, where p_1 is a positive number. Since the second piece is traveling north, its momentum will have the form $(0, p_2)$, where p_2 is a positive number. If a third momentum is to be added to these and the result is to be $(0,0)$, then the third momentum must be $(-p_1, -p_2)$. Since its east-west and north-south components are both negative, the momentum of the third piece must point south west and so the velocity must be south west. The answer is D.

Assess: It makes sense that the third piece would need to travel southwest. It needs a western component of momentum to cancel the eastern component of the first piece and it needs a southern component to cancel the northern component of the second piece.

Q9.9. Reason: Since both carts are stationary after the collision the final total momentum of the two-cart system is zero. By conservation of momentum, the total momentum of the system must have also been zero before the collision. Therefore the momentum of the 3 kg cart must have been the same magnitude (and opposite direction) as the momentum of the 2 kg cart. Since the momentum of the 2 kg cart was 6 kg m/s, then the speed of the 3 kg cart must have been 2 m/s.
Assess: If the carts stick together the collision is inelastic.

Q9.11. Reason: The ball must change speed from its original speed to zero whether you wear the glove or not. So $\Delta \vec{p}$ is the same in either case. The impulse-momentum theorem also tells us $\vec{F}_{avg} \Delta t = \Delta \vec{p}$.

Given that the right side is the same in either case, the left side must also be the same in either case. But if we can increase Δt then F will be decreased correspondingly. The padding of the glove increases the collision time, thereby decreasing the force.
Assess: This is also how air bags work in car collisions.

Q9.13. Reason: Looking at the asteroid coming toward us, let us suppose we direct our laser to the asteroid's right side, which we will call the $+x$ direction. The asteroid initially has no momentum in the $+x$ direction. As the surface is heated, material is ejected from the surface (as in small explosions) off to the right. That ejected material will now have momentum in $+x$. But since the asteroid and all its material is an isolated system, the total momentum cannot change. Thus the remaining asteroid must have some momentum to the left, such that $\Delta p_x = \Delta p_{ejected, x} + \Delta p_{asteroid, x} = 0$.

Assess: The asteroid would be so massive that its change in velocity would be small. Humans would need to begin the process very early on so that even a small deflection could cause it to miss Earth.

Q9.15. Reason: The impulse on each object is equal to the object's change in momentum. During the collision total momentum is conserved. The change in momentum of the rubber ball must be exactly equal to the negative of the change in momentum of the steel ball since the total change in momentum must be zero. Both balls receive the same impulse.
Assess: This can also be seen from the fact that the duration of the collision is the same for both objects, and the forces on each are action/reaction pairs. So the impulses are equal and opposite.

Q9.17. Reason: You should throw the rubber ball rather than the beanbag to increase your chances of knocking down the bowling pin. This is because the rubber ball will exert more impulse on the pin. The impulse on the bowling pin equals the negative of the impulse on the projectile, which in turn equals the negative of the change in momentum of the projectile: $J_{on\ pin} = -\Delta p_{proj}$. The rubber ball and beanbag would start with the same momentum, but the rubber ball would have a greater change in momentum because it would bounce off and so its direction would change. The beanbag would continue to move in the same direction. Thus the rubber ball would exert a greater impulse on the bowling pin.
Assess: This result make sense because we see that a collision with the rubber ball is more violent than a collision with the beanbag in that the former collision turns the projectile around.

Q9.19. Reason: Consider the Monica-platform system to be isolated (hence the frictionless axle) so that the angular momentum is conserved. As Monica walks toward the center the moment of inertia of the system decreases so the angular velocity increases because $L = I\omega$ stays constant. Once she passes the center and gets closer to the opposite edge the moment of inertia increases until it is the same value as initially. So the angular velocity also then decreases until reaching is initial value.
Assess: How much the moment of inertia decreases as she approaches the center depends on her mass relative to the mass of the platform.

Q9.21. Reason: Let us suppose an observer is off to one side such that "rotating forward" corresponds to clockwise motion of your body. If you want the bulk of your body (your torso and legs) to remain upright, rotate your arms in the clockwise direction, quickly. If you are already tipping forward/clockwise, then you have some clockwise angular momentum, and that angular momentum will be conserved for your body as a whole. But if your arms rotate clockwise very quickly, then most of the angular momentum will be in them, and not your torso and legs.

Assess: If you exert forces (causing torque) on your arms to make them rotate clockwise, then must exert forces causing a counterclockwise torque, and stopping you from rotating clockwise.

Q9.23. Reason: Let us call the rockets A and B. We know from the definition of impulse that

$$J_A = J_B \Rightarrow F_A \Delta t_A = F_B \Delta t_B \Rightarrow \Delta t_B = F_A \Delta t_A / F_B = (6 \text{ N})(2 \text{ s}) / (4 \text{ N}) = 3 \text{ s}.$$

So the correct answer is C.

Assess: Since the second rocket provides a smaller thrust, that thrust must persist longer for the second rocket. So we expect a time that is greater than 2 s.

Q9.25. Reason: Momentum is conserved, so the momentum before the collision must equal the momentum after the collision. The momentum conservation equation gives

$$m_1 v_i = (m_1 + m_2) v_f$$

Solving for the mass of the second ball,

$$m_2 = \frac{m_1 v_i - m_1 v_f}{v_f} = \frac{(1.0 \text{ kg})(2.0 \text{ m/s}) - (1.0 \text{ kg})(1.2 \text{ m/s})}{1.2 \text{ m/s}} = 0.67 \text{ kg}$$

The correct choice is A.

Assess: This result makes sense. The second ball must have small mass relative to the first ball, since the velocity of the pair is not that much less than the initial velocity of the first ball.

Q9.27. Reason: Neglect the drag force from the water on the canoe. Consider the two people, ball, and canoe as the system. The initial momentum of the system is zero. The person in the front throws the ball to the person in the rear. The ball has a negative momentum, so the rest of the system must gain a positive momentum. The canoe and people move forward. When the other person catches the ball, the canoe, people, and ball move with a common velocity. Since the momentum of the system must remain zero, the final velocity of the system must be zero. The correct choice is A.

Assess: An important thing to note here is that when the person in the rear catches the ball, the entire system will move with a common velocity.

Q9.29. Reason: The two-disk system is isolated, so the angular momentum is conserved: $L_f = L_i$. The friction force is not an external force on the system, but within the system. The moment of inertia of the identical disks is the same: call it I.

$$I\omega_1 + I\omega_2 = I(\omega_1 + \omega_2) = I_f \omega_f$$

Where I_f is the combined moment of inertia of the disks stuck together; it equals twice the moment of inertia of one disk.

$$I(\omega_1 + \omega_2) = 2I\omega_f \Rightarrow \omega_f = \tfrac{1}{2}(\omega_1 + \omega_2) = \tfrac{1}{2}(30 \text{ rpm} + 20 \text{ rpm}) = 25 \text{ rpm}$$

So the answer is C.

Assess: Intuition would have guided us to this answer. The faster disk speeds up the slower one and the slower one slows down the faster one.

Problems

P9.1. Strategize: This problem deals with the momentum of moving objects. Model the bicycle and its rider as a particle. Also model the car as a particle.
Prepare: We will use Equations 9.7 for momentum.
Solve: From the definition of momentum,

$$p_{car} = p_{bicycle} \Rightarrow m_{car} v_{car} = m_{bicycle} v_{bicycle} \Rightarrow v_{bicycle} = \frac{m_{car}}{m_{bicycle}} v_{car} = \left(\frac{1500 \text{ kg}}{100 \text{ kg}}\right)(1.0 \text{ m/s}) = 15 \text{ m/s}$$

Assess: This is a reasonable speed. This problem shows the importance of mass in comparing two momenta.

P9.3. Strategize: This problem deals with a change in momentum from an external force acting over some time interval (this is the impulse).
Prepare: We can use Equation 9.6 to calculate the change in momentum of the ball, and then use Equation 9.8 to find the impulse.
Solve: Since the ball is initially at rest, the change in the momentum of the ball is

$$\Delta p = p_f - p_i = p_f = m v_f = (0.057 \text{ kg})(45 \text{ m/s}) = 2.6 \text{ kg} \cdot \text{m/s}$$

The change in momentum of the ball is equal to the impulse on the ball, so $J = 2.6 \text{ kg} \cdot \text{m/s}$.

Assess: Note that since the ball starts from rest, it is hit at the top of its vertical motion in the serve.

P9.5. Strategize: This problem deals with a change in momentum from an external force acting over some time interval (this is the impulse).
Prepare: From Equations 9.5, 9.2, and 9.8, Newton's second law can be profitably rewritten as

$$\vec{F}_{avg} = \frac{\Delta \vec{p}}{\Delta t}$$

Solve: This allows us to find the force on the snowball. By Newton's third law we know that the snowball exerts a force of equal magnitude on the wall.

$$\vec{F}_{avg} = \frac{\Delta \vec{p}}{\Delta t} = \frac{m\vec{v}_f - m\vec{v}_i}{\Delta t} = \frac{m(\vec{v}_f - \vec{v}_i)}{\Delta t} = \frac{(0.12 \text{ kg})(0 \text{ m/s} - 7.5 \text{ m/s})}{0.15 \text{ s}} = -6.0 \text{ N}$$

where the negative sign indicates that the force on the snowball is opposite its original momentum. So the force on the wall is also 6.0 N.
Assess: This is not a large force, but the snowball has low mass, a moderate speed, and the collision time is fairly long.

P9.7. Strategize: This problem deals with the impulse of a force that is changing in time. We will use graphical methods.
Prepare: To get time in seconds, we note that $1 \text{ ms} = 10^{-3} \text{ s}$.
Solve: The impulse as defined through Equation 9.1 is

$$J_x = \text{Area under the } F_x(t) \text{ curve} \quad \text{between } t_i \text{ and } t_f \Rightarrow 6.0 \text{ N} \cdot \text{s} = \frac{1}{2}(F_{max})(8 \text{ ms}) \Rightarrow F_{max} = 1500 \text{ N}$$

Assess: Note that the impulsive force shown is a linear function of time. Therefore, the impulsive force can also be written as $0.5 F_{max}(6 \text{ ms}) + 0.5 F_{max}(2 \text{ ms})$ by looking at the first 6 ms and the next 2 ms separately. Equating this to the change in momentum should also give the same answer, $F_{max} = 1500 \text{ N}$.

P9.9. Strategize: This problem asks about a kinematic situation, but can be solved using concepts of impulse and momentum.

Prepare: Represent the stone as a particle and use Equation 9.1 and the impulse-momentum theorem, Equation 9.8. Note that the falling object is acted on by the gravitational force, whose magnitude is *mg*.

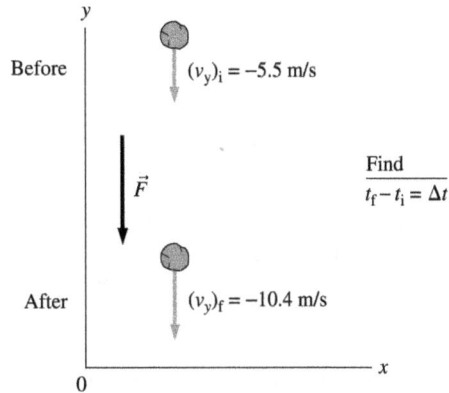

Solve: Using the impulse-momentum theorem,

$$(p_y)_f - (p_y)_i = J_y \Rightarrow m(v_y)_f - m(v_y)_i = -mg\Delta t \Rightarrow \Delta t = \frac{(v_y)_i - (v_y)_f}{g} = \frac{-5.5 \text{ m/s} - (-10.4 \text{ m/s})}{9.8 \text{ m/s}^2} = 0.50 \text{ s}$$

Assess: Acceleration due to gravity is 9.8 m/s^2 or $(9.8 \text{ m/s})/\text{s}$. That is, speed changes by 9.8 m/s every second. A speed change of 4.9 m/s will thus take only 0.5 seconds, as obtained above.

P9.11. Strategize: This problem deals with the impulse of a force that is changing in time. We will use graphical methods.

Prepare: Model the tennis ball as a particle, and its interaction with the wall as a collision. The force increases to F_{max} during the first two ms, stays at F_{max} for two ms, and then decreases to zero during the last two ms. The graph shows that F_x is positive, so the force acts to the right.

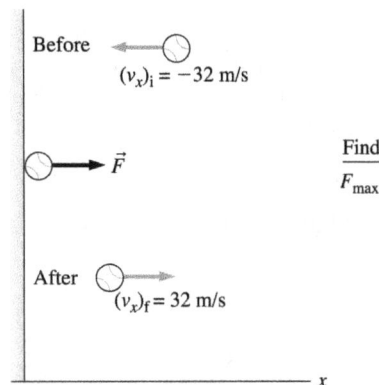

Solve: Using the impulse-momentum theorem $(p_f)_x = (p_i)_x + J_x$,

$$(0.06 \text{ kg})(32 \text{ m/s}) = (0.06 \text{ kg})(-32 \text{ m/s}) + \text{area under force graph}$$

Now,

$$\text{area under force curve} = \frac{1}{2}F_{max}(0.002 \text{ s}) + F_{max}(0.002 \text{ s}) + \frac{1}{2}F_{max}(0.002 \text{ s}) = (0.004 \text{ s})F_{max}$$

$$\Rightarrow F_{max} = \frac{(0.06 \text{ kg})(32 \text{ m/s}) + (0.06 \text{ kg})(32 \text{ m/s})}{0.004 \text{ s}} = 960 \text{ N}$$

P9.13. Strategize: This problem involves a change in momentum of a system because of an external force acting over some time interval (giving an impulse to the system).

Prepare: From Equations 9.5, 9.2, and 9.8, Newton's second law can be profitably rewritten as

$$\vec{F}_{avg} = \frac{\Delta \vec{p}}{\Delta t}$$

Solve: This allows us to find the force on the child and sled.

$$(F_x)_{ave} = \frac{\Delta p_x}{\Delta t} = \frac{m(v_x)_f - m(v_x)_i}{\Delta t} = \frac{m((v_x)_f - (v_x)_i)}{\Delta t} = \frac{(35 \text{ kg})(0 \text{ m/s} - 1.5 \text{ m/s})}{0.50 \text{ s}} = -105 \text{ N} \approx -110 \text{ N}$$

where the negative sign indicates that the force is in the direction opposite the original motion, as stated in the problem. So the *amount* (magnitude) of the average force you need to exert is 110 N.

Assess: This result is neither too large nor too small. In some collision problems Δt is quite a bit shorter and so the force is correspondingly larger.

P9.15. Strategize: This problem involves the change of momentum of two cars under two different circumstances. In the two cases, the impulses are the same. The impulse depends on both force and time.

Prepare: From Equations 9.5, 9.2, and 9.8, Newton's second law can be profitably rewritten as

$$\vec{F}_{avg} = \frac{\Delta \vec{p}}{\Delta t}$$

Solve: This allows us to find the average force on the cars.

(a) Water barrels $(F_x)_{avg} = \dfrac{\Delta p_x}{\Delta t} = \dfrac{m(v_x)_f - m(v_x)_i}{\Delta t} = \dfrac{(m((v_x)_f - (v_x)_i))}{\Delta t} = \dfrac{(1400 \text{ kg})(0 \text{ m/s} - 20 \text{ m/s})}{1.5 \text{ s}} = -19\,000 \text{ N}$

(b) Concrete barrier $(F_x)_{avg} = \dfrac{\Delta p_x}{\Delta t} = \dfrac{m(v_x)_f - m(v_x)_i}{\Delta t} = \dfrac{m((v_x)_f - (v_x)_i)}{\Delta t} = \dfrac{(1400 \text{ kg})(0 \text{ m/s} - 20 \text{ m/s})}{0.10 \text{ s}} = -280\,000 \text{ N}$

where the negative sign indicates that the force is in the direction opposite the original motion.

Assess: We clearly see that a shorter collision time dramatically affects the magnitude of the average force. From a practical standpoint, find something like water barrels or a haystack if you have to crash your car.

P9.17. Strategize: This problem deals with an impulse being delivered to a baseball, causing a change in the baseball's momentum. We will consider the baseball to by our system.

Prepare: We can use Equations 9.6 and 9.8 to calculate the impulse and Equation 9.2 to calculate the average force.

Solve: (a) Consider the ball to be moving along the positive x-axis before it is hit. Then $(v_x)_i = +15.0 \text{ m/s}$. After the collision, the velocity of the ball is $(v_x)_f = -20.0 \text{ m/s}$. The impulse on the ball is given by Equation 9.8:

$$J_x = (p_x)_f - (p_x)_i = (0.145 \text{ kg})(-20.0 \text{ m/s}) - (0.145 \text{ kg})(15.0 \text{ m/s}) = -5.08 \text{ kg} \cdot \text{m/s}$$

The magnitude is $5.08 \text{ kg} \cdot \text{m/s}$.

(b) Given the impulse and duration of the collision, Equation 9.2 gives the average force on the ball.

$$(F_x)_{avg} = \frac{J_x}{\Delta t} = \frac{-5.08 \text{ kg} \cdot \text{m/s}}{0.0015 \text{ s}} = -3400 \text{ N}$$

The magnitude is 3400 N.

Assess: Note that impulse and momentum are vectors. The impulse and force are directed in the negative x direction.

P9.19. Strategize: This problem involves momentum in a collision. We must define our system and consider whether or not we expect the system's momentum to be constant.

Prepare: We'll call the system the two carts and consider it an isolated system so we can apply the law of conservation of momentum. The action all takes place in one dimension, so we don't need y-components. Let the subscript 1 stand for the small cart and 2 for the large.

Solve:

$$(P_x)_i = (P_x)_f$$

$$(p_{1x})_i + (p_{2x})_i = (p_{1x})_f + (p_{2x})_f$$

$$m_1(v_{1x})_i + m_2(v_{2x})_i = m_1(v_{1x})_f + m_2(v_{2x})_f$$

We want to know $(v_{2x})_f$ so we solve for it. Also recall that $(v_{2x})_i = 0$ m/s, so the middle term in the following numerator drops out. The small cart recoils, which means its velocity after the collision is negative.

$$(v_{2x})_f = \frac{m_1(v_{1x})_i + m_2(v_{2x})_i - m_1(v_{1x})_f}{m_2} = \frac{(0.100 \text{ kg})(1.20 \text{ m/s}) - (0.100 \text{ kg})(-0.850 \text{ m/s})}{1.00 \text{ kg}} = 0.205 \text{ m/s}$$

Assess: The large cart does not move quickly, but the answer is reasonable because of the greater mass of the large cart.
We have followed the significant figure rules and kept three significant figures.

P9.21. Strategize: This problem can be solved using momentum conservation during this collision.
Prepare: This is a problem with no external forces acting on the system of the bullet and block, so we can use the law of conservation of momentum.
Solve: The total momentum before the bullet hits the block equals the total momentum after the bullet passes through the block so we can write

$$m_b(v_b)_i + m_{bl}(v_{bl})_i = m_b(v_b)_f + m_{bl}(v_{bl})_f \Rightarrow$$

$$(3.0 \times 10^{-3} \text{ kg})(500 \text{ m/s}) + (2.7 \text{ kg})(0 \text{ m/s}) = (3.0 \times 10^{-3} \text{ kg})(220 \text{ m/s}) + (2.7 \text{ kg})(v_{bl})_f.$$

We can solve for the final speed of the block: $(v_{bl})_f = 0.31$ m/s.

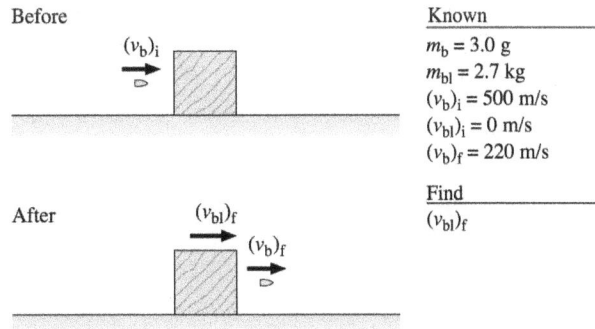

Before

$(v_b)_i$

After

$(v_{bl})_f$

$(v_b)_f$

Known

$m_b = 3.0$ g
$m_{bl} = 2.7$ kg
$(v_b)_i = 500$ m/s
$(v_{bl})_i = 0$ m/s
$(v_b)_f = 220$ m/s

Find

$(v_{bl})_f$

Assess: This is reasonable since the block is about one thousand times more massive than the bullet and its change in speed is about one thousand times less.

P9.23. Strategize: When the gravel is dropped into the car, this is a collision. The gravel and car exert forces on each other, but there is no outside force acting on the system of the gravel and the train car. So we expect the momentum of this system to be conserved.
Prepare: We will choose car + gravel to be our system. The initial x-velocity of the car is 2 m/s and that of the gravel is 0 m/s. To find the final x-velocity of the system, we will apply the momentum conservation Equation 9.15.

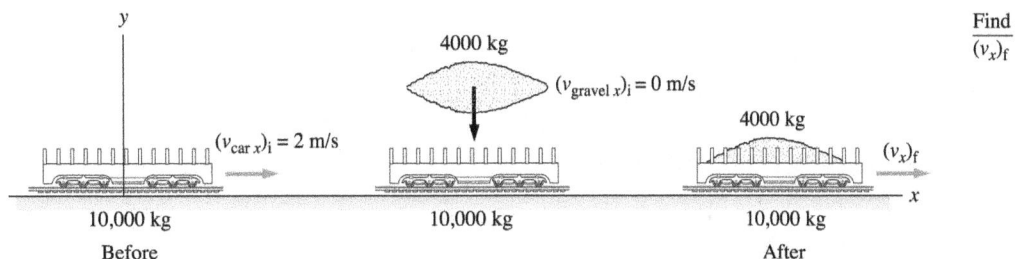

y

4000 kg

$(v_{\text{gravel }x})_i = 0$ m/s

$(v_{\text{car }x})_i = 2$ m/s

10,000 kg

Before

4000 kg

10,000 kg

$(v_x)_f$

After

Find

$(v_x)_f$

x

Solve: There are no *external* forces on the car + gravel system, so the horizontal momentum is conserved. This means $(p_x)_f = (p_x)_i$. Hence,

$$(10\,000\text{ kg} + 4000\text{ kg})(v_x)_f = (10\,000\text{ kg})(2.0\text{ m/s}) + (4000\text{ kg})(0.0\text{ m/s}) \Rightarrow (v_x)_f = 1.4\text{ m/s}$$

Assess: The motion of railroad has to be on a level track for conservation of linear momentum to hold. As we would have expected, the final speed is smaller than the initial speed.

P9.25. Strategize: This problem involves the explosive separation of a bullet from a gun in the hands of a hunter. Since no external forces are exerted on this system, we expect the total momentum of the gun, bullet, and hunter to be constant.

Prepare: The ice is frictionless. We make the hunter, gun and bullet our system, such that momentum will be conserved as discussed above. Gravity does act on all systems, but the effect of gravity during the extremely short duration of the separation will be negligible; also, we are interested in the momentum in the horizontal direction. Assume the hunter shoots to our right.

Solve: The momentum of the system before the man fires the gun is zero because everything is at rest. The momentum of the entire system must also be zero after the man fires the gun since momentum is conserved in the system. See the diagram below.

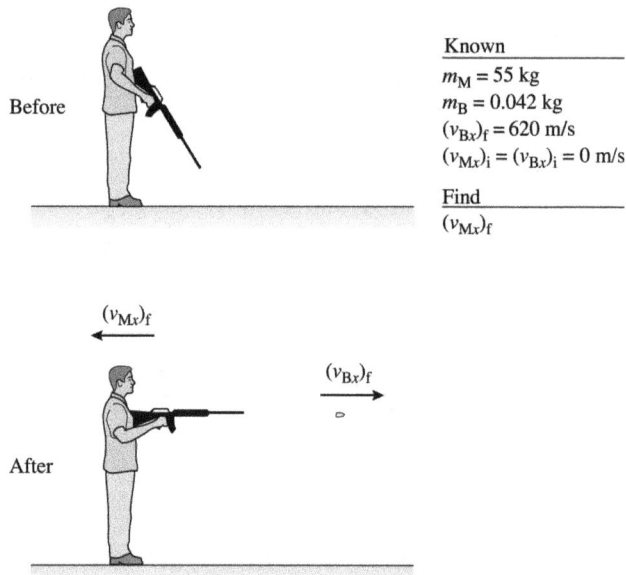

Before

After

Known

$m_M = 55$ kg
$m_B = 0.042$ kg
$(v_{Bx})_f = 620$ m/s
$(v_{Mx})_i = (v_{Bx})_i = 0$ m/s

Find

$(v_{Mx})_f$

$(v_{Mx})_f$

$(v_{Bx})_f$

Writing the momentum of the man and gun as $m_M(v_{Mx})$, and the momentum of the bullet as $m_B(v_{Bx})$, the momentum conservation equation is

$$m_M(v_{Mx})_f + m_B(v_{Bx})_f = m_M(v_{Mx})_i + m_B(v_{Bx})_i = 0$$

Solving for $(v_{Mx})_f$,

$$(v_{Mx})_f = -\left(\frac{m_B}{m_M}\right)(v_{Bx})_f = -\left(\frac{0.042\text{ kg}}{55\text{ kg}}\right)(620\text{ m/s}) = -0.47\text{ m/s}$$

The man moves to the left with a speed of 0.47 m/s.

Assess: This result seems reasonable. Though the bullet has a high speed, the mass of the bullet is much smaller than the mass of the man.

P9.27. Strategize: We will define our system to be bird + bug. There are no external forces in the horizontal direction acting on this system. So we expect the horizontal component of the momentum to be conserved.

Prepare: This is the case of an inelastic collision because the bird and bug move together after the collision.

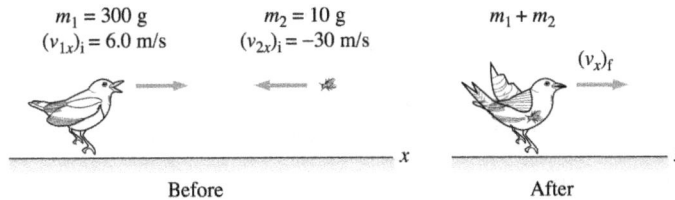

$m_1 = 300$ g	$m_2 = 10$ g	$m_1 + m_2$
$(v_{1x})_i = 6.0$ m/s	$(v_{2x})_i = -30$ m/s	$(v_x)_f$

Before   After

Solve: The conservation of momentum equation $p_{fx} = p_{ix}$ is

$$(m_1 + m_2)(v_x)_f = m_1(v_{1x})_i + m_2(v_{2x})_f$$

$$\Rightarrow (300 \text{ g} + 10 \text{ g})(v_x)_f = (300 \text{ g})(6.0 \text{ m/s}) + (10 \text{ g})(-30 \text{ m/s}) \Rightarrow (v_x)_f = 4.8 \text{ m/s}$$

Assess: We left masses in grams, rather than convert to kilograms, because the mass units cancel out from both sides of the equation. Note that $(v_{2x})_i$ is negative. As would have been expected, the final speed is a little lower than the initial speed because (1) the bug has finite mass and (2) the bug has relatively large speed compared to the bird.

P9.29. Strategize: Consider the player and the ball as the system. There is no external force in the horizontal direction on the system. Thus momentum is conserved in the horizontal direction. Gravity does act on the system from outside, but the impulse from gravity in the extremely short time of the collision is negligible, and we concern ourselves with horizontal momentum.

Prepare: At the highest point in his leap the player has no velocity, such that the initial horizontal component of momentum of the system is just that due to the baseball.

Solve: The momentum of the system before the catch is entirely due to the motion of the ball. Momentum is conserved in the system. See the following diagram.

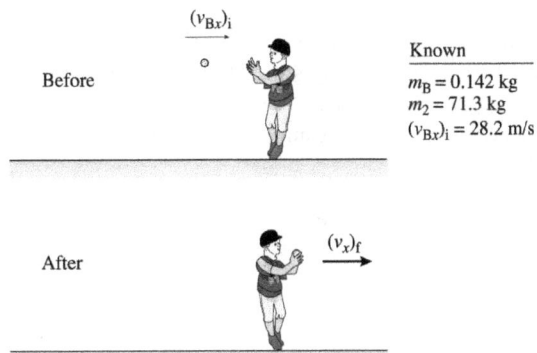

Before

$(v_{Bx})_i$

Known
$m_B = 0.142$ kg
$m_2 = 71.3$ kg
$(v_{Bx})_i = 28.2$ m/s

After $(v_x)_f$

After the player catches the ball, they move together with a common final velocity. Writing the mass of the player as m_p, and the momentum of the ball as $m_B(v_{Bx})$, the momentum conservation equation is

$$(m_p + m_B)(v_x)_f = m_B(v_{Bx})_i$$

Solving for $(v_x)_f$,

$$(v_x)_f = \frac{m_B(v_{Bx})_i}{m_p + m_B} = \frac{(0.140 \text{ kg})(28 \text{ m/s})}{71 \text{ kg} + 0.140 \text{ kg}} = 5.5 \text{ cm/s}$$

The player moves with a speed of 5.5 cm/s in the same direction the ball was originally moving.

Assess: This result seems reasonable, since the mass of the ball is so small relative to the mass of the player.

P9.31. Strategize: Even though this is an inelastic collision, momentum is still conserved during the short collision if we choose the system to be spitball plus carton.

Prepare: Let SB stand for the spitball, CTN the carton, and BOTH be the combined object after impact (we assume the spitball sticks to the carton). We are given $m_{SB} = 0.0030$ kg, $m_{CTN} = 0.020$ kg, and $(v_{BOTHx})_f = 0.30$ m/s.

Solve:

$$(P_x)_i = (P_x)_f$$

$$(p_{SBx})_i + (p_{CTNx})_i = (p_{BOTHx})_f$$

$$m_{SB}(v_{SBx})_i + m_{CTN}(v_{CTNx})_i = (m_{SB} + m_{CTN})(v_{BOTHx})_f$$

We want to know $(v_{SBx})_i$ so we solve for it. Also recall that $(v_{CTNx})_i = 0$ m/s so the last term in the following numerator drops out.

$$(v_{SBx})_i = \frac{(m_{SB} + m_{CTN})(v_{BOTHx})_f - m_{CTN}(v_{CTNx})_i}{m_{SB}} = \frac{(0.0030 \text{ kg} + 0.020 \text{ kg})(0.30 \text{ m/s})}{0.0030 \text{ kg}} = 2.3 \text{ m/s}$$

Assess: The answer of 2.3 m/s is certainly within the capability of an expert spitballer.

P9.33. Strategize: Let the system be made up of the two blocks. Since there are no external forces, we can use the law of conservation of momentum.

Prepare: This is a perfectly inelastic collision, meaning the two blocks stick together and move as one unit after the collision.

Solve: The sum of the momenta of the two blocks before the collision equals the momentum of the coupled blocks after the collision so we write

$$m_1(v_{1x})_i + m_2(v_{2x})_i = (m_1 + m_2)(v_x)_f \Rightarrow$$
$$(2.0 \text{ kg})(1.0 \text{ m/s}) + m_2(4.0 \text{ m/s}) = (2.0 \text{ kg} + m_2)(2.0 \text{ m/s}).$$

We can solve this last equation for the second mass and obtain $m_2 = 1.0$ kg.

Assess: It is reasonable that the second mass is less than the first because the final speed, 2.0 m/s, is closer to the initial speed of the first block, 1.0 m/s, than it is to the initial speed of the second block, 4.0 m/s.

P9.35. Strategize: This problem deals with the conservation of momentum in two dimensions. Let the system be made up of the two balls of clay. Presumably gravity still acts on them, but the effects of the impulse from gravity acting over the very short duration of the collision itself are negligible.

Prepare: This is a perfectly inelastic collision, Meaning the balls stick together and more as one unit after the collision. We will thus use Equation 9.14.

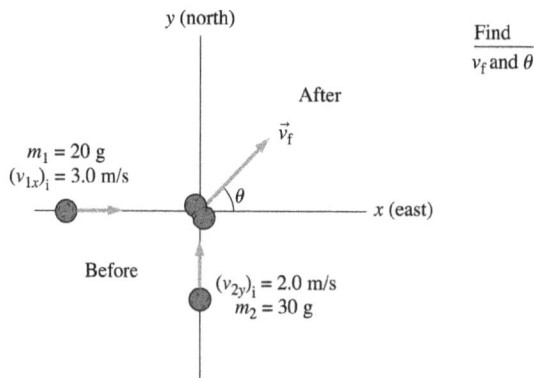

Solve: The conservation of momentum equation $\vec{p}_{before} = \vec{p}_{after}$ is

$$m_1(v_{1x})_i + m_2(v_{2x})_i = (m_1 + m_2)(v_x)_f \qquad m_1(v_{1y})_i + m_2(v_{2y})_i = (m_1 + m_2)(v_y)_f$$

Substituting in the given values,

$$(.02 \text{ kg})(3.0 \text{ m/s}) + 0 \text{ kg} \cdot \text{m/s} = (.02 \text{ kg} + .03 \text{ kg})v_f \cos \theta$$

$$0 \text{ kg m/s} + (.03 \text{ kg})(2.0 \text{ m/s}) = (.02 \text{ kg} + .03 \text{ kg})v_f \sin \theta \Rightarrow v_f \cos \theta = 1.2 \text{ m/s}$$

$$v_f \sin \theta = 1.2 \text{ m/s} \Rightarrow v_f = \sqrt{(1.2 \text{ m/s})^2 + (1.2 \text{ m/s})^2} = 1.7 \text{ m/s}$$

$$\theta = \tan^{-1} \frac{v_y}{v_x} = \tan^{-1}(1) = 45°$$

The ball of clay moves 45° north of east at 1.7 m/s.

Assess: Noting that the *magnitude* of momenta of both particles have the same value 0.6 kg · m/s. This information, along with the fact that they are coming at each other at 90 degrees, may be used to reason out that the outgoing path will be right in the middle, that is, at 45 degrees as shown above.

P9.37. Strategize: We are not told about any external forces acting on the system of the two objects. We assume that the momentum is conserved in the collision. This is a collision in two dimensions.

Prepare: We can apply the conservation of momentum to each direction (*x* and *y*) separately, using the given information.

Solve: The conservation of momentum Equation 9.14 yields

$$(p_{1x})_f + (p_{2x})_f = (p_{1x})_i + (p_{2x})_i \Rightarrow (p_{1x})_f + 0 \text{ kg} \cdot \text{m/s} = 2 \text{ kg} \cdot \text{m/s} - 4 \text{ kg} \cdot \text{m/s} \Rightarrow (p_{1x})_f = -2 \text{ kg} \cdot \text{m/s}$$

$$(p_{1y})_f + (p_{2y})_f = (p_{1y})_i + (p_{2y})_i \Rightarrow (p_{1y})_f - 1 \text{ kg} \cdot \text{m/s} = 2 \text{ kg} \cdot \text{m/s} + 1 \text{ kg} \cdot \text{m/s} \Rightarrow (p_{1y})_f = 4 \text{ kg} \cdot \text{m/s}$$

The final momentum vector of particle 1 that has the above components is shown below.

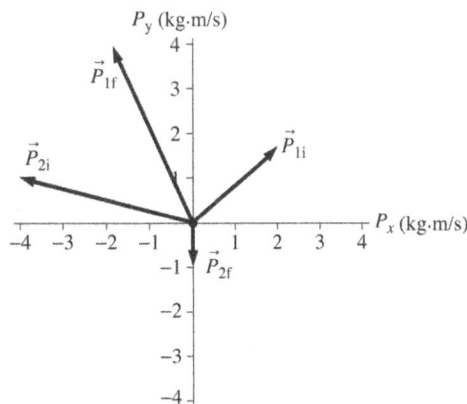

P9.39. Strategize: Our system is comprised of three coconut pieces that are modeled as particles. During the blow up or "explosion," there are no significant external forces acting on this system, so the total momentum of the system is conserved in the *x*-direction and the *y*-direction.

Prepare: This problem deals with an explosive separation, so we can thus apply Equation 9.14. A diagram will help us organize our solution.

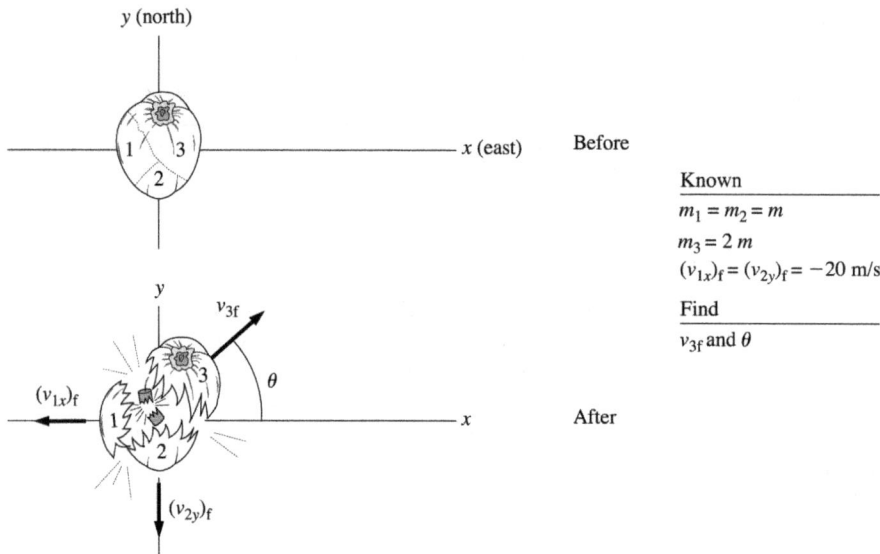

Known

$m_1 = m_2 = m$

$m_3 = 2m$

$(v_{1x})_f = (v_{2y})_f = -20$ m/s

Find

v_{3f} and θ

After

Solve: The initial momentum is zero. From $(p_x)_f = (p_x)_i$ we get

$$+m_1(v_{1x})_f + m_3(v_{3f})\cos\theta = 0 \text{ kg m/s} \Rightarrow (v_{3f})\cos\theta = \frac{-m_1(v_{fx})_1}{m_3} = \frac{-m(-20 \text{ m/s})}{2m} = 10 \text{ m/s}$$

From $(p_y)_f = (p_y)_i$, we get

$$+m_2(v_{2y})_f + m_3(v_{3f})\sin\theta = 0 \text{ kg m/s} \Rightarrow (v_{3f})\sin\theta = \frac{-m_2(v_{fy})_2}{m_3} = \frac{-m(-20 \text{ m/s})}{2m} = 10 \text{ m/s}$$

$$\Rightarrow (v_{3f}) = \sqrt{(10 \text{ m/s})^2 + (10 \text{ m/s})^2} = 14 \text{ m/s} \qquad \theta = \tan^{-1}(1) = 45°$$

Assess: The obtained speed of the third piece is of similar order of magnitude as the other two pieces, which is physically reasonable.

P9.41. Strategize: The disk is a rotating rigid body. The angular momentum of a rigid body depends on how fast it is spinning and on the moment of inertia. Different rigid bodies have different expressions for their moments of inertia, because mass is distributed differently.

Prepare: The angular velocity ω is 600 rpm = $600 \times 2\pi/60$ rad/s = 20π rad/s. rad/s. From Table 7.1, the moment of inertial of the disk about its center is $(1/2)\,MR^2$, which can be used with $L = I\omega$ to find the angular momentum.

Solve:

$$I = \frac{1}{2}MR^2 = \frac{1}{2}(2.0 \text{ kg})(0.020 \text{ m})^2 = 4.0 \times 10^{-4} \text{ kg} \cdot \text{m}^2$$

Thus, $L = I\omega = (4.0 \times 10^{-4} \cdot \text{kg m}^2)(20\pi \text{ rad/s}) = 0.025 \text{ kg m}^2/\text{s}$.

Assess: For objects of size ~1 m and mass ~1 kg spinning at ~1 rad/s, an answer on order of ~1 kg· m²/s is reasonable.

P9.43. Strategize: We neglect any small frictional torque the ice may exert on the skater and apply the law of conservation of angular momentum.

Prepare: We can apply Equation 9.22:

$$L_i = L_f$$
$$I_i \omega_i = I_f \omega_f$$

Even though the data for $\omega_i = 5.0$ rev/s is not in SI units, it's okay because we are asked for the answer in the same units. We are also given $I_i = 0.80$ kg·m² and $I_f = 3.2$ kg·m².

Solve:

$$\omega_f = \frac{I_i \omega_i}{I_f} = \frac{(0.80 \text{ kg}\cdot\text{m}^2)(5.0 \text{ rev/s})}{3.2 \text{ kg}\cdot\text{m}^2} = 1.25 \text{ rev/s} \approx 1.3 \text{ rev/s}$$

Assess: I increased by a factor of 4, so we expect ω to decrease by a factor of 4.

P9.45. Strategize: This problem involves the calculation of angular momentum, which depends on an object's rotational speed, and on the distribution of mass around an axis of rotation.

Prepare: The moon is treated as a particle. The period of the moon is $T = 27.3$ days $= 2.36 \times 10^6$ s, its orbit radius is $r = 3.8 \times 10^8$ m, and its mass is $m = 7.4 \times 10^{22}$ kg. A particle moving in a circular orbit of radius r with velocity v has an angular momentum $L = mvr$. While m and r are given, we will determine v using: $v = 2\pi r/T$.

Solve:

$$L = mvr = m\frac{2\pi}{T}r^2 = 7.4 \times 10^{22} \text{ kg} \frac{2\pi}{2.36 \times 10^6 \text{ sec}} (3.8 \times 10^8 \text{ m})^2 = 2.8 \times 10^{34} \text{ kg} \frac{\text{m}^2}{\text{s}}$$

Assess: Note that the rotation of moon or the earth about their respective axes does not come into these calculations.

P9.47. Strategize: This problem involves a time dependent force causing a change in momentum. Because these things are related through the concept of impulse, we have sufficient information to determine the duration over which the force acts.

Prepare: Model the glider cart as a particle, and its interaction with the spring as a collision. The initial and final speeds of the glider are shown on the velocity graph and the mass of the glider is known. We can thus find the momentum change of the glider, which is equal to the impulse. Impulse is also given by the area under the force graph, which we will find from the force graph in terms of Δt.

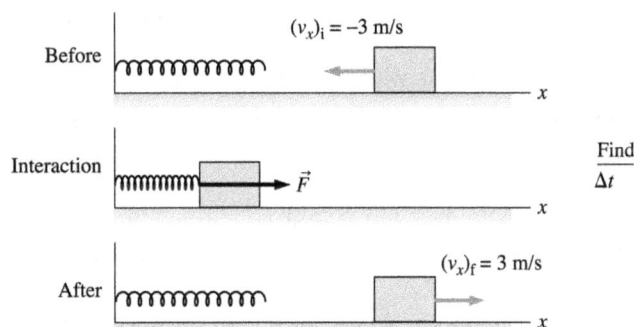

Solve: Using the impulse-momentum theorem $(p_x)_f - (p_x)_i = J_x$,

$$(0.60 \text{ kg})(3 \text{ m/s}) - (0.60 \text{ kg})(-3 \text{ m/s}) = \text{area under force curve} = \tfrac{1}{2}(36 \text{ N})(\Delta t) \Rightarrow \Delta t = 0.2 \text{ s}$$

Assess: You can solve this problem using kinematics to check your answer. From the graph you have the average force during compression to be 18 N, and therefore the average acceleration to be 30 m/s². Now calculate the time taken for the velocity to go from 3 m/s to 0 m/s, and twice this time should match the 0.2 s found in the previous figure.

P9.49. Strategize: This problem involves a change in momentum caused by a time-dependent force. We can relate the two through impulse.

Prepare: Model the ball as a particle that is subjected to an impulse when it is in contact with the floor. We will also use constant-acceleration kinematic equations.

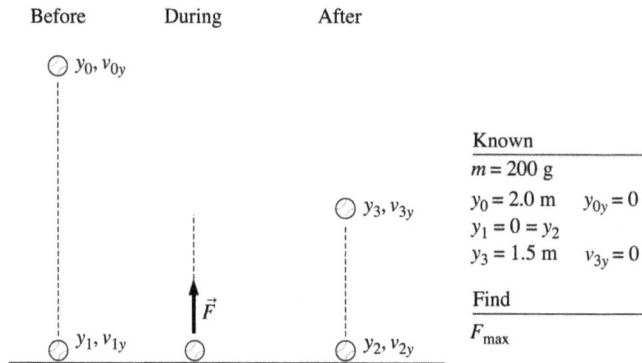

Before   During   After

○ y_0, v_{0y}

                                        ○ y_3, v_{3y}

Known
────────
$m = 200$ g
$y_0 = 2.0$ m $y_{0y} = 0$
$y_1 = 0 = y_2$
$y_3 = 1.5$ m $v_{3y} = 0$

↑$\vec{F}$

Find
────────
F_{max}

○ y_1, v_{1y} ○   ○ y_2, v_{2y}

Solve: To find the ball's velocity just before and after it hits the floor:

$$v_{1y}^2 = v_{0y}^2 + 2a_y(y_1 - y_0) = 0 \text{ m}^2/\text{s}^2 + 2(-9.8 \text{ m/s}^2)(0 - 2.0 \text{ m}) \Rightarrow v_{1y} = -6.261 \text{ m/s}$$

$$v_{3y}^2 = v_{2y}^2 + 2a_y(y_3 - y_2) \Rightarrow 0 \text{ m}^2/\text{s}^2 = v_{2y}^2 + 2(-9.8 \text{ m/s}^2)(1.5 \text{ m} - 0 \text{ m}) \Rightarrow v_{2y} = 5.422 \text{ m/s}$$

The force exerted by the floor on the ball can be found from the impulse-momentum theorem:

$$J_y = \text{area under the force curve} = \Delta p_y = mv_{2y} - mv_{1y}$$

or

$$\frac{1}{2}F_{max}\Delta t = mv_{2y} - mv_{1y}$$

so that

$$F_{max} = \frac{2m(v_{2y} - v_{1y})}{\Delta t} = \frac{2(0.20 \text{ kg})(5.42 \text{ m/s} - (-6.26 \text{ m/s}))}{5.0 \times 10^{-3} \text{ s}} = 930 \text{ N}$$

Assess: A force of 930 N exerted by the floor is typical of such collisions.

P9.51. Strategize: This problem deals with a change in a woman's vertical momentum due to a time-dependent force. We can relate the force (and the duration over which it is exerted) to her change in momentum through the concept of impulse.

Prepare: We combine Equation 9.1 $J = F_{avg}\Delta t$ with the impulse-momentum theorem in the y-direction $J_y = \Delta p_y$.

See Example 9.1. This tells us that Δp_y is the area under the curve of the net force in the vertical direction vs. time.

And if we know the change in the woman's vertical momentum we can figure out the speed with which she leaves the ground; to do this last step, however, we'll need her mass.

Look at the first part of the graph while the force exerted by the floor is constant. During that time she isn't accelerating, so the force the floor exerts must be equal in magnitude to her weight; so she weighs 600 N and her mass is 600 N/(9.8 m/s²) = 61.2 kg.

It should also be clear from the graph that she leaves the floor at $t = 0.5$ s when the force of the floor on her is zero. The graph we are given is not the graph of the net force. The hint warns us that the upward force of the floor is not the only force on the woman. We have just concluded that the earth is exerting a downward gravitational force of 600 N (her weight) on her. Therefore a graph of the *net* vertical force on her vs. time would simply be the same graph only 600 N lower on the force axis.

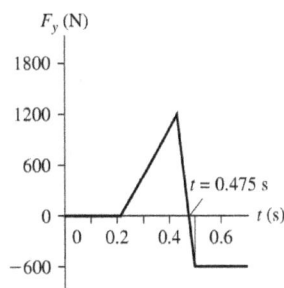

Note carefully that the graph now crosses the t-axis at $t = 0.475\,\text{s}$.

Solve: What is the area under the new graph? We'll take the area of the triangle above the t-axis to be positive and then subtract the area of the smaller triangle below the t-axis. The general formula for the area of a triangle is

$$A = \tfrac{1}{2} \times \text{height} \times \text{base}$$
$$A_{\text{big}} = \tfrac{1}{2}(1800\ \text{N})(0.275\ \text{s}) = 247.5\ \text{N} \cdot \text{s}$$
$$A_{\text{small}} = \tfrac{1}{2}(600\ \text{N})(0.025\ \text{s}) = 7.5\ \text{N} \cdot \text{s}$$

The total area (with the triangle above the axis positive and the triangle below the axis negative) is $247.5\ \text{N} \cdot \text{s} - 7.5\ \text{N} \cdot \text{s} = 240\ \text{N} \cdot \text{s}$; this is the vertical impulse on the woman, and is also equal to her change in vertical momentum. Since $\Delta p_y = m\Delta v_y$ we simply divide by her mass to find her change in velocity:

$$\Delta v_y = \Delta p_y/m = (240\ \text{N} \cdot \text{s})/(61.2\ \text{kg}) = 3.9\ \text{m/s}$$

Because she started from rest this value is also her final speed, just as she leaves the ground.

Assess: We assumed the ability to read the data from the graph to two significant figures. If we were not confident in this we would report the result to just one significant figure: $(v_y)_{\text{f}} \approx 4\ \text{m/s}$. The answer of $\approx 4\ \text{m/s}$ does seem to be in the reasonable range.

It is worth following the units in the last equation to see the answer end up in m/s.

P9.53. Strategize: This problem involves a change in momentum in three dimensions. We can apply the impulse-momentum theorem in each direction separately to determine the vector components of the impulse.

Prepare: To find the impulse delivered by the bat to the ball, we need to know the change in the ball's momentum and use $J = \Delta p$. Since the direction of the ball changes, we need to use vector components. The x-component of the ball's final velocity is

$$(v_x)_{\text{f}} = (-55\ \text{m/s})\cos(25°) = -49.8\ \text{m/s} \quad \text{and the } y\text{-component is} \quad (v_y)_{\text{f}} = (-55\ \text{m/s})\cos(25°) = -23.2\ \text{m/s}$$

Solve: The initial velocity of the ball is $(v_x)_{\text{i}} = 35\ \text{m/s}$ and its initial momentum is obtained by multiplying by the mass of the ball:

$$(p_x)_{\text{i}} = (0.140\ \text{kg})(35\ \text{m/s}) = 4.90\ \text{kg} \cdot \text{m/s} \qquad (p_y)_{\text{i}} = 0\ \text{kg} \cdot \text{m/s}$$

The initial final momentum is the final velocity of the ball times its mass:

$$(p_x)_{\text{f}} = (0.140\ \text{kg})(-49.8\ \text{m/s}) = -6.97\ \text{kg} \cdot \text{m/s}$$
$$(p_y)_{\text{f}} = (0.140\ \text{kg})(23.2\ \text{m/s}) = 3.25\ \text{kg} \cdot \text{m/s}$$

Finally, the impulse on the ball equals the change in the ball's momentum:

$$J_x = (p_x)_{\text{f}} - (p_x)_{\text{i}} = -11.9\ \text{kg} \cdot \text{m/s} \qquad J_y = (p_y)_{\text{f}} - (p_y)_{\text{i}} = 3.25\ \text{kg} \cdot \text{m/s}$$

The magnitude of the impulse can be obtained from the Pythagorean theorem: $J = 12\ \text{kg} \cdot \text{m/s}$ and we can find the angle, θ, above the horizontal using inverse tangent: $\theta = \tan^{-1}(3.25/11.9) = 15°$. The direction is to the left and $15°$ above the horizontal.

Before

$\vec{v}_i$

Known
$m_{ball} = 145$ g
$\vec{v}_i = (35$ m/s, 0 m/s$)$
$\vec{v}_f = 55$ m/s, 25° above the horizontal
$\theta = 25°$

Find
$\vec{J}$

$\vec{v}_f$

After θ

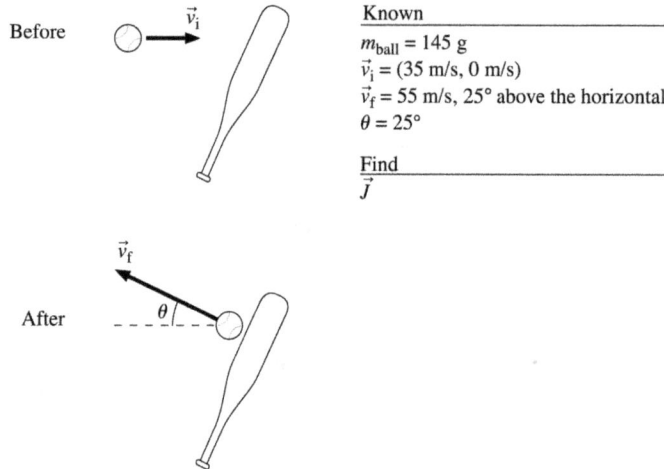

Assess: The angle 15° makes sense because the ball comes in at 0° with the horizontal and leaves the bat at 25° above the horizontal. We expect the force, and therefore the impulse, exerted by the bat to have an angle intermediate to these two.

P9.55. Strategize: Use the particle model for the ball of clay (C) and the 1.0 kg block (B). The two objects are a system and it is a case of a perfectly inelastic collision.
Prepare: Since no significant external forces act on the system in the x-direction during the collision, momentum is conserved along the x-direction.

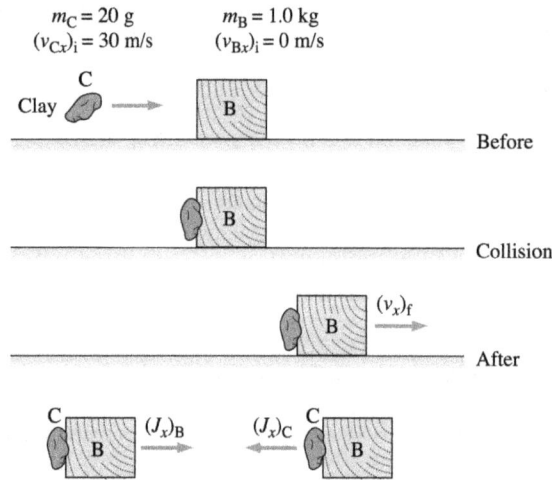

$m_C = 20$ g $m_B = 1.0$ kg
$(v_{Cx})_i = 30$ m/s $(v_{Bx})_i = 0$ m/s

C

Clay   B

Before

B

Collision

$(v_x)_f$

B

After

C $(J_x)_B$ $(J_x)_C$ C

B   B

Solve: (a) The conservation of momentum equation $(p_x)_f = (p_x)_i$ is

$$(1.0 \text{ kg} + 0.02 \text{ kg})(v_x)_f = (0.02 \text{ kg})(30 \text{ m/s}) + (1.0 \text{ kg})(0 \text{ m/s}) \Rightarrow (v_x)_f = 0.588 \text{ m/s} = 0.59 \text{ m/s}$$

(b) The impulse of the ball of clay on the block is calculated as follows:

$$(p_{Bx})_f = (p_{Bx})_i + (J_{Bx}) \Rightarrow (J_{Bx}) = m_B(v_{Bx})_f - m_B(v_{Bx})_i = (1.0 \text{ kg})(v_x)_f - 0 \text{ N} \cdot \text{s} = 0.588 \text{ N} \cdot \text{s} = 0.59 \text{ kg} \cdot \text{m/s}$$

(c) The impulse of the block on the ball of clay is calculated as follows:

$$(p_{Cx})_f = (p_{Cx})_i + (J_{Cx}) \Rightarrow (J_{Cx}) = m_C(v_{Cx})_f - m_C(v_{Cx})_i = (0.02 \text{ kg})(0.588 \text{ m/s}) - (0.02 \text{ kg})(30 \text{ m/s}) = -0.59 \text{ N} \cdot \text{s}$$

(d) Yes, $J_{Bx} = -J_{Cx}$
Assess: During the collision, the ball of clay and the block exert equal and opposite forces on each other for the same time. Impulse is therefore also equal in magnitude but opposite in direction.

P9.57. Strategize: This problem deals with conservation of momentum of a system of objects, as well as the concept of relative motion.

Prepare: We can find the speed of the cart using the law of conservation of momentum. What makes this tricky is that Ethan's speed is given relative to the cart but we need to find the speed of the cart relative to the ground. When using the conservation of momentum, all the velocities must be relative to the same observer. Ethan's velocity relative to the ground is the sum of his velocity relative to the cart and the velocity of the cart relative to the ground. When he has reached his top speed, we have: $(v_{Eg})_f = 8.0 \text{ m/s} + (v_{cg})_f$.

Solve: Initially, Ethan and the cart are at rest, so their total momentum is zero. We can now write down the equation for the conservation of momentum relative to the ground:

$$0 \text{ kg} \cdot \text{m/s} = m_E(v_{Eg})_f + m_c(v_{cg})_f \Rightarrow$$
$$0 \text{ kg} \cdot \text{m/s} = (80 \text{ kg})(8.0 \text{ m/s} + (v_{cg})_f) + (500 \text{ kg})(v_{cg})_f.$$

This equation can be solved for the velocity of the cart: $(v_{cg})_f = -1.1 \text{ m/s}$. The speed of the cart when Ethan has reached his top speed is 1.1 m/s.

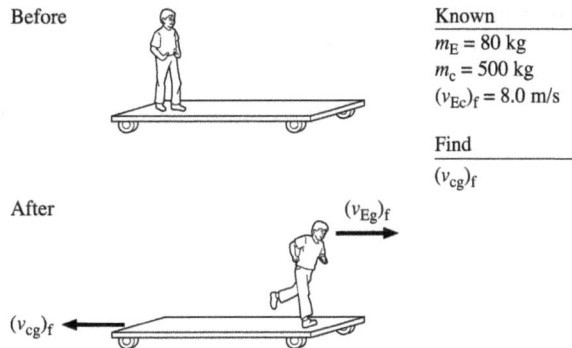

Before

Known
$m_E = 80$ kg
$m_c = 500$ kg
$(v_{Ec})_f = 8.0$ m/s

Find
$(v_{cg})_f$

After $(v_{Eg})_f$

$(v_{cg})_f$

Assess: Relative to the ground, Ethan's speed is 6.9 m/s. It is reasonable that Ethan is moving about six times as fast as the cart because the cart is about six times as massive as Ethan.

P9.59. Strategize: This problem involves conservation of momentum. Specifically, the football players collide in mid-air. No forces other than gravity act on the system made up of the two football players. Although gravity does act during this time, its effect over the small time interval of the collision will be small, and it can only affect the momentum in the vertical direction.

Prepare: We arbitrarily pick the direction the linebacker was running as the positive x-direction since he was mentioned first. If, after we solve the conservation of momentum equation for v_f, the answer is positive, then we know the linebacker ends up moving forward; if v_f is negative, then the quarterback ends up moving forward. We will use subscripts l for linebacker and q for quarterback.

Known
$m_l = 110$ Kg
$m_q = 82$ Kg
$(v_{lx}) = 2.0$ m/s
$(v_{qx})_i = -3.0$ m/s
Find
v_f

We employ the conservation of momentum. Since the collision is inelastic (the linebacker grabs and holds onto the quarterback) the final momentum will be $P_f = (m_l + m_q)v_f$, where v_f is the answer we seek.

Solve:

$$\vec{P}_f = \vec{P}_i$$

$$(m_1 + m_q)v_f = m_1(v_{1x})_i \, m_q(v_{qx})_i$$

$$v_f = \frac{m_1(v_{1x})_i + m_q(v_{qx})_i}{m_1 + m_q} = \frac{(110\text{ kg})(2.0\text{ m/s}) + (82\text{ kg})(-3.0\text{ m/s})}{110\text{ kg} + 82\text{ kg}} = -0.14\text{ m/s}$$

The answer is negative; this indicates that the quarterback ends up moving forward after the hit. (The linebacker gets knocked backward.)

Assess: If all we want to know is the sign of the answer then we do not really need to compute or divide by the denominator—the total mass will certainly be positive and will not affect the sign of the answer. So we could have done a simple mental calculation of the numerator $(82 \times 3 > 110 \times 2)$ to figure out which football players "wins."

P9.61. Strategize: Let us ignore frictional forces and drag, such that our system of two skaters is isolated and the total momentum of that system will be conserved. Note that technically normal forces and gravity still act on our system from outside, but cancel pairwise.

Prepare: Apply conservation of momentum in the x direction. Call the 75-kg skater skater 1, and the lighter skater call skater 2.

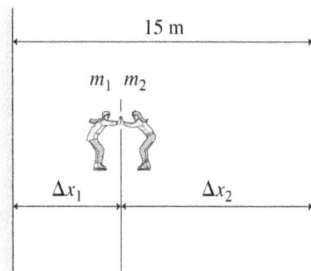

Known
$$m_1 = 75\text{ Kg}$$
$$m_2 = 55\text{ Kg}$$
$$\Delta x_1 + \Delta x_2 = 15\text{ m}$$
Find
$$\Delta x_1$$

Solve:

$$\Sigma(p_x)_i = \Sigma(p_x)_f$$

$$0 = -m_1(v_1)_f + m_2(v_2)_f$$

$$\frac{(v_2)_f}{(v_1)_f} = \frac{m_1}{m_2}$$

Now use $v = \Delta x / \Delta t$ and also that the time intervals for both skaters is the same so Δt cancels.

$$\frac{\Delta x_2}{\Delta x_1} = \frac{(v_2)_f}{(v_1)_f} = \frac{m_1}{m_2} \Rightarrow \Delta x_2 = \frac{m_1}{m_2}\Delta x_1$$

We still have two unknowns in the equation, so we employ another equation:

$$\Delta x_1 + \Delta x_2 = 15\text{ m}$$

Substitute in our expression for Δx_2.

$$\Delta x_1 + \frac{m_1}{m_2}\Delta x_1 = 15\text{ m} \Rightarrow \Delta x_1 = \frac{15\text{ m}}{1 + \frac{m_1}{m_2}} = \frac{15\text{ m}}{1 + \frac{75\text{ kg}}{55\text{ kg}}} = 6.3\text{ m}$$

Assess: A similar computation for Δx_2 gives 8.7 m so the total is 15 m, as expected. We expected the heavier skater to move slower and cover less distance than the lighter skater in the same amount of time.

P9.63. Strategize: This problem deals with conservation of momentum in two dimensions. Let us consider our system to be made up of the two balls. As long as external forces on this system are zero or negligibly small, then we can apply conservation of momentum in the east-west and north-south directions, separately.

Prepare: The billiard balls will be modeled as particles. The two balls in our system are m_1 (moving east) and m_2 (moving west). This is an isolated system because any frictional force during the brief collision period is going to be insignificant. Within the impulse approximation, the momentum of our system will be conserved in the collision. Note that $m_1 = m_2 = m$.

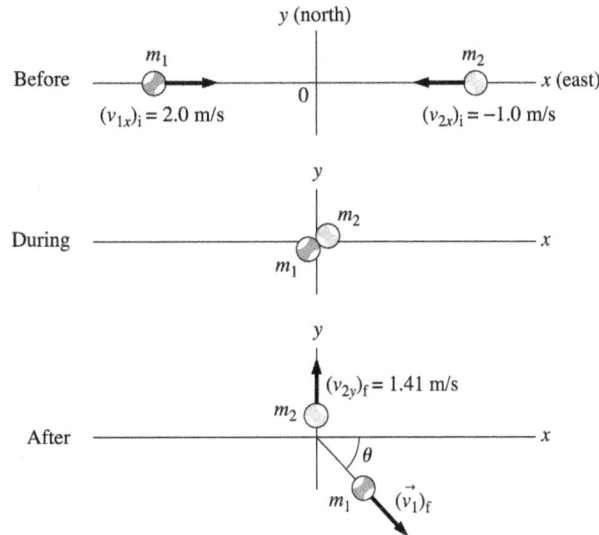

Solve: The equation $(p_x)_f = (p_x)_i$ yields:

$$m_1(v_{1x})_f + m_2(v_{2x})_f = m_1(v_{1x})_i + m_2(v_{2x})_i \Rightarrow m_1(v_1)_f \cos\theta + 0 \text{ kg} \cdot \text{m/s} = m_1(v_{1x})_i + m_2(v_{2x})_i$$

$$\Rightarrow (v_1)_f \cos\theta = (v_{1x})_i + (v_{2x})_i = 2.0 \text{ m/s} - 1.0 \text{ m/s} = 1.0 \text{ m/s}$$

The equation $(p_y)_f = (p_y)_i$ yields:

$$m_1(v_1)_f \sin\theta + m_2(v_{2y})_f = 0 \text{ kg} \cdot \text{m/s} \Rightarrow (v_1)_f \sin\theta = -(v_{2y})_f = -1.41 \text{ m/s}$$

$$\Rightarrow (v_1)_f = \sqrt{(1.0 \text{ m/s})^2 + (-1.41 \text{ m/s})^2} = 1.7 \text{ m/s}$$

$$\theta = \tan^{-1}\left(\frac{1.41 \text{ m/s}}{1.0 \text{ m/s}}\right) = 55°$$

The angle is below $+x$ axis.

Assess: The speed obtained is of the same order of magnitude as the other speeds as would be expected.

P9.65. Strategize: Let us consider our system to be the mosquito and the raindrop. If external forces acting our system are negligible, then we can use conservation of momentum.

Prepare: Gravity does act on our system, but we can consider its effect on the mosquito to be negligible compared to the very large effect on the mosquito from the raindrop. The force of gravity on the raindrop is no negligible, but must be cancelled out by drag if the droplet is falling at a constant speed. Use conservation of momentum in the y direction. Because it is hovering, the initial y velocity of the mosquito is 0 m/s.

Solve: (a)

$$\Sigma(p_y)_i = \Sigma(p_y)_f$$

$$m_d(v_d)_i = (m_d + m_m)v_f$$

$$v_f = \frac{m_d(v_d)_i}{m_d + m_m} = \frac{(40m_m)(8.2 \text{ m/s})}{40m_m + m_m} = \frac{40}{41}(8.2 \text{ m/s}) = 8.0 \text{ m/s}$$

(b) Use the definition of acceleration.

$$a = \frac{\Delta v}{\Delta t} = \frac{8.0 \text{ m/s}}{8.0 \text{ ms}} = 1000 \text{ m/s}^2 = (1000 \text{ m/s}^2)\left(\frac{1g}{9.8 \text{ m/s}^2}\right) = 100g$$

Assess: This is an impressive acceleration to survive.

P9.67. Strategize: This is a three part problem. During the explosion itself, gravity and other external forces are negligible, and we can apply conservation of momentum in the vertical direction. But during the vertical rise and the later fall gravity is certainly not negligible.

Prepare: In the first part, we will use kinematics equations to find the vertical position where the rocket breaks into two pieces. In the second part, we will apply conservation of momentum along the y direction to the system (that is, the two fragments) in the explosion. The momentum conservation "applies" because the forces involved during the explosion are much larger than the external force due to gravity during the small period that the explosion lasts. In the third part, we will again use kinematics equations to find the velocity of the heavier fragment just after the explosion.

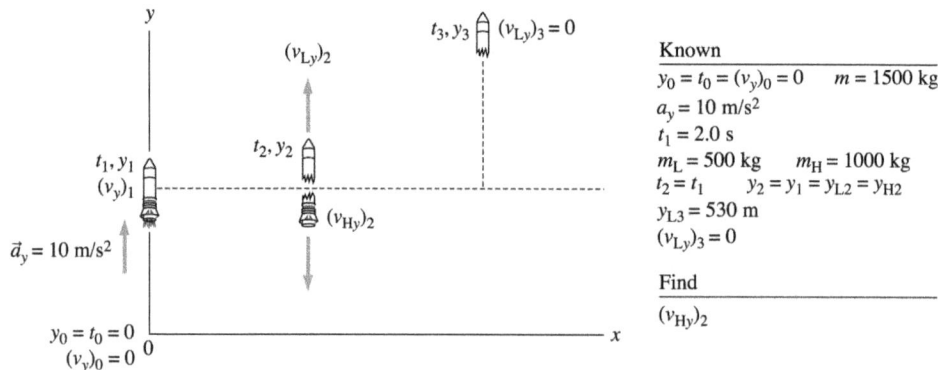

Solve: The rocket accelerates for 2.0 s from rest, so

$$(v_y)_1 = (v_y)_0 + a_y(t_1 - t_0) = 0 \text{ m/s} + (10 \text{ m/s}^2)(2 \text{ s} - 0 \text{ s}) = 20 \text{ m/s}$$

$$y_1 = y_0 + (v_y)_0(t_1 - t_0) + \frac{1}{2}a_y(t_1 - t_0)^2 = 0 \text{ m} + 0 \text{ m} + \frac{1}{2}(10 \text{ m/s}^2)(2 \text{ s})^2 = 20 \text{ m}$$

At the explosion the equation $(p_y)_f = (p_y)_i$ is

$$m_L(v_{Ly})_2 + m_H(v_{Hy})_2 = (m_L + m_H)(v_y)_1 \Rightarrow (500 \text{ kg})(v_{Ly})_2 + (1000 \text{ kg})(v_{Hy})_2 = (1500 \text{ kg})(20 \text{ m/s})$$

To find $(v_{Hy})_2$ we must first find $(v_{Ly})_2$, the velocity after the explosion of the upper section. Using kinematics,

$$(v_{Ly})_3^2 = (v_{Ly})_2^2 + 2(-9.8 \text{ m/s}^2)((y_L)_3 - (y_L)_2) \Rightarrow (v_{Ly})_2 = \sqrt{2(9.8 \text{ m/s}^2)(530 \text{ m} - 20 \text{ m})} = 99.98 \text{ m/s}$$

Now, going back to the momentum conservation equation we get

$$(500 \text{ kg})(99.98 \text{ m/s}) + (1000 \text{ kg})(v_{Hy})_2 = (1500 \text{ kg})(20 \text{ m/s}) \Rightarrow (v_{Hy})_2 = -20 \text{ m/s}$$

The negative sign indicates downward motion.

P9.69. Strategize: Model the two blocks (A and B) and the bullet (L) as particles. This is a two-part problem. First, we have a collision between the bullet and the first block (A). Momentum is conserved since no external force acts on the system (bullet and block A). The second part of the problem involves a perfectly inelastic collision between the bullet and block B. Momentum is again conserved for this system (bullet and block B).

Prepare: Let us call the direction in which the bullet is fired the $+x$ direction. We apply conservation of momentum in the $+x$ direction.

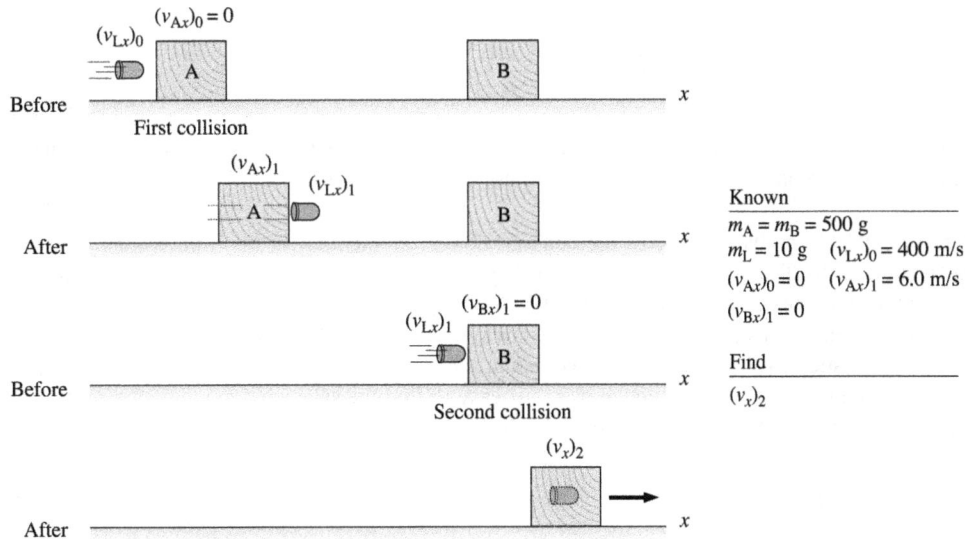

Known
$m_A = m_B = 500$ g
$m_L = 10$ g $(v_{Lx})_0 = 400$ m/s
$(v_{Ax})_0 = 0$ $(v_{Ax})_1 = 6.0$ m/s
$(v_{Bx})_1 = 0$

Find
$(v_x)_2$

Solve: For the first collision the equation $(p_x)_f = (p_x)_i$ is

$$m_L(v_{Lx})_1 + m_A(v_{Ax})_1 = m_L(v_{Lx})_0 + m_A(v_{Ax})_0$$

$$\Rightarrow (0.01\,\text{kg})(v_{Lx})_1 + (0.500\,\text{kg})(6\,\text{m/s}) = (0.01\,\text{kg})(400\,\text{m/s}) + 0\,\text{kg m/s} \Rightarrow (v_{Lx})_1 = 100\,\text{m/s}$$

The bullet emerges from the first block at 100 m/s. For the second collision the equation $(p_x)_f = (p_x)_i$ is

$$(m_L + m_B)(v_x)_2 = m_L(v_{Lx})_1 \Rightarrow (0.01\,\text{kg} + 0.5\,\text{kg})(v_x)_2 = (0.01\,\text{kg})(100\,\text{m/s}) \Rightarrow (v_x)_2 = 2.0\,\text{m/s}$$

Assess: This problem involves repeated application of the law of conservation of momentum. Also note that the actual value of 2 m for the separation between the blocks is not necessary for our calculations.

P9.71. Strategize: Treating the rocket as our system, we will assume it is in deep space far from any strong gravitational forces such that the system is truly isolated. Then its momentum will be conserved.

Prepare: Momentum is a *vector* quantity, so the direction of the initial velocity vector $\vec{v}_i$ establishes the direction of the momentum vector. The final momentum vector, after the explosion, must still point in the $+x$-direction. The two known pieces continue to move along this line and have no y-components of momentum. The missing third piece cannot have a y-component of momentum if momentum is to be conserved, so it must move along the x-axis–either straight forward or straight backward. From the conservation of mass, the mass of piece 3 is $m_3 = m_{total} - m_1 - m_2 = 7.0 \times 10^5$ kg.

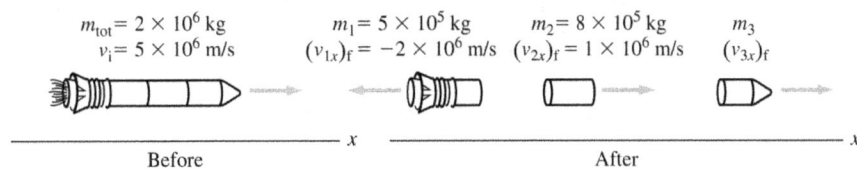

Solve: To conserve momentum along the x-axis, we require

$$[(p_x)_i = m_{total}(v_x)_i] = [(p_x)_f = (p_{1x})_f + (p_{2x})_f + (p_{3x})_f = m_1(v_{1x})_f + m_2(v_{2x})_f + (p_{3x})_f] \Rightarrow$$
$$(p_{3x})_f = m_{total}(v_x)_i - m_1(v_{1x})_f - m_2(v_{2x})_f = +1.02 \times 10^{13} \text{ kg} \cdot \text{m/s}$$

Because $p_{3f} > 0$, the third piece moves in the $+x$-direction, that is, straight forward. Because we know the mass m, we can find the velocity of the third piece as follows:

$$(v_{3x})_f = \frac{(p_{3x})_f}{m_3} = \frac{1.02 \times 10^{13} \text{ kg} \cdot \text{m/s}}{7.0 \times 10^5 \text{ kg}} = 1.5 \times 10^7 \text{ m/s}$$

Assess: Since this event is taking place in outer space, we don't have to worry about any external forces, and naturally the total momentum has to be constant regardless of the direction.

P9.73. Strategize: This problem involves the conservation of momentum in two dimensions. Let the system be made up of the three balls. Presumably gravity acts during the collision, but the effect of gravity during the very short time interval of the collision itself will be negligible, such that we can apply momentum conservation.
Prepare: Model the three balls of clay as particle 1 (moving north), particle 2 (moving west), and particle 3 (moving southeast). The three stick together during their collision, which is perfectly inelastic. All the three masses and the three velocities before the collision are known; it is thus easy to find the speed and the direction of the resulting blob using momentum conservation equations in two dimensions.

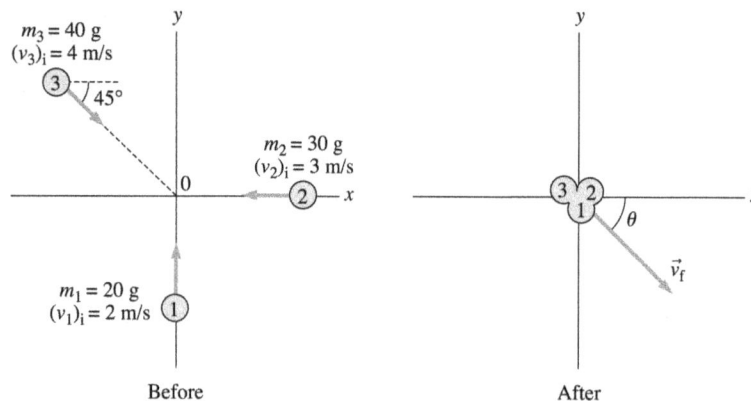

Before   After

Solve: The three initial momenta are

$$(\vec{p}_1)_i = m_1(\vec{v}_1)_i = (0.02 \text{ kg})(2 \text{ m/s}) \text{ along} + x = (0.04 \text{ kg} \cdot \text{m/s, along} + x)$$

$$(\vec{p}_2)_i = m_2(\vec{v}_2)_i = (0.03 \text{ kg})(3 \text{ m/s}) \text{ along} - x = (0.09 \text{ kg} \cdot \text{m/s, along} - x)$$

$$(\vec{p}_3)_i = m_3(\vec{v}_3)_i = [(0.04 \text{ kg})(4 \text{ m/s})\cos 45°, \text{ along} + x] + [(0.04 \text{ kg})(4 \text{ m/s})\sin 45°, \text{ along} - x]$$
$$= (0.113 \text{ kg} \cdot \text{m/s, along} + x) + (0.113 \text{ kg} \cdot \text{m/s, along} - x)$$

Since $\vec{p}_f = \vec{p}_i = (\vec{p}_1)_i + (\vec{p}_2)_i + (\vec{p}_3)_i$, we have

$$(m_1 + m_2 + m_3)\vec{v}_f = (0.023 \text{ kg} \cdot \text{m/s, along} + x) + (0.073 \text{ kg} \cdot \text{m/s, along} - x)$$
$$\Rightarrow \vec{v}_f = (0.256 \text{ m/s, along} + x) + (-0.811 \text{ m/s, along} + x)$$
$$\Rightarrow v_f = \sqrt{(0.256 \text{ m/s})^2 + (-0.811 \text{ m/s})^2} = 0.85 \text{ m/s}$$
$$\theta = \tan^{-1}\frac{|v_{fy}|}{v_{fx}} = \tan^{-1}\frac{0.811}{0.256} = 72° \text{ below} + x$$

Assess: The final speed is of the same order of magnitude as the initial speeds, as one would expect.

P9.75. Strategize: Because there's no friction or other tangential forces, the angular momentum of puck is conserved.

Prepare: Because the mass of the puck is concentrated at the end of the string and the string has negligible mass, the moment of inertia for the puck is mr^2.

Solve: The conservation of angular momentum equation $L_i = L_f$ is

$$I_i \omega_i = I_f \omega_f$$

$$\omega_f = \frac{I_i}{I_f} \omega_i = \frac{mr_i^2}{mr_f^2} \omega_i = \frac{(20 \text{ cm})^2}{(10 \text{ cm})^2} 100 \text{ rpm} = 400 \text{ rpm}$$

Assess: The increase in the angular speed by a factor of 4 comes because the radius is squared in the calculation of the moment of inertia.

P9.77. Strategize: This problem involves the conservation of angular momentum. If we take our system to be made up of the merry-go-round and Joey, then there is no external torque acting to change the angular momentum.

Prepare: We can use the conservation of angular momentum to solve this problem, as follows:

$$(I_1)_f (\omega_1)_f + (I_2)_f (\omega_2)_f + \ldots = (I_1)_i (\omega_1)_i + (I_2)_i (\omega_2)_i + \ldots$$

Treat Joey as a particle and the merry-go-round as a disk. We need to use the moment of inertia of a rotating disk $I_{disk} = \frac{1}{2} MR^2$ in order to write the final angular momentum of the merry-go-round. The moment of inertia of Joey is $I_J = (I_J)_f = (I_J)_i = m_J R$.

Before

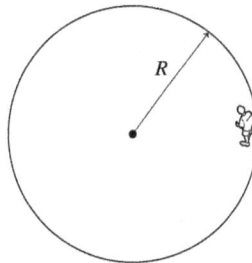

Known
$m = 36$ kg
$(v_J)_f = 5.0$ m/s
$M = 200$ kg
$R = 2.0$ m

Find
ω_f

After

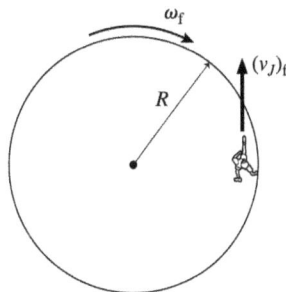

Solve: Before Joey begins running, both he and the merry-go-round are at rest, so the total angular momentum is 0. Let us say that he runs counterclockwise so that his final angular momentum is positive. Then the merry-go-round must rotate clockwise so that its final angular momentum is negative. In this way the angular momenta of Joey and the merry-go-round will still add to 0. The equation for conservation of angular momentum is

$$(I_m)_f(\omega_m)_f + (I_J)_f(\omega_J)_f = (I_m)_i(\omega_m)_i + (I_J)_i(\omega_J)_i \Rightarrow$$

$$(\omega_m)_f = \frac{(I_m)_i(\omega_m)_i + (I_J)_i(\omega_J)_i - (I_J)_f(\omega_J)_f}{(I_m)_f}$$

$$= \frac{(I_m)_i(0 \text{ rad/s}) + (I_J)_i(0 \text{ rad/s}) - (I_J)_f(\omega_J)_f}{(I_m)_f}$$

$$= \frac{-(I_J)_f(\omega_J)_f}{(I_m)_f} = \frac{-(m_J R^2)\left(\frac{v_J}{R}\right)}{\frac{1}{2}m_m R^2} = \frac{-(36 \text{ kg})(2.0 \text{ m})^2\left(\dfrac{5.0 \text{ m/s}}{2.0 \text{ m}}\right)}{\frac{1}{2}(200 \text{ kg})(2.0 \text{ m})^2} = -0.90 \text{ rad/s}$$

Assess: The negative sign is as expected considering that the total angular momentum is zero. For this to be true, the two bodies, the merry-go-round and Joey, must move in opposite directions around the axis.

P9.79. Strategize: Define the system to be Disk A plus Disk B so that during the time of the collision there are no net torques on this isolated system. This allows us to use the law of conservation of angular momentum.
Prepare: After the collision the combined object will have a moment of inertia equal to the sum of the moments of inertia of Disk A and Disk B and it will have one common angular speed ω_f (which we seek).

We must remember to call counterclockwise angular speeds positive and clockwise angular speeds negative.

Known
$M_A = 2.0 \text{ kg}$
$R_A = 0.40 \text{ m}$
$(\omega_A)_i = -30 \text{ rev/s}$
$M_B = 2.0 \text{ kg}$
$R_B = 0.20 \text{ m}$
$(\omega_B)_i = 30 \text{ rev/s}$

Find
ω_f

Preliminarily compute the moments of inertia of the disks:

$$I_A = \frac{1}{2}M_A R_A^2 = \frac{1}{2}(2.0 \text{ kg})(0.40 \text{ m})^2 = 0.16 \text{ kg} \cdot \text{m}^2$$

$$I_B = \frac{1}{2}M_B R_B^2 = \frac{1}{2}(2.0 \text{ kg})(0.20 \text{ m})^2 = 0.040 \text{ kg} \cdot \text{m}^2$$

Solve: Since they stick together afterwards they will have a common angular speed.

$$((I_A)_f + (I_B)_f)(\omega_{tot})_f = (I_A)_i(\omega_A)_i + (I_B)_i(\omega_B)_i \Rightarrow$$

$$(\omega_{tot})_f = \frac{(I_A)_i(\omega_A)_i + (I_B)_i(\omega_B)_i}{(I_A)_f + (I_B)_f}$$

$$= \frac{(0.16 \text{ kg} \cdot \text{m}^2)(-30 \text{ rev/s}) + (0.040 \text{ kg} \cdot \text{m}^2)(30 \text{ rev/s})}{0.16 \text{ kg} \cdot \text{m}^2 + 0.040 \text{ kg} \cdot \text{m}^2} = -18 \text{ rev/s}$$

That is, the angular speed is 18 rev/s and the direction is clockwise.

Assess: Since the two disks were rotating in opposite directions at the same speed we expect the angular speed afterwards to be less than the original speeds, and indeed it is.

P9.81. Strategize: This problem is effectively asking whether or not the system is isolated during the collision, and thus whether or not we can apply the concept of momentum conservation.

Prepare: In this problem we make all the usual assumptions: that the collision happens quickly enough that we can ignore any net external forces and consider the system (club plus ball) to be isolated.

Solve: Since the system is isolated the momentum of the system is conserved.

The correct choice is B.

Assess: Momentum is conserved in *all* collisions if we make the usual assumptions.

P9.83. Strategize: This is a straightforward application of the conservation of momentum. During the short time of the collision, external forces cause a negligible change in momentum.

Prepare: We are asked to find $(v_{Cx})_f - (v_{Cx})_i$.

Solve:

$$(P_x)_i = (P_x)_f$$
$$(p_{Cx})_i + (p_{Bx})_i = (p_{Cx})_f + (p_{Bx})_f$$
$$m_C(v_{Cx})_i + m_B(v_{Bx})_i = m_C(v_{Cx})_f + m_B(v_{Bx})_f$$

Rearranging terms allows us to solve for $(v_{Cx})_f - (v_{Cx})_i$:

$$(v_{Cx})_f - (v_{Cx})_i = \frac{m_B[(v_{Bx})_i - (v_{Bx})_f]}{m_C} = \frac{0.0460\,\text{kg}(0.0\,\text{m/s} - 60.0\,\text{m/s})}{0.200\,\text{kg}} = -14\,\text{m/s}$$

The correct choice is C. The negative sign indicates that the club slowed down.

Assess: Examining our equation for $(v_{Cx})_f - (v_{Cx})_i$ confirms that we had a different (and maybe easier) approach from the very beginning. If the momentum is conserved for a two-body system, then the change in momentum of the club must be equal in magnitude (and opposite in direction) to the change in momentum of the ball. The numerator above is the change in momentum of the ball, so simply dividing by the mass of the club gives the difference in the club's velocity.

P9.85. Strategize: One can solve this problem using impulse and momentum.

Prepare: The dancer's initial speed is zero, and therefore her initial momentum is zero. We are asked to find the time at which she is again, momentarily, at zero momentum. Between those two times the net impulse (area under the curve we are given) must be zero.

Solve: The dancer just begins descending at time 0.20 s. At time 0.40 s, the area under the curve is definitely negative. After 0.70 s, the area under the curve is definitely positive, and that time is labeled in the figure as the time the dancer leaves the ground. By a process of elimination, the net impulse can only be zero at some time between 0.40 s and 0.70 s. The correct answer is C.

Assess: By eye, it looks like the area might be zero under the curve after a time of about 0.52 s.

P9.87. Strategize: This problem involves relating impulse to the change in momentum.

Prepare: Impulse and momentum are related by $\vec{J} = \Delta\vec{p} = m\Delta\vec{v}$. In this case, since the motion is vertical and the dancer starts from rest, we obtain $J_y = mv_{f,y}$.

Solve: Rearranging and inserting numbers, we find $v_{f,y} = J_y / m = (30\,\text{kg}\cdot\text{m/s})/(42\,\text{kg}) = 0.71\,\text{m/s}$. So to the nearest m/s, the dancer's speed as she leaves the ground is 1 m/s. A is the correct answer.

Assess: This is a reasonable speed for a person who will only be in the air for a moment.

10

ENERGY AND WORK

Q10.1. Reason: The brakes in a car slow down the car by converting its kinetic energy to thermal energy in the brake shoes through friction. Cars have large kinetic energies, and all of that energy is converted to thermal energy in the brake shoes, which causes their temperature to increase greatly. Therefore they must be made of material that can tolerate very high temperatures without being damaged.
Assess: This is an example of an energy conversion. All of the car's kinetic energy is converted to thermal energy through friction. To get an appreciation of how much kinetic energy is absorbed by the brake shoes, consider instead the energy explicit in stopping the car by hitting a stationary object instead!

Q10.3. Reason: Here we must increase potential energy without increasing kinetic energy. Consider lifting an object at constant speed. Consider the object plus the earth as the system. The force does work that increases the gravitational potential energy of the object, while the kinetic energy does not increase because the velocity of the object remains the same. Another possibility is the compression of a spring by an applied force at constant velocity. Note that constant velocity is not necessary for the change in kinetic energy between the beginning and end of a process to be zero. Lifting an object or compressing a spring in any way, as long as the initial velocity is equal to the final velocity at the end of the process leads to no change in net kinetic energy. Any kinetic energy gained during the process is lost when the object is brought the rest.
Assess: Kinetic energy does not change if an object has the same velocity at the beginning and end of a process.

Q10.5. Reason: We need a process that converts kinetic energy to work without any change in potential energy. Consider a block sliding on level ground, to which is attached a cord you are holding on to. As the block slides, it exerts a force on your hand by virtue of its kinetic energy. As the block pulls your hand, it is doing work on you. The kinetic energy of the block will decrease as it continues to exert the force on your hand.
Assess: To have a change in gravitational potential energy you must have a change in height.

Q10.7. Reason: We need a process that converts work totally into thermal energy without any change in the kinetic or potential energy. Consider moving a block of wood across a horizontal rough surface at constant speed. Because the surface is horizontal there is no change in potential energy, and because the speed is constant there is no change in kinetic energy. All the work done in moving the block across the rough horizontal surface is transferred into thermal energy.
Assess: To have a change in gravitational energy you must have a change in height and to have a change in kinetic energy you must have a change in speed.

Q10.9. Reason: We need a process that converts kinetic energy totally into thermal energy without changing the gravitational potential energy. Consider a wood block sliding across a rough horizontal surface and slowing to a stop. The gravitational potential energy is not changing and all the kinetic energy is being transferred into thermal energy. All the decrease in kinetic energy becomes an increase in thermal energy.
Assess: Gravitational potential energy does not change because there is a change in height of the block. The kinetic energy decreases as the block slows to a stop.

Q10.11. Reason: Consider all the forms of energy involved: gravitational potential energy, kinetic energy, and incoherent motion of water molecules that we treat as thermal energy. All objects involved in the storage/transformation of these energies must be included in our system. Thus we include Earth, the diver, and the water in the pool.

Assess: The gravitational potential energy is related to the mutual attraction of the diver and Earth. The only kinetic energy is that of the diver. When the diver stops due to resistance from the water, the thermal energy of the water increases. Thus all objects involved in the energy transformations are contained in our choice of system. Since no external forces do work adding or subtracting energy to/from our system, it is isolated.

Q10.13. Reason: (a) The work done is $W = Fd$. Both particles experience the same force and move the same distance, so the work done on them is the same. From the work-kinetic energy theorem we know that the kinetic energy of each puck has changed by the same amount. They both started from rest, so they end up with the same kinetic energy.

(b) The kinetic energies are equal, but the speeds are not.

$$\tfrac{1}{2}m_A v_A^2 = \tfrac{1}{2}m_B v_B^2 \Rightarrow \frac{v_A}{v_B} = \sqrt{\frac{m_B}{m_A}} = \sqrt{\frac{2m_A}{m_A}} = \sqrt{2}$$

So the speed of puck A is 1.41 times the speed of puck B.

Assess: This makes intuitive sense.

Q10.15. Reason: The ball of mass m has an initial potential energy of mgh and the ball of mass $2m$ has an initial potential energy if $2mgh$. The ball of mass $2m$ has twice the initial potential energy of the ball of mass m. There is no initial kinetic energy. Since energy is conserved, as the balls hit the floor the ball of mass $2m$ still has twice as much energy as the ball of mass m. As the balls hit the floor the energy is totally kinetic. So the ball of mass $2m$ hits the floor with twice the kinetic energy of the ball of mass m.

Assess: No energy was lost, rather it was changed from potential to kinetic energy.

Q10.17. Reason: (a) If the car is to go twice as fast at the bottom, its kinetic energy, proportional to v^2, will be *four times* as great. You thus need to give it four times as much gravitational potential energy at the top. Since gravitational potential energy is linearly proportional to the height h, you'll need to increase the height of the track by a factor of four. **(b)** Using considerations of conservation of energy, as in part (a), we see that the speed of the car at the bottom depends only on the height of the track, not its shape.

Assess: Kinetic energy is proportional to the *square* of the velocity.

Q10.19. Reason: A quarterback and a baseball player have comparable abilities to exert forces and do work on an object. So to a good approximation, it should be possible for the football and baseball to leave a player's hands with equal kinetic energies. That is $\frac{1}{2}m_f v_f^2 = \frac{1}{2}m_b v_b^2$. But since $m_f > m_b$, this equality of energy means $v_f < v_b$. Thus we expect the football to have a lower speed.

Assess: Kinetic energy depends on both mass and speed. Thus a given person can throw a ball with small mass faster than he/she can throw a ball with greater mass.

Q10.21. Reason: By the time the blocks reach the ground, they have transformed identical amounts of gravitational potential energy into translational kinetic energy of the blocks and rotational kinetic energy of the cylinders. But the moment of inertia of a hollow cylinder is higher than that of a solid cylinder of the same mass, so more of the energy of the system is in the form of rotational kinetic energy for the hollow cylinder than for the solid one. This leaves less energy in the form of translational kinetic energy for the hollow cylinder. But it is the translational kinetic energy that determines the speed of the block. So the block moves more slowly for the system with the hollow cylinder, and so its block reaches the ground last.

Assess: The energy is shared between the rotating cylinder and the falling block. The more energy the cylinder has, the less is available for the block.

Q10.23. Reason: If the springy part of the fern were attached to one seed and the stem, then when the elastic potential energy of a springy part is converted to kinetic energy, some of the kinetic energy would be on the stem. In fact, if the stem has very little mass, then it would move a greater distance than the seed and according to $W = F\Delta x$, more work would be done on the stem than on the seed. If there are two seeds, then each will leave with the same kinetic energy and no energy is wasted on unwanted motion of the stem.

Assess: If a compressed springy part of a plant is attached to a seed and a lightweight stem, then released, the stem would move out of the way so immediately, that the seed would not move very far. Since work depends on force and displacement, this small displacement of the seed would mean not much of the energy would be transformed into kinetic energy of the seed. By placing two seeds opposite each other, nature solves this problem.

Q10.25. Reason: When the coaster is at the top $U = mgy$ relative to the ground. That amount of energy equals the kinetic energy at the bottom. Halfway down (or up) the potential energy is half of what it was at the top, so the kinetic energy must also be half of what it is at the bottom. If v' is the speed at the halfway height, then

$$\frac{\frac{1}{2}m(v')^2}{\frac{1}{2}m(v)^2} = \frac{1}{2} \Rightarrow \left(\frac{v'}{v}\right)^2 = \frac{1}{2} \Rightarrow \frac{v'}{v} = \sqrt{\frac{1}{2}} \Rightarrow v' = \frac{\sqrt{2}}{2}(30 \text{ m/s}) = 21 \text{ m/s}$$

So the correct choice is C.
Assess: Even though the height is half the total, the speed is not half of 30 m/s.

Q10.27. Reason: Since kinetic energy is proportional to the square of the velocity of an object, an object with twice the velocity will have four times the amount of kinetic energy. In this question, all the kinetic energy is converted to elastic potential energy in the spring. The potential energy stored in a spring is proportional to the square of the compression from its equilibrium position. Since we start with four times the kinetic energy, four times as much energy is stored in the spring. But since the energy stored in the spring is proportional to the *square* of the compression, the compression is only twice the compression previously, or $2(2.0 \text{ cm}) = 4.0 \text{ cm}$. The correct choice is C.
Assess: Kinetic energy is proportional to the *square* of the velocity of an object and the potential energy of a spring is proportional to the *square* of the displacement from the equilibrium position.

Q10.29. Reason: As the ball falls, energy is conserved since the only force doing work is the force of gravity. Since gravity is conserved we may write

$$\Delta E = 0$$

or

$$\Delta K + \Delta U_g = 0$$

As the ball falls we have

$$\Delta K = mv^2/2 \text{ and } \Delta U_g = -mgh = -mg(L - L\cos 30°) = -mgL(1 - \cos 30°)$$

Combining these obtain

$$mv^2/2 = mgL(1 - \cos 30°)$$

or

$$v = \sqrt{2gL(1 - \cos 30°)} = 3.6 \text{ m/s}$$

The correct response is C.
Assess: The key to the problem is to realize that energy is conserved and then find the change in kinetic and potential energy. A speed of 3.6 m/s is reasonable.

Q10.31. Reason: The power in this case is given by $P = W / \Delta t = Fd / \Delta t = Fv$. Doubling the speed doubles the force required, but the speed itself also shows up in the calculation. Thus, if $P_i = F_i v_i$, then $P_f = F_f v_f = (4F_i)(2v_i) = 8F_i v_i = 8P_i$. The correct response is C.

Assess: Increasing cycling speed increases the power required by the cyclist in two different ways: increasing the drag force and increasing the speed itself. Both show up in the equation $P = Fv$.

Problems

P10.1. Strategize: Assume you lift the book steadily, so that the force exerted on the book is constant. Since there is a component of the lifting force in the direction of the displacement, we expect the work done by your hand to be positive. There is a component of the gravitational force opposite the direction of the displacement. Thus we expect the work done by gravity to be negative.

Prepare: Equation 10.6 gives the work done by a force $\vec{F}$ on a particle. The work is defined as $W = Fd\cos(\theta)$, where d is the particle's displacement.

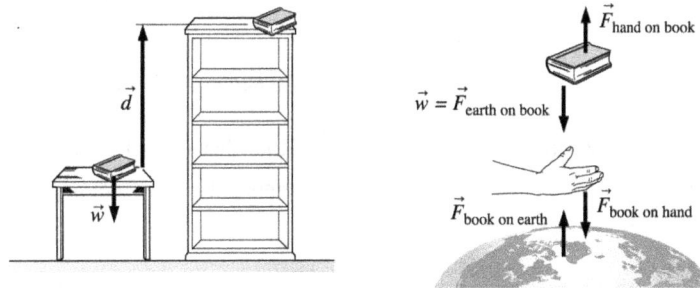

Solve: **(a)** Refer to the diagram. We are assuming the book does not accelerate, so the force you exert on the book is exactly equal to the force of gravity on the book. $\vec{F}_{\text{hand on book}} = -\vec{F}_{\text{gravity on book}}$

The total displacement of the book is $2.3 \text{ m} - 0.75 \text{ m} = 1.55 \text{ m}$ (keeping one extra significant figure for this intermediate result).

The work done by gravity is then

$$W_{\text{gravity on book}} = wd\cos(\theta) = (2.0 \text{ kg})(9.80 \text{ m/s}^2)(1.55 \text{ m})\cos(180°) = -30 \text{ J}$$

(b) The work done by hand is $W_{\text{hand on book}} = F_{\text{hand on book}} d\cos(\theta)$.

$$\Rightarrow W_{\text{hand on book}} = (2.0 \text{ kg})(9.80 \text{ m/s}^2)(1.55 \text{ m})\cos(0°) = +30 \text{ J}$$

Assess: Note that the only difference is in the sign of the answer. This is because the two forces are equal, but act in opposite directions. The work done by gravity is negative because gravity acts opposite to the displacement of the book. Your hand exerts a force in the same direction as the displacement, so it does positive work. We should expect the total work to be zero from Equation 10.4 since energy is conserved in this process. Referring to the results, we see that the work by your hand cancels the work done by gravity and the total work is zero as expected.

P10.3. Strategize: Note that not all forces act in the same direction as the displacement. The tensions in the two ropes have components that are in the same direction as the displacement, so we expect the work done by those forces to be positive. The force of friction is opposite the displacement, so we expect the work done by friction to be negative.

Prepare: We must use Equation 10.6: $W = Fd\cos(\theta)$ for each force. W is the work done by a force of magnitude F on a particle and d is the particle's displacement. The crate is moving directly to the right. We assume all forces are given to three significant figures, including the 500 N force since the other two forces are given to three significant figures.

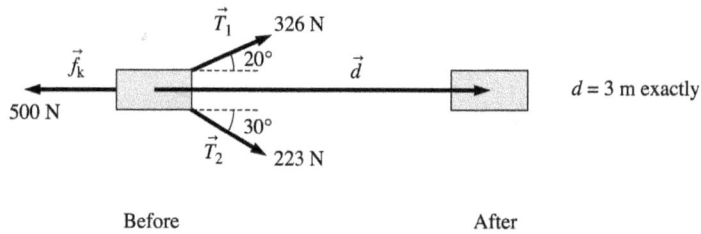

Before   After

Solve: For the force $\vec{f}_k$, the displacement is exactly opposite the force, so

$$W_{f_k} = f_k d \cos(180°) = (500 \text{ N})(3 \text{ m})(-1) = -1.50 \text{ kJ}$$

For the tension $\vec{T}_1$:

$$W_{T_1} = T_1 d \cos(20°) = (326 \text{ N})(3 \text{ m})(0.9397) = 0.919 \text{ kJ}$$

For the tension $\vec{T}_2$:

$$W_{T_2} = T_2 d \cos(30°) = (223 \text{ N})(3 \text{ m})(0.8660) = 0.579 \text{ kJ}$$

Assess: Negative work done by the force of kinetic friction $\vec{f}_k$ means that 1.50 kJ of energy has been transferred *out* of the crate and converted to heat. The other two forces have components along the displacement, and therefore do positive work to move the crate.

P10.5. Strategize: In this problem the force on the kite and the displacement of the kite are not parallel.
Prepare: Equation 10.6 is the definition of work when the force and displacement are not parallel.
Solve: (a) The boy is standing still in this case, so the displacement is zero. $W = Fd\cos(\theta) = (F)(0 \text{ m})\cos(\theta) = 0 \text{ J}$. The work done on the boy by the string is exactly zero Joules.
(b) The displacement is non-zero in this case, so we expect the work done to be non-zero. Refer to the following figure. The angle between the force and the displacement is $180° - 30° = 150°$.

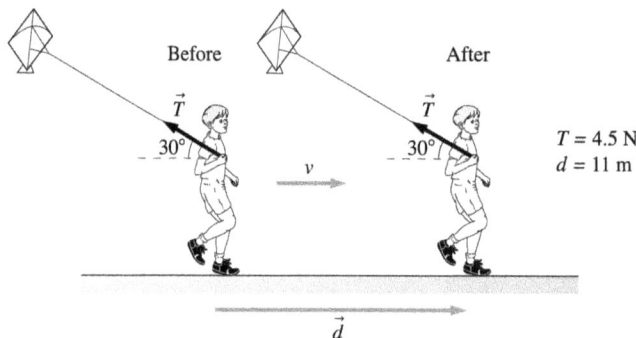

The work is

$$W = Fd\cos(\theta) = (4.5 \text{ N})(11 \text{ m})\cos(150°) = -43 \text{ J}$$

(c) The angle between the force and displacement in this case is $30°$ (Look at the figure and imagine the direction of the displacement vector is reversed). The work is

$$W = Fd\cos(\theta) = (4.5 \text{ N})(11 \text{ m})\cos(30°) = 43 \text{ J}$$

Assess: For there to be work done, the displacement must not be zero. If there is no displacement there is no work done. Note that the answers to parts (b) and (c) have opposite signs. This is because the displacement is exactly opposite in those cases for the same direction of the force.

P10.7. Strategize: The work done by Paige can be easily calculated using $W = Fd\cos(\theta)$.

Prepare: Note that both the displacement is down the ramp, while the force is horizontal, with a significant component up the ramp. The angle between them is $160°$.

Solve: Inserting values, we find $W = Fd\cos(\theta) = (68\text{ N})(3.5\text{ m})\cos(160°) = -220\text{ J}$

Assess: Since the system is losing gravitational potential energy and not gaining any kinetic energy, we expect that some negative work must be done on it. Our answer above fits our expectation.

P10.9. Strategize: Kinetic energy depends only on mass and speed. We can equate the kinetic energies of the two vehicles, and then solve for the unknown quantity.

Prepare: Use the definition of kinetic energy, Equation 10.8, to set up an equation such that the kinetic energy of the car is equal to that of the truck.

$$m_C = 1000\text{ kg}$$
$$v_C$$

$$m_T = 20{,}000\text{ kg}$$
$$v_T = 25\text{ km/h}$$

Solve: For the kinetic energy of the compact car and the kinetic energy of the truck to be equal,

$$K_C = K_T \Rightarrow \frac{1}{2}m_C v_C^2 = \frac{1}{2}m_T v_T^2 \Rightarrow v_C = \sqrt{\frac{m_T}{m_C}}v_T = \sqrt{\frac{20\,000\text{ kg}}{1000\text{ kg}}}(25\text{ km/h}) = 110\text{ km/h}$$

To match the kinetic energy of the truck, the car needs a velocity of 110 km/h (to two significant figures).

Assess: Note that the smaller mass needs a greater velocity for its kinetic energy to be the same as that of the larger mass. Though the truck has 20 times the mass, the car only needs about four times the velocity of the truck to have the same kinetic energy. This is because kinetic energy is proportional to the mass, but proportional to the *square* of the velocity.

P10.11. Strategize: Kinetic energy depends on mass and speed. Here mass is constant, and we only change speed.

Prepare: In order to work this problem, we need to know that the kinetic energy of an object is given by $K = mv^2/2$.

Solve: The problem may be solved in a qualitative manner or in a quantitative manner. Since some students think one way and some the other, we will use both methods.

(a) First, in a qualitative manner. Since the kinetic energy depends on the square of the speed, the kinetic energy will be doubled if the speed is increases by a factor of $\sqrt{2}$. This is true because $(\sqrt{2})^2 = 2$. Then the new speed is $\sqrt{2}(10\text{ m/s}) = 14\text{ m/s}$.

Second, in a more quantitative manner. Use a subscript 1 for the present case where the speed is 10 m/s and a subscript 2 for the new case where the speed is such that the kinetic energy is doubled.

$$K_1 = mv_1^2/2 \quad \text{and} \quad K_2 = mv_2^2/2$$

We want

$$K_2 = 2K_1$$

Inserting expressions for K_1 and K_2 obtain

$$\frac{mv_2^2}{2} = 2\frac{mv_1^2}{2} \quad \text{or} \quad v_2 = \sqrt{2}v_1 = 14\text{ m/s}$$

(b) First, in a qualitative manner, if the speed is doubled and the kinetic energy depends on the square of the speed, the kinetic energy will increase by a factor of four.

Second, in a more quantitative manner, use a subscript 1 for the present case and a subscript 2 for the new case where the speed is doubled.

$$K_1 = mv_1^2/2 \quad \text{and} \quad K_2 = mv_2^2/2$$

We want $v_2 = 2v_1$. Inserting v_2 into K_2, we obtain

$$K_2 = \frac{mv_2^2}{2} = \frac{m(2v_1)^2}{2} = 4\left(\frac{mv_1^2}{2}\right) = 4K_1$$

This expression clearly shows that the kinetic energy is increased by a factor of four when the speed is doubled.
Assess: The key to the problem is to know that the kinetic energy depends on the speed squared. After that, we can approach the problem in a qualitative or a quantitative manner. You may prefer one method over the other, but you should be able to work in either mode.

P10.13. Strategize: Kinetic energy depends only on mass and speed. We can equate the kinetic energies of the two objects, and then solve for the unknown quantity.

Prepare: Use the definition of kinetic energy, Equation 10.8: $K = \frac{1}{2}mv^2$. Apply this to both objects.

Solve: The man and the bullet have the same kinetic energy.

$$\tfrac{1}{2}m_m v_m^2 = \tfrac{1}{2}m_b v_b^2 \Rightarrow$$

$$v_m = \sqrt{\frac{m_b}{m_m}}\, v_b = \sqrt{\frac{8.0 \text{ g}}{80 \text{ kg}}}(400 \text{ m/s}) = 4.0 \text{ m/s}$$

Assess: We expected the man to need much less speed than the bullet to have the same kinetic energy.

P10.15. Strategize: We will assume that all the work Sam does goes into stopping the boat.
Prepare: We can use conservation of energy as expressed in Equation 10.7 to calculate the work done from the change in kinetic energy.

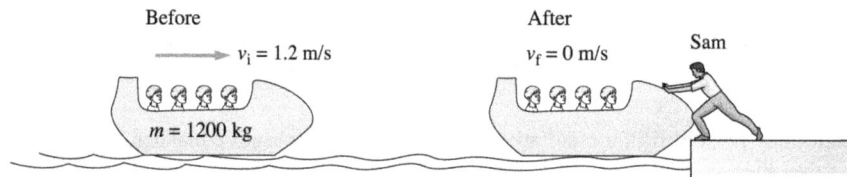

Solve: Refer to the before and after representation of Sam stopping a boat. Equation 10.7 becomes

$$\frac{1}{2}mv_f^2 - \frac{1}{2}mv_i^2 = W$$

Since the boat is at rest at the end of the process, $v_f = 0$ m/s. Therefore, the final kinetic energy is zero. The work done on the boat is then

$$W = -\frac{1}{2}mv_i^2 = -\frac{1}{2}(1200 \text{ kg})(1.2 \text{ m/s})^2 = -0.86 \text{ kJ}$$

The amount of work is 0.86 kJ.
Assess: Note that the work done by Sam on the boat is negative. This is because the force Sam exerts on the boat must be opposite to the direction of motion of the boat to slow it down.

P10.17. Strategize: Since the turntable has no translational motion, we need only consider rotational kinetic energy.
Prepare: The definition of kinetic energy for objects rotating around a stationary axis is given by Equation 10.9. In Equation 10.9, the rotational velocity should be in units of radians/second.

Solve: The turntable turns once every 4.0 s. So its rotational velocity is

$$\omega = \left(\frac{1 \text{ rev}}{4.0 \text{ s}}\right)\left(\frac{2\pi \text{ rad}}{\text{rev}}\right) = 1.57 \text{ rad/s}$$

This should be reported as 1.6 rad/s to two significant figures. We keep an extra significant figure for substitution in the next step:

$$K_{\text{rot}} = \frac{1}{2}I\omega^2 = \frac{1}{2}(0.040 \text{ kg} \cdot \text{m}^2)(1.57 \text{ rad/s})^2 = 0.049 \text{ J}$$

Assess: This is a reasonable result for such a low rotational velocity and moment of inertia.

P10.19. Strategize: The rotational kinetic energy depends on the moment of inertia and the rotational velocity. Since rotational kinetic energy and rotational velocity (after some conversion) are given, we can use the definition of rotational kinetic energy to find the moment of inertia.
Prepare: The rotational kinetic energy is given by Equation 10.9. In this equation, units for rotational velocity must be rad/s.
Solve: Using Equation 10.9,

$$K_{\text{rot}} = \frac{1}{2}I\omega^2, \text{ so } I = \frac{2K_{\text{rot}}}{\omega^2}$$

We need to convert ω to proper units, radians/s. Since $\omega = 20\,000$ rev/min and there are 2π rad/rev and 60 s/min,

$$\omega = \left(20\,000\frac{\text{rev}}{\text{min}}\right)\left(\frac{1 \text{ min}}{60 \text{ s}}\right)\left(\frac{2\pi \text{ rad}}{\text{rev}}\right)$$

So.

$$I = \frac{(2)(4.0 \times 10^6 \text{ J})}{\left[\left(20\,000\frac{\text{rev}}{\text{min}}\right)\left(\frac{1 \text{ min}}{60 \text{ s}}\right)\left(\frac{2\pi \text{ rad}}{\text{rev}}\right)\right]^2} = 1.8 \text{ kg} \cdot \text{m}^2$$

Assess: The flywheel can store this large amount of energy even though it has a low moment of inertia because of its high rate of rotation.

P10.21. Strategize: We want to relate a known kinetic energy to an unknown gravitational potential energy. We are not given the mass of the runner, but we recall that mass shows up in expressions for both kinetic and gravitational potential energy. We proceed, expecting that mass may drop out of our calculation.
Prepare: We need to know how to find kinetic energy $(K = mv^2/2)$ and gravitational potential energy $(U_g = mgh)$ and be aware that we want the potential energy to change by an amount equal to the kinetic energy.
Solve: Equate the potential energy to the kinetic energy

$$\frac{mv^2}{2} = mgh$$

And solve for h to obtain

$$h = \frac{v^2}{2g} = \frac{(11 \text{ m/s})^2}{2(9.8 \text{ m/s}^2)} = 6.2 \text{ m}$$

Assess: This is as high as a two-story building.

P10.23. Strategize: We are told the length of the cable and the angle from vertical. We know the change in gravitational potential energy depends on the change in height, which we can find from the given information.

Prepare: The wrecking ball will increase its gravitational potential energy by an amount $\Delta U_g = mgh$, where h is the change in the vertical position.

Solve: The increase in gravitational potential energy of the wrecking ball is determined by

$$\Delta U_g = mgh = mg(L - L\cos 25°) = mgL(1 - \cos 25°) = 14\text{ kJ}$$

Assess: This is a reasonable change in gravitational potential energy for such a massive object.

P10.25. Strategize: Assume an ideal spring that obeys Hooke's law. We can rearrange this law to solve for the unknown.

Prepare: Equation 10.16 gives the potential energy stored in an ideal spring. The elastic potential energy of a spring is defined as $U_s = \frac{1}{2}kx^2$, where x is the magnitude of the stretching or compression relative to the unstretched or uncompressed length.

Solve: We have $x = 20\text{ cm} = 0.20\text{ m}$ and $k = 500\text{ N/m}$. This means

$$U_s = \frac{1}{2}kx^2 = \frac{1}{2}(500\text{ N/m})(0.20\text{ m})^2 = 10\text{ J}$$

Assess: Since x is squared, U_s is positive for a spring that is either compressed or stretched. U_s is zero when the spring is in its equilibrium position.

P10.27. Strategize: We will assume the knee extensor tendon behaves according to Hooke's Law and stretches in a straight line. We can use this to find the spring potential energy for both athletes and non-athletes, and then find the difference.

Prepare: The elastic energy stored in a spring is given by Equation 10.16, $U_s = \frac{1}{2}kx^2$.

Solve: For athletes,

$$U_{s,\text{athlete}} = \frac{1}{2}kx^2 = \frac{1}{2}(33\,000\text{ N/m})(0.041\text{ m})^2 = 27.7\text{ J}$$

For non-athletes,

$$U_{s,\text{non-athlete}} = \frac{1}{2}kx^2 = \frac{1}{2}(33\,000\text{ N/m})(0.033\text{ m})^2 = 18.0\text{ J}$$

The difference in energy stored between athletes and non-athletes is therefore 9.7 J.

Assess: Notice the energy stored by athletes is over 1.5 times the energy stored by non-athletes.

P10.29. Strategize: Since the gravitational potential energy and the kinetic energy of the car do not change, all the work Mark does on the car goes into thermal energy.

Prepare: We must use the equation for work done by a constant force: $W = Fd\cos(\theta)°$, and equate this to the change in thermal energy.

Solve: The thermal energy created in the tires and the road may be determined by:

$$\Delta E_t = W_{\text{Mark}} = F_{\text{Mark}}d\cos 0° = (110\text{ N})(150\text{ m})\cos 0° = 16.5\text{ kJ}$$

Assess: All the work Mark does in pushing the car, becomes thermal energy of the tires and road. Since Mark is pushing in the direction the car is moving, the angle between the direction of F and d is $50°$

P10.31. Strategize: There is no change in kinetic energy, so all the work done on the crate by gravity goes into thermal energy.

Prepare: The work done by gravity depends on the angle between the force of gravity and the displacement of the crate, which is 55 degrees, and also on the distance over which the force acts.

Solve: The thermal energy created may be determined by:

$$\Delta E_t = W_{gravity} = F_{gravity} L \cos 55° = (900 \text{ N})(12 \text{ m}) \cos 55° = 6.2 \text{ kJ}$$

Assess: This seems like a lot of thermal energy, however the crate is heavy (about 200 lbs) and the ramp is long so we should expect a large number. We could also solve this problem as follows:

$$\Delta E_{Th} = W_{gravity} = -\Delta U_{gravity} = -(-mgh) = F_{gravity} L \sin 35° = F_{gravity} L \cos 55°$$

P10.33. Strategize: The force of gravity and the force of friction are doing work on the child. Since the child slides at a constant speed, the net work (which is the change in kinetic energy) is zero. We can write expressions for the work done by the force of gravity and by the force of friction in terms of the frictional force. Equating the net force to zero will allow us to solve for the unknown frictional force.

Reason: We can write:

$$W_g + W_f = \Delta K = 0 \text{ or } W_f = -W_g$$

Knowing how the work done by gravity is related to the change in gravitational potential energy and how the work done by friction is related to the force of friction, we can determine the force of friction.

Solve: Writing expressions for the work done by friction obtain

$$W_f = -W_g = -(-\Delta U_g) = \Delta U_g = Mgh \text{ and } W_f = F_f L \cos 180° = -F_f L$$

Combining these and solving for the force of friction obtain

$$F_f = -Mgh/L = -(25 \text{ kg})(9.8 \text{ m/s}^2)(3.0 \text{ m})/(7.0 \text{ m}) = -1.0 \times 10^2 \text{ N}$$

The minus reminds us that the force of friction opposes the motion of the object, so the magnitude is 100 N.

Assess: A 100 N force of friction for a child sliding down a playground slid is a reasonable number.

P10.35. Strategize: This is a case of free fall, so the sum of the kinetic and gravitational potential energy does not change as the ball rises and falls. We have sufficient information to write the initial kinetic and gravitational potential energies, except for the mass. We note that mass shows up in every expression for gravitational potential energy and kinetic energy at any time, so we intuitively expect mass to drop out of our calculation.

Prepare:

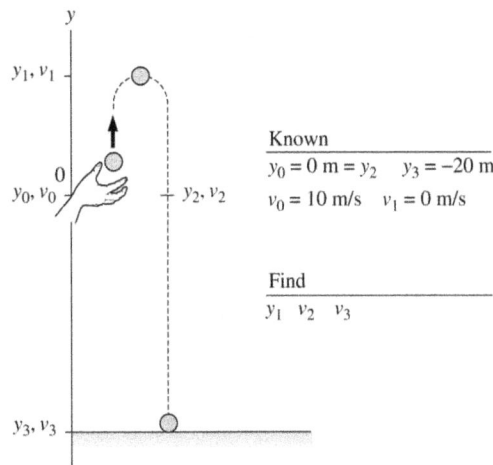

The figure shows a ball's before-and-after pictorial representation for the three situations in parts (a), (b) and (c).

Solve: The quantity $K + U_g$ is the same during free fall: $K_f + U_{gf} = K_i + U_{gi}$. At the top of its trajectory where the ball turns around the ball's velocity is 0 m/s. We have

(a) $\dfrac{1}{2}mv_1^2 + mgy_1 = \dfrac{1}{2}mv_0^2 + mgy_0$

$$\Rightarrow y_1 = (v_0^2 - v_1^2)/2g = [(10 \text{ m/s})^2 - (0 \text{ m/s})^2]/(2 \times 9.80 \text{ m/s}^2) = 5.1 \text{ m}$$

5.1 m is therefore the maximum height of the ball above the window. This is 25 m above the ground.

(b) $\dfrac{1}{2}mv_2^2 + mgy_2 = \dfrac{1}{2}mv_0^2 + mgy_0$

Since $y_2 = y_0 = 0$, we get for the magnitudes $v_2 = v_0 = 10 \text{ m/s}$.

(c) $\dfrac{1}{2}mv_3^2 + mgy_3 = \dfrac{1}{2}mv_0^2 + mgy_0 \Rightarrow v_3^2 + 2gy_3 = v_0^2 + 2gy_0 \Rightarrow v_3^2 = v_0^2 + 2g(y_0 - y_3)$

$$\Rightarrow v_3^2 = (10 \text{ m/s})^2 + 2(9.80 \text{ m/s}^2)[0 \text{ m} - (-20 \text{ m})]$$

Taking the square root, the magnitude of v_3 is equal to 22 m/s.

Assess: Note that the ball's speed as it passes the window on its way down is the same as the speed with which it was tossed up, but in the opposite direction.

P10.37. Strategize: Since the ramp is frictionless, the sum of the puck's kinetic and gravitational potential energy does not change during its sliding motion. We will use conservation of energy. We also note that the final speed can be taken to be zero, since we are asked for the minimum initial speed.
Prepare: Use Equation 10.4 for the conservation of energy.

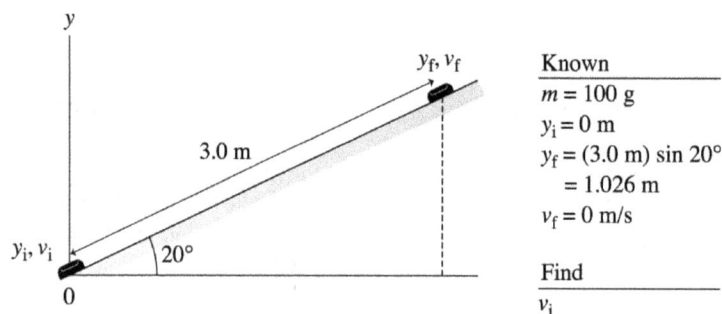

Solve: The quantity $K + U_\text{g}$ is the same at the top of the ramp as it was at the bottom. The energy conservation equation $K_\text{f} + U_\text{gf} = K_\text{i} + U_\text{gi}$ is

$$\frac{1}{2}mv_\text{f}^2 + mgy_\text{f} = \frac{1}{2}mv_\text{i}^2 + mgy_\text{i} \Rightarrow v_\text{i}^2 = v_\text{f}^2 + 2g(y_\text{f} - y_\text{i})$$
$$\Rightarrow v_\text{i}^2 = (0 \text{ m/s})^2 + 2(9.80 \text{ m/s}^2)(1.03 \text{ m} - 0 \text{ m}) = 20.1 \text{ m}^2/\text{s}^2 \Rightarrow v_\text{i} = 4.5 \text{ m/s}$$

Assess: An initial push with a speed of 4.5 m/s $\approx$ 10 mph to cover a distance of 3.0 m up a 20° ramp seems reasonable. Note that a ramp of *any* angle to the same final height would lead to the same final velocity for the puck. Note that the mass cancels out in the equation since both kinetic energy and gravitational potential energy are proportional to mass.

P10.39. Strategize: Assume there is zero rolling friction since friction is not mentioned in the problem. The sum of the kinetic and gravitational potential energy, therefore, does not change during the car's motion. We can calculate the total (gravitational potential and kinetic) energy before the hill, then compare it to how much gravitational potential energy is required to be at the top of the hill. Comparison will tell us whether or not this is possible.
Prepare: Consider the other side of the hill to be the zero for gravitational potential energy.

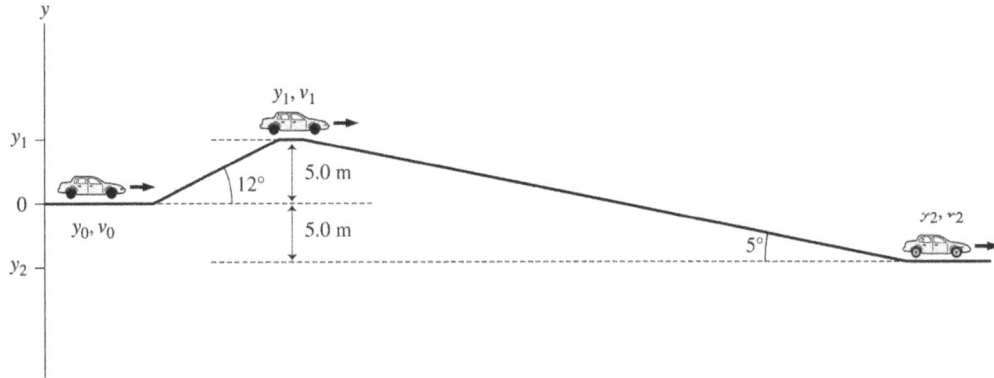

Solve: **(a)** The initial energy of the car is

$$K_0 + U_{g0} = \frac{1}{2}mv_0^2 + mgy_0 = \frac{1}{2}(1500 \text{ kg})(10 \text{ m/s})^2 = 7.5 \times 10^4 \text{ J}$$

The energy of the car at the top of the hill is

$$K_1 + U_{g1} = K_1 + mgy_1 = K_1 + (1500 \text{ kg})(9.80 \text{ m/s}^2)(5.0 \text{ m}) = K_1 + 7.4 \times 10^4 \text{ J}$$

If the car just wants to make it to the top, then $K_1 = 0$. That is, the car has no velocity at the top of the hill. In other words, a minimum energy of 7.4×10^4 J is needed to get to the top. Since this energy is less than the available energy of 7.5×10^4 J, the car can make it to the top.

(b) The conservation of energy equation $K_0 + U_{g0} = K_2 + U_{g2}$ is

$$7.5 \times 10^4 \text{ J} = \frac{1}{2}mv_2^2 + mgy_2 \Rightarrow 7.5 \times 10^4 \text{ J} = \frac{1}{2}(1500 \text{ kg})v_2^2 + (1500 \text{ kg})(9.80 \text{ m/s}^2)(-5.0 \text{ m})$$

$$\Rightarrow v_2 = 14 \text{ m/s}$$

Assess: A higher speed on the other side of the hill is reasonable because the car has increased its kinetic energy and lowered its potential energy compared to its starting values. Note that the shape of the hill is irrelevant because gravitational potential energy depends only on height.

P10.41. Strategize: We must consider gravitational potential energy, kinetic energy, and thermal energy, since it is evident that conversions between these types are occurring in this problem. Since total energy is conserved, we can compare the sum of gravitational potential energy and kinetic energy in the initial and final cases. The difference will be accounted for by a change in thermal energy.

Prepare: We will use conservation of energy, Equation 10.4, to calculate the increase in thermal energy. Assume the initial velocity of the fireman to be zero.

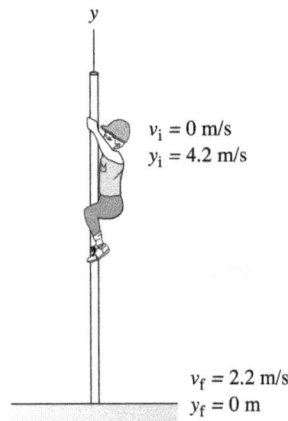

Solve: Consider the figure. Using ground level as the reference for gravitational potential energy, the conservation of energy equation becomes

$$\Delta K + \Delta U_g + \Delta E_{th} = 0 \Rightarrow \Delta E_{th} = -(\Delta K + \Delta U_g)$$

The change in his gravitational potential energy is

$$\Delta U_g = mgy_f - mgy_i = 0 - (80 \text{ kg})(9.80 \text{ m/s}^2)(4.2 \text{ m}) = -3.3 \text{ kJ}$$

The change in his kinetic energy is

$$\Delta K = K_f - K_i = 0 \text{ J} - \frac{1}{2}(80 \text{ kg})(2.2 \text{ m/s})^2 = 0.19 \text{ kJ}$$

So

$$\Delta E_{th} = -(190 \text{ J} - 3300 \text{ J}) = +3.1 \text{ kJ}.$$

Assess: Note that most of the gravitational potential energy is converted to thermal energy.

P10.43. Strategize: The only force doing work on the puck is friction. Knowing that the change in kinetic energy of the puck is equal to the work done on the puck, we can determine how far the puck will slide before coming to rest.
Prepare: We will use the definition of kinetic energy and Equation 10.6 to describe the work done by friction.
Solve: The change in kinetic energy of the puck is

$$\Delta K = K_f - K_i = -mv_i^2/2$$

The work done on the puck by the force of friction is
$$W_f = F_f d \cos 180° = -\mu N d = -\mu mg d$$

Knowing that
$$W_f = \Delta K$$

obtain

$$-\mu mg d = -mv_i^2/2$$

or

$$d = v_i^2/2\mu g = (5.0 \text{ m/s})^2/2(0.05)(9.8 \text{ m/s}^2) = 26 \text{ m}$$

Assess: This is a reasonable distance for a puck traveling 5 m/s to slide across the ice before coming to rest.

P10.45. Strategize: Call the system the tube, rider, and slope (but not the rope or rope puller). The snow is not frictionless, and the change in thermal energy due to friction will be shared between the tube and the slope, so both of those must be in our system. The rope and rope puller have some electrical or chemical potential energy driving it. Since we have no information to account for that energy, it makes the most sense to leave them out of our system and treat the rope as doing external work.
Prepare: Assume the tow rope is parallel to the slope. Use the work-energy equation and $W = Fd$.
Solve: Since there are no springs or chemical reactions involved the work-energy equation is

$$W = \Delta K + \Delta U_g + \Delta E_{th}$$

We are told the towing takes place at a constant speed so $\Delta K = 0$. Solve for the change in thermal energy.

$$\Delta E_{th} = W - \Delta U_g = Fd - mg\Delta y = (340 \text{ N})(120 \text{ m}) - (80 \text{ kg})(9.8 \text{ m/s}^2)(30 \text{ m}) = 17 \text{ kJ}$$

Assess: W is the work done on the system by the rope.

P10.47. Strategize: Call the system the bike, rider, and Earth. With that choice, the only external force that is doing work is that from drag.

Prepare: Use the work-energy equation and $W = Fd$. Assume the cyclist and air do not heat up: $\Delta E_{th} = 0$.

Solve: Since there are no springs or chemical reactions involved the work-energy equation is

$$K_f + (U_g)_f + \Delta E_{th} = K_i + (U_g)_i + W$$

$$\tfrac{1}{2}mv_f^2 + 0 + 0 = \tfrac{1}{2}mv_i^2 + mgh - F_D L$$

$$v_f^2 = v_i^2 + 2gh - \frac{2F_D L}{m} = (12 \text{ m/s})^2 + 2(9.8 \text{ m/s}^2)(30 \text{ m}) - \frac{2(12 \text{ N})(450 \text{ m})}{70 \text{ kg}}$$

$$v_f = 24 \text{ m/s}$$

Assess: 24 m/s is quite fast (54 mph) for a cyclist, but the drag was small.

P10.49. Strategize: Choose the system to be the woman and Earth. Since there is no friction and no significant drag mentioned, total mechanical energy is conserved. It might seem as though tension (an external force) can do work on our system. But note that the woman will move in a circular arc, in which her motion is always tangential and the tension is always radial. Since the force and motion are always orthogonal, no work is done by tension.

Prepare: We can use simple geometry to find her height above the far edge of the ravine. Using the law of conservation of energy will allow us to relate her final gravitational potential energy to her initial kinetic energy. For simplicity, we choose the zero point of gravitational potential energy to be on level ground where she is initially running.

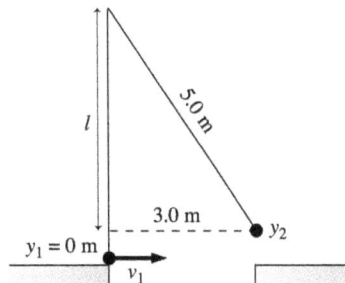

Solve: (a) As she swings, her height above the cliff increases since the rope doesn't stretch. Her initial kinetic energy is being converted to gravitational potential energy during the swing.

(b) Refer to the diagram. When she is directly above the opposite side of the ravine, she has moved 3.0 m horizontally while the rope is also swinging her upwards. Using the Pythagorean theorem, we can find the distance between the branch and her new height. $l = \sqrt{(5.0 \text{ m})^2 - (3.0 \text{ m})^2} = 4.0 \text{ m}$. Therefore, she is $5.0 \text{ m} - 4.0 \text{ m} = 1.0 \text{ m}$ above the cliff.

(c) To make it to the other side of the ravine, she must have enough kinetic energy to be converted to the equivalent gravitational potential energy of her additional 1.0 m of height. The minimum initial velocity she will need will be when she just makes it to the other side of the ravine with no kinetic energy (all her initial kinetic energy being converted to potential energy of her height above the cliff). Using the cliff as the reference for gravitational potential energy, the conservation of energy equation reads

$$K_1 + (U_g)_1 = K_2 + U_g(U_g)_2$$

$$\frac{1}{2}mv_1^2 = mgy_2$$

$$v_1 = \sqrt{2gy_2} = \sqrt{2(9.80 \text{ m/s}^2)(1.0 \text{ m})} = 4.4 \text{ m/s}$$

Assess: We calculated for the case where all her initial kinetic energy is converted to gravitational potential energy at the other side of the ravine. Note she could start with a greater initial kinetic energy, which would also get her to the other side of the ravine. In this case, when she's above the other side of the ravine, she will have some additional kinetic energy instead of just making it to the other side.

P10.51. Strategize: For an energy diagram, the sum of the kinetic and potential energy is a constant.

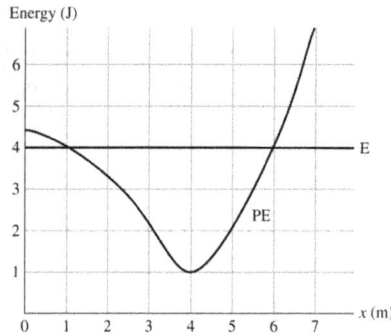

Prepare: The particle is released from rest at $x = 1.0$ m. That is, $K = 0$ at $x = 1.0$ m. Since the total energy is given by $E = K + U$, and is constant, we can draw a horizontal total energy line through the point of intersection of the potential energy curve (PE) and the $x = 1.0$ m line. The distance from the PE curve to the energy line is the particle's kinetic energy. These values are transformed as the position changes, causing the particle to speed up or slow down, but the sum $K + U$ does not change.

Solve: (a) We have $E = 4.0$ J and this energy is a constant. For $x < 1.0, U > 4.0$ J and, therefore, K must be negative to keep E the same (note that $K = E - U$ or $K = 4.0$ J $- U$). Since negative kinetic energy is unphysical, the particle cannot move to the left. That is, the particle will move to the right of $x = 1.0$ m.

(b) The expression for the kinetic energy is $E - U$. This means the particle has maximum speed or maximum kinetic energy when U is minimum. This happens at $x = 4.0$ m. Thus,

$$K_{max} = E - U_{min} = (4.0 \text{ J}) - (1.0 \text{ J}) = 3.0 \text{ J} \qquad \frac{1}{2}mv_{max}^2 = 3.0 \text{ J} \Rightarrow v_{max} = \sqrt{\frac{2(3.0 \text{ J})}{m}} = \sqrt{\frac{8.0 \text{ J}}{0.020 \text{ kg}}} = 17.3 \text{ m/s}$$

The particle possesses this speed at $x = 4.0$ m.

(c) The energy line intersects the potential energy (PE) curve at $x = 1.0$ m and $x = 6.0$ m. These are the turning points of the motion.

Assess: It is reasonable that one of the turning points is at the location the particle was released. If the particle oscillated back and went further to the left, it would need to have more energy than it started with.

P10.53. Strategize: This question relates to the concept of equilibrium. We need to determine at what separation of atoms the system is in stable equilibrium.

Prepare: A system is in stable equilibrium (or at least metastable equilibrium) when at a local energy minimum.

Solve: We can read from the plot that the minimum in energy occurs at approximately $x = 0.07$ nm. Estimates as high as $x = 0.08$ nm might be acceptable given the coarse grain of the plot.

Assess: Since the Bohr radius of a hydrogen atom is around 0.05 nm, this is a reasonable length scale for a bond between two such atoms. It turns out the exact bond length is 0.074 nm.

P10.55. Strategize: This is a one-dimensional collision that obeys the conservation laws of momentum. Since the collision is perfectly elastic, mechanical energy is also conserved.

Prepare: Equation 10.22 applies to perfectly elastic collisions.

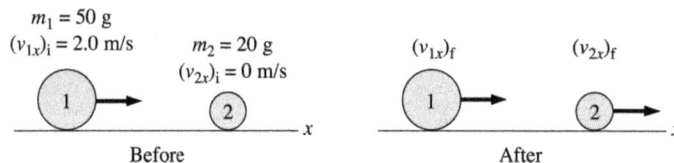

Solve: Using Equation 10.22,

$$(v_{1x})_f = \frac{m_1 - m_2}{m_1 + m_2}(v_{1x})_i = \frac{50 \text{ g} - 20 \text{ g}}{50 \text{ g} + 20 \text{ g}}(2.0 \text{ m/s}) = 0.86 \text{ m/s}$$

$$(v_{1x})_f = \frac{2m_1}{m_1 + m_2}(v_{1x})_i = \frac{2(50 \text{ g})}{50 \text{ g} + 20 \text{ g}}(2.0 \text{ m/s}) = 2.9 \text{ m/s}$$

Assess: These velocities are of a reasonable magnitude. Since both these velocities are positive, both balls move along the positive x direction. This makes sense since ball 1 is more massive than ball 2 and ball 2 is initially at rest.

P10.57. Strategize: For a perfectly inelastic collision, the two colliding objects stick together after the collision and energy is not conserved. Since we are given no information about outside forces acting during the collision, we will assume there are none and that momentum is conserved. Knowing these two pieces of information we can solve the problem.

Reason: We can use conservation of momentum to determine the final speed of the two objects after the collision. We can use the definition of kinetic energy to determine the initial and final kinetic energies. We can equate the decrease in kinetic energy to the increase in thermal energy.

Solve: Conserving momentum we obtain the velocity of the compound object: $mv = 2mV$ or $V = v/2$

The initial kinetic energy (the kinetic energy of the incident glider) is

$$K_i = mv^2/2$$

The final kinetic energy (the kinetic energy of the combined two gliders) is

$$K_f = (2m)V^2/2 = (2m)(v/2)^2/2 = mv^2/4$$

The final kinetic energy is some fraction f of the initial kinetic energy or $K_f = fK_i$

Solving for the fraction f, obtain

$$f = \frac{K_f}{K_i} = \frac{(mv^2/4)}{(mv^2/2)} = \frac{1}{2}$$

Knowing that the final kinetic energy is one-half the initial kinetic energy, we may conclude that one-half the first glider's kinetic energy is transformed into thermal energy during the collision.

Assess: After a quick first glance at this problem, one might conclude that nothing is given and that the problem can not be solved. After thinking about the concepts involved, the problem can be solved and a numerical value obtained even though no values are given.

P10.59. Strategize: If we treat the system as the block only, then two outside forces are doing work on it: friction and our push. Since there is no change in kinetic energy (which in this case is all the energy in our system) the work done by the two forces must be equal and opposite. Therefore we find the work done by friction. Once we have the work and the time, the power is easy to calculate.

Prepare: We can use the definition of work, Equation 10.5, to calculate the work you do in pushing the block. The displacement is parallel to the force, so we can use $W = Fd$. Since the block is moving at a steady speed, the force you exert must be exactly equal and opposite to the force of friction.

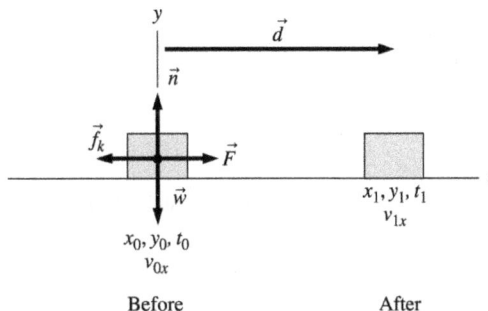

Solve: (a) The work done on the block is $W = Fd$ where d is the displacement. We will find the displacement using kinematic equations. The displacement in the x-direction is

$$d = (1.0 \text{ m/s})(3.0 \text{ s}) = 3.0 \text{ m}$$

We will find the force using Newton's second law of motion. Consider the preceding diagram. The equations for Newton's second law along the x and y components are

$$(F)_y = n - w = 0 \text{ N} \Rightarrow n = w = mg = (10 \text{ kg})(9.80 \text{ m/s}^2) = 98.0 \text{ N}$$

$$(F)_x = \vec{F} - \vec{f}_k = 0 \text{ N} \Rightarrow F = f_k = \mu_k n = (0.60)(98 \text{ N}) = 58.8 \text{ N}$$

$$\Rightarrow W = Fd = (58.8 \text{ N})(3.0 \text{ m}) = 176 \text{ J}, \text{ which should be reported as } 1.8 \times 10^2 \text{ J to two significant figures.}$$

An extra significant figure has been kept in intermediate calculations.

(b) The power required to do this much work in 3.0 s is

$$P = \frac{W}{t} = \frac{176 \text{ J}}{3.0 \text{ s}} = 59 \text{ W}$$

Assess: This seems like a reasonable amount of power to push a 10 kg block at 1.0 m/s. Note that this power is almost what a standard 60 W light bulb requires!

P10.61. Strategize: If we treat the system as the elevator and Earth, then only the elevator motor is doing work on it. Since the speed is constant, the work being done is used solely to increase the elevator's gravitational potential energy.
Prepare: We can apply the work-energy equation to calculate the work done and then use this to calculate the power supplied by the motor using Equation 10.24.
Solve: The tension in the cable does work on the elevator to lift it. Because the cable is pulled by the motor, we say that the motor does the work of lifting the elevator.

(a) The energy conservation equation is $K_i + U_i + W = K_f + U_f + \Delta E_{th}$. Using $K_i = 0$ J, $K_f = 0$ J, and $\Delta E_{th} = 0$ J gives

$$W = (U_f - U_i) = mg(y_f - y_i) = (1000 \text{ kg})(9.80 \text{ m/s}^2)(100 \text{ m}) = 9.8 \times 10^5 \text{ J}$$

(b) The power required to give the elevator this much energy in a time of 50 s is

$$P = \frac{W}{\Delta t} = \frac{9.8 \times 10^5 \text{ J}}{50 \text{ s}} = 2.0 \times 10^4 \text{ W}$$

Assess: Since 1 horsepower (hp) is 746 W, the power of the motor is 27 hp. This is a reasonable amount of power to lift a mass of 1000 kg to a height of 100 m in 50 s.

P10.63. Strategize: The work done on the car while it is accelerating from rest to the final speed is the change in kinetic energy. Knowing the work done and the time to do this work we can determine the power associated with this work.
Prepare: This is a straightforward application of the work-energy equation (Equation 10.2) and the definition of kinetic energy.
Solve: The change in kinetic energy of the car is

$$W = \Delta K = K_f - K_i = \frac{1}{2}mv_f^2 = \frac{1}{2}(1000 \text{ kg})(30 \text{ m/s})^2 = 4.5 \times 10^5 \text{ J}$$

since the initial kinetic energy is zero.
The power associated with this work is

$$P = \frac{W}{\Delta t} = \frac{4.5 \times 10^5 \text{ J}}{10 \text{ s}} = 45 \text{ kW}$$

Assess: This is reasonable. In most cars only a small fraction of the work done by the engine goes into propelling the car.

P10.65. Strategize: If the system is the cyclist and Earth, then work is being done to increase the gravitational potential energy only. Note that the kinetic energy is constant.

Prepare: Use the definition of power: $P = \dfrac{\Delta E}{\Delta t}$, and solve for the time interval.

Solve: The change in energy is the change in gravitational potential energy.

$$P = \frac{\Delta E}{\Delta t} \Rightarrow \Delta t = \frac{\Delta E}{P} = \frac{mg\Delta y}{P} = \frac{(85 \text{ kg})(9.8 \text{ m/s}^2)(1100 \text{ m})}{450 \text{ W}} = 2036 \text{ s} \approx 34 \text{ min}$$

Assess: This is a reasonable length of time for a cyclist in a climbing stage in the Tour de France.

P10.67. Strategize: Since the car is traveling at a constant speed, the force that the car's engine provides to move the car forward must equal the total force opposing the car's motion. This is related to energy-work concepts, but since the forces act over the same distances, it can be reduced to a force problem.

Prepare: We will use Equation 10.25 to calculate the power, since we are given the velocity of the car and will calculate the force.

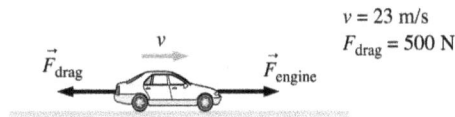

Solve: **(a)** Refer to the diagram. Since the force provided by the engine must equal the drag force,

$$F_{\text{engine}} = F_{\text{drag}} = 500 \text{ N}$$

(b) The power is

$$P = Fv = (500 \text{ N})(23 \text{ m/s}) = 1.2 \times 10^4 \text{ W}$$

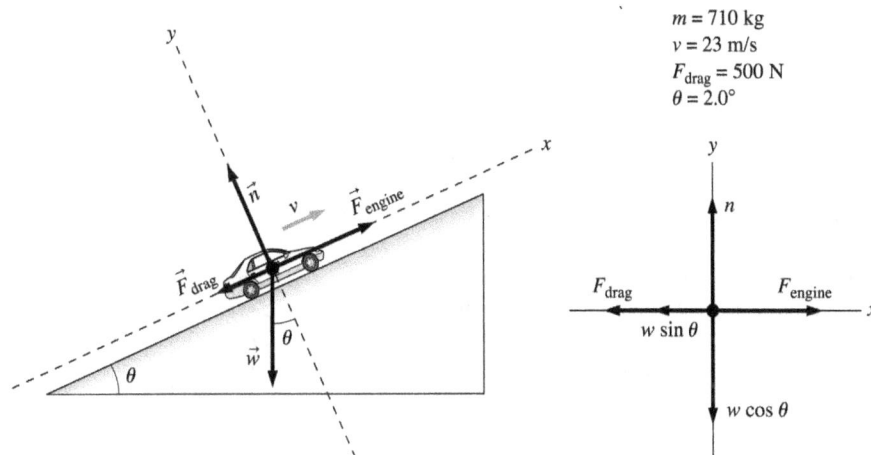

Refer to the diagram. Now in addition to the drag force, there is a component of the weight of the car that is opposing the motion of the car. From the diagram, the component of the weight which is opposite the direction of motion of the car is

$$F_{\text{gravity, along the slope}} = mg\sin(\theta) = (710 \text{ kg})(9.80 \text{ m/s}^2)\sin(2°) = 243 \text{ N}$$

The total force the engine must overcome is now $500 \text{ N} + 243 \text{ N} = 743 \text{ N}$.
The power required from the engine for this is

$$P = Fv = (743 \text{ N})(23 \text{ m/s}) = 1.7 \times 10^4 \text{ W}$$

Assess: Note that this form of the definition of power is more convenient to use since the velocity is given.

P10.69. Strategize: The two forces acting on the elevator are its weight and the force $\vec{F}$ due to the motor. Since the elevator is moving with constant velocity, the net force on the elevator is zero.

Prepare: We simply write the sum of all forces in the vertical direction and equate it to the zero acceleration. Once we know the force being applied, we determine the power using $P = Fv$.

Solve: Since the net force on the elevator is zero, $\vec{F} + \vec{w} = \vec{0}$ N. So

$$F = -w = 2500 \text{ N}$$

The power due to this force acting on the elevator moving with constant velocity can be calculated using Equation 10.25.

$$P = Fv = (2500 \text{ N})(8.0 \text{ m/s}) = 2.0 \times 10^4 \text{ W}$$

Assess: One horsepower (hp) is 746 W, so the power of the motor is 26.8 hp. This is a reasonable amount of power to lift an elevator.

P10.71. Strategize: The elevator is not moving with constant velocity, so there is a non-zero net force. The tension in the cable is greater than the weight of the elevator. If the system is the elevator and Earth, then only the elevator motor is doing work and increasing both the kinetic and gravitational potential energies.

Prepare: We use kinematics to determine the final speed at the end of this 15 m rise. This allows us to calculate the final kinetic and gravitational potential energies. We then use the work-energy equation (Equation 10.2).

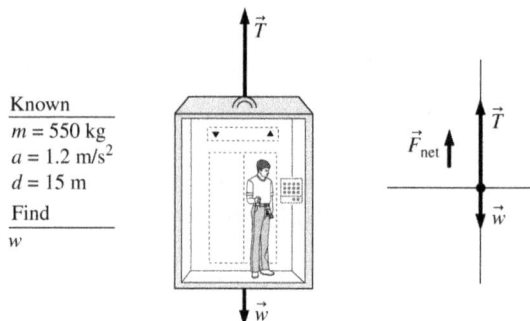

Solve: We need to find the tension in the cable and use that in the work equation

$$W = Fd = Td = m(a + g)d = (550 \text{ kg})(1.2 \text{ m/s}^2 + 9.8 \text{ m/s}^2)(15 \text{ m}) = 91 \text{ kJ}$$

Assess: This seems to be a reasonable amount of work on a large elevator.

P10.73. Strategize: We can choose the system to be the box, the ramp, and Earth. That way friction causes a change in internal thermal energy and only the external pull does work on the system.

Prepare: Use the definitions of kinetic energy (Equation 10.8), gravitational potential energy (Equation 10.14), work (Equation 10.5), and conservation of energy (Equation 10.3) in this problem.

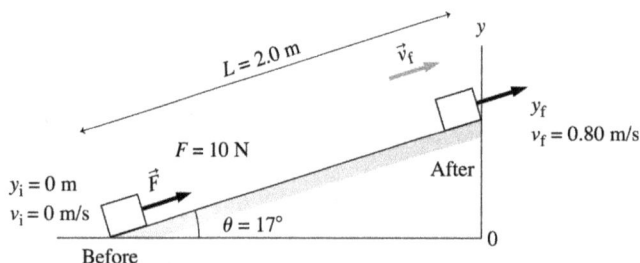

Solve: (a) The work done by the force can be calculated with Equation 10.5 since the force is parallel to the displacement of the box.

$$W = Fd = (10 \text{ N})(2.0 \text{ m}) = 20 \text{ J}$$

(b) We know the box starts from rest, so $K_i = 0$ J. We are given that the speed of the box at the top of the ramp is 0.80 m/s, so we can calculate the change in kinetic energy.

$$\Delta K = K_f - K_i = \frac{1}{2}mv_f^2 = \frac{1}{2}(2.3 \text{ kg})(0.80 \text{ m/s})^2 = 0.74 \text{ J}$$

(c) Taking the reference for gravitational potential energy to be the bottom of the ramp, $U_i = 0$ J. We need to find the final height of the box to calculate the final gravitational potential energy. Refer to the preceding diagram. Since the ramp is 2.0m long and at an angle of $17°$ the final height of the box is

$$y_f = (2.0 \text{ m})\sin(17°) = 0.58 \text{ m}$$

So the change in gravitational potential energy is

$$\Delta U_g = (U_g)_f - (U_g)_i = mgy_f = (2.3 \text{ kg})(9.80 \text{ m/s}^2)(0.58 \text{ m}) = 13 \text{ J}$$

(d) From the conservation of energy equation we have

$$\Delta K + \Delta U_g + \Delta E_{th} = W \Rightarrow \Delta E_{th} = W - \Delta K - \Delta U_g = 20 \text{ J} - 0.74 \text{ J} - 13 \text{ J} = 6.3 \text{ J}$$

Assess: Note that much of the work goes into overcoming gravity and the friction in the ramp, giving a relatively small increase in kinetic energy.

P10.75. Strategize: We will take the system to be the flea plus the earth. Then the only external force that does work on the system is drag.
Prepare: Right after the flea jumps, it has kinetic energy. This is transformed to potential energy and thermal energy as it moves upward.
Solve: (a) If there is no air resistance, none of the initial kinetic energy is transformed to thermal energy, so we can write $K_i + U_i = K_f + U_f$. We want the initial point to be when the flea takes off; at this point, the kinetic energy $K_i = \frac{1}{2}mv^2$ is a maximum. We take the zero of potential energy to be at the ground, so $U_i = 0$. We will take the final point to be the highest point of the motion. Here, $K_f = 0$, and the potential energy $U_f = mgh$ is a maximum. Solving, we find

$$K = \frac{1}{2}mv_i^2 = mgh = (5.0 \times 10^{-4} \text{ kg})(9.80 \text{ m/s}^2)(0.40 \text{ m}) = 2.0 \text{ mJ}$$

(b) If some of the initial kinetic energy is transformed to thermal energy, the final potential energy, and the final height, will be less. When air resistance is a factor, there is a loss to thermal energy and the final height is half the height with no air resistance. As potential energy is proportional to height above the ground, half the energy, 50%, is lost to thermal energy. So only half the initial kinetic energy has been converted to potential energy.
Assess: As shown here, air resistance and wind speed are a real concern for a flea.

P10.77. Strategize: Let the system be the girl and Earth. Assume the chain to be massless. In the absence of frictional and air-drag effects, the sum of the kinetic and gravitational potential energy does not change during the swing's motion.
Prepare: We use geometry to calculate the change in height during the swing. Since the system is isolated, this will tell us the change in kinetic energy, and therefore the initial kinetic energy.

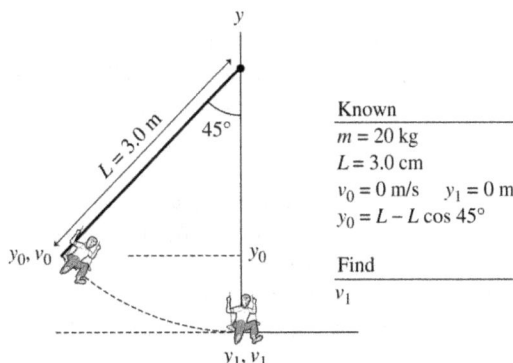

Known
$m = 20$ kg
$L = 3.0$ cm
$v_0 = 0$ m/s $y_1 = 0$ m
$y_0 = L - L \cos 45°$

Find
v_1

Solve: The quantity $K + U_g$ is the same at the highest point of the swing as it is at the lowest point. That is, $K_0 + U_{g0} = K_1 + U_{g1}$. The preceding equation is

$$\frac{1}{2}mv_0^2 + mgy_0 = \frac{1}{2}mv_1^2 + mgy_1 \Rightarrow v_1^2 = v_0^2 + 2g(y_0 - y_1)$$

$$\Rightarrow v_1^2 = (0 \text{ m/s})^2 + 2g(y_0 - 0 \text{ m}) \Rightarrow v_1 = \sqrt{2gy_0}$$

We see from the pictorial representation that

$$y_0 = L - L\cos 45° = (3.0 \text{ m}) - (3.0 \text{ m})\cos 45° = 0.88 \text{ m}$$

$$\Rightarrow v_1 = \sqrt{2gy_0} = \sqrt{2(9.80 \text{ m/s}^2)(0.88 \text{ m})} = 4.2 \text{ m/s}$$

Assess: We did not need to know the swing's or the child's mass. Also, a maximum speed of 4.2 m/s is reasonable.

P10.79. Strategize: Let the system be the sledder, sled, and Earth. Since the hill is frictionless, mechanical energy will be conserved during the sledder's trip. To make it over the next hill, the sledder's velocity must be greater than or equal to zero at the top of the hill. Even though the sledder dips down between the two hills, we only need to concern ourselves with the initial and final positions.

Prepare: The sum of the kinetic and potential energies will be the same for the initial and final positions. Since the minimum velocity the sledder can have at the top of the second hill is 0 m/s to just make it over, we will set her final kinetic energy to zero. Since the heights are known, we can determine gravitational potential energies, and solve for the initial speed.

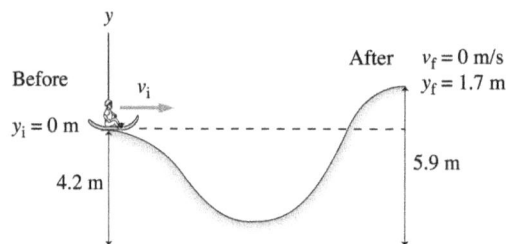

Solve: Consider the before and after pictorial representation. We will use the sledder's initial height as the reference for gravitational potential energy. Since there is no friction, the conservation of energy equation, Equation 10.4 reads

$$\frac{1}{2}mv_i^2 + mgy_i = \frac{1}{2}mv_f^2 + mgy_f \Rightarrow v_i^2 = 2gy_f$$

$$v_i = \sqrt{2gy_f} = \sqrt{2(9.80 \text{ m/s}^2)(1.7 \text{ m})} = 5.8 \text{ m/s}$$

Where we have used $y_i = 0$ m, and $v_f = 0$ m/s for the sledder to just make it over the second hill. Note that since we are using the top of the first hill as the reference of gravitational potential energy, we must use the height of the top of the second hill above the first for y_f, $y_f = 5.9$ m − 4.2 m = 1.7 m.

Assess: Note the shape of the hill doesn't matter, only the difference in height between the first and second hill is needed, as expected for gravitational potential energy. Since the second hill is higher than the first, we expect that the sledder needs the additional kinetic energy at the initial hill to make up for the additional potential energy needed at the top of the second hill.

P10.81. Strategize: We will take the system to be the person plus the earth. When a person drops from a certain height, the initial potential energy is transformed to kinetic energy. When the person hits the ground, if they land rigidly upright, we assume that all of this energy is transformed into elastic potential energy of the compressed leg bones.

Prepare: The maximum energy that can be absorbed by the leg bones is 200 J; this limits the maximum height. We equate the gravitational potential energy at this maximum height to the maximum elastic potential energy, and solve for the height.

Solve: **(a)** The initial potential energy can be at most 200 J, so the height h of the jump is limited by $mgh = 200$ J For $m = 60$ kg, this limits the height to

$$h = 200 \text{ J}/mg = 200 \text{ J}/(60 \text{ kg})(9.80 \text{ m/s}^2) = 0.34 \text{ m}$$

(b) If some of the energy is transformed to other forms than elastic energy in the bones, the initial height can be greater. If a person flexes her legs on landing, some energy is transformed to thermal energy. This allows for a greater initial height.

Assess: There are other tissues in the body with elastic properties that will absorb energy as well, so this limit is quite conservative.

P10.83. Strategize: Assume an ideal spring that obeys Hooke's law. This is a two-part problem. The first part, when the bullet embeds itself in the block, is a perfectly inelastic collision. In a perfectly inelastic collision, the momentum is conserved while energy is not conserved. In the second part of the problem, when the bullet and block hit the spring, there is no friction. Since there is no friction after the bullet enters the block, the mechanical energy of the system (bullet + block + spring) is conserved during that part of the motion.

Prepare: We will equate the initial and final momenta of the bullet and the block together. Initially the block is at rest, and after the collision, they move with the same speed.

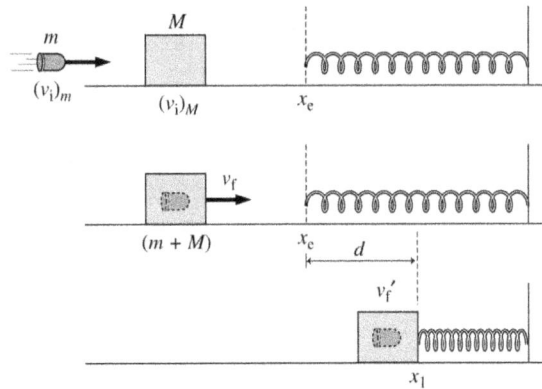

We place the origin of our coordinate system at the end of the spring that is not anchored to the wall.

Solve: **(a)** Momentum conservation for perfectly inelastic collision states $p_f = p_i$. This means

$$(m + M)v_f = m(v_i)_m + M(v_i)_M \Rightarrow (m + M)v_f = mv_B + 0 \text{ kg m/s} \Rightarrow v_f = \left(\frac{m}{m+M}\right)v_B$$

where we have used v_B for the initial speed of the bullet. This is velocity of the bullet and block after the bullet embeds itself in the block. Now, when the bullet and block hit the spring and compress it, mechanical energy is conserved. The mechanical energy conservation equation $K_1 + U_{s1} = K_e + U_{se}$ as the bullet-embedded block compresses the spring is:

$$\frac{1}{2}m(v_f')^2 + \frac{1}{2}k(x_1 - x_e)^2 = \frac{1}{2}(m + M)(v_f)^2 + \frac{1}{2}k(x_e - x_e)^2$$

$$0 \text{ J} + \frac{1}{2}kd^2 = \frac{1}{2}(m + M)\left(\frac{m}{m+M}\right)^2 v_B^2 + 0 \text{ J} \Rightarrow v_B = \sqrt{\frac{(m+M)kd^2}{m^2}}$$

(b) Using the preceding formula with $m = 5.0$ g, $M = 2.0$ kg, $k = 50$ N/m, and $d = 10$ cm,

$$v_B = \sqrt{\frac{(0.0050 \text{ kg} + 2.0 \text{ kg})(50 \text{ N/m})(0.10 \text{ m})^2}{(0.0050 \text{ kg})^2}} = 200 \text{ m/s}$$

which should be reported as 2.0×10^2 m/s to two significant figures.

(c) The fraction of energy lost is (initial energy − final energy)/(initial energy), which is

$$\frac{\frac{1}{2}mv_{\mathrm{B}}^{2}-\frac{1}{2}(m+M)v_{\mathrm{f}}^{2}}{\frac{1}{2}mv_{\mathrm{B}}^{2}}=1-\frac{m+M}{m}\left(\frac{v_{\mathrm{f}}}{v_{\mathrm{B}}}\right)^{2}=1-\frac{m+M}{m}\left(\frac{m}{m+M}\right)^{2}$$

$$=1-\frac{m}{m+M}=1-\frac{0.0050\ \mathrm{kg}}{(0.0050\ \mathrm{kg}+2.0\ \mathrm{kg})}=99.8\%$$

where we have kept an additional significant figure.

Assess: During the perfectly inelastic collision 99.8% of the bullet's energy is lost. The energy is transformed into the energy needed to deform the block and bullet and to the thermal energy of the bullet and block combination.

P10.85. Strategize: If we choose our system to be all three railroad cars, then there are no external forces. But there is an inelastic collision, in which other forms of energy could be transformed into heat (and sound).

Reason: Since this is a perfectly inelastic collision and there are no significant outside forces acting during the collision, momentum is conserved and energy is not.

Solve:

(a) From conservation of momentum we may write $2mv = 3mV$ or $V = 2v/3$

The final speed (V) of the three cars is 2/3 the speed (v) of the two cars before collision, or 1.7 m/s.

$$v_{\mathrm{f}} = \tfrac{2}{3}v_{\mathrm{i}} = \tfrac{2}{3}(2.5\ \mathrm{m/s}) = 1.7\ \mathrm{m/s}$$

(b) The initial kinetic energy is $K_{\mathrm{i}} = (2m)v^{2}/2 = mv^{2}$

The final kinetic energy is $K_{\mathrm{f}} = (3M)V^{2}/2 = (3m)(2V/3)^{2}/2 = 2mv^{2}/3$

Since the final kinetic energy is a fraction f of the initial kinetic energy, we may write

$$K_{\mathrm{f}} = fK_{\mathrm{i}}$$

Or solving for f obtain

$$f = K_{\mathrm{f}}/K_{\mathrm{i}} = (2mv^{2}/3)/mv^{2} = 2/3$$

Knowing that the final kinetic energy is 2/3 the initial kinetic energy, we conclude that 1/3 of the initial kinetic energy is transformed to thermal energy.

Assess: Losing 1/3 the initial kinetic energy to thermal energy is reasonable

P10.87. Strategize: The first collision described is perfectly inelastic, and the second is perfectly elastic. We must carefully decide when we can use conservation of mechanical energy.

Prepare: Momentum is conserved in both inelastic and elastic collisions. Mechanical energy is conserved only in an elastic collision. We will break this problem into separate parts, using conservation of momentum and/or conservation of energy where appropriate.

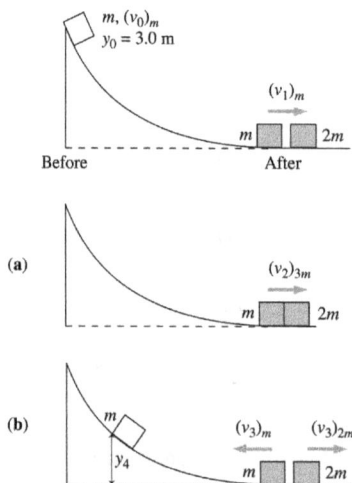

Solve: Consider the motion of the first package down the chute. Mechanical energy is conserved during its slide since the chute is frictionless. For a package with mass m the conservation of energy equation is

$$K_1 + U_{g1} = K_0 + U_{g0} \Rightarrow \frac{1}{2}m(v_1)_m^2 + mgy_1 = \frac{1}{2}m(v_0)_m^2 + mgy_0$$

Using $(v_0)_m = 0$ m/s and $y_1 = 0$ m,

$$\frac{1}{2}m(v_1)_m^2 = mgy_0 \Rightarrow (v_1)_m = \sqrt{2gy_0} = \sqrt{2(9.80 \text{ m/s}^2)(3.0 \text{ m})} = 7.67 \text{ m/s}$$

We will keep an additional significant figure in this intermediate calculation.
(a) For the perfectly inelastic collision the conservation of momentum equation is

$$p_{fx} = p_{ix} \Rightarrow (m + 2m)(v_2)_{3m} = m(v_1)_m + (2m)(v_1)_{2m}$$

Using $(v_1)_{2m} = 0$ m/s, we get

$$(v_2)_{3m} = (v_1)_m / 3 = 2.6 \text{ m/s}$$

(b) For the elastic collision, the mass m package rebounds with velocity

$$(v_3)_m = \frac{m - 2m}{m + 2m}(v_1)_m = -\frac{1}{3}(7.67 \text{ m/s}) = -2.56 \text{ m/s}$$

Where we keep an additional significant figure in this intermediate result for use later.
The negative sign with $(v_3)_m$ shows that the package with mass m rebounds. It will travel back up the chute and goes to the position y_4. During this part of its motion, mechanical energy is conserved. We can determine y_4 by applying the conservation of energy equation as follows. For a package of mass m:

$$K_f + U_{gf} = K_i + U_{gi} \Rightarrow \frac{1}{2}m(v_4)_m^2 + mgy_4 = \frac{1}{2}m(v_3)_m^2 + mgy_3$$

Using $(v_3)_m = -2.56$ m/s, $y_3 = 0$ m, and $(v_4)_m = 0$ m/s, we get

$$mgy_4 = \frac{1}{2}m(-2.56 \text{ m/s})^2 \Rightarrow y_4 = 33 \text{ cm}$$

Assess: The first mass (m) rebounds up the ramp to a height of 33 cm. This is reasonable since the first package (m) has given a lot of its energy to the second mass ($2m$) Note that for part (b), energy is conserved during the whole process.

P10.89. Strategize: Let the system be the elevator only, such that gravity is treated as an external force that can do work. This is a work-energy problem, but since the forces from the motor and gravity do work over the same distance, this can be reduced to a force problem.
Prepare: To go at constant speed the net force must be zero, so the force of the motor must equal the weight of the elevator (and passengers).
Solve: Use Equation 10.25.

$$P = Fv \Rightarrow v = \frac{P}{F} = \frac{15\,000 \text{ W}}{(1200 \text{ kg})(9.8 \text{ m/s}^2)} = 1.3 \text{ m/s}$$

Assess: This seems like a reasonable speed for an elevator.

P10.91. Strategize: This is a kinematics problem. We will use kinematic equations to determine the average acceleration.

Prepare: We will need to consider the acceleration after the ball first hits the floor, starts to compress up to the point at which the ball reaches maximum compression, and is about to rebound.

Solve: During rebound, the ball compresses 6 mm and slows to a speed of 0 m/s^2. Knowing the initial and final speed of the ball and the distance over which this compression takes place, we can determine the acceleration. Since the acceleration is constant, we can use the kinematics we learned in a previous chapter

$$v_f^2 = v_i^2 + 2a\Delta y \text{ so}$$

$$v_i^2 = -2a\Delta y \Rightarrow a = \frac{v_i^2}{2\Delta y} = \frac{-(7.0 \text{ m/s})^2}{(2(-0.006 \text{ m}))} = 4 \times 10^3 \text{ m/s}^2$$

The closest choice is D.

Assess: This is a very large acceleration compared to the acceleration due to gravity. This is due to the large change in velocity over a very short distance.

P10.93. Strategize: We need to calculate the kinetic energies just before and just after the collision. In problem 10.90 the speed just before impact was calculating using conservation of energy. We can apply the same conservation of energy to the ball as it rises up to its new maximum height after the collision, to determine the speed just after the collision.

Prepare: We will need the results of Problem 10.90 and the ball's speed after rebounding.

Solve: From Problem 10.90, the ball's speed before hitting the ground is 7.0 m/s. Its kinetic energy is then

$$K_{before} = \frac{1}{2}m(v_{before})^2 = \frac{1}{2}(0.0575 \text{ kg})(7.0 \text{ m/s})^2 = 1.4 \text{ J}$$

To find the ball's speed after rebounding, use the fact that it rebounds to a height of 1.4 m and use conservation of energy, as in Problem 10.90.

$$K_{after} = mgh_{after} = (0.0575 \text{ kg})(9.80 \text{ m/s}^2)(1.4 \text{ m}) = 0.79 \text{ J}$$

The percentage change in energy is given by

$$\frac{K_{after} - K_{before}}{K_{before}} = \frac{0.79 \text{ J} - 1.4 \text{ J}}{1.4 \text{ J}} = -0.44 = -44\%$$

The choice corresponding to this is B.

Assess: It is reasonable that the ball loses nearly half of its energy in rebounding from the floor. The height it rebounds to is much less than the initial height it was dropped from.

P10.95. Strategize: This is a work-energy problem. We have the force and the distance over which it acts. We can find the work done.

Prepare: We will use $W = Fd$ to determine the work done by the cyclist.

Solve: Since the cyclist is moving with constant speed, the drag force must equal the force exerted by the cyclist, assuming no friction from other forces.
Using Equation 10.25, $W = Fd = (10 \text{ N})(1000 \text{ m}) = 10 \text{ kJ}$.

The correct choice is B.

Assess: This is a reasonable amount of energy for a person to expend.

P10.97. Strategize: This is a work energy problem. We know that the drag depends on the square of the velocity, and we know the distance over which the new drag force will act.

Prepare: Since the cyclist has sped up, the drag force has increased. We will use that proportionality and the equation $W = Fd$.

Solve: The drag force is proportional to the square of the velocity. Use k to denote the constant of proportionality. Then $F = kv^2$.

Before we can use this equation to calculate the new drag force, we need to calculate k. We can use the fact that at $v = 5$ m/s   the force is 10 N. Then

$$k = \frac{F}{v^2} = \frac{10 \text{ N}}{(5 \text{ m/s})^2} = 0.40 \frac{\text{N}}{(\text{m/s})^2}$$

With the new velocity,

$$F = kv^2 = \left(0.40 \frac{\text{N}}{(\text{m/s})^2}\right)(10 \text{ m/s})^2 = 40 \text{ N}$$

Then the work done is $W = Fd = (40.0 \text{N})(1000 \text{ m}) = 40$ kJ.

The correct choice is B.

Assess: Since force is proportional to the square of velocity, as the velocity goes up by a factor of two, and the force goes up by a factor of four. So we would expect the work to also increase by a factor of four, which is what explicit calculation shows.

P10.99. Strategize: This is a power problem. We care about how quickly the rider is doing work.

Prepare: The speed of the wind relative to the rider has changed, so the drag force will change.

Solve: The wind speed is now 5 m/s relative to the ground, and the rider is moving at 5 m/s relative to the ground and into the wind. The wind velocity relative to the rider is then 10 m/s.

The force due to a relative wind velocity of 10 m/s was calculated in Problem 10.91, and the result is $F = 40$ N.

Her velocity is now only 5 m/s, so the power she is exerting is $p = Fv = (40 \text{ N})(5 \text{ m/s}) = 200$ W. The correct choice is B.

Assess: The force is the same as in Problem 10.98 but the speed is half, as a result we expect the power to be half that in Problem 10.98 and that is the case.

11

USING ENERGY

Q11.1. Reason: The friction between your hands increases the kinetic energies in molecules of your hands, and this exhibits itself as an increase in thermal energy of your hands. The temperature of your hands goes up.
Assess: Thermal energy is related to molecular kinetic energies.

Q11.3. Reason: Riding a bike, you do not change your height. As you run you constantly propel yourself off the ground. Extra energy is required for this.

Q11.5. Reason: Every creature has some efficiency with which they can turn chemical potential energy into some form of mechanical energy (kinetic, gravitational potential, etc). The fact that this efficiency is not 100% means that some chemical potential energy is lost to heat. That heat is lost inside the animal, meaning the temperature will rise.
Assess: This is one reason that vigorous exercise makes you feel warm, even if you are in a cold room. It is also the reason why humans (and other creatures) sweat to cool off during vigorous exercise.

Q11.7. Reason: We know from the previous chapter that kinetic energy increases as the square of the velocity. As velocity increases, kinetic energy increases. As the engine turns faster, the rate at which friction creates waste thermal energy also increases. Also, as the speed of the car increases the drag force on the car increases. All of this requires a faster rate of energy consumption of the energy available as chemical energy in the fuel.

Q11.9. Reason: Temperature (thermal energy) can only increase by adding energy in the form of heat transfer or through mechanical work. The space shuttle gets hotter, and its thermal energy increases, but it isn't heated in the sense of being in thermal contact with a hotter object. Instead, it gets hot by the other method of changing the thermal energy: work. As the shuttle slams into air molecules in the upper atmosphere, work is done on them by the shuttle and work is done on the shuttle by them. The atoms in the shuttle jiggle more vigorously.
Assess: Since the air in the upper atmosphere is not especially not, it is reasonable that the shuttle's thermal energy does not increase due to heat transfer, but rather due to work. It is worth noting that energy is still conserved here. The spacecraft initially has a large kinetic energy. As it slows, the kinetic energy drops, but the thermal energy of the spacecraft and surrounding air increases.

Q11.11. Reason: The temperature of the rod remains constant and no work is being done. Some small amount of heat is lost to the environment, but the energy transferred to the ice must be nearly 100 J, from the first law of thermodynamics.

Q11.13. Reason: In cooling a pan of water in the fridge, the thermal energy of the water decreases while the water does no work.

Q11.15. Reason: Expanding a gas in an insulated container will decrease its thermal energy while no heat can be transferred from the system. Expanding a gas does negative work on the gas.

Q11.17. Reason: Decompressing a gas in a non-insulated container will decrease the thermal energy of the gas, since negative work is being done on it. Since container is not insulated, it is possible for some heat to enter the container.

Q11.19. Reason: The fan blades exert forces on the molecules of air, and exert these forces over some distance. The work done on the air molecules initially shows up as coherent motion in a "breeze", but quickly becomes incoherent. One can argue that the increase in energy can only be accounted for by a higher thermal energy, or one can simply recall that temperature is the average kinetic energy of the particles making up the air. And the fan obviously increases the kinetic energy of those particles.
Assess: Fans help a person feel cooler if the surrounding air is below body temperature. If the air is stationary, then your body will start to warm the air around it. A fan brings fresh, cooler air to your body. But a fan does not cool the air; it heats the air slightly.

Q11.21. Reason: The piston does work on the gas which increases the thermal energy; and it does so quickly enough that heat doesn't escape the system fast enough to keep the gas below ignition temperature.
Assess: Diesel engines have a hard time starting in cold weather for this reason.

Q11.23. Reason: Every heat engine has efficiency less than 100%. This means that some energy is converted to a mechanical form (does work) and some energy is lost as waste heat. In your car, there is a lot of waste heat. This is not related to poor manufacturing, but is a fundamental rule in thermodynamics. As long as this waste heat is being spat out by the engine, you may as well use it to feel warmer.
Assess: The same argument is not true of air conditioning. That is a separate engine that doesn't need to be on for the car to function. So having it on does use up additional fuel.

Q11.25. Reason: Often, refrigerators have a greater coefficient of performance, when the hot and cold reservoirs have temperatures that are very similar. In this context, that means having the cold reservoir (the inside of the fridge) and the hot reservoir (the air outside) at temperatures that are very similar will increase efficiency and reduce costs. Since basement air is cooler than air upstairs, putting the fridge in the basement should lower costs.

Assess: You have seen this numerically in the special case of the Carnot refrigerator, which has $\text{COP}_{max} = \dfrac{T_C}{T_H - T_C}$.

Clearly, if T_C and T_H get close the coefficient of performance rises.

Q11.27. Reason: Some of the energy of braking is recovered in the battery and can be used again as electric energy to drive the car. In a gasoline-fueled vehicle the energy of braking is not recovered, but simply heats the brakes and then the air and is not recoverable.
Assess: Flywheel systems recover some energy in braking by storing it as rotational kinetic energy which can then be re-directed to moving the car.

Q11.29. Reason: A 100% efficient electric heater doesn't violate the second law of thermodynamics because entropy increases as expected.
Assess: The law prohibits moving heat from a cold object to a hot object with 100% efficiency, but not the other way around.

Q11.31. Reason: We know that the total energy is conserved. The only way the energy of a system can change is if energy flows into or out of the system (through heat, particle flow, etc). If the container is perfectly sealed and insulated, then no energy can flow into or out of the container and the energy cannot decrease. Free energy is different, in that it takes entropy into account $G = H - TS$: Even if the energy remains the same, the free energy (or Gibbs' free energy) can decrease due to the entropy increase. In fact, we know that for a reaction to proceed spontaneously, the requirement is that $\Delta G < 0$.
Assess: Even though energy cannot change in a perfectly isolated system, we know that something changes in a reaction. So it is entirely plausible that the Gibbs' free energy changes.

Q11.33. Reason: For most activities, humans have about 25% efficiency. That means that we would have to use up four times as much chemical energy as we need to put out in electric/mechanical energy. So we would have to metabolize at a rate of 600 W. The correct answer is D.

Assess: There would also likely be losses in converting from the kinetic energy of bicycle parts to the required electric energy.

Q11.35. Reason: Walking takes less energy for the same distance run. The time it takes to run the same distance as the walk is less. Therefore the both the energy used and power required for the run is greater. The answer is C.

Q11.37. Reason: The efficiency is $e = \dfrac{Q_H - Q_C}{Q_H}$ and the maximum efficiency is $e_{max} = 1 - \dfrac{T_C}{T_H}$. Set these equal to each other and solve for $Q_C \cdot T_C = 20°C = 293\,K$ and $T_H = 450°C = 723\,K$.

$$\frac{Q_H - Q_C}{Q_H} = 1 - \frac{T_C}{T_H}$$

$$1 - \frac{Q_C}{Q_H} = 1 - \frac{T_C}{T_H} \quad \Rightarrow \quad Q_C = Q_H \frac{T_C}{T_H} = (100\,\text{MJ})\left(\frac{293\,K}{723\,K}\right) = 40\,\text{MJ}$$

So the answer is D.

Assess: Power plants do heat up rivers like this.

Problems

P11.1. Strategize: We will use conservation of energy to calculate the energy generated by the engine that is converted entirely to kinetic energy and then use the definition of efficiency to calculate the total energy generated by the engine.

Prepare: The conservation of energy states that the work done by the car engine is equal to the change in the kinetic energy. Thus $W_{out} = \Delta K = \frac{1}{2}mv^2$

Solve: Using the definition of thermal efficiency,

$$e = \frac{W_{out}}{Q_H} \Rightarrow Q_H = \frac{W_{out}}{e} = \frac{\frac{1}{2}(1500\ \text{kg})(15\ \text{m/s})^2}{0.10} = 1.7 \times 10^6\,\text{J}$$

That is, the burning of gasoline transfers into the engine 1.7×10^6 J of energy.

Assess: Note that the vast majority of the energy generated by the car's engine is converted to heat. This is typical for engines in cars.

P11.3. Strategize: We consider the definition of efficiency and identify what we "get out" and what we had to "put in".

Prepare: Efficiency is given by Equation 11.2

$$e = \frac{\text{what you get}}{\text{what you had to pay}}$$

In this case the 4.0×10^{-3} W of electrical energy is "what you get" as visible light, and the 1.2×10^{-1} W of light energy is "what you had to pay."

Solve: The efficiency calculation gives

$$e = \frac{4.0 \times 10^{-3}\ \text{W}}{1.2 \times 10^{-1}\ \text{W}} = 0.033 = 3.3\%$$

Assess: Photovoltaic (PV) cells, also known as solar cells, are notoriously inefficient, and this is a typical value for traditional cells. However, advances in technology have been made and efficiencies of up to 20% are available; some researchers are aiming for 40% efficiency.

P11.5. Strategize: We consider the definition of efficiency to determine the power effectively radiated by the LED, and then compare that power to that of an incandescent bulb.
Prepare: $e = 0.20$
Solve: Each LED provides $(1.0\,W)(0.20) = 0.20\,W$ of visible light. We therefore need eight of them to give 1.6 W of visible light power. Since each of them uses 1.0 W then the total power necessary is 8.0 W.
Assess: This is a factor of five better than the incandescent bulb.

P11.7. Strategize: We take the required rate of power and multiply by time to find the total energy requirement. From, there it is a simple unit conversion.
Prepare: As detailed in Table 11.3, the energy per second needed to carry on basic life processes totals to about 100 W for a 68 kg individual (we'll assume that is the mass of an average human). That means they'll need 100 J of energy every second. We need to figure out how many seconds in a day and then multiply by 100 J/s to give the total number of joules needed $(\text{energy} = \text{power} \times \text{time})$. Then we'll convert joules to calories and then to kcal (which is the nutritional Cal).
Solve:

$$\text{daily intake} = (1\,d)\left(\frac{24\,h}{1\,d}\right)\left(\frac{60\,\min}{1\,h}\right)\left(\frac{60\,s}{1\,\min}\right)\left(\frac{100\,J}{1\,s}\right)\left(\frac{1\,cal}{4.19\,J}\right)\left(\frac{1\,kcal}{1000\,cal}\right)$$
$$= 2060\,Cal \approx 2100\,Cal$$

Assess: So a person of average weight (mass) needs just over two thousand Calories per day to maintain basic life processes. As the intermediate calculation showed (before the last two factors), that's about $8\,600\,000$ joules per day. Active people of the same mass will need more caloric intake. Mass matters too: Larger people need greater daily caloric intake. On the other hand, if you regularly eat more energy than you use in basic life processes and activities then your body will store it as fat.

P11.9. Strategize: Various fuels and the corresponding energy in 1 g of each are listed in Table 11.1. We use that information and then carry out simple unit conversions.
Prepare: Carbohydrates in foods such as "energy bars" have an energy content of 17 kJ per gram. Since we have 22 g of carbohydrates in our energy bar, we simply need to multiply 22 g by 17 kJ/g.
Solve: Keep one extra digit in intermediate calculations but report the answers to only two significant figures.

$$22\,g\left(\frac{17\,kJ}{1\,g}\right) = 374\,kJ \approx 370\,kJ$$

$$374\,kJ\left(\frac{1000\,J}{1\,kJ}\right) = 374000\,J \approx 370000\,J$$

$$374000\,J\left(\frac{1\,cal}{4.19\,J}\right) = 89300\,cal \approx 89000\,cal$$

$$89300\,cal\left(\frac{1\,kcal}{1000\,cal}\right) = 89.3\,kcal = 89.3\,Cal \approx 89\,Cal$$

Assess: Comparing with Table 11.2 shows that the carbohydrates in the energy bar do not provide as much energy as a fried egg; however, the energy bar may also contain fat (see Problem 11.7) in addition, and fat provides more energy per gram than carbohydrates do, so the total number of food calories in the energy bar might be 150.
A 68 kg person needs just over 2000 Cal for basic life processes, so they would need to eat about 15 energy bars per day if that is all they ate.
You may have learned in a health or nutrition class that 1 g of carbohydrates provides about 4 Cal of energy. Our calculations above bear this out: 89.3 Cal/22 g = 4.0 Cal/g.

P11.11. Strategize: We use the definition of power and the duration of the sleep to determine the total energy used.

Prepare: $P = \dfrac{\Delta E}{\Delta t}$

Solve: Keep one extra digit in intermediate calculations but report the answers to only two significant figures.

$$\Delta E = P\Delta t = (71\,\text{W})(8.0\,\text{h})\left(\frac{1\,\text{J/s}}{1\,\text{W}}\right)\left(\frac{3600\,\text{s}}{1\,\text{h}}\right)\left(\frac{1\,\text{cal}}{4.19\,\text{J}}\right)\left(\frac{1\,\text{Cal}}{1000\,\text{cal}}\right) = 490\,\text{Cal}$$

Assess: Since a reasonable number of Calories to eat in a day would be about 2000 Cal, it seems that one could burn almost 500 Cal sleeping at night.

P11.13. Strategize: Various fuels and the corresponding energy in 1 g of each are listed in Table 11.1, and the metabolic power needs of a 68 kg human for various activities are listed in table 11.4. We will use the energy content of the bars and equate that to the power times the duration of the walk.

Prepare: Carbohydrates in foods such as "energy bars" have an energy content of 17 kJ per gram. Since we have 22 g of carbohydrates in our energy bar, we simply need to multiply 22 g by 17 kJ/g.

Table 11.4 tells us that a 68 kg person needs to expend 380 J/s to walk at a speed of 5 km/h.

Solve: The energy in the carbohydrates in the bar is

$$22\,\text{g}\left(\frac{17\,\text{kJ}}{1\,\text{g}}\right) = 370\,\text{kJ}\left(\frac{1000\,\text{J}}{1\,\text{kJ}}\right) = 370000\,\text{J} = 3.7 \times 10^5\,\text{J}$$

The time that the chemical energy will last at the rate of 380 J/s is

$$\Delta t = \frac{\Delta E_{\text{chem}}}{P} = \frac{3.7 \times 10^5\,\text{J}}{380\,\text{W}} = 970\,\text{s} = 16\,\text{min} = 0.27\,\text{h}$$

And the distance that can be covered during this time at 5 km/h is

$$\Delta x = v\Delta t = (5\,\text{km/h})(0.27\,\text{h}) \approx 1.4\,\text{km}$$

Assess: The answer seems to be in the right ball park; we didn't get an answer of just a few cm nor an answer of many km—either of which we would be suspicious of given just one energy bar.

P11.15. Strategize: We will assemble data from the tables in the chapters so that we can relate the required metabolic power during the swim to the total energy in the fast food meal.

Prepare: We consult Table 11.2 to see that we will need to assume a *large* size meal of burger, fries, and drink which has an energy content of $1350\,\text{Cal} = 5660\,\text{kJ}$. In addition, we consult Table 11.4 to see that swimming at a fast crawl requires about 800 W of metabolic power for a 68 kg individual.

Solve: Use energy = power × time:

$$\text{time} = \frac{\text{energy}}{\text{power}} = \frac{(5660\,\text{kJ})}{800\,\text{J/s}} = 7080\,\text{s} = 118\,\text{min} = 2.0\,\text{h}$$

Assess: Fast swimming is strenuous exercise and uses up the metabolic energy fairly quickly, but it would still take 2 h of hard swimming to burn off such a large meal.

P11.17. Strategize: We calculate the energy in the candy bar, in Joules. Using the typical efficiency for a human, we determine what change in gravitational potential energy this could provide.

Prepare: A typical efficiency for climbing stairs is about 25%, so we can assume that 25% of the chemical energy in the candy bar is transformed to increased potential energy.

$$\Delta U_{\text{g}} = (0.25)(400\,\text{Cal})\left(\frac{1\,\text{kcal}}{1\,\text{Cal}}\right)\left(\frac{1000\,\text{cal}}{1\,\text{kcal}}\right)\left(\frac{4.2\,\text{J}}{1\,\text{cal}}\right) = 4.2 \times 10^5\,\text{J}$$

Solve: Since $\Delta U_g = mg\Delta y$, the height gained is

$$\Delta y = \frac{\Delta_g}{mg} = \frac{4.2 \times 10^5 \text{ J}}{(60 \text{ kg})(9.8 \text{ m/s}^2)} = 710 \text{ m}$$

If we assume that each flight of stairs has a height of 2.7 m (as is done in Example 11.7), this gives

$$\text{Number of flights} = \frac{710 \text{ m}}{2.7 \text{ m}} \approx 260 \text{ flights}$$

Assess: This is more than enough to get to the top of the Empire State Building twice—all fueled by one candy bar! This is a remarkable result.

P11.19. Strategize: We calculate the energy used by the body to accommodate each lift, including efficiency in our consideration, and then equate some unknown number of such lifts to the energy content of a piece of pizza.
Prepare: In weightlifting, a barbell curl is an exercise in which the barbell is held down at arms' length against the thighs and then raised in semi-circular motion until the forearms touch the biceps. We'll assume that the weightlifter expends metabolic energy when he lifts the 30 kg bar, but not as he lowers it. We'll also assume 25% efficiency. Table 11.2 tells us that a typical slice of pizza has a metabolic energy content of 300 Cal or 1260 kJ.
The weightlifter will use the 1260 kJ (at 25% efficiency) to lift the 30 kg bar, increasing its potential energy. In one repetition he'll increase the potential energy by $\Delta U_g = mg\Delta y = mg(0.060 \text{ m})$, and in n repetitions by nmg (0.060 m), where n is what we want to know.
Solve:

$$\text{(energy from pizza) (efficiency)} = nmg\Delta y$$
$$(1.26 \times 10^6 \text{ J})(0.25) = n(30 \text{ kg})(9.80 \text{ m/s}^2)(0.60 \text{ m})$$

Solving for n gives 1790 repetitions, which should be reported to two significant figures as 1800 repetitions.
Assess: That's a lot of curls! Exercising in this way to "burn off" an extra slice of pizza is almost impossible; people can't do 1800 reps in a row, and there isn't time before the next meal anyway. And n would be four times larger if the weightlifter were 100% efficient!

P11.21. Strategize: The work done in one repetition is the force (the weight of the barbell) multiplied by the distance lifted h, since the force and displacement vectors are in the same direction. To figure the energy needed for 20 repetitions we need to multiply the answer from part **(a)** by 20.
However, we also need to take into account that the weightlifter only uses the energy in her food at 25% efficiency.
Prepare: The expression for the gravitational potential energy required for each lift is $\Delta U_g = mg\Delta h$.
The expression for efficiency is, generally,

$$e = \frac{\text{what you get}}{\text{what you had to pay}}$$

where "what you get" is the total work done on the barbell in 20 reps, and "what you had to pay" is the total energy expended (which is what we want to know).
Solve: (a) $W = mgh = (40 \text{ kg})(9.8 \text{ m/s}^2)(0.5 \text{ m}) = 196 \text{ J} \approx 200 \text{ J}$

(b) what you had to pay $= \dfrac{\text{what you get}}{25\%} = \dfrac{(196 \text{ J/rep})(20 \text{ reps/day})}{0.25} = 15\,680 \text{ J/day} \approx 16\,000 \text{ J/day}$

(c) The number of 400 Calorie donuts needed to supply $15\,680 \text{ J}$ is simply $15\,680 \text{ J}/(400 \text{ Cal/donut})$.

$$\frac{15\,680 \text{ J/day}}{400 \text{ Cal/donut}} \left(\frac{1 \text{ Cal}}{1 \text{ kcal}} \right) \left(\frac{1 \text{ kcal}}{1000 \text{ cal}} \right) \left(\frac{1 \text{ Cal}}{4.19 \text{ J}} \right) = 0.0094 \text{ donuts/day}$$

Assess: Straightforward problems like part (a) help develop our intuition about how big a joule is. 200 J is not a lot of energy compared to the chemical energy consumed in a meal.
16 000 J is still not a large number of joules compared to the 420 000 J of chemical energy in a fried egg.
About a hundredth of a donut is enough to provide the energy for 20 reps of a 40 kg bench press.

P11.23. Strategize: We will determine the energy content of one gallon of gasoline in Joules. We will look up the power requirement for cycling and equate the energy used for cycling some unknown period of time to the energy content of the gasoline.

Prepare: Table 11.4 tells us that a 68 kg person (we'll assume this is your mass) needs to expend 480 J/s to pedal a bicycle at a speed of 15 km/h.

Table 11.1 helps us calculate the chemical energy stored in one gallon of gasoline (which has a mass of 3.2 kg).

$$E_{chem} = 3.2 \, kg \left(\frac{1000 \, g}{1 \, kg} \right) \left(\frac{44 \, kJ}{1 \, g} \right) \left(\frac{1000 \, J}{1 \, kJ} \right) \approx 1.4 \times 10^8 \, J$$

Solve: The time that the chemical energy will last at the rate of 480 J/s is

$$\Delta t = \frac{\Delta E_{chem}}{P} = \frac{1.4 \times 10^8 \, J}{480 \, W} = 2.93 \times 10^5 \, s = 81 \, h$$

And the distance that can be covered during this time at 15 km/h is

$$\Delta x = v \Delta t = (15 \, km/h)(81 \, h) \approx 1200 \, km$$

to two significant figures.

Assess: The driving distance from Dallas, Texas to Denver, Colorado is just over 1200 km. This seems far for one gallon of gasoline, but you are going much slower than a car would (which increases your efficiency by decreasing the drag) and you are taking a lot less mass. Also remember that a car's efficiency is probably less than 10% as shown in Integrated Example 11.19, while the efficiency of a human cycling is 20%–30%.

To put it in units of mpg to compare to your car, you would be able to cycle 760 miles with the energy supplied by that one gallon of gas, which is 30 times better than a car that gets 25 mpg.

P11.25. Strategize: We are given the required metabolic power. We need only determine the total energy and convert to g of carbohydrates by consulting Table 11.1.

Prepare: Table 11.1 tells that a single gram of fat contains 37 kJ of energy. We will determine the time of flight using $\Delta t = d / v$, and use this time in $\Delta E = P \Delta t = Pd / v$ to determine the energy requirement.

Solve: Inserting given values, we find

$$\Delta E = Pd / v = \frac{(12 \, W)(1,000 \, m)}{(11 \, m/s)} = 1091 \, J.$$

This is equivalent to $(1091 \, J) \left(\frac{1 \, g \, fat}{37,000 \, J} \right) = 2.9 \times 10^{-2} \, g.$

Assess: Although this is a relatively small mass, the starling itself has a very small mass. Also, we expect a smaller mass of fat than we found of carbohydrate in the previous problem, since fat has a higher energy per gram. So this is reasonable.

P11.27. Strategize: This is a simple conversion between temperature scales.
Prepare: Use the temperature conversion formula from Equation 11.6.
Solve:

$$T(^{\circ}C) = T(K) - 273 = 4.2 - 273 = -269 \, ^{\circ}C$$

$$T(^{\circ}F) = \frac{9}{5} T(^{\circ}C) = 32^{\circ} = \frac{9}{5}(-269 \, ^{\circ}C) + 32^{\circ} = -452 \, ^{\circ}F$$

Assess: This is just 4.2 K above absolute zero, so it is reasonable that are answers are so low.

P11.29. Strategize: We write out the expression for the conversion to the Kelvin scale for the initial and final temperatures and calculate the ratio.

Prepare: The Celsius temperature doubles from $100\,°C$ to $200\,°C$. Change both of these to the Kelvin scale. $100\,°C = 373\,K$, $200\,°C = 473\,K$

Solve:

$$\frac{473\,K}{373\,K} = 1.3$$

The Kelvin temperature has increased by a factor of 1.3.

Assess: To make good mathematical sense when doubling a temperature, we should use the Kelvin scale.

P11.31. Strategize: This is a straightforward application of the first law of thermodynamics, which is essentially energy conservation.

Prepare: We will use Equation 11.9.

Solve: The first law of thermodynamics is

$$\Delta E_{th} = W + Q \Rightarrow -200\,J = 500\,J + Q \Rightarrow Q = -700\,J$$

The negative sign means a transfer of energy from the system to the environment.

Assess: Because $W > 0$ means a transfer of energy into the system, Q must be less than zero and larger in magnitude than W so that $E_{th\,f} < E_{th\,i}$.

P11.33. Strategize: This is a straightforward application of the first law of thermodynamics, which is essentially energy conservation.

Prepare: Equation 11.9 gives the thermal energy change $\Delta E_{th} = W + Q$, and Figure 11.15 helps us figure out the signs. The 600 J of heat energy transferred to the system will be a positive Q while the 400 J of work that the system does means that W will be negative.

Solve:

$$\Delta E_{th} = (-400\,J) + 600\,J = 200\,J$$

Assess: We must remember that when the system does work W is negative.

P11.35. Strategize: This is a straightforward application of the first law of thermodynamics, which is essentially energy conservation.

Prepare: Equation 11.9 gives the thermal energy change $\Delta E_{th} = W + Q$, and Figure 11.15 helps us figure out the signs. The 10 J of heat removed the system (the gas sample) means Q is negative, while the 20 J of work that the piston does on the system is a positive W.

Solve:

$$\Delta E_{th} = 20\,J + (-10\,J) = 10\,J$$

Since the change in thermal energy is positive, the temperature of the gas increases.

Assess: We must remember that when work is done on the system W is positive.

P11.37. Strategize: We use the first law of thermodynamics to determine the heat input. Once we know the work we got out and the heat we put in we can apply the definition of efficiency.

Prepare: To calculate the energy removed from the hot reservoir, we use $W_{out} = Q_H - Q_C$. To determine the efficiency, we use $e = \dfrac{W_{out}}{Q_H}$ or equivalently $e = 1 - \dfrac{Q_C}{Q_H}$.

Solve: During each cycle, the work done by the engine is $W_{out} = 20\,J$ and the engine exhausts $Q_C = 30\,J$ of heat energy. Because $W_{out} = Q_H - Q_C$,

$$Q_H = W_{out} + Q_C = 20\,J + 30\,J = 50\,J$$

Thus, the efficiency of the engine is

$$e = 1 - \frac{Q_C}{Q_H} = 1 - \frac{30\,J}{50\,J} = 0.40$$

Assess: This makes sense, since about half the energy of the engine is lost as heat.

P11.39. Strategize: The maximum possible efficiency for a heat engine depends only on the absolute temperature of the hot and cold reservoirs.

Prepare: The maximum efficiency is $e_{max} = 1 - \frac{T_C}{T_H}$. Here T_C and T_H must be in Kelvin.

Solve: Inserting the given values, we find.

$$e_{max} = 1 - \frac{T_C}{T_H} = 1 - \frac{(273 + 20)\,K}{(273 + 600)\,K} = 0.6644$$

Because the heat engine is running at only 30% of the maximum efficiency, $e = (0.30)e_{max} = 0.20$. The amount of heat that must be extracted is

$$Q_H = \frac{W_{out}}{e} = \frac{1000\,J}{0.20} = 5000\,J$$

Assess: This is reasonable, since most of the energy in this engine is lost to the cold reservoir.

P11.41. Strategize: We use the known efficiency expression for a maximally efficient heat engine, and insert the given temperatures.

Prepare: The temperatures must be converted to the Kelvin scale. $650\,°C = 923\,K$ and $30\,°C = 303\,K$.

Solve:

$$e_{max} = 1 - \frac{T_C}{T_H} = 1 - \frac{303\,K}{923\,K} = 0.67 = 67\%$$

Assess: 67% is quite respectable (better than photovoltaic cells) but requires a high T_H.

P11.43. Strategize: We determine the maximum theoretical efficiency using the temperatures in the absolute scale. The actual efficiency is determined through the ratio of useful electric energy to heat energy put into the system.

Prepare: The power plant is treated as a heat engine.

Solve: **(a)** The maximum possible thermal efficiency of the power plant is

$$e_{max} = 1 - \frac{T_C}{T_H} = 1 - \frac{303\,K}{573\,K} = 0.471 \approx 47\%$$

(b) The plant's actual efficiency is

$$e = \frac{W_{out}}{Q_H} = \frac{700 \times 10^6\,J/s}{2000 \times 10^6\,J/s} = 0.35 = 35\%$$

Assess: If the plant were operating at the maximum possible thermal efficiency, the output would be $(2000\,MJ)(0.471) = 942\,MJ$. There is a loss of 242 MJ due to inefficiencies in the plant.

P11.45. Strategize: We use the definition of the coefficient of performance for cooling, and we employ the first law of thermodynamics to determine the unknown heat removed from the cold reservoir.

Prepare: The COP of a refrigerator is given by the equation before Equation 11.13.

Solve: The coefficient of performance of the refrigerator is

$$COP = \frac{Q_C}{W_{in}} = \frac{Q_H - W_{in}}{W_{in}} = \frac{50\,J - 20\,J}{20\,J} = 1.5$$

P11.47. Strategize: We use the definition of the coefficient of performance for cooling to find the relationship between the heat removed from the cold reservoir and the work done. We then employ the first law of thermodynamics to determine the unknown heat exhausted to the hot reservoir.
Prepare: The COP of a refrigerator is given by the equation just before Equation 11.13.
Solve: (a) The heat extracted from the cold reservoir is calculated as follows:

$$\text{COP} = \frac{Q_C}{W_{in}} \Rightarrow 4.0 = \frac{Q_C}{50\,\text{J}} \Rightarrow Q_C = 200\,\text{J}$$

(b) The heat exhausted to the hot reservoir is

$$Q_H = Q_C + W_{in} = 200\,\text{J} + 50\,\text{J} = 250\,\text{J}$$

P11.49. Strategize: We use the known expression for the theoretical maximum coefficient of performance for heating.
Prepare: The heat pump's job is to heat the inside of the house by pumping thermal energy from the (colder) outside to the (warmer) inside.
The temperatures of the hot and cold sides must be expressed in Kelvin. In this problem $T_C = -20°C = 253\,\text{K}$ and $T_H = 20°C = 293\,\text{K}$.
Solve: We use Equation 11.15 to compute the maximum coefficient of performance as follows:

$$\text{COP}_{max} = \frac{T_H}{T_H - T_C} = \frac{293\,\text{K}}{293\,\text{K} - 253\,\text{K}} = 7.325 \approx 7.3$$

Assess: The COP describes the ratio of thermal energy pumped to electrical power consumption; therefore a coefficient of performance of 7.3 means that we pump 7.3 J of thermal energy for an energy cost of 1 J. This value is the maximum possible for the given temperatures. In practice a typical heat pump has a COP of about three. An electrical resistance heater has a COP of one. Can you see why?

P11.51. Strategize: We are told the power output of the students, and all of that must be removed in order for the temperature of the auditorium to remain the same. We equate this student power output to $Q_C / \Delta t$, the rate at which heat must be removed from the cold reservoir.
Prepare: The amount of energy that needs to be pumped out of the cold reservoir is
$Q_C = (250 \text{ students})(125\,\text{J}) = 31,250\,\text{J}$ every second.
Solve:

$$\text{COP} = \frac{Q_C}{W_{in}} \Rightarrow W_{in} = \frac{Q_C}{\text{COP}} = \frac{31250\,\text{J}}{5} = 6250\,\text{J}$$

That is the work needed per second. We round to two significant figures, so the power needed is 6.3 kW.
Assess: The cost for this cooling isn't too much for one hour, but it adds up over many hours.

P11.53. Strategize: We check the accounting of energy in each diagram to make sure the first law is satisfied. To make sure the second law is satisfied, we must determine the maximum theoretical coefficient of performance and check whether or not the proposed process exceeds that maximum.
Prepare: For a refrigerator $Q_H = Q_C + W_{in}$, and the coefficient of performance and the maximum coefficient of performance are

$$\text{COP} = \frac{Q_C}{W_{in}} \quad \text{COP}_{max} = \frac{T_C}{T_H - T_C}$$

Solve: Please refer to Figure P11.53. **(a)** For refrigerator (a) $Q_H = Q_C + W_{in}$ $(60\,\text{J} = 40\,\text{J} + 20\,\text{J})$, so the first law of thermodynamics is obeyed. For refrigerator (b) $50\,\text{J} = 40\,\text{J} + 10\,\text{J}$, so the first law of thermodynamics is obeyed. For the refrigerator (c) $40\,\text{J} \neq 30\,\text{J} + 20\,\text{J}$, so the first law of thermodynamics is violated.

(b) For the three refrigerators, the maximum coefficient of performance is

$$COP_{max} = \frac{T_C}{T_H - T_C} = \frac{300\,K}{400\,K - 300\,K} = 3$$

For refrigerator (a),

$$COP = \frac{Q_C}{W_{in}} = \frac{40\,J}{20\,J} = 2 < COP_{max}$$

so the second law of thermodynamics is obeyed. For refrigerator (b),

$$COP = \frac{Q_C}{W_{in}} = \frac{40\,J}{10\,J} = 4 > COP_{max}$$

so the second law of thermodynamics is violated. For refrigerator (c),

$$COP = \frac{30\,J}{20\,J} = 1.5 < COP_{max}$$

so the second law is obeyed.

Assess: Both the first and second laws of thermodynamics must be obeyed by a refrigerator. The only refrigerator which does not violate either is refrigerator (a).

P11.55. Strategize: We compare the energy content of some unknown number of pieces of pizza to that required by expending the metabolic power for walking over a period of 1.0 h. We take care to convert to SI units.

Prepare: We see from Table 11.2 that a typical slice of pizza has an energy content of $300\,Cal = 1260\,kJ$.

In addition, we consult Table 11.4 to see that walking at 5 km/h requires about 380 W of metabolic power for a 68 kg individual (hence that assumption for your mass as well). The speed of 5 km/h (a typical walking speed) won't directly enter the rest of the calculation; we only needed it to correspond with the entry in Table 11.4.

We will also estimate your efficiency at 25%, as is typical of walking, running, and cycling.

It goes without saying (but we will anyway) that you could nudge these estimates either way if you want to do the computation for walking speeds faster or slower than 5 km/h, or for a person of mass slightly greater or less than 68 kg, or for pizza slices that are larger or smaller than average. You should still be able to produce an answer that is in the right ballpark, correct to one or two significant figures.

Solve: Use energy = power × time. Call the number of pizza slices (what we want to know) n. The total energy you eat is n times 1.26 kJ.

A couple of simple pre-conversions will make the following clearer and cleaner. $380\,W = 380\,J/s$ and $1h = 3600\,s$.

$$energy = power \times time$$
$$n(1.26 \times 10^6\,J) = 380\,J/s \times 3600\,s$$

Solving for n:

$$n = \frac{380\,J/s \times 3600\,s}{(1.26 \times 10^6\,J)} = 1.1\,slices$$

Assess: One slice of pizza sounds about right for a 1-h walk, doesn't it?

P11.57. Strategize: We will use the rate of increase of the gravitational potential energy as well as the 25% efficiency of a typical human.

Prepare: We convert the time to seconds: 37.5 min = 2250 s.

Solve:

(a)

$$\Delta U_g = mg\Delta y = (65\,kg)(9.8\,m/s^2)(1100\,m) = 701\,kJ$$

This is true for the winner and any other 60 kg person who finished the race.

We'll use the efficiency formula where "what you get" is the change in potential energy of the person, and "what you had to pay" is the total energy Marco expended during the race.

$$\text{what you had to pay} = \frac{\text{what you get}}{25\%} = \frac{701\,\text{kJ}}{0.25} = 2.80\,\text{MJ} = 670\,\text{Cal}$$

(b) To compute Marco's metabolic power we need to divide the total metabolic energy "burned up" by the time it took.

$$P = \frac{\Delta E}{\Delta t} = \frac{2.80\,\text{MJ}}{2250\,\text{s}} = 1200\,\text{W}$$

Assess: This is just a bit more than the metabolic power of a 68 kg individual running at 15 km/h.

P11.59. Strategize: We assume we can add the metabolic power required to walk horizontally to the metabolic power needed to increase the gravitational potential energy. We assume the body's efficiency is 25%.
Prepare: The increase in the hiker's gravitational potential energy as she ascends a height Δy is $mg\Delta y$. Because the body's efficiency is 25%, the metabolic power required to ascend this height is four times as much, or $(4.0)mg\Delta y$.
Solve: The total metabolic power is

$$P = 380\,\text{W} + \frac{4mg\Delta y}{\Delta t} = (4.0)mg\frac{\Delta y}{\Delta t} 380\,\text{W} + (4.0)mgv\tan 7^{\circ}$$

$$= 380\,\text{W} + (4.0)(68\,\text{kg})(9.8\,\text{m/s}^2)(5\,\text{km/h})\left(\frac{1\,\text{h}}{3600\,\text{s}}\right)\left(\frac{1000\,\text{m}}{1\,\text{km}}\right)\tan 7^{\circ} = 830\,\text{W}$$

Assess: We see that it takes an extra 450 W to go up the incline over what it would take to walk horizontally.

P11.61. Strategize: This is a simple application of the definition of efficiency. The only trick is that there are two processes, each with a different efficiency.
Prepare: The typical efficiency for a human is 25%.
Solve:
(a) power output = (power input)$(e_{\text{human}})(e_{\text{generator}})$ = $(400\,\text{W})(0.25)(0.8) = 80\,\text{W}$

(b) $\dfrac{400\,\text{W}}{80\,\text{W/person}} = 5$ people

Assess: If both efficiencies were 100% then it would only take one person.

P11.63. Strategize: This is a simple ratio exercise.
Prepare: We use ratios to solve the problem.
Solve:

(a) $(68\,\text{kg})\left(\dfrac{70000\,\text{Cal}}{5000\,\text{kg}}\right) = 952\,\text{Cal} \approx 950\,\text{Cal}$

(b) $P = \dfrac{E}{\Delta t} = \dfrac{952\,\text{Cal}}{1\,\text{d}}\left(\dfrac{1\,\text{d}}{24\,\text{h}}\right)\left(\dfrac{1\,\text{h}}{3600\,\text{s}}\right)\left(\dfrac{1000\,\text{cal}}{1\,\text{Cal}}\right)\left(\dfrac{4.19\,\text{J}}{1\,\text{cal}}\right) = 46\,\text{W}$

This is 54 W less than, or just under half the 100 W stated in the chapter.
Assess: The 46 W is less than a human, as we expect.

P11.65. Strategize: We can write out the expression for the maximum theoretical efficiency in terms of reservoir temperatures and apply it twice. We assume the temperature of the cold reservoir is unchanged and solve for the temperature of the hot reservoir that would provide each stated efficiency.
Prepare: The maximum efficiency of a heat engine depends only on the temperatures of the hot and cold reservoirs.
Solve: The maximum efficiency of a heat engine is $e_{\text{max}} = 1 - T_C/T_H$. This can be increased by either increasing the hot-reservoir temperature or decreasing the cold-reservoir temperature. For a 40% Carnot efficiency and a cold-reservoir temperature of 7°C,

$$e_{\text{max}} = 0.40 = 1 - \frac{280\,\text{K}}{T_H} \Rightarrow T_H = 470\,\text{K}$$

A higher efficiency of 60% can be obtained by raising the hot-reservoir temperature to T_H'. Thus,

$$e_{max} = 0.60 = 1 - \frac{280\,\text{K}}{T_H'} \Rightarrow T_H' = 700\,\text{K}$$

The reservoir temperature must be increased by 230 K to cause the efficiency to increase from 40% to 60%.
Assess: Note this is a large temperature change for a relatively small increase in efficiency.

P11.67. Strategize: The first part of the problem is a simple conversion using density. For the second part we use the definition of efficiency and the given values for the power output and the energy in the coal.
Prepare: The power plant is to be treated as a heat engine.
Solve: (a) Every hour 300 metric tons or $3 \times 10^5\,\text{kg}$ of coal is burnt. The volume of coal is

$$\frac{3.00 \times 10^5\,\text{kg}}{1\,\text{hour}} \times \frac{1.5\,\text{m}^3}{1000\,\text{kg}} \times 24\,\text{hour} = 1.08 \times 10^4\,\text{m}^3$$

The height of the room will be 110 m.
(b) The thermal efficiency of the power plant is

$$e = \frac{W_{out}}{Q_H} = \frac{2.7 \times 10^{12}\,\text{J}}{(3.00 \times 10^5\,\text{kg}) \times \dfrac{28 \times 10^6\,\text{J}}{\text{kg}}} = 32\%$$

Assess: An efficiency of 32% is typical of power plants.

P11.69. Strategize: This is an estimation problem, so a small range of answers may be possible. We use the equation for the theoretically maximal COP for cooling, and convert to a SEER rating using the given conversion rate.
Prepare: Say the outside temperature is $95°\text{F} = 35°\text{C} = 308\,\text{K}$, and the desired indoor temperature is $75°\text{F} = 24°\text{C} = 297\,\text{K}$.
Solve:

$$\text{COP}_{max} = \frac{T_C}{T_H - T_C} = \frac{297\,\text{K}}{308\,\text{K} - 297\,\text{K}} = 27$$

$$\text{SEER}_{max} = 3.4 \times \text{COP}_{max} = (3.4)(27) = 92$$

Assess: Your answer may vary if you assumed different inside and outside temperatures.

P11.71. Strategize: In part **(a)** we simply need to multiply the *rate* of energy consumption (i.e., power) by the time.
In part **(b)** we use the definition of efficiency.
Prepare: The efficiency is given by the equation:

$$e = \frac{\text{what you get}}{\text{what you had to pay}}$$

"What you get" is the energy needed by the 250 000 houses. "What you had to pay" is the energy collected by an area A of solar cells (at 15% efficiency), where A is what we want to know.
Solve: (a) $E = P\Delta t = (1\,\text{kW})(24\,\text{h})\left(\dfrac{3600\,\text{s}}{1\,\text{h}}\right) = 8.64 \times 10^7\,\text{J} = 86.4\,\text{MJ} \approx 86\,\text{MJ}$

(b) $e = 0.15\% = \dfrac{(250\,000\text{ houses})(86.4\,\text{MJ/house})}{A(20\,\text{MJ/m}^2)}$

Then solve for A:

$$A = \frac{(250\ 000\ \text{houses})(86.4\,\text{MJ/house})}{(0.15)(20\,\text{MJ/m}^2)} = 7.2 \times 10^7\,\text{m}^2 = 2.8\,\text{mi}^2$$

Assess: 86.4 MJ sounds like a lot, but one joule isn't a lot of energy. 86.4 MJ is only ten times as much as an average person consumes in metabolic energy in a day.

An area of $2.8\,\text{mi}^2$ is a lot, but that's for $250\ 000$ houses; for one house it is $29\,\text{m}^2$.

Of course there are complicating factors such as clouds and storage of the energy for use at night, etc. And so we'll need much more efficient solar cells (also known as photovoltaic cells) before they'll become a viable factor in the mix of sources of electrical energy.

This is an active area of research. Solar cells with efficiencies in the teens are now available, and there is some speculation that the efficiencies may get up to 40% eventually.

P11.73. Strategize: Consider the graph of oxygen uptake as a function of speed. Note the differences between the curve for the kangaroo and for the human.

Prepare: As mentioned, kangaroo hops are not as efficient at slow speeds as they are at high speeds, so it will take less energy to go 1 km at high speed.

Looked at another way, since the necessary power is approximately the same for low and high speeds it would take approximately the same amount of energy for the kangaroo to hop for equal amounts of *time*, but it could cover the 1 km distance in a lot less time at a fast speed.

Solve: A. A faster speed requires less total energy.

Assess: This result may be somewhat surprising, at least compared to humans. For humans, running twice as fast requires about twice as much power, so the energy expended over the 1 km distance would be about the same either way. Not so for kangaroos.

P11.75. Strategize: Examine the graph and look for a speed at which the oxygen uptake for a human is about half that for a kangaroo.

Prepare: Figure P11.72 shows oxygen uptake versus speed, and it mentions that oxygen uptake is a measure of energy use per second, or power.

So we are simply looking on the graph for the speed at which the value on the line for humans is half the value on the line for kangaroos; it appears that it happens at about 3 m/s.

Solve: A. 3 m/s.

Assess: That's the speed at which a human would use half the power of an equal-mass kangaroo.

P11.77. Strategize: We determine the metabolic power requirement, then use the time to determine the total energy requirement.

Prepare: We use the same metabolic efficiency as humans (25%), meaning that 3,000 W of metabolic power are required. We will determine the corresponding energy for one hour of work using $\Delta E = P\Delta t$.

Solve: Inserting the given values, we have $\Delta E = P\Delta t = (3{,}000\,\text{W})(3600\,\text{s}) = 1.08 \times 10^7\,\text{J} \approx 11\,\text{MJ}$. So the correct answer is D.

Assess: If the horse is eating carbohydrates, this is about 0.64 kg of food, which is reasonable.

P11.79. Strategize: Let us assume that the maximum power output of the sled dog is fixed.

Prepare: We know that $P = Fv$, and we also know the maximum power output of the sled dog is 132 W.

Solve: $v = P/F = (132\,\text{W})/(120\,\text{N}) = 1.1$ m/s. So the correct answer is A.

Assess: This means a dog can run at 1 m/s with a force pulling back on it equivalent to about 12 kg, or about 27 lbs, which is reasonable.

CONSERVATION LAWS

PptII.1. Reason: The water molecules that become the snowflake become more ordered, so the entropy of the water decreases. The answer is A.
Assess: But the entropy of the entire universe increases. The entropy of the air increases by more than the entropy of the water decreases.

PptII.3. Reason: When energy is transferred between hot and cold (as in convection cells) the entropy increases, so the process is not reversible. The answer is B.
Assess: The entropy of the universe never decreases.

PptII.5. Reason: The rubber bands slow down the rider over a longer period of time so the force is reduced. $F_{\text{net}}\Delta t = \Delta p$ shows this. The answer is D.
Assess: This is the same reason it doesn't hurt to land on a trampoline from a large height.

PptII.7. Reason: Solve $\frac{1}{2}kx^2 = mgh$ for h, so the height the rider will go is inversely proportional to the mass (given that that spring energy in the numerater is the same) $h = \frac{1}{2}kx^2/mg$, so the rider will go twice as high with half the mass. The answer is C.
Assess: Four meters is dangerously high.

PptII.9. Reason: Since $F = kx$ reducing the spring constant will reduce the force on the rider. And since $\frac{1}{2}kx^2 = mgh$ reducing the spring constant will also reduce the final height (all else remaining equal). So the answer is A.
Assess: It makes sense that a smaller force would produce a lower jump height.

PptII.11. Reason: The force is what makes the ball change momentum and bounce back up, so if it doesn't bounce as high it must be because the force wasn't as great. The answer is C.
Assess: The force must also be less to slow the ball down over a longer time.

PptII.13. Reason: Use conservation of momentum, with the initial momentum of the system zero.

$$m_s v_s = -m_w v_w \quad \Rightarrow \quad v_s = \frac{-m_w v_s}{m_s} = \frac{-(0.30 \text{ kg})(10 \text{ m/s})}{4.0 \text{ kg}} = -0.75 \text{ m/s}$$

The squid's speed is thus 0.75 m/s, so the answer is D.
Assess: Because the water's mass is less than the squid's, we expect the squid's speed to be less than the water's.

PptII.15. Reason: Compute the average force from

$$F_{net} = \frac{\Delta p}{\Delta t} = \frac{m\Delta v}{\Delta t} = \frac{(0.30 \text{ kg})(10 \text{ m/s})}{0.10 \text{ s}} = 30 \text{ N}$$

So the answer is B.
Assess: This force is not very big.

PptII.17. Reason:

$$\Delta p = m\Delta v = (0.046 \text{ kg})(63 \text{ m/s}) = 2.9 \text{ kg} \cdot \text{m/s}$$

The answer is C.
Assess: Momentum is conserved for the club-ball system.

PptII.19. Reason: See if the kinetic energy is the same before and after.

$$K_i = \frac{1}{2}m_c(v_c)_i^2 = \frac{1}{2}(0.30 \text{ kg})(40 \text{ m/s})^2 = 240 \text{ J}$$

$$K_f = \frac{1}{2}m_c(v_c)_f^2 + \frac{1}{2}m_b(v_b)_f^2 = \frac{1}{2}(0.30 \text{ kg})(30.3 \text{ m/s})^2 + \frac{1}{2}(0.046 \text{ kg})(63 \text{ m/s})^2 = 229 \text{ J}$$

Kinetic energy is not conserved, so the answer is B.
Assess: We would be worried if $K_f > K_i$.

PptII.21. Reason:

$$P = \frac{\Delta E}{\Delta t} = \frac{\frac{1}{2}mv^2}{\Delta t} = \frac{\frac{1}{2}(80 \text{ kg})(11 \text{ m/s})^2}{4.1 \text{ s}} = 1180 \text{ W} \approx 1200 \text{ W}$$

The answer is D.
Assess: It is hard to select between the choices based on ballpark expectations because they are all in a reasonable range.

PptII.23. Reason: Momentum is conserved in this inelastic collision.

$$m_1(v_1)_i + m_2(v_2)_i = (m_1 + m_2)v_f$$

$$v_f = \frac{m_1(v_1)_i + m_2(v_2)_i}{m_1 + m_2} = \frac{(100 \text{ kg})(6.0 \text{ m/s, E}) + (130 \text{ kg})(5.0 \text{ m/s, W})}{230 \text{ kg}} = (0.22 \text{ m/s, W})$$

Assess: Since they are about equal in momentum the final speed is small.

PptII.25. Reason:
(a) Use conservation of momentum.

$$m_L(v_L)_i = (m_L + m_s)v_f \quad \Rightarrow \quad v_f = \frac{m_L(v_L)_i}{m_L + m_s} = \frac{(30 \text{ kg})(4.0 \text{ m/s})}{40 \text{ kg}} = 3.0 \text{ m/s}$$

(b)

$$F = \frac{\Delta p}{\Delta t} = \frac{(10 \text{ kg})(3.0 \text{ m/s})}{0.25 \text{ s}} = 120 \text{ N}$$

(c) Use $K = \frac{1}{2}mv^2$.

$$K_i = \frac{1}{2}(30 \text{ kg})(4.0 \text{ m/s})^2 = 240 \text{ J} \qquad K_f = \frac{1}{2}(40 \text{ kg})(3.0 \text{ m/s})^2 = 180 \text{ J}$$

$$\Delta K = K_f - K_i = 180 \text{ K} - 240 \text{ J} = -60 \text{ J}$$

This energy was dissipated as thermal energy.
Assess: In part **(c)** it seems reasonable to "lose" a quarter of the energy.

THERMAL PROPERTIES OF MATTER

Q12.1. Reason: The mass of a mole of a substance in grams equals the atomic or molecular mass of the substance. Since neon has an atomic mass of 20, a mole of neon has a mass of 20 g. Since N_2 has a molecular mass of 28, a mole of N_2 has a mass of 28 g. Thus a mole of N_2 has more mass than a mole of neon.

Assess: Even though nitrogen *atoms* are lighter than neon atoms, nitrogen molecules are more massive, so a mole of nitrogen has more mass than a mole of neon.

Q12.3. Reason: Since there is almost pure helium in a helium balloon and almost no helium in the outside air, helium tends to diffuse out of the balloon. Similarly, with almost no oxygen or nitrogen in the balloon initially and high concentrations of oxygen and nitrogen in the air, these molecules tend to diffuse into the balloon. However, since helium atoms travel about three times faster than oxygen or nitrogen molecules and since helium atoms are smaller, they diffuse much faster so gas leaves the balloon faster than it enters. An air-filled balloon has the same particles inside as out and so the stated effect does not contribute to the deflation of such balloons. Instead, there is a weaker effect which is also at work for helium balloons: Higher pressure inside the balloon than outside makes the interior air molecules diffuse faster.

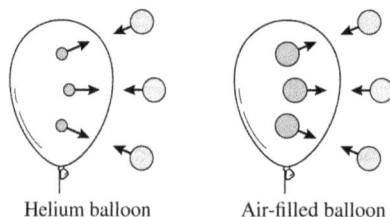

Helium balloon   Air-filled balloon

Assess: Because of the advantages helium atoms have in diffusion, helium balloons deflate faster than air-filled ones.

Q12.5. Reason: Note that N/V and v_{rms} are the same for both gases.
(a) In the process of deriving the ideal gas law we saw that

$$p = \frac{1}{3}\frac{N}{V}mv_{rms}^2$$

or that $p \propto m$, so given the conditions above, the gas with the more massive molecules (gas 2) will have the higher pressure.

$$p_2 > p_1$$

(b) The ideal gas law can be rearranged as

$$p = \frac{N}{V} k_B T$$

which shows that $p \propto T$, so given the answer to part **(a)** the temperature of gas 2 must be greater than the temperature of gas 1.

$$T_2 > T_1$$

Assess: We conclude that, other things being equal, the gas with the more massive molecules will have a greater pressure and a greater temperature.

Q12.7. Reason: Since the bottles have the same temperature, the average kinetic energies of the atoms that are in each bottle is the same. Since both bottles have the same amount of atoms, the total thermal energy must be the same also.
Assess: Thermal energy in a gas is directly proportional to the number of atoms in the gas and its temperature.

Q12.9. Reason: Equation 12.15 applies. The number of molecules in the gas is constant since the container is sealed. Equation 12.15 can be written as $p = Nk_B(T/V)$.
(a) If the volume is doubled and the temperature tripled, the pressure increases by a factor of 3/2.
(b) If the volume is halved and the temperature tripled, the pressure increases by a factor of six.
Assess: This makes sense. Increasing the temperature increases the pressure in a gas as does decreasing the volume of the container.

Q12.11. Reason: Thermal expansion will make your sun-drenched tape longer than the shaded tape, so if you and your coworker measure the same object yours will read a smaller value.
Assess: To the extent that this is noticeable, this would be an error in the measurements made with the hot tape, as they are calibrated to be correct at room temperature.

Q12.13. Reason: From Table 12.3, we see that water has a significantly higher coefficient of thermal expansion than steel—about six times as much. As the water and steel get hotter, the water expands six times more than the steel. Thus the water will overflow out of the steel container.
Assess: This seems reasonable since we expect gases to expand more than liquids and liquids to expand more than solids when the temperature increases.

Q12.15. Reason: You are heating both containers (each with n moles of nitrogen gas) and thereby increase the internal energy of each by Q, but the temperatures do not rise by the same amount. See Equations 12.28 and 12.29.

$$Q = nC_V \Delta T_A = nC_p \Delta T_B = 10 \text{ J}$$

Consulting Table 12.6 shows that $C_p > C_V$; therefore $\Delta T_A > \Delta T_B$ and since $(T_A)_i = (T_B)_i$, then $(T_A)_f > (T_B)_f$.
Assess: Some of the energy in container B is used as work done in changing the volume. In container A the volume did not change, so no work was done and all of the energy went into changing the temperature.

Q12.17. Reason: Consider the iron and Earth to have roughly fixed volume, such that any changes in entropy can only be from changing the temperature of the iron or Earth around it. A crystal is more ordered than liquid iron, so as iron crystallizes, its entropy decreases. The entropy of an isolated system cannot decrease, so the entropy of the Earth around the iron (with which it is in good thermal contact) must increase. The only way to do this at constant volume is to increase the thermal energy and temperature of the Earth around the iron.
Assess: Changes of phase are often associated with absorbing or giving off heat. This is the principle behind many kinds of warming/cooling packs used by humans.

Q12.19. Reason: When water changes from a vapor to a solid it gives up heat (the same amount as the heat of vaporization). This warms the environment and slows the cooling process.

Assess: Clearly water is the relevant substance here because water vapor naturally exists in the air. But we note that this argument would be true for any substance changing phase from vapor to liquid. Water is especially effective at this because of its high heat of vaporization.

Q12.21. Reason: The chocolate starts at some temperature. As it's heated the temperature rises to a point where the chocolate changes phase. Assuming the chocolate started as a solid, the chocolate melts during the second portion of the graph. During melting, the temperature is constant. After all the chocolate melts, the temperature rises as the liquid chocolate is heated.
Assess: Compare to Figure 12.21.

Q12.23. Reason: Heat transfer processes like convection and conduction are very effective in water, but ineffective in air. This means animals that live in water will quickly equilibrate with their environment whereas land animals can maintain a temperature difference with the environment for some time.
Assess: Some aquatic animals try to reduce the effective thermal conductivity by maintaining a thick layer of insulating fat on their bodies.

Q12.25. Reason: Part (a) of the figure represents a constant pressure, or isobaric, expansion of the gas. Part (b) represents a constant volume reduction of pressure of the gas. During part (b), the temperature also decreases, from Equation 12.16. Part (c) represents a decrease in volume along with an increase in pressure. However, part (c) is not isothermal since the graph is a straight line. Isothermal processes are hyperbolae on pV diagrams.
Assess: A correct diagram would look like the following figure.

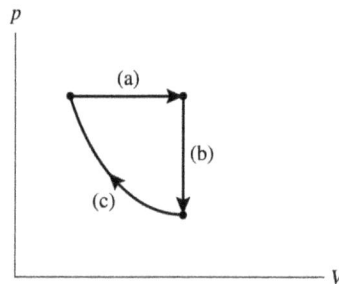

Q12.27. Reason: The trees help prevent the energy from being radiated out into space on a cold clear night; the trees reflect back down some of the infrared radiation and keep the ground under them warmer. In contrast, the open ground radiates its thermal energy into space without the "blanket" of the trees or clouds to keep the energy in.
Assess: Gardeners in northern climes know to cover their plants on clear fall nights to keep the radiation in and keep the plants from freezing. These early first frosts in the fall are even called radiation frosts. They take place on clear nights with calm winds. Another type, called advective freeze, occurs when very cold air moves in (by convection); advective freezes can take place with winds and clouds present, and are much harder to protect plants against.

Q12.29. Reason: Fick's law (Equation 12.42) tells us that the rate of diffusion depends on the difference in concentrations. Thus, if there is a large difference in the oxygen concentration in the lungs as compared to the adjacent blood vessels, the rate of diffusion increases. Air is only about 21% oxygen (with much of the remainder being nitrogen). Thus, breathing pure oxygen makes the gradient in your lungs much larger and increases oxygen uptake.
Assess: This explains why people use oxygen tanks when they have difficulties with their lungs.

Q12.31. Reason: Apply the ideal gas law in Equation 12.15: $pV = Nk_BT$. We are told that T is kept constant and assume that V is also. So $p \propto N$; if the number of atoms is doubled then the pressure is also.
The correct answer is D.
Assess: makes sense because with twice as many atoms there will be twice as many atomic collisions in a given amount of time, so the pressure is doubled.

Q12.33. Reason: Since the energy deceases by half and the energy is directly proportional to temperature, the temperature must be halved also. The answer is B.

Q12.35. Reason: We'll use Equation 12.25 $Q = Mc\Delta T$ and solve for M. We are given $\Delta T = 20°C = 20\,K$ and Table 12.4 provides $c = 4190$ J/kg·K.

$$Q = \text{energy} = \text{power} \times \text{time} = (100\text{ W})(1\text{ min})\left(\frac{60\text{ s}}{1\text{ min}}\right) = 6000\text{ J}$$

$$M = \frac{Q}{c\Delta T} = \frac{6000\text{ J}}{(4190\text{ J/kg}\cdot\text{K})(20\text{ K})} = 0.072\text{ kg} = 72\text{ g}$$

The correct answer is A.
Assess: For the water (with a large specific heat) to rise in temperature by that much in one minute the mass must be pretty small (72 g of water has a volume of about 1/3 of a cup).
To boil a liter (just over a quart, and almost 14 times the answer to this question) of water on your stove in just a few minutes you need to deliver much more than 100 W to it.

Q12.37. Reason: Choice A can't be right because the question explicitly states both the steam and liquid water are at 100°C. But as the steam condenses on the skin it transfers a lot of heat to the skin; this causes the more severe burn. The correct answer is B.
Assess: The specific heat of steam is actually less than that of liquid water.

Problems

P12.1. Strategize: We will use the molar mass of each substance to convert from grams to moles, and then compare.
Prepare: For each element, one mole of atoms has a mass in grams equal to the atomic mass number; for example, since the atomic mass number of carbon is 12 then there is one mole of carbon atoms in 12 grams; likewise, there is one mole of argon atoms in 40 grams of argon. See Table 12.1.
Solve: The only catch is that hydrogen gas is diatomic, so one mole of diatomic hydrogen gas molecules has a mass of 2 g.
For hydrogen: (10 g)(1 mol/2 g) = 5 mol
For carbon: (100 g)(1 mol/12 g) = 8.3 mol
For lead: (500 g)(1 mol/207 g) = 2.4 mol
The answer is that 100 g of carbon has the most moles.
Assess: Notice that we count atoms for the solids, but molecules for the diatomic gas.

P12.3. Strategize: We'll first compute how many moles of hydrogen peroxide molecules there are in 100 g and then use Equation 12.1 to find how many particles that is.
Prepare: The molecular mass number for H_2O_2 is $2 \times 1 + 2 \times 16 = 34$.
Solve: (100 g)(1 mol/34 g) = 2.94 mol of hydrogen peroxide molecules. However, there are two hydrogen atoms in each molecule of hydrogen peroxide, so there are 2×2.94 mol = 5.88 mol of hydrogen atoms.

$$N = nN_A = (5.88\text{ mol})(6.02 \times 10^{23}\text{ mol}^{-1}) = 3.5 \times 10^{24}$$

Assess: Three trillion hydrogen atoms is a lot, but one gets used to huge numbers in these types of problems.

P12.5. Strategize: This is a simple calculation of volume of a box from the given dimensions.
Prepare: The volume is clearly the product of the three length measurements; the issue is converting the units. First multiply $L \times W \times H$ to get the number of cm^3, then convert to m^3.
Solve:

$$V = (200\text{ cm})(40\text{ cm})(3.0\text{ cm}) = 24{,}000\text{ cm}^3$$

Now remember that while $1\,\text{m} = 100\,\text{cm}$, $1\,\text{m}^3 \neq 100\,\text{cm}^3$. Instead, $1\,\text{m}^3 = 1{,}000{,}000\,\text{cm}^3$.

$$24{,}000\,\text{cm}^3 = (24{,}000\,\text{cm}^3)\left(\frac{1\,\text{m}^3}{1{,}000{,}000\,\text{cm}^3}\right) = 0.024\,\text{m}^3$$

Assess: The answer is small—not a very big fraction of one cubic meter; however, this is reasonable given the small height. The conversion factor comes from $(1\,\text{m}/100\,\text{cm})^3$.

P12.7. Strategize: To solve this problem, we must understand that temperature (on the absolute Kelvin scale) is equivalent to a description of the average kinetic energy of the particles of the gas.
Prepare: Doubling the kinetic energy corresponds to doubling the absolute temperature, so we find the equivalent on the Kelvin scale.
Solve:

$$20\,^{\circ}\text{C} = 293\,\text{K}$$

Double this to get $586\,\text{K}$ and subtract 273 to get back to $313\,^{\circ}\text{C}$.

Assess: Don't double the Celsius temperature. It seems that $313\,^{\circ}\text{C}$ is more than double $20\,^{\circ}\text{C}$ but it works right when converted to the absolute Kelvin scale.

P12.9. Strategize: To solve this problem, we must understand that temperature (on the absolute Kelvin scale) is equivalent to a description of the average kinetic energy of the particles of the gas.
Prepare: We must convert the temperature to the Kelvin scale before reducing it by 10%. $20\,^{\circ}\text{C} = 293\,\text{K}$.
Solve:

$$(293\,\text{K})(0.90) - 273 = -9.3\,^{\circ}\text{C}$$

Assess: It is important to do these calculations in an absolute temperature scale.

P12.11. Strategize: We can use expressions that relate temperature to the kinetic energy of an ideal gas.
Prepare: Solve $\Delta E_{\text{th}} = \frac{3}{2} N k_{\text{B}} \Delta T$ for ΔT.

Solve:

$$\Delta T = \frac{2}{3}\frac{\Delta E_{\text{th}}}{N k_{\text{B}}} = \frac{2}{3}\frac{-4.3\,\text{J}}{(2.2 \times 10^{22})(1.38 \times 10^{-23})} = -9.4\,\text{K} = -9.4\,^{\circ}\text{C}$$

$$T_{\text{f}} = T_{\text{i}} + \Delta T = 20\,^{\circ}\text{C} - 9.4\,^{\circ}\text{C} = 11\,^{\circ}\text{C}$$

Assess: We expected the temperature to drop from the removed thermal energy.

P12.13. Strategize: Pressure is the force per unit area. We can use this knowledge and the known area to determine the force.
Prepare: Equation 12.12 gives the force due to a pressure applied over an area. The preliminary calculation is to compute the cross section area of the tube.

$$A = \pi R^2 = \pi \left(\frac{0.015\,\text{m}}{2}\right)^2 = 1.77 \times 10^{-4}\,\text{m}^2$$

We are given $p = 6.0\,\text{kPa}$.
Solve:

$$F = pA = (6.0\,\text{kPa})(1.77 \times 10^{-4}\,\text{m}^2) = 1.1\,\text{N}$$

Assess: 1.1 N is not a large force, but it is pushing a light dart, so the dart achieves a respectable acceleration.

P12.15. Strategize: The absolute pressure is the gauge pressure plus one atmosphere at sea level.
Prepare: We are asked to give our answer is psi, so we will use $1\,\text{atm} = 14.7\,\text{psi}$.
Solve:

$$p = p_\text{g} + 1\,\text{atm} = 35.0\,\text{psi} + 14.7\,\text{psi} = 49.7\,\text{psi}$$

Assess: The difference between p and p_g is due to the fact that your tire gauge measures pressure *differences*.

P12.17. Strategize: We will use the known relationship between temperature and average kinetic energy to determine the rms speed.
Prepare: We need the mass of a mole of CO_2. Since carbon has an atomic mass of 12 and oxygen has an atomic mass of 16, the molecular mass of CO_2 is 44. Hence a mole of CO_2 has a mass of 44 g or 0.044 kg.
Solve: We use Equation 12.13 for rms speed but modify it by multiplying numerator and denominator by Avogadro's number.

$$v_\text{rms} = \sqrt{3kT/m} = \sqrt{3RT/(m_\text{mol})} = \sqrt{3(8.315\,\text{J/(mol·K)})(210\,\text{K})/(44\times10^{-3}\,\text{kg/mol})} = 350\ \text{m/s}$$

Assess: This is a typical speed for a gas molecule with a temperature in the hundreds of Kelvins. For example, the rms speed of an O_2 molecule at room temperature is about 480 m/s.

P12.19. Strategize: We will use the ideal gas equation.
Prepare: The take care to convert the temperature to the Kelvin scale: 293 K.
Solve: Solve the equation for N.

$$N = \frac{pV}{k_\text{B}T} = \frac{(4.0\times10^{-11}\,\text{Pa})(0.090\,\text{m}^3)}{(1.38\times10^{-23}\,\text{J/K})(293\,\text{K})} = 8.9\times10^8\ \text{molecules}$$

Assess: This sounds like a lot of molecules, and, indeed, the vacuums we can achieve in the laboratory are not nearly as good as outer space.

P12.21. Strategize: We can use the ideal gas equation to determine the unknown volume.
Prepare: The temperature in kelvin is 273 K, the pressure is $1\,\text{atm} = 101.3\,\text{kPa}$, and the number of helium molecules is $125\ \text{g}\left(\dfrac{1\,\text{mol}}{4\,\text{g}}\right)\left(\dfrac{6.02\times10^{23}\,\text{molecules}}{1\,\text{mol}}\right) = 1.88\times10^{25}\ \text{molecules}$.
Solve: Solve the equation for V.

$$V = \frac{Nk_\text{B}T}{p} = \frac{(1.88\times10^{25}\,\text{molecules})(1.38\times10^{-23}\,\text{J/K})(273\,\text{K})}{101.3\,\text{kPa}} = 0.70\ \text{m}^3 = 700\ \text{L}$$

Assess: From 1.0 L in liquid form to 700 L in gaseous form sounds reasonable.

P12.23. Strategize: We use the known relationship between thermal energy and temperature.
Prepare: Use $E_\text{th} = \dfrac{3}{2}Nk_\text{B}T$ and $\Delta E = P\Delta t$ to find ΔT.
Solve:

$$\Delta T = \frac{2}{3}\frac{\Delta E}{Nk_\text{B}} = \frac{2}{3}\frac{P\Delta t}{Nk_\text{B}} = \frac{2}{3}\frac{(125\,\text{W})(10\,\text{min})\left(\dfrac{60\,\text{s}}{1\,\text{min}}\right)}{(6.0\times10^{26})(1.38\times10^{-23}\,\text{J/K})} = 6.0\ \text{K}$$

This ΔT is equal to $6.0\,°\text{C}$.

Assess: The temperature wouldn't really go up that much in real life because thermal energy would be transferred to the walls instead of merely raising the temperature of the air.

P12.25. Prepare: The carbon dioxide in the cube is an ideal gas and we will use the ideal gas Equation 12.16 at STP with $n = M/M_{mol}$.

Solve: Using the ideal gas equation

$$pV = nRT \Rightarrow V = \frac{nRT}{p} = \frac{MRT}{pM_{mol}}$$

The molar mass of CO_2 is 44 g/mol or 0.044 kg/mol. Thus,

$$V = \frac{(10,000 \text{ kg})(8.31 \text{ J}/(\text{mol} \cdot \text{K}))(293 \text{ K})}{(1.013 \times 10^5 \text{ Pa})(0.044 \text{ kg/mol})} = 5.5 \times 10^3 \text{ m}^3$$

Assess: This is just under two hot air balloons' worth of gas.

P12.27. Strategize: The gas is assumed to be ideal. We can use the ideal gas equation applied to the initial and final states and relate them through the fixed number of particles.

Prepare: As a general rule, we must convert all quantities into SI units. In the present case, however, we will be dealing with the ratio of the final and the initial value of V, so we do not have to convert L into m^3.

Solve: The before-and-after relationship of an ideal gas is

$$\frac{p_1 V_1}{T_1} = \frac{p_2 V_2}{T_2} \Rightarrow p_2 = p_1 \frac{V_1}{V_2} \cdot \frac{T_2}{T_1} = (2.4 \text{ atm}) \left(\frac{3.0 \text{ L}}{9.0 \text{ L}}\right) \left(\frac{600 \text{ K}}{300 \text{ K}}\right) = 1.6 \text{ atm}$$

P12.29. Strategize: In an isochoric process, the volume of the container stays unchanged. Argon gas in the container is assumed to be an ideal gas. We can use the ideal gas equation twice: once for the initial state and once for the final state of the gas.

Prepare: We must first convert the volumes and temperatures to SI units with $V_1 = 50 \text{ cm}^3 = 50 \times 10^{-6} \text{ m}^3$, $T_1 = 20°C = (273 + 20) \text{ K} = 293 \text{ K}$, and $T_2 = 300°C = (300 + 273) \text{ K} = 573 \text{ K}$.

Solve: **(a)** The container has only argon inside with $n = 0.1$ mol. The pressure before heating is

$$p_1 = \frac{nRT}{V_1} = \frac{(0.10 \text{ mol})(8.31 \text{ J}/(\text{mol} \cdot \text{K}))(293 \text{ K})}{50 \times 10^{-6} \text{ m}^3} = 4.87 \times 10^6 \text{ Pa} = 4870 \text{ kPa}$$

An ideal gas process has $p_2 V_2/T_2 = p_1 V_1/T_1$. Isochoric heating to a final temperature T_2 has $V_2 = V_1$, so the final pressure is

$$p_2 = \frac{V_1}{V_2} \frac{T_2}{T_1} p_1 = 1 \times \frac{573 \text{ K}}{293 \text{ K}} \times 4870 \text{ kPa} = 9500 \text{ kPa}$$

(b)

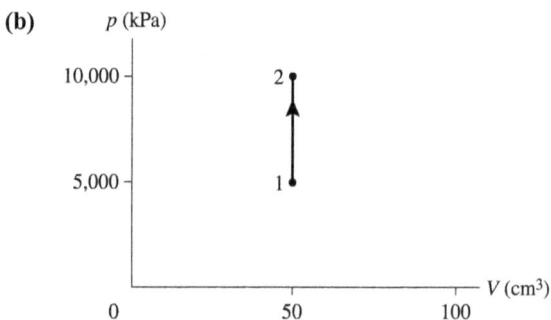

Assess: Note that it is essential to express temperatures in kelvins. Increase in temperature at a constant volume leads to increased pressure, as would be expected.

P12.31. Strategize: The isobaric heating means that the pressure of the argon gas stays unchanged. Argon gas in the container is assumed to be an ideal gas. We can apply the ideal gas law twice: once for the initial state and once for the final state of the gas.

Prepare: We must first convert the volumes and temperatures to SI units with

$V_1 = 50 \text{ cm}^3 = 50 \times 10^{-6} \text{ m}^3$, $T_1 = 20°C = (273 + 20) \text{ K} = 293 \text{ K}$, and $T_2 = 300°C = (300 + 273) \text{ K} = 573 \text{ K}$.

Solve: (a) The container has only argon inside with $n = 0.10$ mol. This produces a pressure

$$p_1 = \frac{nRT_1}{V_1} = \frac{(0.10 \text{ mol})(8.31 \text{ J/(mol} \cdot \text{K)})(293 \text{ K})}{50 \times 10^{-6} \text{ m}^3} = 4.87 \times 10^6 \text{ Pa} = 4870 \text{ kPa}$$

An ideal gas process has $p_2 V_2 / T_2 = p_1 V_1 / T_1$. Isobaric heating to a final temperature $T_2 = 300°C = 573 \text{ K}$ has $p_2 = p_1$, so the final volume is

$$V_2 = \frac{p_1}{p_2} \frac{T_2}{T_1} V_1 = 1 \times \frac{573}{293} \times 50 \text{ cm}^3 = 98 \text{ cm}^3$$

(b)

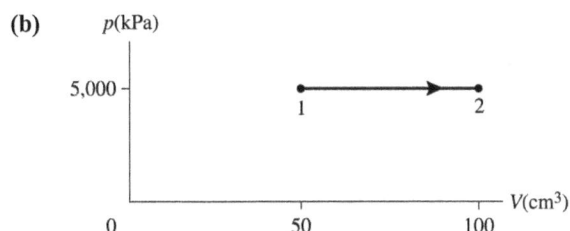

P12.33. Strategize: Assume the gas to be an ideal gas. We examine what quantity remains fixed throughout the process and then use the ideal gas equation.

Prepare: We will make use of the following conversions: $1 \text{ atm} = 1.013 \times 10^5 \text{ Pa}$ and $1 \text{ cm}^3 = 1 \times 10^{-6} \text{ m}^3$.

Solve: (a) Because the volume stays unchanged, the process is isochoric.

(b) The ideal-gas law $p_1 V_1 = nRT_1$ gives

$$T_1 = \frac{p_1 V_1}{nR} = \frac{(3 \times 1.013 \times 10^5 \text{ Pa})(100 \times 10^{-6} \text{ m}^3)}{(0.0040 \text{ mol})(8.31 \text{ J/(mol} \cdot \text{K)})} = 914 \text{ K} \approx 910 \text{ K}$$

The final temperature T_2 is calculated as follows for an isochoric process:

$$\frac{p_1}{T_1} = \frac{p_2}{T_2} \Rightarrow T_2 = T_1 \frac{p_2}{p_1} = (914 \text{ K}) \left(\frac{1 \text{ atm}}{3 \text{ atm}} \right) = 300 \text{ K}$$

P12.35. Strategize: Assume that the gas is an ideal gas. Consider what quantity might be held fixed in such a curve. Then we can use the ideal gas equation.

Prepare: We will make use of the following conversions: $1 \text{ atm} = 1.013 \times 10^5 \text{ Pa}$ and $1 \text{ cm}^3 = 1 \times 10^{-6} \text{ m}^3$.

Solve: (a) Because the process is at a constant pressure, it is isobaric.

(b) For an ideal gas at constant pressure,

$$\frac{V_2}{T_2} = \frac{V_1}{T_1} \Rightarrow T_2 = T_1 \frac{V_2}{V_1} = [(273 + 900) \text{ K}] \frac{100 \text{ cm}^3}{300 \text{ cm}^3} = 391 \text{ K} \approx 120°C$$

(c) Using the ideal-gas law $p_2 V_2 = nRT_2$,

$$n = \frac{p_2 V_2}{RT_2} = \frac{(3 \times 1.013 \times 10^5 \text{ Pa})(100 \times 10^{-6} \text{ m}^3)}{(8.31 \text{ J/(mol} \cdot \text{K)})(391 \text{ K})} = 9.4 \times 10^{-3} \text{ mol}$$

P12.37. Strategize: Treat the air like it is an ideal gas, and we are told to assume this is an isothermal expansion.

Prepare: We are given the volume and gauge pressure at the bottom of the sea: $V_b = 1.0$ cm^3 and $p_{bg} = 1.5$ atm. The absolute pressure is: $p_b = p_{bg} + 1$ atm $= 2.5$ atm. Since the process is isothermal, from the ideal gas law, pV is constant.

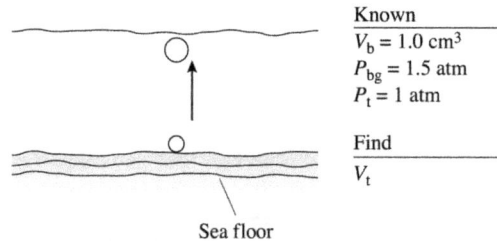

Sea floor

Solve: When the bubble reaches the surface, its pressure is 1 atm. Since pV is constant, we can equate its value at the bottom of the sea with its value at the top as follows:

$$(2.5 \text{ atm})(1.0 \text{ cm}^3) = (1.0 \text{ atm})V_t$$

This equations can be solved to yield: $V_t = 2.5$ cm^3.

(b) As the bubble rises it expands. If no heat were exchanged with the sea, it would cool by adiabatic expansion. Since it does not cool down, we know heat is flowing into the bubble.

Assess: We can understand why heat flows into the bubble if we realize that as the bubble expands it does work on its surroundings and does this work at the expense of its thermal energy. The sea responds to the loss in thermal energy of the bubble by giving it heat. You can imagine the bubble doing an infinitesimal amount of work, which causes its temperature to decrease infinitesimally, which in turn causes heat to flow from the sea to the slightly cooler bubble until they have the same temperature again.

P12.39. Strategize: Assume the gas is ideal such that we can apply the ideal gas equation. We will apply it twice: once to the initial state of the gas, and once to the final state.

Prepare: We will use the ideal gas Equation 12.17 and assume that the volume of the tire and that of the air in the tire is constant. That is, the gas undergoes an isochoric (constant-volume) process. Because the gas equation needs absolute rather than gauge pressure, a gauge pressure of 30 psi corresponds to an absolute pressure of $(30 \text{ psi}) + (14.7 \text{ psi}) = 44.7$ psi.

Solve: Using the before-and-after relationship of an ideal gas for an isochoric process,

$$\frac{p_i}{T_i} = \frac{p_f}{T_f} \Rightarrow p_f = \frac{T_f}{T_i}p_i = \left(\frac{273 + 45}{273 + 15}\right)(44.7 \text{ psi}) = 49.4 \text{ psi}$$

Your tire gauge will read a gauge pressure $p_f = 49.4 \text{ psi} - 14.7 \text{ psi} = 34.7 \text{ psi}$, which is to be reported as 35 psi.

Assess: A 5 psi increase in gauge pressure due to an increase in temperature by 30°C is reasonable.

P12.41. Strategize: Assume the gas is ideal such that we can apply the ideal gas equation. We will apply it twice: once to the initial state of the gas, and once to the final state.

Prepare: For a gas in a sealed container (n is constant) we use Equation 12.17.

$$\frac{p_f V_f}{T_f} = \frac{p_i V_i}{T_i}$$

In our constant volume case V_i and V_f cancel.

The 35 psi is the gauge pressure. We take the usual steps of converting the gauge pressure to absolute pressure and converting the temperatures to absolute temperatures.

$$p_i = 35 \text{ psi} + 14.7 \text{ psi} = 49.7 \text{ psi}$$
$$T_i = 20°C = 293 \text{ K}$$
$$T_f = 0°C = 273 \text{ K}$$

Solve: Solve the top equation for p_f.

$$p_f = \left(\frac{T_f}{T_i}\right) p_i = \left(\frac{273 \text{ K}}{293 \text{ K}}\right) 49.7 \text{ psi} = 46.3 \text{ psi}$$

This answer is the absolute final pressure. To report the answer as a gauge pressure, we subtract $1 \text{ atm} = 14.7 \text{ psi}$. $46.3 \text{ psi} - 14.7 \text{ psi} = 31.6 \text{ psi}$, which should be reported to two significant figures as 32 psi.

Assess: We expected the pressure to decrease with decreasing temperature. On the absolute scale the temperature didn't decrease a lot, however, so neither did the pressure.

We did not need to convert all the pressure data to SI units because we wanted the answer in the same units as the original data, and the formula shows the ratio in any units (using absolute pressure, not gauge pressure) is the ratio of the absolute temperatures.

P12.43. Strategize: This problem deals with linear expansion when a beam is heated. We will use the given data to find the initial length (before heating).
Prepare: We are given much of the data needed in Equation 12.22 except the coefficient of linear expansion for steel, which we look up in Table 12.3. $\Delta L = 0.73 \text{ mm}$, $\Delta T = 13 \text{ K}$, $\alpha_{steel} = 12 \times 10^{-6} \text{ K}^{-1}$.

We want to know the original length, so we solve Equation 12.22 for that quantity.
Solve:

$$L_i = \frac{\Delta L}{\alpha \Delta T} = \frac{(0.73 \text{ mm})}{(12 \times 10^{-6} \text{ K}^{-1})(13 \text{ K})} \left(\frac{1 \text{ m}}{1000 \text{ mm}}\right) = 4.7 \text{ m}$$

Assess: This seems to be a reasonable answer—in the realm of daily life, about the width of a room. It would have taken an aluminum beam only about half that long to produce the same ΔL under the same ΔT.

P12.45. Strategize: This problem deals with linear expansion when a beam is heated. We will use the given data to find the final length of the segment after heating. The gap must be sufficient to accommodate this difference in lengths.
Prepare: The gaps must be at least the length of the thermal expansion in the rails. Equation 12.22 and Table 12.3 apply.
Solve: The expansion of a single rail during this temperature change is

$$\Delta L = \alpha L_i \Delta T = (12 \times 10^{-6} \text{ K}^{-1})(12 \text{ m})(50 \text{ K} - 16 \text{ K}) = 4.9 \times 10^{-3} \text{ m}$$

The rails should be laid with gaps of about half a centimeter.
Assess: This result makes sense. Gaps on real tracks are about this size or larger.

P12.47. Strategize: This problem involves the thermal expansion of the volume of a solid as it is heated. We will use the known relationship for volume expansion.
Prepare: We need to rearrange Equation 12.2 to give the fractional volume expansion (the percentage expansion is then just 100 times that fractional expansion).
Look up $\beta_{Al} = 69 \times 10^{-6} \text{ K}^{-1}$ in Table 12.3. $\Delta T = 120 \text{ K}$.
Solve:

$$\frac{\Delta V}{V_i} = \beta \Delta T = (69 \times 10^{-6} \text{ K}^{-1})(120 \text{ K}) = 0.0083 = 0.83\%$$

Assess: The expansion is small, but greater than it would have been for steel. While 120°C is a larger temperature swing than we might see on a daily basis, there are situations (engines, etc.) where there are significant temperature differences, and people who design and build precise things must take such expansions into account.

P12.49. Strategize: We use the concepts of specific heat to determine the energy required to yield a specific temperature change.

Prepare: The heat needed to change an object's temperature by ΔT is $Q = Mc\Delta T$. The mass of the ice cube is 0.200 kg and its specific heat from Table 12.4 is $c_{ice} = 2090$ J/(kg·K).

Solve:

$$Q = (0.200 \text{ kg})(2090 \text{ J/(kg·K)})(243 \text{ K} - 273 \text{ K}) = -12,500 \text{ J}$$

Thus, the energy removed from the ice block is 13 kJ.

Assess: The negative sign with Q means loss of energy because removal of heat from the ice reduces its thermal energy and its temperature.

P12.51. Strategize: We will use the specific heat of mercury (and later of water) to relate an increase of thermal energy to the increase in temperature.

Prepare: The mass of the mercury is $20 \text{ g} = 2.0 \times 10^{-2}$ kg and its specific heat from Table 12.4 is $c_{Hg} = 140$ J/kg K. The mass of the water is $20 \text{ g} = 2.0 \times 10^{-2}$ kg and its specific heat from Table 12.4 is $c_{water} = 4190$ J/kg K. We will use Equation 12.24 to obtain the needed heats.

Solve: **(a)** The heat needed to change the mercury's temperature is

$$Q = Mc_{Hg}\Delta T \Rightarrow \Delta T = \frac{Q}{Mc_{Hg}} = \frac{100 \text{ J}}{(0.020 \text{ kg})(140 \text{ J/(kg·K)})} = 35.7 \text{ K} = 35.7°C, \text{ which will be reported as } 36°C.$$

(b) The amount of heat required to raise the temperature of the same amount of water by the same number of degrees is

$$Q = Mc_{water}\Delta T = (0.020 \text{ kg})(4190 \text{ J/(kg·K)})(35.7 \text{ K}) = 3000 \text{ J}$$

Assess: Q is directly proportional to c_{water} and the specific heat for water is much higher than the specific heat for mercury. This explains why $Q_{water} > Q_{mercury}$.

P12.53. Strategize: We will use the value given in Example 12.18 for the effective latent heat of vaporization in the process of sweating.

Prepare: We'll compute the energy necessary to evaporate 3.5 L and then divide by an hour to get the rate. One liter of water has a mass of one kilogram.

Solve:

$$Q = ML_f = (3.5 \text{ kg})(24 \times 10^5 \text{ J/K}) = 8.4 \times 10^6 \text{ J}$$

The rate is the energy divided by the time.

$$\text{rate} = \frac{Q}{\Delta t} = \frac{8.4 \times 10^6 \text{ J}}{1 \text{ h}}\left(\frac{1 \text{ h}}{3600 \text{ s}}\right) = 2300 \text{ W}$$

Assess: That is an impressive power output, but necessary to keep cool in tropical climates. Notice that the value given for L_f at body temperature is different from the one given in Table 12.5 for standard temperature (0°C).

P12.55. Strategize: We will use the specific heat to relate the energy absorbed to the increase in temperature.

Prepare: We are asked to find the time it takes for the alligator's temperature to rise from 25°C to 30°C, that is by 5 K. It is absorbed at a rate of 1200 W. We use the mammalian specific heat: $c = 3400$ J/(kg·K). We will need to use the definition of power, which in this case can be written as $P = Q/\Delta t$, as well as the formula for heat absorbed in connection with an increase in temperature, $Q = Mc\Delta T$.

Solve: The time needed for the alligator to reach its final temperature can be obtained by solving the power equation: $\Delta t = Q/P$. This, coupled with the formula for heat gives us the following:

$$\Delta t = \frac{Q}{P} = \frac{Mc\Delta T}{P} = \frac{(300 \text{ kg})(3400 \text{ J/(kg·K)})(5 \text{ K})}{1200 \text{ W}} = 4250 \text{ s} = 71 \text{ min}$$

This is just over 1 hour.

Assess: This seems like a reasonable length of time. The alligator will still have most of the day at the warmer body temperature to catch prey.

P12.57. Strategize: We will use the known effective latent heat of vaporization for sweat to relate the energy losses to the quantity of moisture converted to vapor.

Prepare: Water in the body is converted to water vapor. Equation 12.25 applies.

Solve: Each breath converts 25 mg of water to water vapor. The heat required for this is

$$Q = ML_v = (2.5 \times 10^{-5} \text{ kg})(24 \times 10^5 \text{ J/kg}) = 60 \text{ J}$$

At 12 breaths/min, there are 0.2 breaths/s. Multiplying by the above result of 60 J/breath we obtain that $P = 12$ J/s or $P = 12$ W is the rate of heat loss.

Converting to Calories/day, we have

$$P = (12 \text{ W})\left(\frac{1 \text{ Calorie}}{4190 \text{ J}}\right)\left(\frac{60 \text{ s}}{\text{min}}\right)\left(\frac{60 \text{ min}}{\text{h}}\right)\left(\frac{24 \text{ h}}{\text{d}}\right) = 250 \text{ Cal/d}$$

Assess: This seems reasonable since it is a small fraction of a person's daily caloric intake of about 2000 Calories.

P12.59. Strategize: We will use the specific heat to relate thermal energy to a change in temperature for a human body.

Prepare: We need to find the temperature increase and we are given the metabolic power $P = 1000$ W, the time of exercise $\Delta t = 30 \text{ min} = (30 \text{ min})(60 \text{ s})/(1 \text{ min}) = 1800 \text{ s}$, and the man's mass $m = 70$ kg. We need the formula for power, which in this case is $P = \Delta E_{th}/\Delta t$ and Equation 12.23 for heat absorbed. Here the heat produced by the exercise goes to increasing the thermal energy of the man, so we can write the following: $\Delta E_{th} = Mc\Delta T$. For the specific heat, we use the mammalian specific heat from Table 12.4: $c = 3400$ J/kg·K.

Solve: First we will find the thermal energy produced by the exercise by solving the power equation.

$$\Delta E_{th} = P\Delta t = (1000 \text{ W})(1800 \text{ s}) = 1.8 \times 10^6 \text{ J}$$

We can solve the equation $\Delta E_{th} = Mc\Delta T$ for ΔT.

$$\Delta T = \Delta E_{th}/Mc = \frac{(1.8 \times 10^6 \text{ J})}{(70 \text{ kg})(3400 \text{ J/(kg·K)})} = 7.56 \text{ K}$$

To two significant figures, his temperature increases by 7.6 K which is 14°F.

Assess: Even though he only exercised for 30 min, the man's temperature has increased by 14°F. This would bring a temperature of 98.6°F up to about 113°F, which would be very dangerous.

P12.61. Strategize: This is a basic calorimetry problem. We will assume that all energy that leaves the coffee goes to heating the aluminum.

Prepare: We are given the mass and initial temperature of the aluminum $m_{Al} = 200$ g and $T_{i \, Al} = -20°C$, as well as the mass and initial temperature of the coffee $m_c = 500$ g and $T_{i \, c} = 85°C$. We obtain the specific heats from Table 12.4, according to which $c_{Al} = 900$ J/(kg·K) and $c_c = 4190$ J/(kg·K).

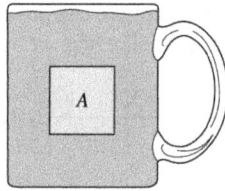

Known
$m_{A1} = 200$ g
$T_{iA1} = -20° C$
$C_{A1} = 900$ J/(kg.K)
$m_C = 500$ g
$T_{iC} = 85° C$
$c_C = 4190$ J/(kg.K)

Find
T_f

Solve: The basic calorimetry equation becomes $m_{Al}c_{Al}\Delta T_{Al} + m_c c_c \Delta T_c = 0$ which we can solve for the final temperature:

$$(0.200 \text{ kg})(900 \text{ J/kg})(T_f - -20°C) + (0.500 \text{ kg})(4190 \text{ J/(kg·K)})(T_f - 85°C) = 0 \text{ J}$$
$$\Rightarrow (2280 \text{ J/K})T_f - 174000 \text{ J} = 0$$
$$\Rightarrow T_f = 76°C$$

Assess: Notice that even though the aluminum was comparable in mass to the coffee, its temperature change of $96°C$ was far more than the temperature change of the coffee, $9°C$. This is a result of water having a much higher specific heat than metal.

P12.63. Strategize: This is a basic calorimetry problem. We will assume all energy that leaves the copper pellets goes into the water.
Prepare: We have a thermal interaction between the copper pellets and the water. The initial temperatures of both copper pellets and the water are known and we denote the common final temperature by T_f. The specific heats of copper and water from Table 12.4 are as follows: $c_c = 385$ J/(kg·K) and $c_w = 4190$ J/(kg·K). While the mass of copper pellets is $M_c = 0.030$ kg, the mass of water is obtained from

$M_w = V\rho$: $M_w = (100 \text{ mL})(10^{-6} \text{ m}^3/1 \text{ mL})(1000 \text{ kg/m}^3) = 0.10$ kg. We can determine the common final temperature using Equations 12.25 and 12.28.
Solve: The conservation of energy equation $Q_c + Q_w = 0$ is

$$M_c c_c (T_f - 300°C) + M_w c_w (T_f - 20°C) = 0 \text{ J}$$

Solving this equation for the final temperature T_f gives

$$T_f = \frac{M_c c_c (300°C) + M_w c_w (20°C)}{M_c c_c + M_w c_w} = \frac{(0.030 \text{ kg})(385 \text{ J/(kg·K)})(300°C) + (0.10 \text{ kg})(4190 \text{ J/(kg·K)})(20°C)}{(0.030 \text{ kg})(385 \text{ J/(kg·K)}) + (0.10 \text{ kg})(4190 \text{ J/(kg·K)})} = 28°C$$

The final temperature of the water and the copper is 28°C.
Assess: Due to the large specific heat of water compared to copper and three times more water in our system, we expected a small temperature increase.

P12.65. Strategize: This is a basic calorimetry problem. We will assume that all energy that leaves the copper block goes into the water.
Prepare: A thermal interaction between the copper block and the water leads to a common final temperature denoted by T_f. The initial temperatures of both the copper block and the water are known. The specific heats of copper and water from Table 12.4 are as follows: $c_c = 385$ J/(kg·K) and $c_w = 4190$ J/(kg·K). While the mass of the water is known, we can determine the mass of the copper block using Equations 12.25 and 12.28.
Solve: The conservation of energy equation $Q_{copper} + Q_{water} = 0$ J is

$$M_{copper} c_{copper} (T_f - T_{i \text{ copper}}) + M_{water} c_{water} (T_f - T_{i \text{ water}}) = 0 \text{ J}$$

Both the copper and the water reach the common final temperature $T_f = 25.5°C$. Thus

$$M_{copper}(385 \text{ J/(kg·K)}(25.5°C - 300°C) + (1.00 \text{ kg})(4190 \text{ J/(kg·K))}(25.5°C - 20.0°C) = 0 \text{ J}$$

$$\Rightarrow M_{copper} = 0.218 \text{ kg}$$

Assess: Due to the large specific heat of water compared to copper, a smaller value obtained for the mass of the copper block is reasonable.

P12.67. Strategize: If we assume that all heat leaving the patient's body goes into melting the ice, then this is effectively a calorimetry problem.
Prepare: We need the formula for heat absorbed in melting $Q_{ice} = m_{ice}L_f$, and the formula for the heat lost by the patient $Q_p = m_p c_p \Delta T$. For the specific heat, we will use the value for mammals from Table 12.4, $c_p = 3400 \text{ J/kg·K}$. We need to solve for m_{ice}.
Solve: The calorimetry equation is as follows:

$$m_{ice}(3.33 \times 10^5 \text{ J/kg}) + (60 \text{ kg})(3400 \text{ J/(kg·K))}(39°C - 40°C) = 0 \text{ J} \Rightarrow$$
$$m_{ice} = \frac{(3400 \text{ J/(kg·K))}}{(3.33 \times 10^5 \text{ J/kg})}(60 \text{ kg})(1 \text{ K}) = 0.61 \text{ kg} = 610 \text{ g}$$

It takes 610 g of ice to reduce the fever by $1°C$.

Assess: Since the ratio of the mammalian specific heat to the latent heat of fusion of ice is about $\frac{1}{100}K^{-1}$, it follows that a rule of thumb is that the amount of ice needed to cool a person by $1°C$ is one hundredth of their body mass. This is seen to be the case here.

P12.69. Strategize: This problem requires you to use the concept of specific heat in two contexts. At constant volume, all heat energy goes into changing the temperature of a gas; at constant pressure, the gas can expand and do work, so the heat input is split up between increasing the temperature and doing work
Prepare: The heating processes are isobaric (in part **(a)**) and isochoric (in part **(b)**). O_2 is a diatomic ideal gas. We will use Equations 12.29 and 12.30 to find the heat needed at constant volume and constant pressure. Since these equations involve the number of moles of the gas, we will calculate it from the mass of the gas and its molar mass. From Table 12.6, $C_p = 29.2 \text{ J/(mol·K)}$ and $C_V = 20.9 \text{ J/(mol·K)}$. Note that the change in temperature on the Kelvin scale is the same as the change in temperature on the Celsius scale.
Solve: (a) The number of moles of oxygen is

$$n = \frac{M}{M_{mol}} = \frac{1.0 \text{ g}}{32 \text{ g/mol}} = 0.03125 \text{ mol}$$

For the isobaric process,

$$Q = nC_p\Delta T = (0.03125 \text{ mol})(29.2 \text{ J/(mol·K))}(100°C) = 91 \text{ J}$$

(b) For the isochoric process,

$$Q = nC_V\Delta T = 91.2 \text{ J} = (0.03125 \text{ mol})(20.9 \text{ J/(mol·K))} \Delta T \Rightarrow \Delta T = 140°C$$

P12.71. Strategize: We can write out the heat energy required for this process at constant volume, and then use the same heat energy in an expression for adding heat at constant pressure. Relating the two expressions will allow us to find a temperature change in the latter case.
Prepare: In the first part the "rigid container" means constant volume, so $Q = \frac{3}{2}nR\Delta T$. That Q will be the same as the Q in the second constant pressure part of the question for which $Q = \frac{5}{2}nR\Delta T'$.

Solve: Solve for $\Delta T'$.

$$\frac{5}{2}nR\Delta T' = \frac{3}{2}nR\Delta T \Rightarrow \Delta T' = \frac{\frac{3}{2}}{\frac{5}{2}}\Delta T = \frac{3}{5}(15\text{ K}) = 9.0°\text{ C}.$$

Assess: We expected the change in temperature to be smaller at constant pressure than at constant volume. The number of moles was irrelevant.

P12.73. Strategize: This problem involves the flow of heat through conduction. We ignore other mechanisms of heat exchange like radiation and convection.
Prepare: The rate of conduction across a temperature difference is given in Equation 12.35.

$$\frac{Q}{\Delta t} = \left(\frac{kA}{L}\right)\Delta T$$

where $A = 4.0\text{ m} \times 5.5\text{ m} = 22\text{ m}^2$ is the area, $L = 0.018\text{ m}$ is the thickness of the flooring, and $k = 0.2\text{ W/(m}\cdot\text{K})$ is the thermal conductivity of wood given in Table 12.7. $\Delta T = 19.6°\text{C} - 16.2°\text{C} = 3.4°\text{C}$.
Solve:

$$\frac{Q}{\Delta t} = \left(\frac{kA}{L}\right)\Delta T = \frac{(0.2\text{ W/(m}\cdot\text{K}))(22\text{ m}^2)}{0.018\text{ m}}(3.4°\text{C}) = 830\text{ J/s} = 830\text{ W}$$

Assess: 830 W is about as much as a dozen incandescent light bulbs. In the winter when you are trying to keep the room warm this energy is being wasted; you could do drastic things like increase the thickness of the wood, or simpler, cheaper things like cover the floor with carpet, which has a much smaller k. In the summer you might be grateful to have this energy conducted from the room if the subfloor can stay at a cooler temperature.

P12.75. Strategize: This problem involves heat transfer due to radiation. We will take into account both the radiation from the seal and the radiation from the environment to obtain the net rate of heat loss.
Prepare: The rate of net energy loss by radiation is given by Equation 12.37.

$$\frac{Q_{\text{net}}}{\Delta t} = e\sigma A(T^4 - T_0^4)$$

where T_0 is the termperature of the surroundings.
We are given $T = 30°\text{C} = 303\text{ K}$, $T_0 = -10°\text{C} = 263\text{ K}$, and $A = 0.030\text{ m}^2$. We are told to assume the emissivity of seal skin is the same as human skin; the text gives this vaule as $e = 0.97$.
The textbook gives Stefan's constant as $\sigma = 5.67 \times 10^{-8}\text{ W/(m}^2\cdot\text{K}^4)$.
Solve:

$$\frac{Q_{\text{net}}}{\Delta t} = e\sigma A(T^4 - T_0^4) = (0.97)(5.67 \times 10^{-8}\text{ W/(m}^2\cdot\text{K}^4))(0.030\text{ m}^2)[(303\text{ K})^4 - (263\text{ K})^4] = 6.0\text{ W}$$

Assess: 6 W isn't a lot, but it is sufficient to cool the seal when the surroundings are very cool. If there were no thermal windows the seal would have difficulty regulating its temperature.

P12.77. Strategize: This problem involves radiative energy loss. The temperature of the filament is so high, that we can ignore radiative absorption from the environment.
Prepare: The rate of energy loss by radiation is given by Equation 12.36.

$$\frac{Q}{\Delta t} = e\sigma AT^4$$

We are given $e = 0.23$, $T = 1500°\text{C} = 1773\text{ K}$, and $Q/\Delta t = 60\text{ W}$. We are asked to find A.

The textbook gives Stefan's constant as $\sigma = 5.67 \times 10^{-8}\text{ W/(m}^2\cdot\text{K}^4)$.

Solve: Solve the equation for A.

$$A = \frac{Q/\Delta t}{e\sigma T^4} = \frac{60 \text{ W}}{(0.23)(5.67\times10^{-8} \text{ W/(m}^2\cdot\text{K}^4))(1773 \text{ K})^4} = 4.7\times10^{-4} \text{ m}^2$$

Assess: We knew that light bulb filaments have a small surface area, so we are not concerned to get a small answer. The units work out properly.

P12.79. Strategize: This problem has to do with heat loss through radiation. There is some degree of estimation involved in the surface area of the human that effectively radiates. So a small range of answers might be acceptable.
Prepare: Since you are lying on the ground, your back does not emit radiation or absorb radiation from the sky. We might guess that a little more than half of a person's area is off the ground. A typical person's surface area from the book is 1.8 m^2. So we will use 1 m^2 for the area in contact with the air. We will use $e = 0.97$. The temperature of your clothing is $T_c = 303$ K and the temperature of the sky is $T_s = 233$ K.
Solve: The net rate that your body loses energy to the sky is given by Equation 12.36.

$$\frac{Q_{\text{net}}}{\Delta t} = e\sigma A(T_c^4 - T_s^4) = (0.97)(5.67\times10^{-8} \text{ W/(m}^2\cdot\text{K}^4))(1 \text{ m}^2)((303 \text{ K})^4 - (233 \text{ K})^4) = 300 \text{ W}$$

Assess: This is about three times the net rate of heat loss you would experience if you were in a room at room temperature.

P12.81. Strategize: This problem involves diffusion across a membrane.
Prepare: We can use Equation 12.40: $\Delta t = \dfrac{L^2}{6D}$ to determine the diffusion time. We will use the diffusion coefficient given: $D = 2.0\times10^{-11} \text{ m}^2 / \text{s}$.

Solve: Inserting the given values, we have $\Delta t = \dfrac{L^2}{6D} = \dfrac{\left(1.0\times10^{-6} \text{ m}\right)^2}{6\left(2.0\times10^{-11} \text{ m}^2 / \text{s}\right)} = 8$ ms.

Assess: It makes sense that this is substantially shorter than the time that the time to traverse the alveolus.

P12.83. Strategize: Assume that the compressed air in the cylinder is an ideal gas and that the volume of the air in the cylinder is a constant.
Prepare: We will use Equation 12.17 to calculate the new pressure in atm and compare it with the maximum pressure (in atm) of the compressed gas that the cylinder can withstand.
Solve: Using the before-and-after relationship of an ideal gas,

$$\frac{p_f V_f}{T_f} = \frac{p_i V_i}{T_i} \Rightarrow p_f = p_i \frac{T_f}{T_i} \frac{V_i}{V_f} = (25 \text{ atm})\left(\frac{1223\text{K}}{293\text{K}}\right)\frac{V_i}{V_i} = 104 \text{ atm}$$

where we have converted to the Kelvin temperature scale. Because the pressure does not exceed 110 atm, the compressed air cylinder does not blow.

P12.85. Strategize: The work done by a gas is the area under the p-versus-V curve.
Prepare: The gas is expanding, so the work done *by* the gas is positive or the work done *on* the gas is negative.
Solve: The area under the pV curve is the area of the rectangle and triangle. The work done *by* the gas is as follows:

$$(200\times10^{-6} \text{ m}^3)(200\times10^3 \text{ Pa}) + \frac{1}{2}(200\times10^{-6} \text{ m}^3)(200\times10^3 \text{ Pa}) = 60 \text{ J}$$

P12.87. Strategize: For the first part, we will use the given equation to determine the work done on the fluid (the blood). For the second part, we can simply use the relationship between work and power.
Prepare: We are given the pressure difference $\Delta p = 16$ kPa, and the volume $V = 5.0$ L $= 5.0\times10^{-3} \text{m}^3$. For the second part, we know $P = \Delta E / \Delta t = W / \Delta t$.

Solve: (a) Inserting the given numbers, we have

$$W = V\Delta p = (5.0 \times 10^{-3} \text{ m}^3)(16{,}000 \text{ N/m}^2) = 80 \text{ J}$$

(b) Using the time of 1.0 min = 60 s, we have

$$P = W / \Delta t = (80 \text{ J}) / (60 \text{ s}) = 1.3 \text{ W}$$

Assess: Given the millions of Joules of food energy that humans take in each day, this requirement of 80 J per minute just for the heart is very reasonable.

P12.89. Strategize: Assume the gravitational potential energy is transformed completely into heat energy.
Prepare: We will use the specific heat of water to relate the change in thermal energy to a change in termperature.
Solve:

$$\Delta U_g = -Q$$

$$Mg\Delta y = -Mc\Delta T$$

Solve for the change in temperature.

$$\Delta T = \frac{Mg\Delta y}{-Mc} = \frac{(9.8 \text{ m/s}^2)(-51 \text{ m})}{-4190 \text{ J/kg} \cdot \text{K}} = 0.12 \text{ K}$$

We expect the water to be 0.12°C warmer at the bottom.
Assess: This is not a big difference in temperature, but it should be measurable. We expected this result to depend on g and the distance the water falls, but not M.

P12.91. Strategize: This problem requires us to relate the waste heat taken up by water to the change in temperature of the water using the specific heat.
Prepare: There are two interacting systems: the nuclear reactor and the water. The heat generated by the nuclear reactor is used to raise the water temperature. For the closed reactor–water system, energy conservation per second requires $Q = Q_{\text{reactor}} + Q_{\text{water}} = 0$ J. The heat from the reactor in $\Delta t = 1$ s is $Q_{\text{reactor}} = -2000$ MJ $= -2.0 \times 10^9$ J and we will use Equation 12.24 for Q_{water}.
Solve: The heat absorbed by the water is

$$Q_{\text{water}} = m_{\text{water}}c_{\text{water}}\Delta T = m_{\text{water}}(4190 \text{ J/(kg} \cdot \text{K)})(12 \text{ K})$$
$$\Rightarrow -2.0 \times 10^9 \text{ J} + m_{\text{water}}(4190 \text{ J/(kg} \cdot \text{K)})(12 \text{ K}) = 0 \text{ J} \Rightarrow m_{\text{water}} = 3.98 \times 10^4 \text{ kg}$$

Each second, 3.98×10^4 kg of water is needed to remove heat from the nuclear reactor. Thus, the water flow per minute is

$$3.98 \times 10^4 \frac{\text{kg}}{\text{s}} \times \frac{60 \text{ s}}{\text{min}} = 2.4 \times 10^6 \text{ kg/min}$$

P12.93. Strategize: We will look up the metabolic power of this cyclist and require that all waste heat be taken up through perspiration.
Prepare: From Table 11.4, the metabolic power of a 68 kg cyclist is 480 W. We assume that 25% of this goes to propelling the cyclist and the other 75%, or 360 W, becomes heat which serves to evaporate perspiration. Equation 12.29 gives the heat needed to evaporate a liquid. From the discussion following Table 12.5, a good value for the latent heat of vaporization of sweat is $L_v = 2.4 \times 10^6$ J/kg.
Solve: We solve Equation 12.29 for m and get $m = Q/L_v$. This is the mass of perspiration which heat Q could evaporate. We know $\dfrac{Q}{\Delta T} = 360$ W and we can combine this with the preceding equation to obtain the following:

$$\frac{m}{\Delta t} = \frac{Q/L_v}{\Delta t} = \frac{Q/\Delta t}{L_v} = \frac{360 \text{ J/s}}{2.4 \times 10^6 \text{ J/kg}} = 1.5 \times 10^{-4} \text{ kg/s} = 0.54 \text{ kg/h}$$

Assess: A kilogram of water has a volume of about one liter, so this is about half a liter per hour. A value on the order of one liter per hour seems reasonable.

P12.95. Strategize: We equate the metabolic rate of the elephant to the rate at which heat must be taken up by evaporating water in a process equivalent to perspiration.
Prepare: We need Equation 12.29 which gives the amount of heat need to evaporate a mass of liquid. We are given the rate that heat must be absorbed by the water as follows: $Q/\Delta t = 2500$ W. We can use the latent heat of vaporization given in the discussion following Table 12.5, $L_v = 2.4 \times 10^6$ J/kg.
Solve: Solving Equation 12.29 for the mass gives us: $m = Q/L_v$ and dividing both sides by Δt gives our answer

$$\frac{m}{\Delta t} = \frac{Q/L_v}{\Delta t} = \frac{Q/\Delta t}{L_v} = \frac{2500 \text{ J/s}}{2.4 \times 10^6 \text{ J/kg}} \left(\frac{3600 \text{ s}}{1 \text{ hr}} \right) = 3.8 \text{ kg/h}$$

Assess: This works out to 3.8 L/h, which seems reasonable. During vigorous exercise, humans sweat a couple of liters per hour. An elephant at rest has about twice the metabolic rate of a human exercising vigorously (using the data from Table 11.4), so we would expect a value of around 4 L/h.

P12.97. Strategize: Heating the material increases its thermal energy. Sometimes the thermal energy goes into increasing the temperature and other times it goes into a phase change. We can determine the thermal energy required per Kelvin of temperature increase from the slope of the line where the temperature is changing. We can determine the heat of vaporization from the energy input required at constant temperature.
Prepare: The material melts at 300°C and undergoes a solid-liquid phase change. The material's temperature increases from 300°C to 1500°C. Boiling occurs at 1500°C and the material undergoes a liquid-gas phase change. We will use Equations 12.24 and 12.25 to determine the specific heat and the heat of vaporization of the liquid.
Solve: (a) In the liquid phase, the specific heat of the liquid can be obtained as follows:

$$\Delta Q = Mc\Delta T \Rightarrow c = \frac{1}{M}\frac{\Delta Q}{\Delta T} = \left(\frac{1}{0.200 \text{ kg}} \right) \left(\frac{20 \text{ kJ}}{1200 \text{ K}} \right) = 83 \text{ J/(kg} \cdot \text{K)}$$

(b) The latent heat of vaporization is

$$L_v = \frac{Q}{M} = \frac{40 \text{ kJ}}{(0.200 \text{ kg})} = 2.0 \times 10^5 \text{ J/kg}$$

Assess: The values obtained are of the same order of magnitude as in Tables 12.4 and 12.5 for a few materials.

P12.99. Strategize: We will use the first law of thermodynamics to describe this isobaric process.
Prepare: W_{gas} is negative because the gas is compressed. Compression transfers energy into the system, that is, work done *on* the gas is positive. Also, 100 J of heat energy is transferred out of the gas, that is $Q = -60$ J.
Solve: The first law of thermodynamics is

$$\Delta E_{th} = -W_{gas} + Q = -p\Delta V + Q = -(4.0 \times 10^5 \text{ Pa})(200 - 600) \times 10^{-6} \text{ m}^3 - 100 \text{ J} = 60 \text{ J}$$

Thermal energy increases by 60 J.

P12.101. Strategize: The monatomic gas is an ideal gas, which is subject to isobaric and isochoric processes.
Prepare: We will use the ideal gas Equation 12.16 and Equations 12.28 and 12.29.
Solve: (a) For this isobaric process, $p_1 = 4.0$ atm, $V_1 = 800 \times 10^{-6}$ m^3, $p_2 = 4.0$ atm, and $V_2 = 1600 \times 10^{-6}$ m^3. The temperature T_1 of the gas is obtained from the ideal-gas equation as

$$T_1 = \frac{p_1 V_1}{nR} = 390 \text{ K}$$

where $n = 0.10$ mol. Also,

$$T_2 = T_1 \frac{V_2}{V_1} = T_1 \left(\frac{1600 \times 10^{-6} \, \text{m}^3}{800 \times 10^{-6} \, \text{m}^3} \right) = 2T_1 = 780 \, \text{K}$$

Thus, the heat required for the process $1 \rightarrow 2$ is

$$Q = nC_p(T_2 - T_1) = (0.10 \, \text{mol})(20.8 \, \text{J/(mol} \cdot \text{K})(390 \, \text{K}) = 811 \, \text{J}$$

which is 810 J in two significant figures.
This is heat transferred to the gas.
(b) For the isochoric process, $V_2 = V_3 = 1600 \times 10^{-6}$ m³, $p_2 = 4.0$ atm, $p_3 = 2.0$ atm, and $T_2 = 780°$K. T_3 can be obtained from the ideal gas equation as follows:

$$\frac{p_2 V_2}{T_2} = \frac{p_3 V_3}{T_3} \Rightarrow T_3 = T_2 (p_3/p_2) = (780 \, \text{K}) \left(\frac{2.0 \, \text{atm}}{4.0 \, \text{atm}} \right) = 390 \, \text{K}$$

The heat required for the process $2 \rightarrow 3$ is

$$Q = nC_V(T_3 - T_2) = (0.10 \, \text{mol})(12.5 \, \text{J/mol K})(390 \, \text{K} - 780 \, \text{K}) = -488 \, \text{J}$$

which is -490 J in two significant figures.
Because of the negative sign, this is the amount of heat removed from the gas.
(c) The change in the thermal energy of the gas is

$$\Delta E_{\text{th}} = (Q_{1 \rightarrow 2} + Q_{2 \rightarrow 3}) - (W_{1 \rightarrow 2} + W_{2 \rightarrow 3}) = 811 \, \text{J} - 488 \, \text{J} - W_{1 \rightarrow 2} - 0 \, \text{J} = 324 \, \text{J} - p\Delta V$$
$$= 324 \, \text{J} - (4.0 \times 1.013 \times 10^5 \, \text{Pa})(1600 \times 10^{-6} \, \text{m}^3 - 800 \times 10^{-6} \, \text{m}^3) = 0 \, \text{J}$$

Assess: This result was expected since $T_3 = T_1$.

P12.103. Strategize: We will apply the first law of thermodynamics and the definition of efficiency.
Prepare: The heat engine follows a closed cycle.
Solve: (a) The work done by the gas per cycle is the area inside the closed p-versus-V curve. We get

$$W_{\text{out}} = \frac{1}{2}(300 \, \text{kPa} - 100 \, \text{kPa})(600 \, \text{cm}^3 - 200 \, \text{cm}^3) = \frac{1}{2}(200 \times 10^3 \, \text{Pa})(400 \times 10^{-6} \, \text{m}^3) = 40 \, \text{J}$$

The heat exhausted is $Q_C = 180 \, \text{J} + 100 \, \text{J} = 280 \, \text{J}$. So, the heat extracted from the hot reservoir is $Q_H = 280 \, \text{J} + 40 \, \text{J} = 320 \, \text{J}$.
(b) The thermal efficiency of the engine is

$$\eta = \frac{W_{\text{out}}}{Q_H} = \frac{40 \, \text{J}}{320 \, \text{J}} = 0.125$$

or 0.13 to two significant figures.

P12.105. Strategize: In this problem we are asked to calculate heat losses due to two processes: radiation and conduction. We have equations that describe both in terms of given variables. In the end we compare the heat being metabolized to the body as waste heat to the total losses to determine the person's comfort.
Prepare: Heat loss by conduction can be calculated with Equation 12.34. Equation 12.36 applies to heat lost by radiation.
Solve: (a) The dead layer of air separates and insulates you from the air in the room. The conduction through the air layer is

$$\frac{Q}{\Delta t} = \left(\frac{kA}{L} \right) \Delta T = \left(\frac{(0.026 \, \text{W/(m} \cdot \text{K})(1.8 \, \text{m}^2)}{0.005 \, \text{m}} \right) (9 \, \text{K}) = 84 \, \text{W}$$

(b) The heat lost through radiation is given by Equation 12.24. Body temperature is $T = 34 + 273 = 307\,\text{K}$. The temperature of the walls is $T = 17 + 273 = 290\,\text{K}$.

$$\frac{Q_{net}}{\Delta t} = e\sigma A(T^4 - T_0^{\,4}) = (.97)(5.67\times10^{-8}\ \text{W/(m}^2\cdot\text{K}^4))(1.8\ \text{m}^2)((307\ \text{K})^4 - (290\ \text{K})^4) = 180\ \text{W}$$

(c) The heat lost to radiation is greater.
(d) If the person is metabolizing food at a rate of 155 W, he feels chilly because he is producing heat at a rate of 155 W and losing heat at a rate of $84\ \text{W} + 180\ \text{W}$ or 260 W.
Assess: Putting some clothes on would decrease the heat lost by radiation, convection, and conduction.

P12.107. Strategize: We will relate a change in temperature to a change in thermal energy through the specific heat of water.
Prepare: We can use Equation 12.24. Table 12.4 lists the specific heat of water.
Solve: The mass of the water in the top layer is $M = \rho d_i A$. The area of the top level of the oceans is

$$A = (3.6\times10^8\ \text{km}^2)\left(\frac{10^3\ \text{m}}{\text{km}}\right)^2 = 3.6\times10^{14}\ \text{m}^2$$

The heat required to change the temperature by one degree Celsius is

$$Q = Mc\Delta T = \rho d_i Ac\Delta T = (1000\ \text{kg/m}^3)(500\ \text{m})(3.6\times10^{14}\ \text{m}^2)(4190\ \text{J/(kg}\cdot\text{K})(1\,\text{K}) = 7.5\times10^{23}\ \text{J}$$

To one significant figure, this is 1×10^{24} J. The correct choice is A.
Assess: This result makes sense. The density and specific heat of water are relatively large.

P12.109. Strategize: We carefully consider the definition of each mechanism of heat transfer.
Prepare: Conduction is the transfer of heat directly through physical material. Radiation is energy transfer through electromagnetic waves. Evaporation transfers energy through removal of molecules with high thermal energy.
Solve: Since there is no mixing in the warmer surface water, convection does not happen effectively. The correct choice is B.
Assess: Note that the water is heated by the light from the sun, so it is heated through radiation.

P12.111. Strategize: This is an estimation problem, so we will make our best estimate of the area enclosed by the two paths of the PV diagram.
Prepare: We can estimate an upper bound by considering the maximum/minimum pressures and volumes. We can then refine our estimate by doing a count of small grid boxes, each of which has an area of $\Delta P\Delta V = \left(2.0\times10^3\ \text{Pa}\right)\left(0.5\times10^{-3}\text{m}^3\right) = 1.0\ \text{J}$.

Solve: Counting the boxes completely enclosed and trying to keep track of those partially enclosed, I count approximately 16 J. This is closest to option B, so we select option B.
Assess: If we expand the area to a square with vertices at the maxima and minima of each parameter, we can obtain an upper bound. We check to make sure our answer is below this. Inserting the numerical values for these minima and maxima, we see the energy cannot possibly be more than

$$\left(P_{max} - P_{min}\right)\left(V_{max} - V_{min}\right) = \left(\left(4.0\times10^3\ \text{Pa}\right) - \left(-3.0\times10^3\ \text{Pa}\right)\right)\left(\left(4.0\times10^{-3}\ \text{m}^3\right) - \left(1.0\times10^{-3}\ \text{m}^3\right)\right) = 21\ \text{J}$$

This is consistent with the value we obtained.

FLUIDS

Q13.1. Reason: Density does not depend on the volume. That is, 1 g of mercury would have the same density as 1000 g of mercury, and 1 g of water would have the same density as 1000 g of water.

Table 13.1 shows the density of mercury to be $13\,600$ kg/m^3 and that of water to be only 1000 kg/m^3.

The density of 1 g of mercury is 13.6 times as much as the density of 1000 g of water.

Assess: It is important to get used to the idea that density is a ratio of mass to volume, so different samples of the same substance would have the same density.

Q13.3. Reason: Density is given by Equation 13.1.

(a) The size and shape of the two objects is the same, so the volume of the two objects is the same. The second object has twice the mass as the first so its density is $\rho_2 = 2m/V = 2\rho_1$. The second object has twice the density of the first, or 8000 kg/m^3.

(b) The third object has a size in all three dimensions twice that of the first object. Its volume is eight times that of the first object. Since its mass is the same, the density of the third object is $\rho_3 = m/8V = \frac{1}{8}\rho_1$. The density of the third object is one eighth that of the first, or 500 kg/m^3.

Assess: To make the volume increase in part **(b)** more concrete, consider increasing the length of each side of a cube by a factor of two. The volume increases by a factor of $2 \times 2 \times 2 = 8$.

Q13.5. Reason: The pressure at a depth of 10 m is

$$p = p_0 + \rho g h = 1.013 \times 10^5 \,\text{Pa} + (1000 \,\text{kg/m}^3)(9.80 \,\text{m/s}^2)(10 \,\text{m}) = 2 \times 10^5 \,\text{Pa}$$

This is almost twice atmospheric pressure. But the air pressure in the hose will be only slightly higher than atmospheric pressure, because the density of air is so low. So Tom will have great difficulty breathing in the low-pressure air when the large pressure of the water is pressing in on his chest.

Assess: Note that atmospheric pressure is approximately the gauge pressure at a depth of 10 m of water.

Q13.7. Reason: Air comes out of a high pressure tank at the same pressure as the water around the diver (metered by the regulator). This means the lungs are inflated with highly-pressurized gas. This does not adversely affect the diver when deep underwater, because the entire environment around the diver is at a similarly high pressure. If the diver suddenly surfaces, the air in the alveoli in the lungs will still be at high pressure, but the air around the diver will be at a low pressure. The gas in the diver's lungs will expand and can burst the alveoli.

Assess: It is important to breathe several times as you ascend to make sure as much of the high-pressure air is cleared from the lungs before surfacing.

Q13.9. Reason: The pressure only depends on the depth from the opening. Since point D is the deepest and point E the highest then

$$p_D > p_F > p_E$$

Assess: A point halfway between point E and point B would have a pressure about the same as the pressure at point D.

Q13.11. Reason: (a) The pressure at the bottom of either tank is given by Equation 13.5. The pressure at the bottom of each tank will be the same, since the height of water in each tank is the same. The area of the bottom of tank A is larger than the area of the bottom of tank B. From Equation 13.3, the force on the bottom of tank A will be larger.
(b) The pressure at each height in both tanks is the same, since the depth of water is the same. Since the area of the sides indicated in the diagram in Figure Q13.11 is the same for each tank, the force on the side of each tank is the same also.
Assess: This makes sense. Since there's more water in tank A, the force of the water on the bottom of tank A is larger.

Q13.13. Reason: The density of the heated water is less, but water increases in height so the pressures are the same. The pressure at the bottom of the beaker can be calculated with Equation 13.5. Consider a beaker whose cross-sectional area is A. The volume of water in the cool beaker is $V = Ah_1$. For the cool water the pressure at the bottom of the beaker is,

$$p_1 = p_0 + \rho g h_1 = p_0 + \left(\frac{m}{Ah_1}\right) g h_1 = p_0 + \left(\frac{m}{A}\right) g$$

For the hot water,

$$p_2 = p_0 + \rho g h_2 = p_0 + \left(\frac{m}{Ah_2}\right) g h_2 = p_0 + \left(\frac{m}{A}\right) g$$

The pressure at the bottom of the beaker is the same for the hot water and cold water.
Assess: This result makes sense, since the mass of the water is the same. The total weight of water exerting force on the bottom of the beaker is the same, so the pressure will be the same.

Q13.15. Reason: The fluid exerts an upward force—the buoyant force—on the ball. Thus, by Newton's third law, the ball exerts a downward force on the fluid. This downward force raises the scale reading, just as would happen if you pushed down on the beaker itself.
Assess: There's another way to look at the problem. Without the ball, the downward force on the beaker and water is just their weight; this force must be balanced by the upward force of the scale. When the ball is lowered in, if we take the system to be the beaker, water, *and* ball, the downward force is now greater by the weight of the ball. But there's also an *upward* force, the tension force of the string. But, just as in Example 13.5, this tension force is less than the weight of the ball, leading to an overall greater downward force. This increases the scale's reading.

Q13.17. Reason: For objects that are completely submerged the buoyant force is proportional to the volume of the object. Since the two blocks are the same size (volume), then the buoyant force is the same on both blocks.
Assess: A refresher of Example 13.5 would be very valuable here.

Q13.19. Reason: Archimedes' principle states that the buoyant force on an object is equal to the weight of the fluid displaced by the object. Each object displaces exactly the same amount of fluid since each is the same volume. So the buoyant force on all three objects is the same.
Assess: Note that the buoyant force does not depend on the mass or location of the object.

Q13.21. Reason: Adding salt to water increases its density. How high a person floats in the water depends on the ratio of the density of the person to the density of the water. The density of the person doesn't change. With denser water, however, the person would float higher in the water.
Assess: This makes sense. An egg consists essentially of water mixed with the substance of the embryo, which makes the density of the egg slightly larger than that of water.

Q13.23. Reason: Let us assume that freshwater fish and saltwater fish use their swim bladders in the same way and have the same requirements for changes in depth and predator evasion. That would mean that both types of fish would need to be able to produce comparable upward buoyant forces. Since saltwater is about 3% denser than freshwater, a given buoyant force would require about 3% smaller volume in saltwater (since the fluid being displaced by a given volume is heavier). Since saltwater fish can get away with a slightly smaller volume changes, it makes sense that they would have slightly smaller swim bladders than freshwater fish.
Assess: There are bound to be exceptions to this since many saltwater fish will have different ranges of depth, and different evasion requirements from freshwater fish.

Q13.25. Reason: When the blood is released gravity pulls it downward and the buoyant force acts upward on it. The gravitational force depends on the mass of material in the droplet, whereas the buoyant force depends on the volume of the blood that displaces the solution. This means denser blood will experience a higher gravitational force for a given volume (fixed buoyant force). Thus, denser blood with higher levels of hemoglobin should experience a larger net downward force and should accelerate faster. If the descent takes too long the density is too low, and there is not enough hemoglobin.
Assess: This analysis fits the test criteria given in the problem.

Q13.27. Reason: The density of air decreases with increasing temperature. The hot air inside the balloon has low density and takes up a fixed volume (the volume of the balloon envelope). We are told there are limits to how low the density can become. But the buoyant force depends on the weight of the volume of gas displaced by the balloon. If the balloon were in a denser medium, then the buoyant force would increase. Cool morning air has a greater density, so the buoyant force will be greatest when the balloon is in that cool air.
Assess: This is easy to understand if you consider trying to submerge a (smaller) balloon in a much denser medium, such as water. The buoyant force is much greater than when the balloon is in the air, illustrating the fact that the buoyant force increases with the density of the medium being displaced.

Q13.29. Reason: The volume flow rate must be the same at all points. So if the water is moving more slowly in the direction of the current, there must be a larger cross-sectional area of water moving, compared to a region with a small cross-sectional area and quick water speed. Since the river does not get wider or narrower, this larger cross-section can only come from depth. Thus, where the water is slow-moving, it is deep.
Assess: This is similar to the case of a garden hose. When there is no nozzle on the end and water simply flows out of the end, the water moves rather slowly. When a nozzle is attached that forces the water through a small cross-sectional area, it speeds up.

Q13.31. Reason: The pressure is reduced at the chimney due to the movement of the wind above. Thus, the air will flow in the window and out the chimney.
Assess: Prairie dogs ventilate their burrows this way; a small breeze above their mound lowers the pressure there and allows the air in the burrow to move between openings of different types or heights.

Q13.33. Reason: Since both blocks are more dense than water they will both submerge and not float (so D is not the answer). However, while the two blocks have the same mass, they do not have the same volume. The aluminum (with lower density) will have a larger volume. For submerged blocks, the one with the larger volume (the aluminum) will experience a greater buoyant force.
Since the string is massless and the pulley is massless and frictionless then the tension T in the string is the same everywhere. Now draw a free-body diagram for each block. For each block the downward weight force is the same and the upward force of tension in the string is the same. But because the buoyant force is not the same for the two blocks then there is a net upward force on the aluminum block and a net downward force on the copper block.
The correct answer is A.
Assess: To the extent that the buoyant force of air is not negligible the same effect would occur out of water.

Q13.35. Reason: Since the question talks about the "extra pressure" we will ignore the air pressure above the water; you have an equal amount of air pressure on the inside. The 7 N quoted is the "increased pressure." Therefore the equation we need is Equation 13.5 without the p_0 term,

$$p = \rho g d$$

where we want to solve for d. We will also use $p = F/A$ and round g to 10 m/s^2 for one significant figure accuracy.

$$d = \frac{p}{\rho g} = \frac{F/A}{\rho g} = \frac{7 \text{ N}/(7 \times 10^{-5} \text{ m}^2)}{(1000 \text{ kg/m}^3)(10 \text{ m/s}^2)} = 10 \text{ m}$$

The correct answer is D.
Assess: This one significant figure calculation can easily be done in the head without a calculator. The answer seems plausible, and the other choices seem too small. The units cancel appropriately to leave the answer in m.

Q13.37. Reason: Archimedes' principle says the magnitude of the buoyant force equals the weight of the fluid displaced by the object. The amount of water in the collecting tray is the amount of fluid displaced. If the block had sunk it would displace its volume of water, but since it floated it displaced only its weight of water. The answer is C.
Assess: If the block were dense enough to sink the answer would have been B.

Q13.39. Reason: The density of the water remains the same (water is not very compressible). However, the density of the manatee can change slightly, since the small volume of air in its lungs can be compressed. When the manatee is compressed, it does not displace as much water, such that the buoyant force is somewhat reduced. This means the buoyant force will be somewhat less than the weight force. The correct answer is A.
Assess: If the air volume was initially small, the compression is not likely to make a large difference in the volume of the manatee. It will still be easy for the manatee to swim.

Q13.41. Reason: Since the object is floating, the buoyant force equals the weight of the water displaced. The volume of water displaced is 75% of the volume of the object, so $\rho g V = \rho_{\text{water}} g((0.75)V)$. So $\rho = (0.75)\rho_{\text{water}} = 750 \text{ kg/m}^3$. The correct choice is B.
Assess: This makes sense. The density of the object must be less than the density of water in order for the object to float.

Q13.43. Reason: The question says to ignore viscosity, so we do not need Poiseuille's equation; Bernoulli's equation should suffice. And because the pipe is horizontal we can drop the $\rho g y$ terms (because they will be the same on both sides).

$$\Delta p = p_2 - p_1 = \frac{1}{2}\rho(v_1^2 - v_2^2)$$

We are given $r_1 = 0.040 \text{ m}, r_2 = 0.02 \text{ m}$, and $v_1 = 1.3 \text{ m/s}$.
We will also use the equation of continuity to solve for v_2.

$$v_2 = \frac{A_1}{A_2}v_1 = \frac{r_1^2}{r_2^2}v_1 = \frac{(0.040 \text{ m})^2}{(0.020 \text{ m})^2}(1.3 \text{ m/s}) = 5.2 \text{ m/s}$$

Putting it all together

$$\Delta p = p_2 - p_1 = \frac{1}{2}\rho(v_1^2 - v_2^2) = \frac{1}{2}(1000 \text{ kg/m}^3)[(1.3 \text{ m/s})^2 - (5.2 \text{ m/s})^2] = -12\,700 \text{ Pa}$$

The magnitude of this is 12 700 Pa, so the correct answer is D.
Assess: Since $\text{Pa} = \text{N/m}^2$ the units work out.

Problems

P13.1. Strategize: We will use the definition of density, and convert to base SI units.
Prepare: In SI Units $1 \text{ L} = 10^{-3} \text{ m}^3$ and $1 \text{ g} = 10^{-3} \text{ kg}$.
Solve: The density of the liquid is

$$\rho = \frac{m}{V} = \frac{0.120 \text{ kg}}{100 \text{ mL}} = \frac{0.120 \text{ kg}}{100 \times 10^{-3} \times 10^{-3} \text{ m}^3} = 1200 \text{ kg/m}^3$$

Assess: The liquid's density is a little more than that of water (1000 kg/m^3) and is a reasonable number.

P13.3. Strategize: Convert the inches to metric units and compute the volume of the bar.
Prepare: We use the fact that 1.0 in = 2.54 cm, for conversion. We then use the definition of density: $\rho = m/V$.

$$V = (7.00 \text{ in})(3.63 \text{ in})(1.75 \text{ in})\left(\frac{2.54 \text{ cm}}{1 \text{ in}}\right)^3\left(\frac{1 \text{ m}}{100 \text{ cm}}\right)^3 = 7.287 \times 10^{-4} \text{ m}^3$$

Solve: Using the equation for density,

$$\rho = \frac{m}{V} \Rightarrow m = \rho V = (19\,300 \text{ kg/m}^3)(7.287 \times 10^{-4} \text{ m}^3) = 14.1 \text{ kg}$$

Assess: For the size of the bar this is a large mass, but gold is very dense, so this makes sense.

P13.5. Strategize: We will use the definition of density, and the fact that the mass is unchanged, to determine the new density. We don't know the initial dimensions of the sphere, but we only need to know how the volume of a sphere depends on the radius.
Prepare: The volume of the sphere will be reduced by a factor of 8 when its radius is halved. We will use the definition of mass density $\rho = m/V$.
Solve: The new density is

$$\rho' = \frac{m}{V/8} = 8\frac{m}{V} = 8\rho = 8(1.4 \text{ kg/m}^3) = 11 \text{ kg/m}^3$$

Assess: If the mass is constant and the volume is reduced by a factor of 8, the density will increases by a factor of 8.

P13.7. Strategize: This is a hydrostatics problem; we can use the known expression for the pressure at a given depth, noting that the pressure at the surface is atmospheric pressure.
Prepare: The density of seawater is 1030 kg/m³. Also, $1 \text{ atm} = 1.013 \times 10^5 \text{ Pa}$.
Solve: The pressure below sea level can be found from Equation 13.5 as follows:

$$p = p_0 + \rho gd = 1.013 \times 10^5 \text{ Pa} + (1030 \text{ kg/m}^3)(9.80 \text{ m/s}^2)(1.1 \times 10^4 \text{ m})$$
$$= 1.013 \times 10^5 \text{ Pa} + 1.1103 \times 10^8 \text{ Pa} = 1.11 \times 10^8 \text{ Pa} = 1100 \text{ atm}$$

Assess: The pressure deep in the ocean is very large.

P13.9. Strategize: Assume both liquids are incompressible. We have an expression for the pressure at a given depth in a fluid of a given density. We will apply this twice: once for the pressure change through each liquid.
Prepare: The densities are given in Table 13.1: 1000 kg/m³ for water and 900 kg/m³ for oil. The pressure at the bottom of a liquid is due to the air pressure at the top plus the pressure due to the liquid. In this case the pressure at the top of the water is the pressure at the bottom of the oil, so we apply the concept twice. One further point: since we are asked for gauge pressure we will subtract the atmospheric pressure from the absolute pressure.
Solve: The pressure below the surface can be found from Equation 13.5 as follows:

$$p_{\text{guage}} = p - 1 \text{ atm} = (p_0 + \rho_w gd_w + \rho_{\text{oil}} gd_{\text{oil}}) - 1 \text{ atm} = \rho_w gd_w + \rho_{\text{oil}} gd_{\text{oil}}$$
$$= (1000 \text{ kg/m}^3)(9.80 \text{ m/s}^2)(0.25 \text{ m}) + (900 \text{ kg/m}^3)(9.80 \text{ m/s}^2)(0.15 \text{ m})$$
$$= 3.8 \text{ kPa}$$

Assess: If the water and oil had somehow managed to mix together the answer would still be the same.

P13.11. Strategize: This is a hydrostatics problem. We will use Equation 13.5 for pressure.
Prepare: The gauge pressure is $p - p_0 = 0.40 \text{ atm}$.
Solve: The gauge pressure at the bottom of the cylinder is $p - p_0 = \rho gd$. Because the cross-sectional area of the second cylinder is greater by a factor of four, the depth will be smaller by the same factor. As the above relationship indicates, the decrease in depth thus reduces the gauge pressure by the same factor. That is, the gauge pressure at the bottom of the second cylinder is $0.40 \text{ atm}/4 = 0.10 \text{ atm}$.

Assess: We could also look at this as $p = F/A$. The force is the weight of the liquid and it does not change. However since the radius doubles, the area will increase by a factor of four and the pressure will decrease by a factor of four.

P13.13. Strategize: We relate the surface area of the window to the force and pressure. Then we determine at what depth that pressure is reached.

Prepare: The density of seawater $\rho_{seawater} = 1030 \text{ kg/m}^2$.

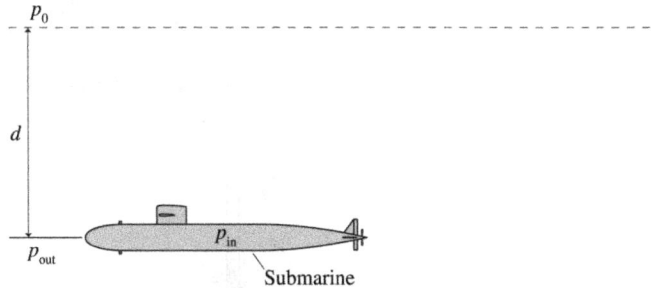

Submarine

Solve: The pressure outside the submarine's window is $p_{out} = p_0 + \rho_{seawater}gd$, where d is the maximum safe depth for the window to withstand a force F. This force is $F/A = p_{out} - p_{in}$, where A is the area of the window. With $p_{in} = p_0$, we simplify the pressure equation to

$$p_{out} - p_0 = \frac{F}{A} = \rho_{seawater}gd \Rightarrow d = \frac{F}{A\rho_{seawater}g} \qquad d = \frac{1.0 \times 10^6 \text{ N}}{\pi(0.10 \text{ m})^2(1030 \text{ kg/m}^2)(9.8 \text{ m/s}^2)} = 3153 \text{ m} = 3.2 \text{ km}$$

Assess: A force of 1.0×10^6 N corresponds to a pressure of

$$p = \frac{F}{A} = \frac{1.0 \times 10^6 \text{ N}}{\pi(0.10 \text{ m})^2} = 310 \text{ atm}$$

A depth of 3.2 km is therefore reasonable.

P13.15. Strategize: This is a hydrostatics problem. We will use our expression for the pressure change through a depth of fluid.

Prepare: The density of mercury $\rho = 13\,600 \text{ kg/m}^3$.

Solve: Equation 13.6 tells us how the pressure at some depth in a fluid compares to the pressure at the top: $p = p_0 + \rho gh$. The pressure at the mercury surface on the right is 1.0 atm, because it is open to the air. At that same height on the left, the pressure of the mercury must be the same (and this is 10 cm below the top of the mercury on the left). Rearranging Equation 13.6, we have

$$p_{gas} + \rho gh = p_{atmos} \Rightarrow p_{gas} + (13\,600 \text{ kg/m}^3)(9.8 \text{ m/s}^2)(0.10 \text{ m}) = 1.013 \times 10^5 \text{ Pa} \Rightarrow p_{gas} = 88\,000 \text{ Pa}$$

Assess: Using 1 atm $= 1.013 \times 10^5$ Pa, the gas pressure is 0.87 atm. This is expected because the mercury level in the left tube is higher than that in the right tube.

P13.17. Strategize: We will use the fact that, in hydrostatics, the pressure change in a fluid of constant density depends only on the depth in the fluid.

Prepare: Oil is incompressible and has a density of 900 kg/m^3.

Solve: (a) The pressure at point A, which is 0.50 m below the open oil surface, is

$$p_A = p_0 + \rho_{oil}g(1.00 \text{ m} - 0.50 \text{ m}) = 101\,300 \text{ Pa} + (900 \text{ kg/m}^3)(9.8 \text{ m/s}^2)(0.50 \text{ m}) = 1.1 \times 10^5 \text{ Pa}$$

(b) The pressure difference between A and B is

$$p_B - p_A = (p_0 + \rho gd_B) - (p_0 + \rho gd_A) = \rho g(d_B - d_A) = (900 \text{ kg/m}^3)(9.8 \text{ m/s}^2)(0.50 \text{ m}) = 4400 \text{ Pa}$$

Pressure depends only on depth, and C is the same depth as B. Thus $p_C - p_A = 4400$ Pa also, even though C isn't directly under A.

Assess: This problem illustrates clearly that the pressure depends only on the depth of the fluid.

P13.19. Strategize: Water and mercury are incompressible and immiscible liquids. The water in the left arm floats on top of the mercury and presses the mercury down from its initial level. If we start at the lowest point in the tube, and look at a point on the left and on the right some height above that lowest point, those two points will have the same pressure as long as we are still in glycerin. Once the substance changes (and the density changes) that statement is no longer true.

Prepare: The diagram below will help us visualize the height difference, and the points at equal pressure.

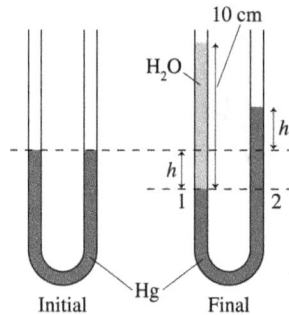

Initial   Final

Solve: The pressure at point 1 is due to water of depth $d_w = 10$ cm.

$$p_1 = p_{atmos} + \rho_w g d_w$$

Because mercury is incompressible, the mercury in the left arm goes down a distance h while the mercury in the right arm goes up a distance h. Thus, the pressure at point 2 is due to mercury of depth $d_{Hg} = 2h$.

$$p_2 = p_{atmos} + \rho_{Hg} g d_{Hg} = p_{atmos} + 2\rho_{Hg} g h$$

Equating p_1 and p_2 gives

$$p_{atmos} + \rho_w g d_w = p_{atmos} + 2\rho_{Hg} g h \Rightarrow h = \frac{1}{2}\frac{\rho_w}{\rho_{Hg}}d_w = \frac{1}{2}\frac{1000 \text{ kg/m}^3}{13\,600 \text{ kg/m}^3}10 \text{ cm} = 3.68 \text{ mm}$$

Assess: The mercury in the right arm rises 3.68 mm above its initial level. This is a reasonable number due to the rather large density of mercury compared to water.

P13.21. Strategize: We know that the barge displaces an amount of seawater that weighs the same as the barge. This will allow us to compute the weight of the barge. Then, as the barge moves into the fresh water we know that it will also displace an amount of water that weighs the same as the barge.

Prepare: Note that the weight of the displaced seawater, the weight of the displaced fresh water, and the weight of the barge are all the same; their mAssess:are also equal, call that value m. Because fresh water is less dense than seawater (see Table 13.1), the barge will displace a greater volume of fresh water, and the barge will ride lower in the water.

Solve: The volume of seawater displaced is

$$V_{sea} = 3.0 \text{ m} \times 20.0 \text{ m} \times 0.80 \text{ m} = 48 \text{ m}^3$$

The mass of that volume of seawater (and therefore also the mass of the barge) is

$$m = \rho_{sea}V_{sea} = (1030 \text{ kg/m}^3)(48 \text{ m}^3) = 49\,440 \text{ kg}$$

Now, the volume of fresh water that the barge displaces is

$$V_{fresh} = \frac{m}{\rho_{fresh}} = \frac{49\,440\,kg}{1000\,kg/m^3} = 49.44\,m^3$$

Lastly, since the area of the barge has not changed, we solve for the new depth.

$$d = \frac{V_{fresh}}{A} = \frac{49.44\,m^3}{3.0\,m \times 20.0\,m} = 0.824\,m \approx 0.82\,m$$

In the fresh water the barge rides 2 cm lower than in the seawater.

Assess: The answer is precisely what we expected: The barge rides a bit (2 cm) lower because the fresh water is less dense than the seawater.

In fact, a shortcut would be to see that seawater is 3% more dense than fresh water, so the ship will ride 3% deeper in the fresh water. This gives exactly the same answer: $0.80\,m \times 103\% = 0.0824\,m$.

P13.23. Strategize: The buoyant force on the wood block is given by Archimedes' principle.

Prepare: The density of water is $1000\,kg/m^3$ and the density of seawater is $1030\,kg/m^3$. A floating object is in static equilibrium.

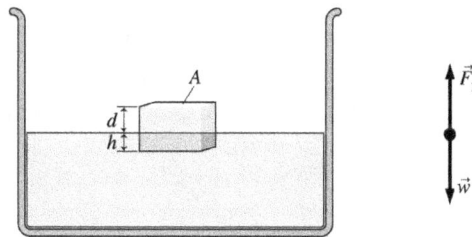

Solve: The volume of displaced fluid is $V_f = Ah$. In static equilibrium,

$$F_B = w \Rightarrow \rho_f V_f g = \rho_{wood} V_{wood} g \Rightarrow \rho_f (Ah) = \rho_{wood} V_{wood}$$

$$\Rightarrow h = \left(\frac{\rho_{wood}}{\rho_f}\right)\left(\frac{V_{wood}}{A}\right) = \left(\frac{\rho_{wood}}{\rho_f}\right)\frac{(0.10 \times 0.10 \times 0.10\,m^3)}{(0.10 \times 0.10)\,m^2} = \left(\frac{\rho_{wood}}{\rho_f}\right)(0.10\,m)$$

For fresh water, $h = (0.7)(0.10\,m) = 0.070\,m$ and for seawater $h = 0.068\,m$. The corresponding values for d are (a) 3.0 cm and (b) 3.2 cm.

Assess: Objects float better in seawater, so the above result is reasonable.

P13.25. Strategize: The buoyant force on the aluminum block is given by Archimedes' principle. We will use Newton's second law to relate the buoyant force to gravity and tension.

Prepare: The density of aluminum and ethyl alcohol are $\rho_{Al} = 2700\,kg/m^3$ and $\rho_{ethyl\,alcohal} = 790\,kg/m^3$. The buoyant force F_B and the tension due to the string act vertically up, and the weight of the aluminum block acts vertically down. The block is submerged, so the volume of displaced fluid equals V_{Al}, the volume of the block.

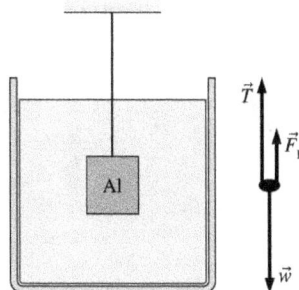

Solve: The aluminum block is in static equilibrium, so

$$\Sigma F_y = F_B + T - w = 0 \text{ N} \Rightarrow \rho_f V_{Al} g + T - \rho_{Al} V_{Al} g = 0 \text{ N} \Rightarrow T = V_{Al} g(\rho_{Al} - \rho_f)$$
$$T = (100 \times 10^{-6} \text{ m}^3)(9.80 \text{ m/s}^2)(2700 \text{ kg/m}^3 - 790 \text{ kg/m}^3) = 1.9 \text{ N}$$

Assess: The weight of the aluminum block is $\rho_{Al} V_{Al} g = 2.7$ N. A similar order of magnitude for T is reasonable.

P13.27. Strategize: Follow Example 13.5. The buoyant forces are calculated using Archimedes' principle.
Prepare: We are told $w_o = 690$ N, where the object in question is the athlete. When the athlete is submerged the scales she stands on read 42 N; this is the upward normal force.

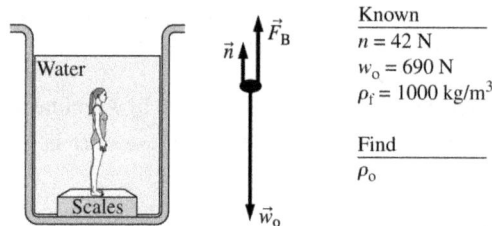

Known
$n = 42$ N
$w_o = 690$ N
$\rho_f = 1000$ kg/m^3

Find
ρ_o

Solve: The submerged athlete is in static equilibrium, so

$$\Sigma F_y = F_B + n - w_o = 0 \text{ N} \Rightarrow F_B = w_o - n = 690 \text{ N} - 42 \text{ N} = 648 \text{ N}$$

The person's weight is $w_o = m_o g = \rho_o V_o g$, so the person's volume is

$$V_o = \frac{w_o}{\rho_o g}$$

Inserting this into Archimedes' principle gives

$$F_B = \rho_f V_f g = \rho_f V_o g = \rho_f \left(\frac{w_o}{\rho_o g}\right) g = \frac{\rho_f}{\rho_o} w_o$$

Solving for the density of the object (person) gives

$$\rho_o = \frac{\rho_f w_o}{F_B} = \frac{(1000 \text{ kg/m}^3)(690 \text{ N})}{648 \text{ N}} = 1065 \text{ kg/m}^3 \approx 1100 \text{ kg/m}^3$$

Assess: The density is greater than regular water, but less than the Dead Sea, so this person would float in the Dead Sea. Once we know the density we can accurately look up the fat percentage in a table. See Problem 13.52.

P13.29. Strategize: We will calculate the buoyant force when the Styrofoam is just barely fully submerged, and related it to the weight of the hanging mass.
Prepare: The buoyant force on the sphere is given by Archimedes' principle.

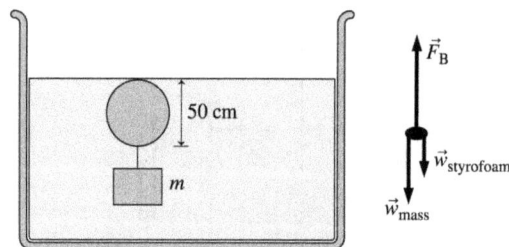

Solve: For the Styrofoam sphere and the mass not to sink, the sphere must be completely submerged and the buoyant force F_B must be equal to the sum of the weight of the Styrofoam sphere and the attached mass. The volume of displaced water equals the volume of the sphere, so

$$F_B = \rho_{water} V_{water} g = (1000 \text{ kg/m}^3) \frac{4\pi}{3} (0.25 \text{ m})^3 (9.80 \text{ m/s}^2) = 641.4 \text{ N}$$

$$w_{Styrofoam} = \rho_{Styrofoam} V_{Styrofoam} g = (300 \text{ kg/m}^3) \left[\frac{4}{3} \pi (0.25 \text{ m})^3 \right] (9.80 \text{ m/s}^2) = 20.5 \text{ N}$$

Because

$$w_{Styrofoam} + mg = F_B,$$

$$m = \frac{F_B - w_{Styrofoam}}{g} = \frac{641.4 \text{ N} - 20.5 \text{ N}}{9.80 \text{ m/s}^2} = 63 \text{ kg}$$

Assess: This large mass allows one to appreciate the importance of the buoyant force.

P13.31. Strategize: We determine the volume of the man and the resulting buoyant force in air.
Prepare: We can calculate the volume of the man using Equation 13.1. For the buoyant force, Equation 13.7 applies.
Solve: The volume of the man is $V = m/\rho_{man} = w/g\rho_{man}$. The buoyant force is

$$F_B = \rho_{air} V g = \rho_{air} \frac{w}{g\rho_{man}} g = \frac{\rho_{air}}{\rho_{man}} w = \frac{1.20 \text{ kg/m}^3}{1000 \text{ kg/m}^3} (800 \text{ N}) = 0.96 \text{ N}$$

Assess: This result makes sense. The weight of the air displaced is very small.

P13.33. Strategize: We can relate the volume flow rate to the speed and cross-sectional area using the continuity equation.
Prepare: To work with SI units, we need the conversion $1 \text{ L} = 10^{-3} \text{ m}^3$.
Solve: The volume flow rate is

$$Q = \frac{300 \text{ L}}{5.0 \text{ min}} = \frac{300 \times 10^{-3} \text{ m}^3}{5.0 \times 60 \text{ s}} = 1.0 \times 10^{-3} \text{ m}^3/\text{s}$$

Using the definition $Q = vA$, we get

$$v = \frac{Q}{A} = \frac{1.0 \times 10^{-3} \text{ m}^3/\text{s}}{\pi (0.01 \text{ m})^2} = 3.2 \text{ m/s}$$

Assess: This is a reasonable speed for water flowing through a 2.0-cm pipe.

P13.35. Strategize: We can relate the volume flow rate to the speed and cross-sectional area in each segment using the continuity equation.
Prepare: Note that A_1, A_2, and A_3 and v_1, v_2, and v_3 are the cross-sectional areas and the speeds in the first, second, and third segments of the pipe.

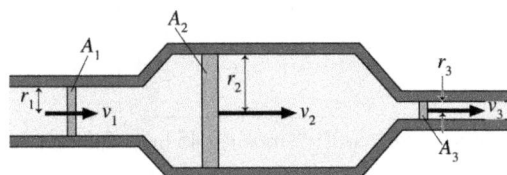

Solve: (a) The equation of continuity is

$$A_1v_1 = A_2v_2 = A_3v_3 \Rightarrow \pi r_1^2 v_1 = \pi r_2^2 v_2 = \pi r_3^2 v_3 \Rightarrow r_1^2 v_1 = r_2^2 v_2 = r_3^2 v_3$$

$$\Rightarrow (0.005 \text{ m})^2(4.0 \text{ m/s}) = (0.01 \text{ m})^2 v_2 = (0.0025 \text{ m})^2 v_3$$

$$\Rightarrow v_2 = \left(\frac{0.005 \text{ m}}{0.01 \text{ m}}\right)^2 (4.0 \text{ m/s}) = 1.0 \text{ m/s} \qquad v_3 = \left(\frac{0.005 \text{ m}}{0.0025 \text{ m}}\right)^2 (4.0 \text{ m/s}) = 16 \text{ m/s}$$

(b) The volume flow rate through the pipe is

$$Q = A_1v_1 = \pi(0.005 \text{ m})^2(4.0 \text{ m/s}) = 3.1 \times 10^{-4} \text{ m}^3\text{/s}$$

Assess: Since most of us do not have a good feel for flow rate in m³/s, let's look at this value in L/s. A flow rate 3.1×10^{-4} m³/s is equal to 0.31 L/s. This is a small but reasonable flow rate for a 0.5-cm diameter pipe.

P13.37. Strategize: We have a moving fluid that experiences changes in speed, elevation and pressure. If we assume that energy density is conserved in the flow (negligible viscosity), we can apply Bernoulli's equation.
Prepare: Treat the oil as an ideal fluid (obeying Bernoulli's equation). Consider the path connecting point 1 in the lower pipe with point 2 in the upper pipe a streamline.
Solve: Bernoulli's equation is

$$p_2 + \frac{1}{2}\rho v_2^2 + \rho g y_2 = p_1 + \frac{1}{2}\rho v_1^2 + \rho g y_1 \Rightarrow p_2 = p_1 + \frac{1}{2}\rho(v_1^2 - v_2^2) + \rho g(y_1 - y_2)$$

Using $p_1 = 200 \text{ kPa} = 200 \times 10^5 \text{ Pa}$, $\rho = 900 \text{ kg/m}^3$, $y_2 - y_1 = 10.0 \text{ m}$ $v_1 = 20.0 \text{ m/s}$, and $v_2 = 3.0 \text{ m/s}$, we get $p_2 = 1.1 \times 10^5$ Pa $= 110$ kPa.

Assess: We expect the pressure at point 2 to be less than the pressure at point 1. If this were not the case, the fluid would not flow from point 1 to point 2.
I also note that the blind solver has used $p = p_0 + \rho g h$.

P13.39. Strategize: Treat the water as an ideal fluid obeying Bernoulli's equation.
Prepare: Bernoulli's equation is given by

$$\rho g(y_2 - y_1) = (p_1 - p_2) + \frac{1}{2}\rho(v_1^2 - v_2^2).$$

Consider a streamline connecting a point at the surface with a point in the hole. The pressure at the two points is the same, so $p_1 - p_2 = 0$. Further assumptions are that the area of the trough is so large that (1) it doesn't matter, and (2) the speed of the water at the top is zero ($v_1 = 0$).
Call $y_2 - y_1 = h = 0.45$ m.
Solve: Since the pressures are equal we have

$$\rho g h = \frac{1}{2}\rho(v_1^2 - v_2^2)$$

Now set $v_1 = 0$ and cancel ρ.

$$gh = \frac{1}{2}v_2^2$$

Solve for v_2.

$$v_2 = \sqrt{2gh} = \sqrt{2(9.8 \text{ m/s}^2)(0.45 \text{ m})} = 3.0 \text{ m/s}$$

Assess: The result is independent of the area of the trough, as long as it is big enough that we can assume $v_1 = 0$. The result $v_2 = \sqrt{2gh}$ is known as Torricelli's theorem.

P13.41. Strategize: This is a hydrodynamics problem in which viscosity plays a major role. We must use an equation that takes viscosity into account and relates pressure differences to flow rates.
Prepare: We can use Equation 13.14.
Solve: The pressure difference required is given by Equation 13.14.

$$\Delta p = 8\pi\eta\frac{Lv_{avg}}{A} = 8\pi\eta\frac{Lv_{avg}}{\pi R^2} = 8\eta\frac{Lv_{avg}}{R^2} = 8(0.7\times10^{-3}\,\text{Pa}\cdot\text{s})\frac{(2.0\,\text{m})(4.0\,\text{m/s})}{(5.0\times10^{-4}\,\text{m})^2} = 1.8\times10^5\,\text{Pa}$$

Assess: This seems like a reasonable pressure for such a narrow tube.

P13.43. Strategize: This is a hydrodynamics problem in which viscosity will play a role. Thus, we use Poiseuille's equation.
Prepare: One end of the hose is open to the air, so the gauge pressure requested is the same as Δp.

Known
$Q = 0.25\,\text{L/s} = 2.5\times10^{-4}\,\text{m}^3/\text{s}$
$R = 1.25\,\text{cm} = 1.25\times10^{-2}\,\text{m}$
$L = 10\,\text{m}$
$\eta = 1.0\times10^{-3}\,\text{Pa}\cdot\text{s}$ at 20°C
Find
Δp

Solve: Solve Poiseuille's equation for Δp.

$$\Delta p = \frac{8\eta L Q}{\pi R^4} = \frac{8(1.0\times10^{-3}\,\text{Pa}\cdot\text{s})(10\,\text{m})(2.5\times10^{-4}\,\text{m}^3/\text{s})}{\pi(1.25\times10^{-2}\,\text{m})^4} = 260\,\text{Pa}$$

Assess: This is just a couple of mm of Hg and sounds reasonable. Isn't it nice how the units cancel to leave Pa?

P13.45. Strategize: This is a hydrodynamics problem in which viscosity will play a role. We use Poiseuille's equation.
Prepare: In applying Equation 13.14 to the water in the syringe, we will need the appropriate viscosity from Table 13.4: $\eta = 1.0\times10^{-3}\,\text{Pa}\cdot\text{s}$.
Solve: Using Equation 13.14,

$$\Delta p = p - p_{atm} = \frac{8\eta L v_{avg}}{R^2} = \frac{8(1.0\times10^{-3}\,\text{Pa}\cdot\text{s})(0.040\,\text{m})(10\,\text{m/s})}{(1\times10^{-3}\,\text{m})^2} = 3200\,\text{Pa}$$

Assess: This result seems reasonable.

P13.47. Strategize: We assume the flow of blood can be ignored, such that this becomes a hydrostatic problem.
Prepare: We apply Equation 13.5: $p = p_0 + \rho g d$. Here $\rho_{blood} = 1{,}060\,\text{kg/m}^3$, from Table 13.1.
Solve: The pressure difference is $p - p_0 = \rho_{blood} g d = (1{,}060\,\text{kg/m}^3)(9.8\,\text{m/s}^2)(0.30\,\text{m}) = 3.1\,\text{kPa}$. One can also express this as 23 mm Hg.
Assess: This is just a few percent of atmospheric pressure. It seems reasonable that this pressure difference can exist in our bodies without causing problems.

P13.49. Strategize: For the first part, we simply use the flow rate and area to find the speed. In the second part we will use Poiseuille's equation to express the pressure difference in terms of flow speed and other given quantities.

Prepare: For the first part, we will use Equation 13.12: $Q = vA$, and the fact that the total cross-sectional area of the venules is $3{,}000 \text{ cm}^2 = 0.30 \text{ m}^2$. In the second part we will use Equation 13.15 and the result from (a). We also use (from Figure 13.37) that the diameter of the average venule is $0.003 \text{ cm} = 3.0 \times 10^{-5} \text{ m}$. The viscosity of blood is $\eta = 3.5 \times 10^{-3} \text{ Pa} \cdot \text{s}$.

Solve: (a) Rearranging, and inserting the given quantities, we have

$$v = Q/A = \frac{\left(5.0 \times 10^{-3} \text{ m}^3\right)}{(60 \text{ s})\left(0.30 \text{ m}^2\right)} = 2.78 \times 10^{-4} \text{ m/s}$$

We note this value out to three significant digits for use in part (b), but our final answer for (a) to two significant digits is 2.8×10^{-4} m/s.

(b) Rearranging Equation 13.15, we find

$$\Delta p = \frac{8 \eta L v_{\text{avg}}}{R^2} = \frac{8\left(3.5 \times 10^{-3} \text{ Pa} \cdot \text{s}\right)(0.01 \text{ m})\left(2.78 \times 10^{-4} \text{ m/s}\right)}{\left(1.5 \times 10^{-5} \text{ m}\right)^2} = 3.5 \times 10^2 \text{ Pa.}$$

Assess: Since the venules are one of the smallest vessels through which blood flows, it is reasonable that the speed would be very slow, and a fraction of a mm/s is reasonable. Similarly, we would expect significant pressure changes along such very narrow vessels, but in order for this to be biologically tractable, we expect the pressure differences to still be much less than atmospheric changes. Thus, our answer is reasonable.

P13.51. Strategize: Although viscosity in blood always plays a role, we can assume that over the short distance of the change described, and in a relatively large artery, we may ignore viscosity. In that case, this becomes a hydrodynamics question in which we can use energy conservation, which means Bernoulli's equation can be used.

Prepare: We ignore any small differences in elevation, and focus on the pressure change due to the change in speed of blood flow. In particular, we use the continuity equation to relate the blood flow before the blockage (1) and after the blockage has reduced the diameter (2): $v_1 A_1 = v_2 A_2 \Rightarrow v_2 = \left(\dfrac{r_1}{r_2}\right)^2 v_1 = \left(\dfrac{1}{0.80}\right)^2 v_1$. We insert this into Bernoulli's equation, which (with negligible height change) reads $p_1 + \dfrac{1}{2}\rho v_1^2 = p_2 + \dfrac{1}{2}\rho v_2^2$.

Solve: Combining equations, we have

$$p_1 + \frac{1}{2}\rho v_1^2 = p_2 + \frac{1}{2}\rho \left(\left(\frac{1}{0.80}\right)^2 v_1\right)^2 \Rightarrow p_1 - p_2 = \frac{1}{2}\rho v_1^2 \left(\frac{1}{(0.80)^4} - 1\right).$$

$$\Delta p = \frac{1}{2}\left(1{,}060 \text{ kg/m}^3\right)(0.15 \text{ m/s})^2 \left(\frac{1}{(0.80)^4} - 1\right) = 17 \text{ Pa}$$

Assess: Although blockages are always a serious concern, one might expect the drop due narrowing of 20% to be relatively small compared to other pressure changes in the circulatory system. Our answer matches this expectation.

P13.53. Strategize: We will use the definition of density and the given volumetric information to determine the mass. We will then use the molar mass of Al and Avogadro's number to determine the number of atoms.

Prepare: $N = (M/M_A)N_A$, where N_A is Avogadro's number. Because the atomic mass number of Al is 27, one mole of Al has a mass of $M_A = 27$ g.

Solve: The volume of the aluminum cube $V = 8.0 \times 10^{-6} \text{ m}^3$ and its mass is

$$M = \rho V = (2700 \text{ kg/m}^3)(8.0 \times 10^{-6} \text{ m}^3) = 0.0216 \text{ kg} = 21.6 \text{ g}$$

One mole of aluminum (^{27}Al) has a mass of 27 g. The number of atoms is

$$N = \left(\frac{6.02 \times 10^{23} \text{ atoms}}{1 \text{ mol}} \right) \left(\frac{1 \text{ mol}}{27 \text{ g}} \right) (21.6 \text{ g}) = 4.8 \times 10^{23} \text{ atoms}$$

Assess: A number slightly smaller than Avogadro's number is expected since we have slightly less than a mole of aluminum.

P13.55. Strategize: We will use hydrostatics to express the pressure as a function of depth into a fluid, and we will use the definition of pressure to relate it to force.

Prepare: Assume that the oil is incompressible and its density is 900 kg/m^3.

Solve: (a) The pressure at depth d in a fluid is $p = p_0 + \rho g d$. Here, pressure p_0 at the top of the fluid is due both to the atmosphere *and* to the weight of the floating piston. That is, $p_0 = p_{\text{atm}} + w_p/A$. At point A,

$$p_A = p_{\text{atm}} + \frac{w_p}{A} + \rho g (1.00 \text{ m} - 0.30 \text{ m})$$

$$= 1.013 \times 10^5 \text{ Pa} + \frac{(10 \text{ kg})(9.8 \text{ m/s}^2)}{\pi (0.02 \text{ m})^2} + (900 \text{ kg/m}^3)(9.8 \text{ m/s}^2)(0.70 \text{ m}) = 185\,460 \text{ Pa}$$

$$\Rightarrow F_A = p_A A = (185\,460 \text{ Pa})\pi(0.10 \text{ m})^2 = 5800 \text{ N}$$

(b) In the same way,

$$p_B = p_{\text{atm}} + \frac{w_p}{A} + \rho g (1.30 \text{ m}) = 190{,}752 \text{ Pa} \Rightarrow F_B = 6000 \text{ N}$$

Assess: F_B is larger than F_A, because p_B is larger than p_A.

P13.57. Strategize: We will use Newton's second law to relate the forces acting on the sphere. Archimedes' principle will help us relate the buoyant force to the density of the sphere.

Prepare: The sphere is in static equilibrium.

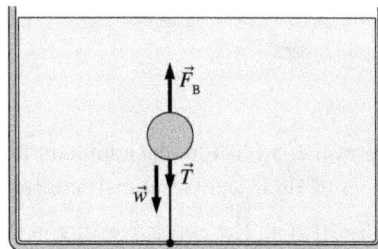

Solve: The free-body diagram on the sphere shows that

$$\Sigma F_y = F_B - T - w = 0 \text{ N} \Rightarrow F_B = T + w = \frac{1}{3}w + w = \frac{4}{3}w$$

$$\Rightarrow \rho_w V_{\text{sphere}} g = \frac{4}{3}\rho_{\text{sphere}} V_{\text{sphere}} g \Rightarrow \rho_{\text{sphere}} = \frac{3}{4}\rho_w = \frac{3}{4}(1000 \text{ kg/m}^3) = 750 \text{ kg/m}^3$$

Assess: We expected the sphere's density to be smaller than the water's because the sphere is tethered to the bottom.

P13.59. Strategize: We will use Newton's second law to relate the tension, weight, and buoyant force. We will then relate the buoyant force to the density of the stone and the fluid using Archimedes principle and the definition of density.

Prepare: A pictorial representation of the situation and the forces on the rock are shown.

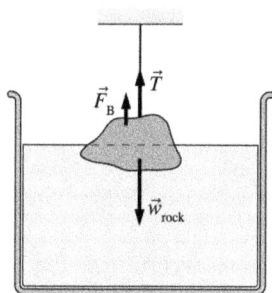

Solve: Because the rock is in static equilibrium, Newton's first law is

$$F_{net} = T + F_B - w_{rock} = 0 \text{ N} \Rightarrow T = \rho_{rock} V_{rock} g - \rho_{water}\left(\frac{1}{2}V_{rock}\right)g$$

$$= \left(\rho_{rock} - \frac{1}{2}\rho_{water}\right)V_{rock}g = \left(\rho_{rock} - \frac{1}{2}\rho_{water}\right)\left(\frac{m_{rock}g}{\rho_{rock}}\right) = \left(1 - \frac{\rho_{water}}{2\rho_{rock}}\right)m_{rock}g$$

Using $\rho_{rock} = 4800 \text{ kg/m}^3$ and $m_{rock} = 5.0 \text{ kg}$, we get $T = 44 \text{ N}$.

Assess: A buoyant force of $(5.0\times9.8 \text{ N} - 44 \text{ N}) \approx 5 \text{ N}$ is reasonable for this problem.

P13.61. Strategize: For the first part, we will assume the fluid is moving very slowly, such that we simply need to equate the pressure applied to the minimum blood pressure inside the patient. The second part is a simple application of the definition of a volume flow rate.

Prepare: The pictorial representation that follows gives the relevant diameters of the syringe.

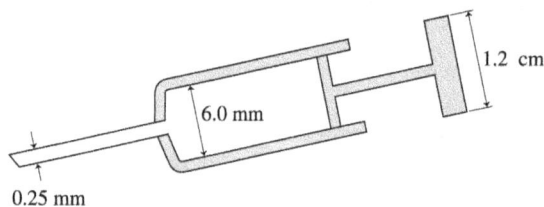

Solve: (a) Because the patient's blood pressure is 140/100, the minimum fluid pressure needs to be 100 mm of Hg above atmospheric pressure. Since 760 mm of Hg is equivalent to 1 atm and 1 atm is equivalent to 1.013×10^5 Pa, the minimum pressure is 100 mm $= 1.333\times10^4$ Pa. The excess pressure in the fluid is due to force F pushing on the internal 6.0-mm-diameter piston that presses against the liquid. Thus, the minimum force the nurse needs to apply to the syringe is

$$F = \text{fluid pressure} \times \text{area of plunger} = (1.333\times10^4 \text{ Pa})[\pi(0.003 \text{ m})^2] = 0.38 \text{ N}$$

(b) The flow rate is $Q = vA$, where v is the flow speed of the medicine and A is the cross-sectional area of the needle. Thus,

$$v = \frac{Q}{A} = \frac{2.0\times10^{-6} \text{ m}^3/2.0 \text{ s}}{\pi(0.125\times10^{-3} \text{ m})^2} = 20 \text{ m/s}$$

Assess: Note that the pressure in the fluid is due to F that is not dependent on the size of the plunger pad. Also note that the syringe is not drawn to scale.

P13.63. Strategize: This is a simple application of the expression for volume flow rate. We take care to take all vessels into account in the cross-sectional area.

Prepare: Knowing the volume flow rate, the number of vessels and the diameter of each vessel, we can use Equation 13.12 to solve the problem. We can either say that each vessel handles 1/2000 of the volume flow rate or we can say the total area for this flow rate is 2000 times the area of each vessel. Either way we will get the same result.

Solve: Starting with the definition of volume flow rate and the knowledge that each vessel handles 1/2000 of this volume, we may write

$$\frac{V/2000}{t} = Av$$

Solving for v

$$v = \frac{(V/t)/2000}{\pi r^2} = \frac{3\times 10^{-8}\ \text{m}^3/\text{s}}{(2000)\pi(5\times 10^{-5}\ \text{m})^2} = 2\times 10^{-3}\ \text{m/s}$$

Assess: At this rate, water could travel to the top of an 18 m tree in a day. That seems reasonable.

P13.65. Strategize: We will treat the water as an ideal fluid obeying Bernoulli's equation.

Prepare: A streamline begins in the bigger size pipe and ends at the exit of the narrower pipe. Let point 1 be beneath the standing column and point 2 be where the water exits the pipe.

Solve: (a) The pressure of the water as it exits into the air is $p_2 = p_{atmos}$.

(b) Bernoulli's equation, Equation 13.13, relates the pressure, water speed, and heights at points 1 and 2.

$$p_1 + \frac{1}{2}\rho v_1^2 + \rho g y_1 = p_2 + \frac{1}{2}\rho v_2^2 + \rho g y_2 \Rightarrow p_1 - p_2 = \frac{1}{2}\rho(v_2^2 - v_1^2) + \rho g(y_2 - y_1)$$

From the continuity equation,

$$v_1 A_1 = v_2 A_2 = (4\ \text{m/s})(5\times 10^{-4}\ \text{m}^2) = v_1(10\times 10^{-4}\ \text{m}^2) = 20\times 10^{-4}\ \text{m}^3/\text{s} \Rightarrow v_1 = 2\ \text{m/s}$$

Substituting into Bernoulli's equation,

$$p_1 - p_2 = p_1 - p_{atmos} = \frac{1}{2}(1000\ \text{kg/m}^3)[(4\ \text{m/s})^2 - (2\ \text{m/s})^2] + (1000\ \text{kg/m}^3)(9.80\ \text{m/s})(4.0\ \text{m})$$
$$= 6000\ \text{Pa} + 39\ 200\ \text{Pa} = 45,200\ \text{Pa}$$

But $p_1 - p_2 = \rho g h$, where h is the height of the standing water column. Thus

$$h = \frac{45\ 200\ \text{Pa}}{(1000\ \text{kg/m}^3)(9.80\ \text{m/s}^2)} = 4.6\ \text{m}$$

Assess: In order to sustain fluid flow, the pressure at point 1 must be greater than the pressure at point 2. As a result we should expect the height h to be greater than 4.0 m.

P13.67. Strategize: The ideal fluid obeys Bernoulli's equation. We will apply it to a point in the wider region of the pipe and to one point in the narrower region of the pipe.

Prepare: There is a streamline connecting point 1 in the wider pipe on the left with point 2 in the narrower pipe on the right. The air speeds at points 1 and 2 are v_1 and v_2 and the cross-sectional area of the pipes at these points are A_1 and A_2. Points 1 and 2 are at the same height, so $y_1 = y_2$.

Solve: The volume flow rate is $Q = A_1 v_1 = A_2 v_2 = 1200\times 10^{-6}\ \text{m}^3/\text{s}$. Thus

$$v_2 = \frac{1200\times 10^{-6}\ \text{m}^3/\text{s}}{\pi(0.0020\ \text{m})^2} = 95.49\ \text{m/s} \qquad v_1 = \frac{1200\times 10^{-6}\ \text{m}^3/\text{s}}{\pi(0.010\ \text{m})^2} = 3.82\ \text{m/s}$$

Now we can use Bernoulli's equation to connect points 1 and 2.

$$p_1 + \frac{1}{2}\rho v_1^2 + \rho g y_1 = p_2 + \frac{1}{2}\rho v_2^2 + \rho g y_2 \Rightarrow p_1 - p_2 = \frac{1}{2}\rho(v_2^2 - v_1^2) + \rho g(y_2 - y_1)$$

$$= \frac{1}{2}(1.20 \text{ kg/m}^3)[95.49 \text{ m/s})^2 - (3.82 \text{ m/s})^2] + 0 \text{ Pa} = 5460 \text{ Pa}$$

Because the pressure above the mercury surface in the right tube is p_2 and in the left tube is p_1, the difference in the pressures p_1 and p_2 is $\rho_{\text{Hg}}gh$. That is,

$$p_1 - p_2 = 5460 \text{ Pa} = \rho_{\text{Hg}}gh \Rightarrow h = \frac{5460 \text{ Pa}}{(13\,600 \text{ kg/m}^3)(9.80 \text{ m/s}^2)} = 4.1 \text{ cm}$$

Assess: Note that the pressure difference (5460 Pa) is small compared to atmospheric pressure (1.013×10^5 Pa). As a result we expect the height of the mercury column to be small compared to 760 mm for atmospheric pressure.

P13.69. Strategize: We are not told to treat the fluid as ideal, and indeed, viscosity becomes very important as tubes become very narrow. The 1.0 mm diameter given is small enough that we should use Poiseuille's equation and take viscosity into account.
Prepare: From Table 13.4, we know the viscosity of water at this temperature is 1.0×10^{-3} Pa·s.
Solve: According to Poiseuille's equation, the pressure drop along each section of the tube is

$$\Delta p = \frac{8\eta L Q}{\pi R^4}$$

Then the total pressure drop across the two sections of the tube is

$$\Delta p = \Delta p_1 + \Delta p_2 = \frac{8\eta L_1 Q_1}{\pi R_1^4} + \frac{8\eta L_2 Q_2}{\pi R_2^4} = \frac{8\eta L Q}{\pi}\left(\frac{1}{R_1^4} + \frac{1}{R_2^4}\right)$$

Here, we have used the fact that the two lengths L and the volume flow rates Q are the same in both sections. Thus

$$\Delta p = \frac{8(1.0 \times 10^{-3} \text{ Pa·s})(1.0 \text{ m})(0.02 \times 10^{-3} \text{ m}^3/\text{s})}{\pi}\left(\frac{1}{(5.0 \times 10^{-4} \text{ m})^4} + \frac{1}{(2.0 \times 10^{-3} \text{ m})^4}\right)$$

$$= 8.2 \times 10^5 \text{ Pa}$$

This is the pressure difference between point P and the open end of the tube, which is at atmospheric pressure. Thus Δp is the difference between p_P and p_{atmos}, that is, it is the gauge pressure at P.

P13.71. Strategize: Let us assume that viscosity can be ignored in the large beaker, but that it cannot be ignored as the fluid moves through the narrow tube. We can determine the pressure in the beaker near the tube and use the resulting pressure difference along the tube in Poiseuille's equation.
Prepare: Table 13.4 tells us that the viscosity of water at this temperature is 1.0×10^{-3} Pa·s. To determine the hydrostatic pressure near the bottom of the beaker, we use $p = p_{\text{atm}} + \rho g h$.
Solve: The pressure difference between the end of the tube connected to the beaker and the open end of the tube is

$$p = p_{\text{atm}} + \rho g h \Rightarrow p - p_{\text{atm}} = \rho g h = (1000 \text{ kg/m}^3)(9.80 \text{ m/s}^2)(0.45 \text{ m}) = 4.4 \times 10^3 \text{ Pa}$$

Poiseuille's equation gives that the volume flow rate in the tube as

$$Q = \frac{\pi R^4}{8\eta L}\Delta p = \frac{\pi(1.5 \times 10^{-3} \text{ m})^4}{8(1.0 \times 10^{-3} \text{ Pa·s})(0.10 \text{ m})}(4.4 \times 10^3 \text{ Pa}) = 8.8 \times 10^{-5} \text{ m}^3/\text{s}$$

Assess: This result is reasonable for such a narrow tube and small pressure difference.

P13.73. Strategize: We will look Poiseuille's equation to determine what changes could increase flow rate with only modest changes to the pressure.

Prepare: Poiseuille's equation tells us $Q = \dfrac{\pi R^4 \Delta p}{8\eta L}$, but of course this only describes the flow rate through a particular region with length L and fixed radius R. Each such region in the circulatory system with contribute to the total blood pressure. We can consider what changes would increase blood flow with the smallest effect on pressure.

Solve: From Poiseuille's equation, we see that option A, B, and C could all increase the blood flow, even without changing the pressure much. But it would be a huge physiological change to reduce the viscosity of blood by a factor of 1/5. However, the radius of blood vessels (as in options B and C) would only need to be increased by $R_f / R_i = \sqrt[4]{5} = 1.5$ to increase blood flow by a factor of 5. Since the aorta is already so large, it would be difficult to increase its diameter by this factor. Also, blood flow through the aorta is not where the largest pressure difference occurs. The largest pressure difference occurs where the radius of the blood vessels is smallest, according to $\Delta p = \dfrac{Q 8 \eta L}{\pi R^4}$. Dilation of the smaller blood vessels is most likely to allow larger flow rates with relatively small increases in pressure. The correct choice is B.

Assess: If we consider the circulatory system like one pipe with one radius and one length, either options B or C seem equally valid. However, if we consider that the flow rate must be the same everywhere, but the radius changes, it becomes clear that there are some regions with large pressure differences per unit length and other regions with smaller pressure differences per unit length. Dilating the smaller blood vessels would most effectively increase the flow rate.

P13.75. Strategize: We will consider Poiseuille's equation in determining how a change in artery radius is likely to affect the pressure.

Prepare: Poiseuille's equation can be written as $\Delta p = \dfrac{Q 8 \eta L}{\pi R^4}$. We are ignoring changes in viscosity and assuming the flow rate is kept largely constant.

Solve: Clearly, a decrease in artery radius increases the pressure difference across the artery, and does so with a very strong (R^{-4}) dependence. Therefore the correct answer is C.

Assess: It is reasonable that pushing fluid through a constricted artery requires more pressure than through a wide artery.

PROPERTIES OF MATTER

PptIII.1. Reason: The scaling law for specific metabolic rate vs. body mass is $M^{-0.25}$. Because the wolf has a mass 16 times as great as the jackrabbit's we expect the specific metabolic rate of the jackrabbit to be $(16)^{0.25} = 2$ times the wolf's. The answer is A.

Assess: Remember, the *specific* metabolic rate is the power used per kilogram of tissue.

PptIII.3. Reason: The specific metabolic rate for a rat is $5\ \text{W/kg}$ so the metabolic rate is $(5\ \text{W/kg})(0.20\ \text{kg}) = 1\ \text{W} = 1\ \text{J/s}$. In a day the rat would use

$$(1\ \text{J/s})(1\ \text{d})\left(\frac{1\ \text{cal}}{4.19\ \text{J}}\right)\left(\frac{1\ \text{Cal}}{1000\ \text{cal}}\right)\left(\frac{24\ \text{h}}{1\ \text{d}}\right)\left(\frac{3600\ \text{s}}{1\ \text{h}}\right) \approx 20\ \text{Cal}$$

The answer is B.

Assess: This is about $1/100$ what a human would use; this makes sense.

PptIII.5. Reason: Jump height is the answer that can be explained by scaling laws. We know that all animals can't jump exactly the same height, but the variation in jump height is tiny compared to the ratio of the masses of the animals. The scaling laws must introduce offsetting factors that leave the jump height approximately the same. The answer is D.

Assess: The other choices are true but don't reflect scaling laws.

PptIII.7. Reason: If 25% of the 480 W is converted to mechanical energy of motion, then the other 75% is converted to thermal energy in his body. So $(0.75)(480\ \text{W}) = 360\ \text{W}$. The answer is B.

Assess: It takes some effort to stay cool on a strenuous bike ride.

PptIII.9. Reason: If the cyclist is unable to get rid of this thermal energy then $Q = Mc\Delta T$ will tell us how much the temperature will go up. For mammalian bodies $c = 3400\ \text{J}(\text{kg}\cdot\text{K})$.

$$\Delta T = \frac{Q}{Mc} = \frac{(360\ \text{J/s})(10\ \text{min})}{(68\ \text{kg})(3400\ \text{J}/(\text{kg}\cdot\text{K}))}\left(\frac{60\ \text{s}}{1\ \text{min}}\right) = 0.9\ \text{K} = 0.9\,^{\circ}\text{C}$$

The answer is C.

Assess: Less than one degree doesn't seem like much, but that is a dangerous temperature rise for humans.

PptIII.11. Reason: The buoyant force needs to be about the same magnitude as the weight force, so we compute the weight force on the balloon by assuming its mass is about the same as the mass of the displaced air. Read from the figure that the density of air at 10 km altitude is 0.4 kg/m^3.

$$F_b = mg = \rho Vg = (0.4 \text{ kg/m}^3)(12 \text{ m}^3)(9.8 \text{ m/s}^2) = 47 \text{ N} \approx 50 \text{ N}$$

The answer is A.

Assess: The weight of the balloon has hardly changed at all (being a little bit farther from the earth), so the buoyant force doesn't need to either.

PptIII.13. Reason: If the temperature were unchanged, then as the pressure halved the volume would double to 8.0 m^3. But because the temperature in the balloon drops, this will somewhat shrink the gas volume, leading to a volume less than 8.0 m^3. The answer is C.

Assess: We do not need to actually compute $2^{\frac{1}{14}}$ to know it is less than 2.

PptIII.15. Reason: Assume the helium is an ideal gas. If $T_f = T_i$ then $p_f V_f = p_i V_i$, so reducing the pressure by a factor of three must triple the volume. The answer is D.

Assess: The ratios confirm our intuition.

PptIII.17. Reason: The buoyancy force is in the opposite direction from the weight, so it is up. The drag force is in the opposite direction from the motion, so it is up too. The answer is A.

Assess: Since it is descending at a constant rate the sum of the three forces is zero.

PptIII.19. Reason: Estimate the area of the diaphragm to be $15 \text{ cm} \times 30 \text{ cm} = 0.045 \text{ m}^2$. Since pressure is force/area then

$$F = PA = (7.0 \text{ kPa})(0.045 \text{ m}^2) = 315 \text{ N} \approx 300 \text{ N}$$

Assess: The estimate is probably good to only one significant figure.

PptIII.21. Reason: (a) Assume the air in the bladder is an ideal gas. We solve the ideal gas equation for n. The pressure at a depth d is $p = p_0 + \rho gd$. 80 ft = 24.384 m. $15°\text{C} = 288$ K. $V = (0.070)(7.0 \text{ L}) = 0.00049 \text{ m}^3$.

$$n = \frac{pV}{RT} = \frac{(p_0 + \rho gd)(V)}{RT} = \frac{[101.3 \text{ kPa} + (1030 \text{ kg/m}^3)(9.8 \text{ m/s}^2)(24.384 \text{ m})](0.00049 \text{ m}^3)}{(8.31 \text{ J/(mol} \cdot \text{K)})(288 \text{ K})} = 0.07113 \text{ mol}$$

Which we report as 0.071 mol to two significant figures.

(b) 50 ft = 15.24 m.

$$V = \frac{nRT}{p} = \frac{(0.07113 \text{ mol})(8.310 \text{ J/(mol} \cdot \text{K)})(288 \text{ K})}{101.3 \text{ kPa} + (1030 \text{ kg/m}^3)(9.8 \text{ m/s}^2)(15.24 \text{ m})} = 0.000667 \text{ m}^3 \approx 6.7 \times 10^{-4} \text{ m}^3$$

(c) Find the new number of moles and then subtract.

$$n = \frac{pV}{RT} = \frac{[101.3 \text{ kPa} + (1030 \text{ kg/m}^3)(9.8 \text{ m/s}^2)(15.24 \text{ m})](0.00049 \text{ m}^3)}{(8.31 \text{ J/(mol} \cdot \text{K)})(288 \text{ K})} = 0.05226 \text{ mol}$$

So $0.07113 \text{ mol} - 0.05226 \text{ mol} = 0.01887 \text{ mol} \approx 0.019 \text{ mol}$ need to be removed.

Assess: We expect the needed moles at 50 ft to be less than at 80 ft.

OSCILLATIONS

Q14.1. Reason: The motion will be periodic. This simply means that the motion will repeat with some characteristic period. The pendulum motion may even look qualitatively like a sine or cosine function. But it will not be exactly sinusoidal, and will not be exactly harmonic. Simple harmonic motion means that there is a single frequency (or period, equivalently) independent of amplitude. Since tripling the angular displacement does not triple the restoring force, the described scenario is outside the small-angle regime in which motion is approximately simple harmonic (often taken to be about $10°$).

Assess: A pendulum swinging out to angles much larger than $10°$ will typically not undergo simple harmonic motion. We were also given force information so that we could be sure the restoring force was NOT linear in angular displacement here.

Q14.3. Reason: We are given the graph of x versus t. However, we want to think about the slope of this graph to answer velocity questions.

(a) When the x versus t graph is increasing, the particle is moving to the right. It has maximum speed when the positive slope of the x versus t graph is greatest. This occurs at 0 s, 4 s, and 8 s.

(b) When the x versus t graph is decreasing, the particle is moving to the left. It has maximum speed when the negative slope of the x versus t graph is greatest. This occurs at 2 s and 6 s.

(c) The particle is instantaneously at rest when the slope of the x versus t graph is zero. This occurs at 1 s, 3 s, 5 s, and 7 s.

Assess: This is reminiscent of material studied in Chapter 2; what is new is that the motion is oscillatory and the graph periodic.

Q14.5. Reason: Synthesis 14.1 shows that the maximum speed is proportional to the amplitude. For small angles doubling the angle corresponds to doubling the amplitude, so this increases her maximum speed by a factor of two.

Assess: The maximum acceleration also doubles.

Q14.7. Reason: The maximum kinetic energy is the same as the total mechanical energy. The total energy and amplitude of an oscillator are related by $E = kA^2/2$, we see that the energy is proportional to the square of the amplitude. If A is doubled, E will increase by a factor of four. That is $E_{\text{New}} = 4E_{\text{Old}} = 4(2\text{ J}) = 8\text{ J}$.

Assess: This question may also be answered using a more quantitative approach as outlined in Question 14.6.

Q14.9. Reason: From the graph the strategy is to determine the period, then use $f = 1/T$. As is done in Figure 14.4, one can measure the period between two crests; in this case it appears to be 2 s.

$$f = \frac{1}{T} = \frac{1}{2\text{ s}} = 0.5\text{ Hz}$$

The amplitude is the maximum distance from the equilibrium position. On this graph it appears that $A = 10$ cm.

Assess: The amplitude is *not* the distance from the maximum to the minimum—that would be $2A$. See Figure 14.6.

Q14.11. Reason: The period of a block oscillating on a spring is given in Equation 14.26, $T = 2\pi\sqrt{m/k}$. We are told that $T_1 = 2.0\,\text{s}$.

(a) In this case the mass is doubled, $m_2 = 2m_1$.

$$\frac{T_2}{T_1} = \frac{2\pi\sqrt{m_2/k}}{2\pi\sqrt{m_1/k}} = \sqrt{\frac{m_2}{m_1}} = \sqrt{\frac{2m_1}{m_1}} = \sqrt{2}$$

So $T_2 = \sqrt{2}\,T_1 = \sqrt{2}\,(2.0\,\text{s}) = 2.8\,\text{s}$.

(b) In this case the spring constant is quadrupled, $k_2 = 2k_1$.

$$\frac{T_2}{T_1} = \frac{2\pi\sqrt{m/k_2}}{2\pi\sqrt{m/k_1}} = \sqrt{\frac{k_1}{k_2}} = \sqrt{\frac{k_1}{2k_1}} = \frac{1}{\sqrt{2}}$$

So $T_2 = \sqrt{2}\,T_1 = \sqrt{2}\,(2.0\,\text{s}) = 2.8\,\text{s}$.

(c) The formula for the period does not contain the amplitude; that is, the period is independent of the amplitude. Changing (in particular, doubling) the amplitude does not affect the period, so the new period is still 2.0 s.

Assess: It is equally important to understand what *doesn't* appear in a formula. It is quite startling, really, the first time you realize it, that the amplitude doesn't affect the period. But this is crucial to the idea of simple harmonic motion. Of course, if the spring is stretched too far, out of its linear region, then the amplitude would matter.

Q14.13. Reason: If it behaves like a mass on a spring, then trimming the wings will reduce the mass, and this will increase the frequency because $f = \frac{1}{2\pi}\sqrt{\frac{k}{m}}$.

Assess: It would be easier to beat a wing quickly if it had less mass.

Q14.15. Reason: Reducing the mass increases the frequency because $f = \frac{1}{2\pi}\sqrt{\frac{k}{m}}$. So you would remove water.

Assess: Try it!

Q14.17. Reason: Since both pendulums are started with such small amplitudes, we can use the expression $\omega = \sqrt{\frac{g}{\ell}}$, which is independent of amplitude. Thus, both pendulums will reach their lowest points simultaneously.

Assess: This can be understood by considering that increasing the angle of deflection from vertical increases the distance the pendulum must cover, but also increases the component of gravity accelerating the pendulum. It is plausible that these two effects could cancel to give a period/frequency independent of amplitude, and indeed they do.

Q14.19. Reason: The leg acts somewhat like a pendulum as it swings forward. By bending their knees to bring their feet up closer to the body, sprinters are shortening the pendulums, which makes them swing faster.

Assess: See if you can notice this effect by first running as fast as you can, then again without bending your knees any higher than necessary to clear the ground.

Q14.21. Reason: Every object has natural frequencies of oscillation with which they will respond when deflected from equilibrium. The marching soldiers appear to have been marching at a pace such that their stomping boots drove the oscillation of the bridge at the same frequency as it naturally responded, its resonant frequency. Pushing an oscillating system at the same rate as it responds is like pushing a child on a swing at just the right time to drive the oscillations larger and larger. The wild, large-amplitude oscillations caused (already faulty) bolts to crack and give way.

Assess: If the soldiers "break step" then they will not be driving the oscillation of the bridge in unison. As one steps down with his/her boot, another lifts his/her boot. The total driving force then has no fixed frequency, but is more random and will not drive the bridge into large-amplitude oscillations.

Q14.23. Reason: (a) The unstretched equilibrium position is 20 cm. When we load the spring with 100 g we establish a new equilibrium position at 30 cm. When we pull the oscillator down to 40 cm (i.e., an additional 10 cm) and release it, it will oscillate with an amplitude of A = 10 cm. The correct choice is B.
(b) Knowing that 100 g stretched the spring 10 cm, we can determine the spring constant.

$$k = mg/x = (0.1 \text{ kg})(9.8 \text{ m/s}^2)/(0.10 \text{ m}) = 9.8 \text{ N/m}$$

Knowing the spring constant and the mass on the spring, we can determine the oscillation frequency as follows:

$$f = \sqrt{k/m}/2\pi = \sqrt{(9.8 \text{ N/m})/(0.10 \text{ kg})}/2\pi = 1.6 \text{ Hz}$$

The correct choice is C.
(c) The frequency of oscillation depends on k and m, neither of which would change on the moon. The correct choice is C.
Assess: These are rather small but acceptable values for the spring constant and frequency.

Q14.25. Reason: Some of the question may be answered by comparing the expression given with the general equation for simple harmonic motion.
(a) The expression given for the displacement is $x = (0.350 \text{ m}) \cos(15.0t)$.
The general expression for the displacement is $x = A\cos\omega t$.
Comparing these two expressions for the displacement we see that the amplitude of oscillation is A = 0.350 m, choice B.
(b) The frequency of oscillation may be determined by $f = \omega/2\pi = (15.0/\text{s})/2\pi = 2.39 \text{ Hz}$, choice B.
(c) The mass attached to the spring may be determined by $m = k/\omega^2 = (200 \text{ N/m})/(15/\text{s})^2 = 0.89 \text{ kg}$, choice B.
(d) The total mechanical energy of the oscillator is equal to its maximum potential energy, which may be determined by $E_{\text{Total}} = U_{\max} = kA^2/2 = (200 \text{ N/m})(0.350 \text{ m})^2/2 = 12.2 \text{ J}$, choice E.
(e) The maximum speed may be determined by $v_{\max} = \omega A = (15.0/\text{s})(0.350 \text{ m}) = 5.25 \text{ m/s}$, choice E.

Assess: Working this problem brings our attention to two things. First, there is a lot of information tucked away in the function given for the displacement of the oscillator. Second, there are a lot of details associated with an oscillation. It is important to know these details and their interconnection.

Q14.27. Prepare: We know that the initial displacement Δx is the same in both experiments, and that the initial elastic potential energy is converted to kinetic energy as the mass passes through the equilibrium position. So we can write for either case $\frac{1}{2}k(\Delta x_i)^2 = \frac{1}{2}mv_f^2 \Rightarrow v_f = \sqrt{\frac{k}{m}}\Delta x$. Writing this for the heavier mass (H) and the lighter mass (L), and taking the ratio, we find

$$v_{\text{H,f}} / v_{\text{L,f}} = \sqrt{\frac{k}{m_{\text{H}}}}\Delta x / \left(\sqrt{\frac{k}{m_{\text{L}}}}\Delta x\right) = \sqrt{\frac{m_{\text{L}}}{m_{\text{H}}}} \Rightarrow$$

$$v_{\text{H,f}} = \sqrt{\frac{m_{\text{L}}}{m_{\text{H}}}}v_{\text{L,f}} = \sqrt{\frac{(0.20 \text{ kg})}{(0.40 \text{ kg})}}(0.28 \text{ m/s}) = 0.20 \text{ m/s}$$

So the correct answer is B.
Assess: Clearly, a spring with a fixed rigidity and fixed displacement will have a tougher time accelerating a large mass than a small mass. It is therefore reasonable that the maximum speed of a greater mass would be smaller than the maximum speed of a small mass.

Q14.29. Reason: We see in Figure 14.26 that the cells on the basilar membrane close to the stapes correspond to higher frequencies.
The correct choice is B.
Assess: People often lose hearing sensitivity in the higher frequencies with age.

Problems

P14.1. Strategize: We use the inverse relationship between frequency and period.
Prepare: We will use Equation 14.1.
Solve: The frequency generated by a guitar string is 440 Hz., hence

$$T = \frac{1}{f} = \frac{1}{440 \text{ Hz}} = 2.3 \times 10^{-3} \text{ s} = 2.3 \text{ ms}$$

Assess: The units of frequency are Hz, that is, cycles per second, or s^{-1}, so the period is in seconds.

P14.3. Strategize: We will use the definition of frequency and the inverse relationship between frequency and period.
Prepare: Your pulse or heartbeat is 75 beats per minute or 75 beats/60 s = 1.25 beats/s. The period is the inverse of the frequency, so we will use Equation 14.1.
Solve: The frequency of your heart's oscillations is

$$f = \frac{75 \text{ beats}}{60 \text{ s}} = 1.25 \text{ beats/s} = 1.3 \text{ Hz}$$

The period is the inverse of the frequency, hence

$$T = \frac{1}{f} = \frac{1}{1.3 \text{ Hz}} = 0.80 \text{ s}$$

Assess: A heartbeat of 1.3 beats per second means that one beat takes a little less than 1 second, which is what we obtained above.

P14.5. Strategize: We employ the small angle approximation, which results in the displacement being linearly proportional to the restoring force.
Prepare: For a small angle pendulum the restoring force is proportional to the displacement because $\sin \theta \approx \theta$.
Solve: If we model this as a Hooke's law situation, then doubling the distance will double the restoring force from 20 N to 40 N.
Assess: This would not work for angles much larger than $10°$.

P14.7. Strategize: The air-track glider attached to a spring is in simple harmonic motion. We can use Equation 14.13 to relate the maximum speed and period to the amplitude. We then apply Equation 14.10 to determine the position as a function of time, and finally at the time specified in part (b).
Prepare: The glider is pulled to the right and released from rest at $t = 0$ s. It then oscillates with a period $T = 2.0$ s and a maximum speed $4v_{\text{max}} = 0 \text{ cm/s} = 0.40 \text{ m/s}$. While the amplitude of the oscillation can be obtained from Equation 14.13, the position of the glider can be obtained from Equation 14.10, $x(t) = A \cos(\frac{2\pi t}{T})$.
Solve: (a)

$$v_{\text{max}} = (2\pi A/T) \Rightarrow A = \frac{v_{\text{max}} T}{2\pi} = \frac{(0.40 \text{ m/s})(2.0 \text{ s})}{2\pi} = 0.127 \text{ m} = 0.13 \text{ m}$$

(b) The glider's position at $t = 0.25$ s is

$$x_{0.25 \text{ s}} = (0.127 \text{ m}) \cos\left[\frac{2\pi(0.25 \text{ s})}{2.0 \text{ s}}\right] = 0.090 \text{ m} = 9.0 \text{ cm}$$

Assess: At $t = 0.25$ s, which is less than one quarter of the time period, the object has not reached the equilibrium position and is still moving toward the left.

P14.9. Strategize: We look at the vertical axis to determine the amplitude and the horizontal axis to determine the period, and from that the frequency.

Prepare: The oscillation is the result of simple harmonic motion. As the graph shows, the time to complete one cycle (or the period) is $T = 4.0$ s. We will use Equation 14.1 to find frequency.

Solve: (a) The amplitude $A = 20$ cm.

(b) The period $T = 4.0$ s, thus

$$f = \frac{1}{T} = \frac{1}{4.0 \text{ s}} = 0.25 \text{ Hz}$$

Assess: It is important to know how to find information from a graph.

P14.11. Strategize: We will use the known relation between the maximum acceleration and the amplitude of motion.

Prepare: Equation 14.17 reads $a_{max} = (2\pi f)^2 A$. We know that the argument of the cosine function is $\omega t = 2\pi f t$, so we can read off the value of the quantity $2\pi f = 30$ rad/s. We can similarly read off the maximum acceleration, since it is the magnitude of the prefactor in front of the cosine function: $a_{max} = 18.0$ m/s^2.

Solve: Inserting the frequency information into Equation 14.17 and rearranging, we find

$$A = \frac{a_{max}}{(2\pi f)^2} = \frac{(18.0 \text{ m/s}^2)}{(30 \text{ rad/s})^2} = 2.0 \times 10^{-2} \text{ m}.$$

Assess: Although we are not given a context for this oscillator, 2.0 cm is certainly a plausible amplitude.

P14.13. Strategize: We will assume that ship and passengers are approximately in simple harmonic motion. Synthesis box 14.1 gives us an expression for the maximum acceleration. For part (b) we simply divide by g.

Prepare: The maximum acceleration for an object in simple harmonic motion is given by $a_{max} = (2\pi f)^2 A$.

$$A = 1 \text{ m and } f = 1/T = 1/15 \text{ s} = 0.067 \text{ Hz}.$$

Solve: (a)

$$a_{max} = (2\pi f)^2 A = (2\pi \times 0.067 \text{ Hz})^2 (1 \text{ m}) = 0.2 \text{ m/s}^2$$

(b) To one significant figure, $g = 10$ m/s^2, so the passenger's acceleration is about $\frac{1}{50} g$.

Assess: This is not a large acceleration, but it can play havoc with some people's stomachs.

P14.15 Strategize: We relate the maximum speed and acceleration to the amplitude and frequency, and solve.

Prepare: Equation 14.17 relates the maximum acceleration to the amplitude and frequency: $a_{max} = (2\pi f)^2 A = (2\pi / T)^2 A$, and Equation 14.24 relates the maximum speed to the amplitude and frequency: $v_{max} = (2\pi f) A = (2\pi / T) A$.

Solve: (a) Inserting the given values, we find $v_{max} = (2\pi / T) A = (2\pi / (0.82 \text{ s}))(8.8 \times 10^{-2} \text{ m}) = 0.67$ m/s.

(b) Inserting the given values, we find $a_{max} = (2\pi / T)^2 A = (4\pi^2 / (0.82 \text{ s})^2)(8.8 \times 10^{-2} \text{ m}) = 5.2$ m/s^2.

Assess: Note that even motion with a small amplitude such as this can involve significant speeds and accelerations, due to the relatively short period.

P14.17. Strategize: We can use Equation 14.16 to relate the maximum speed to the frequency and amplitude.
Prepare: We apply $v_{max} = 2\pi f A$.
Solve: Solve the equation for A.

$$v_{max} = 2\pi f A \Rightarrow A = \frac{v_{max}}{2\pi f} = \frac{2.5 \text{ m/s}}{2\pi(250 \text{ Hz})} = 1.6 \text{ mm}$$

Assess: This seems about the right size for a bumblebee.

P14.19. Strategize: The spring undergoes simple harmonic motion. The elastic potential energy in a spring stretched by a distance x from its equilibrium position is given by Equation 14.19, and the total mechanical energy of the object is the sum of kinetic and potential energies as in Equation 14.20.
Prepare: We relate the energy at a given displacement, $U = \frac{1}{2}k(\Delta x)^2$, to the total energy is simply $E = \frac{1}{2}kA^2$, Equation 14.21.
Solve: (a) When the displacement is $x = \frac{1}{2}A$, the potential energy is

$$U = \frac{1}{2}kx^2 = \frac{1}{2}k\left(\frac{1}{2}A\right)^2 = \frac{1}{4}\left(\frac{1}{2}kA^2\right) = \frac{1}{4}E \Rightarrow K = E - U = \frac{3}{4}E$$

Thus, one quarter of the energy is potential and three-quarters is kinetic.
(b) To have $U = \frac{1}{2}E$ requires

$$U = \frac{1}{2}kx^2 = \frac{1}{2}E = \frac{1}{2}\left(\frac{1}{2}kA^2\right) \Rightarrow x = \frac{A}{\sqrt{2}}$$

P14.21. Strategize: The block attached to the spring is in simple harmonic motion. The period of an oscillating mass on a spring is given by Equation 14.26.
Prepare: According to Equation 14.26, $T = 2\pi\sqrt{\frac{m}{k}}$. We note that this is independent of amplitude.

Solve: The period of an object attached to a spring is

$$T = 2\pi\sqrt{\frac{m}{k}} = T_0 = 2.00 \text{ s}$$

where m is the mass and k is the spring constant.
(a) For mass $= 2m$,

$$T = 2\pi\sqrt{\frac{2m}{k}} = (\sqrt{2})T_0 = 2.83 \text{ s}$$

(b) For mass $\frac{1}{2}m$,

$$T = 2\pi\sqrt{\frac{\frac{1}{2}m}{k}} = T_0/\sqrt{2} = 1.41 \text{ s}$$

(c) The period is independent of amplitude. Thus $T = T_0 = 2.00$ s.
(d) For a spring constant $= 2k$,

$$T = 2\pi\sqrt{\frac{m}{2k}} = T_0/\sqrt{2} = 1.41 \text{ s}$$

Assess: As would have been expected, increase in mass leads to slower simple harmonic motion.

P14.23. Strategize: The oscillating mass is in simple harmonic motion. We will use Equations 14.10 to read off some values, but we will also make use of Equation 14.13, 14.16 and 14.18.

Prepare: The position of the oscillating mass is given by $x(t) = (2.0 \text{ cm})\cos(10t)$, where t is in seconds. We will compare this with Equation 14.10.

Solve: (a) The amplitude $A = 2.0$ cm.

(b) The period is calculated as follows:

$$\frac{2\pi}{T} = 10 \text{ rad/s} \Rightarrow T = \frac{2\pi}{10 \text{ rad/s}} = 0.63 \text{ s}$$

(c) The spring constant is calculated from Equation 14.25 as follows:

$$\frac{2\pi}{T} = \sqrt{\frac{k}{m}} \Rightarrow k = m\left(\frac{2\pi}{T}\right)^2 = (0.050 \text{ kg})(10 \text{ rad/s})^2 = 5.0 \text{ N/m}$$

(d) The maximum speed from Equation 14.16 is

$$v_{max} = 2\pi fA = \left(\frac{2\pi}{T}\right)A = (10 \text{ rad/s})(2.0 \text{ cm}) = 20 \text{ cm/s}$$

(e) The total energy from Equation 14.21 is

$$E = \frac{1}{2}kA^2 = \frac{1}{2}(5.0 \text{ N/m})(0.02 \text{ m})^2 = 1.0 \times 10^{-3} \text{ J}$$

(f) At $t = 0.40$ s, the velocity from Equation 14.12 is

$$v_x = -(20.0 \text{ cm/s})\sin[(10 \text{ rad/s})(0.40 \text{ s})] = 15 \text{ cm/s}$$

Assess: Velocity at $t = 0.40$ s, is less than the maximum velocity, as would be expected.

P14.25. Strategize: The mass attached to the spring oscillates in simple harmonic motion. We will use the inverse relationship between the period and frequency. We will then determine the total energy from the given information, then equate that total energy to the maximum potential energy and maximum kinetic energy to determine the maximum displacement and speed, respectively.

Prepare: The mass oscillates at a frequency of 2.0 Hz. We will need the spring constant k which we will determine using Equation 14.26.

Solve: (a) The period using Equation 14.1 is $T = 1/f = 1/2.0 \text{ Hz} = 0.50 \text{ s}$.

(b) Using energy conservation $\frac{1}{2}kA^2 = \frac{1}{2}kx_0^2 + \frac{1}{2}m(v_x)_0^2$.

Using $x_0 = 5.0$ cm, $(v_x)_0 = -30$ cm/s, and $k = m(2\pi f)^2 = 0.2 \text{ kg}[2\pi(2.0 \text{ Hz})]^2 = 31.58 \text{ N/m}$, we get $A = 5.54$ cm, which is to be reported as 5.5 cm.

(c) The maximum speed from Equation 14.26 is $v_{max} = 2\pi fA = 2\pi(2.0 \text{ Hz})(5.54 \text{ cm}) = 69.6 \text{ cm/s}$, which will be reported as 70 cm/s.

(d) The total energy is $E = \frac{1}{2}mv_{max}^2 = \frac{1}{2}(0.200 \text{ kg})(0.696 \text{ m/s})^2 = 0.049 \text{ J}$.

P14.27. Strategize: Let us assume a small angle of oscillation so there is simple harmonic motion, and let us further assume the string is very light compared to the oscillating mass, such that it is a simple pendulum. Equation 14.27 relates the period of a pendulum to its length and gravity.

Prepare: We will use $T_0 = 2\pi\sqrt{\dfrac{L_0}{g}}$, which we note is independent of mass and amplitude.

Solve: The period of the pendulum is

$$T_0 = 2\pi\sqrt{\frac{L_0}{g}} = 4.00 \text{ s}$$

(a) The period is independent of the mass and depends only on the length. Thus $T = T_0 = 4.00$ s.

(b) For a new length $L = 2L_0$,

$$T = 2\pi\sqrt{\frac{2L_0}{g}} = \sqrt{2}T_0 = 5.66 \text{ s}$$

(c) For a new length $L = L_0/2$,

$$T = 2\pi\sqrt{\frac{L_0/2}{g}} = \frac{1}{\sqrt{2}}T_0 = 2.83 \text{ s}$$

(d) The period is independent of the amplitude as long as there is simple harmonic motion. Thus $T = 4.00$ s.

P14.29. Strategize: Because the angle of displacement is less than $10°$, the small-angle approximation holds and the pendulum exhibits simple harmonic motion. We will use Equation 14.27 and $g = 9.80$ m/s^2.

Prepare: Equation 14.27 tells us $T = 2\pi\sqrt{\dfrac{L}{g}}$. We apply the definition of the period to determine it from the given data.

Solve: The period is $T = 12.0$ s$/10$ oscillations $= 1.20$ s and is given by the formula

$$T = 2\pi\sqrt{\frac{L}{g}} \Rightarrow L = \left(\frac{T}{2\pi}\right)^2 g = \left(\frac{1.20 \text{ s}}{2\pi}\right)^2 (9.80 \text{ m/s}) = 35.7 \text{ cm}$$

Assess: A length of 35.7 cm for the simple pendulum is reasonable.

P14.31. Strategize: Assume a small angle of oscillation so that the pendulum has simple harmonic motion. We will use Equation 14.27.

Prepare: We will write out $T = 2\pi\sqrt{\dfrac{L}{g}}$ twice: once for Mars and once for Earth. The constant quantity is the length. So we can equate the two lengths and solve for g_{Mars}.

Solve: The time periods of the pendulums on the earth and on Mars are

$$T_{\text{earth}} = 2\pi\sqrt{\frac{L}{g_{\text{earth}}}} \quad \text{and} \quad T_{\text{Mars}} = 2\pi\sqrt{\frac{L}{g_{\text{Mars}}}}$$

Dividing these two equations,

$$\frac{T_{\text{earth}}}{T_{\text{Mars}}} = \sqrt{\frac{g_{\text{Mars}}}{g_{\text{earth}}}} \Rightarrow g_{\text{Mars}} = g_{\text{earth}}\left(\frac{T_{\text{earth}}}{T_{\text{Mars}}}\right)^2 = (9.80 \text{ m/s}^2)\left(\frac{1.50 \text{ s}}{2.45 \text{ s}}\right)^2 = 3.67 \text{ m/s}^2$$

Assess: Because $T_{\text{Mars}} > T_{\text{earth}}$, the same length of the pendulum would imply smaller g on Mars, as obtained above.

P14.33. Strategize: We will treat the lower leg as a physical pendulum, and apply Equation 14.28.

Prepare: We are given the frequency and the distance to the center of mass, and we can relate those to the unknown moment of inertia using $f = \dfrac{1}{2\pi}\sqrt{\dfrac{mgd}{I}}$.

Solve: Combining the above expressions and solving for the moment of inertia we obtain

$$I = mgd/(2\pi f)^2 = (5.0 \text{ kg})(9.80 \text{ m/s}^2)(0.18 \text{ m})/[2\pi(1.6 \text{ Hz})]^2 = 8.7\times10^{-2} \text{ kg}\cdot\text{m}^2$$

Assess: NASA determines the moment of inertia of the shuttle in a similar manner. It is suspended from a heavy cable, allowed to oscillate about its vertical axis of symmetry with a very small amplitude, and from the period of oscillation one may determine the moment of inertia. This arrangement is called a torsion pendulum.

P14.35. Strategize: We will model the rope as a simple small-angle pendulum. We want to hang on to the rope for a quarter of a period to get as far out as possible.

Prepare: Equation 14.27 gives the period in terms of the length of the rope: $T = 2\pi\sqrt{\dfrac{L}{g}}$.

Solve:

$$t_{\text{hang}} = \frac{T}{4} = \frac{\pi}{2}\sqrt{\frac{L}{g}} = \frac{\pi}{2}\sqrt{\frac{15 \text{ m}}{9.80 \text{ m/s}^2}} = 1.9 \text{ s}$$

Assess: This seems reasonable for such a long pendulum.

P14.37. Strategize: Model the elephant legs as physical pendulums, using the moment of inertia of a solid rod rotating around one end.

Prepare: We insert $I = \frac{1}{3}mL^2$ into Equation 14.28: $T = 2\pi\sqrt{\dfrac{I}{mgd}}$. The distance from the pivot point to the center of gravity is $d = \frac{1}{2}L$. For a leg to swing forward requires half a period.

Solve:

(a)
$$\frac{T}{2} = \pi\sqrt{\frac{I}{mgd}} = \pi\sqrt{\frac{\frac{1}{3}mL^2}{mg\frac{1}{2}L}} = \pi\sqrt{\frac{2L}{3g}} = \pi\sqrt{\frac{2(2.3 \text{ m})}{3(9.8 \text{ m/s}^2)}} = 1.24 \text{ s} \approx 1.2 \text{ s}$$

(b) The right leg takes 1.24 s to swing forward, then stay planted while the left leg takes 1.24 s to swing forward, so it takes 2.48 s for a whole period of the process. That is, each leg hits the ground $1/2.48$ times per second. There are 4 legs.

$$\frac{1}{2.48 \text{ s}}(4 \text{ legs})\left(\frac{60 \text{ s}}{1 \text{ min}}\right) = 97 \text{ steps/min}$$

Assess: This seems to jibe with the nature shows we've seen.

P14.39. Strategize: The motion is a damped oscillation. We can determine the time constant using Equation 14.29.
Prepare: The maximum displacement or amplitude of the oscillation at time t is given by $x_{\text{max}}(t) = Ae^{-t/\tau}$, where τ is the time constant. Using $x_{\text{max}} = 0.368\,A$ and $t = 10.0$ s, we can calculate the time constant.
Solve: From Equation 14.29.

$$0.368A = Ae^{-10.0 \text{ s}/\tau} \Rightarrow \ln(0.368) = \frac{-10.0 \text{ s}}{\tau} \Rightarrow \tau = -\frac{10.0 \text{ s}}{\ln(0.368)} = 10.0 \text{ s}$$

Assess: The above result says that the oscillation decreases to about 37% of its initial value after one time constant.

P14.41. Strategize: We can use Equation 14.29 to determine the maximum displacement after time t .
Prepare: Assume $A = 1$ in arbitrary units. The object continues to oscillate but x_{max} decreases due to the damping.
Solve: The equation for the graph is $x(t) = e^{-t/(4.0 \text{ s})}\cos(2\pi(1.0\,Hz)t)$.

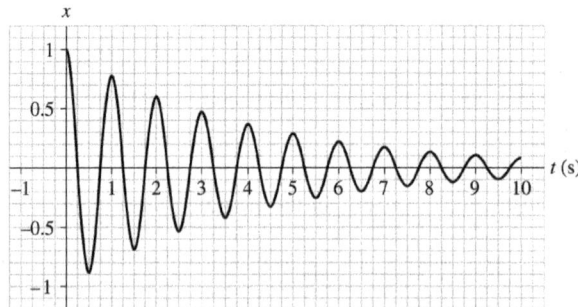

Assess: The oscillations damp out at about the rate we expect for $\tau = 4.0$ s.

P14.43. Strategize: We can count the number of peaks (or troughs) in a certain time interval, and determine the frequency from that. For part (b) we will look at the initial amplitude and the decreased amplitude at some later time, and use Equation 14.29 to relate them to the time constant.

Prepare: We note that between 0 ms and 4.0 ms there are 20 full oscillations. We also note that the initial amplitude is about 5 of the units marked by dashed lines. After 3.0 ms, the amplitude has decayed to just one unit.

Solve: (a) Using the definition of the frequency, we have $f = N / \Delta t = (20) / (4.0 \times 10^{-3} \text{ s}) = 5.0$ kHz.

(b) Rearranging Equation 14.29 and inserting values quoted above, we have

$$\tau = -\Delta t / \ln\left(\frac{x_{\max}(t)}{A} \right) = -(3.0 \times 10^{-3} \text{ s}) / \ln\left(\frac{1}{5} \right) = 4.8 \times 10^{-3} \text{ s}.$$

Assess: A good logical check is that human ears are sensitive to sounds in the kHz range, and humans can hear cricket chirps.

P14.45. Strategize: We read the period off the graph, and use Equation 14.30 to relate the change in amplitude to the time constant.

Prepare: Once the period is read from the graph, we use $f = 1 / T$. The definition of the time constant is the time after which the amplitude has decreased to 37% of its previous value.

Solve: It looks from the graph that the period is 0.5 s so $f = 2$ Hz. Looking at the peaks, it appears the amplitude has decreased to about 37% of its value after only 0.25 s, so that is our guess for τ.

Assess: You want the time constant for damping on your car to be short.

P14.47. Strategize: We will model the child on the swing as a simple small-angle pendulum. To make the amplitude grow large quickly we want to drive (push) the oscillator (child) at the natural resonance frequency. In other words, we want to wait the natural period between pushes.

Prepare: The natural period of a pendulum is given by $T = 2\pi \sqrt{\dfrac{L}{g}}$.

Solve: Inserting the given values, we have

$$T = 2\pi \sqrt{\frac{L}{g}} = 2\pi \sqrt{\frac{2.0 \text{ m}}{9.8 \text{ m/s}^2}} = 2.8 \text{ s}$$

Assess: You could also increase the amplitude by pushing every other time (every $2T$), but that would not make the amplitude grow as quickly as pushing every period.
The mass of the child was not needed; the answer is independent of the mass.

P14.49. Strategize: Equation 14.25 relates the natural frequency to the spring constant and mass.

Prepare: We rearrange $\omega = 2\pi f = \sqrt{k/m} \Rightarrow k = (2\pi f)^2 m$.

Solve: Inserting the given values, we obtain

$$k = (2\pi f)^2 m = [2\pi(29 \text{ Hz})]^2 (7.5 \times 10^{-3} \text{ kg}) = 250 \text{ N/m}$$

Assess: As spring constants go, this is a fairly large value, however the musculature holding the eyeball in the socket is strong and hence will have a large effective spring constant.

P14.51. Strategize: Insert the formula for the period of a mass on a spring into the formula for the maximum speed of a harmonic oscillator.

Prepare: The maximum speed of the ball is given by $v_{\max} = 2\pi f A$, and the frequency is given by $f = \dfrac{1}{2\pi} \sqrt{\dfrac{k}{m}}$.

Solve: Combining these equations, we obtain

$$v_{max} = 2\pi fA = \frac{2\pi}{T}A = \frac{2\pi}{2\pi\sqrt{\frac{m}{k}}}A = \sqrt{\frac{k}{m}}A = \sqrt{\frac{12\,\text{N/m}}{0.40\,\text{kg}}}(0.20\,\text{m}) = 1.1\,\text{m/s}$$

Assess: This problem can also be done with energy conservation considerations.

P14.53. Strategize: The vertical oscillations constitute simple harmonic motion. We employ the definition of the period, and Equation 14.26.

Prepare: A pictorial representation of the spring and the ball is shown in the following figure. The period and frequency of oscillations are

$$T = \frac{20\,\text{s}}{30\,\text{oscillations}} = 0.6667\,\text{s} \quad \text{and} \quad f = \frac{1}{T} = \frac{1}{0.6667\,\text{s}} = 1.50\,\text{Hz}$$

Since k is known, we can obtain the mass m using $f = \frac{1}{2\pi}\sqrt{\frac{k}{m}}$.

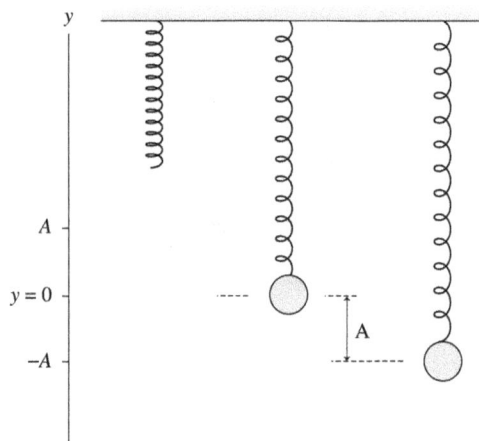

Solve: (a) The mass can be found as follows:

$$f = \frac{1}{2\pi}\sqrt{\frac{k}{m}} \Rightarrow m = \frac{k}{(2\pi f)^2} = \frac{15.0\,\text{N/m}}{[2\pi(1.50\,\text{Hz})]^2} = 0.169\,\text{kg}$$

(b) The maximum speed is given by Equation 14.26, $v_{max} = 2\pi fA = 2\pi(1.50\,\text{Hz})(0.0600\,\text{m}) = 0.565\,\text{m/s}$.

Assess: Both the mass of the ball and its maximum speed are reasonable.

P14.55. Strategize: The buoyant force depends on the volume submerged, which depends on the length of the cylinder submerged. Thus, we have a restoring force that depends linearly on the displacement of the object from equilibrium. Thus, we can treat this like simple harmonic motion. We will use Newton's second law to relate the relevant forces.

Prepare: For this situation, Newton's second law gives $\sum F_y = F_{buoy} - mg = ma_y$. Initially, the cylinder is in equilibrium, such that $\left(\sum F_y\right)_0 = F_{buoy,0} - mg = 0 \Rightarrow \pi r^2 y_0 \rho_w g = mg$. Later, when the block is displaced to some new

$y = y_0 - \Delta y$, we have $\sum F_y(t) = F_{buoy}(t) - mg = ma_y(t) \Rightarrow \pi r^2(y_0 - \Delta y)\rho_w g - mg = m\frac{d^2 y(t)}{dt^2}$, or equivalently

$-\pi r^2 \Delta y \rho_w g = m\frac{d^2 y(t)}{dt^2}$.

Solve: If we collect some of the constants on the left side of the equation above into a variable defined $k = \pi r^2 \rho_w g$, then we have exactly the differential equation solved in the chapter:

$-k\Delta y = m \dfrac{d^2 y(t)}{dt^2}$ such that we know

$$f = \frac{1}{2\pi}\sqrt{\frac{k}{m}} = \frac{1}{2\pi}\sqrt{\frac{\pi r^2 \rho_w g}{m}} = \frac{1}{2\pi}\sqrt{\frac{\pi(0.010 \text{ m})^2 (1{,}000 \text{ kg/m}^3)(9.8 \text{ m/s}^2)}{(0.010 \text{ kg})}} = 2.8 \text{ Hz}.$$

Assess: It is reasonable that a tiny floating bob would oscillate a few times per second when displaced.

P14.57. Strategize: The vertical oscillations constitute simple harmonic motion. We will use Equation 14.26 and Hooke's Law.

Prepare: Newton's second law tells us that $-k\Delta y - mg = 0$, and Equation 14.26 tells us $f = \dfrac{1}{2\pi}\sqrt{\dfrac{k}{m}}$. These two

equations can be combined to determine the unknown gravitational acceleration on Planet X (and the spring constant, although we are not asked for it).

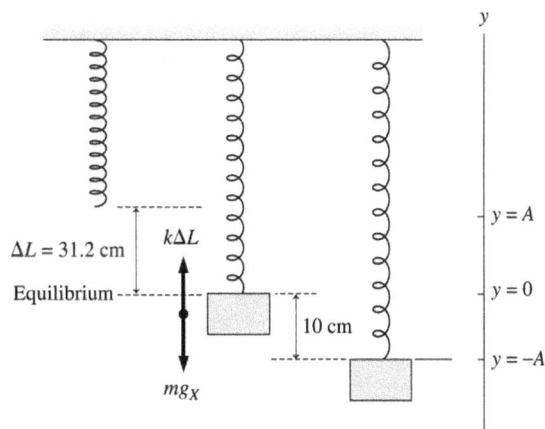

Solve: At the equilibrium position, the net force on mass m on Planet X is the following:

$$F_{\text{net}} = k\Delta L - mg_X = 0 \text{ N} \Rightarrow \frac{k}{m} = \frac{g_X}{\Delta L}$$

For simple harmonic motion, Equation 14.26 yields $k/m = (2\pi f)^2$, thus

$$(2\pi f)^2 = \frac{g_X}{\Delta L} \Rightarrow g_X = \left(\frac{2\pi}{T}\right)^2 \Delta L = \left(\frac{2\pi}{14.5 \text{ s}/10}\right)^2 (0.312 \text{ m}) = 5.86 \text{ m/s}^2$$

Assess: This value of g is of the same order of magnitude as the one for the earth, and would thus seem to be reasonable.

P14.59. Strategize: The vertical mass/spring systems are in simple harmonic motion.

Prepare: We determine the period from the graph and then use $F = 1/T$. We can then use $f = \dfrac{1}{2\pi}\sqrt{\dfrac{k}{m}}$ to relate the

frequencies (or periods) to the spring constants.

Solve: (a) For system A, y is positive for one second as the mass moves downward and reaches maximum negative y after two seconds. It then moves upward and reaches the equilibrium position, $y = 0$, at $t = 3$ seconds. The maximum speed while traveling in the upward direction thus occurs at $t = 3.0$ s. The frequency of oscillation is 0.25 Hz.

(b) For system B, all the mechanical energy is potential energy when the position is at maximum amplitude, which for the first time is at $t = 1.5$ s. The time period of system B is thus 6.0 s.

(c) Spring/mass A undergoes three oscillations in 12 s, giving it a period $T_A = 4.0$ s. Spring/mass B undergoes two oscillations in 12 s, giving it a period $T_B = 6.0$ s. From Equation 14.26, we have

$$T_A = 2\pi\sqrt{\frac{m_A}{k_A}} \quad \text{and} \quad T_B = 2\pi\sqrt{\frac{m_B}{k_B}} \Rightarrow \frac{T_A}{T_B} = \sqrt{\left(\frac{m_A}{m_B}\right)\left(\frac{k_B}{k_A}\right)} = \frac{4.0\text{ s}}{6.0\text{ s}} = \frac{2}{3}$$

If $m_A = m_B$, then

$$\frac{k_B}{k_A} = \frac{4}{9} \Rightarrow \frac{k_A}{k_B} = \frac{9}{4} = 2.25$$

Assess: It is important to learn how to read a graph.

P14.61. Strategize: The ball attached to the spring is in simple harmonic motion. We will use expressions for position and speed as functions of time.
Prepare: The position and velocity at time t are $x_0 = -5$ cm and $v_0 = 20$ cm/s. An examination of Synthesis 14.1 shows that $x(t) = A\cos(2\pi f t)$ and $\frac{v_x(t)}{2\pi f} = -A\sin(2\pi f t)$. Adding the squares of these equations and using the trigonometric relationship $\cos^2\theta + \sin^2\theta = 1$, we have $A = \sqrt{x(t)^2 + \left(\frac{v_x(t)}{2\pi f}\right)^2}$.
Solve: (a) The oscillation frequency is

$$f = \frac{1}{2\pi}\sqrt{\frac{k}{m}} = \frac{1}{2\pi}\sqrt{\frac{2.5\text{ N/m}}{0.10\text{ kg}}} = 0.796\text{ Hz}$$

The amplitude of the oscillation is

$$A = \sqrt{x(t)^2 + \left(\frac{v(t)}{2\pi f}\right)^2} = \sqrt{(-5.00\text{ cm})^2 + \left(\frac{20\text{ cm/s}}{2\pi(0.796\text{ Hz})}\right)^2} = 6.40\text{ cm}$$

(b) We can use the conservation of energy between $x_i = -5$ cm and $x_f = 3$ cm as follows:

$$\frac{1}{2}mv_i^2 + \frac{1}{2}kx_i^2 = \frac{1}{2}mv_f^2 + \frac{1}{2}kx_f^2 \Rightarrow v_f = \sqrt{v_i^2 + \frac{k}{m}(x_i^2 - x_f^2)} = 0.283\text{ m/s} = 28.3\text{ cm/s}$$

Assess: Because k is known in SI units of N/m, the energy calculation *must* be done using SI units of m, m/s, and kg. Both the amplitude and speed are reasonable.

P14.63. Strategize: The compact car is in simple harmonic motion. We can apply Equation 14.27 twice: once empty and once with passengers.
Prepare: The mass on each spring for the empty car is $(1200\text{ kg})/4 = 300$ kg. However, the car carrying four persons means that each spring has, on the average, an additional mass of 70 kg. Equation 14.26 tells us $f = \frac{1}{2\pi}\sqrt{\frac{k}{m}}$, and we will apply this twice as stated above.
Solve: First calculate the spring constant as follows:

$$f = \frac{1}{2\pi}\sqrt{\frac{k}{m}} \Rightarrow k = m(2\pi f)^2 = (300\text{ kg})[2\pi(2.0\text{ Hz})]^2 = 4.74\times10^4\text{ N/m}$$

Now reapply the equation with $m = 370 \text{ kg}$, so

$$f = \frac{1}{2\pi}\sqrt{\frac{k}{m}} \Rightarrow \frac{1}{2\pi}\sqrt{\frac{4.74\times10^4 \text{ N/m}}{370 \text{ kg}}} = 1.8 \text{ Hz}$$

Assess: A small frequency change from the additional mass is reasonable because frequency is inversely proportional to the square root of the mass.

P14.65. Strategize: A completely inelastic collision between the two gliders results in simple harmonic motion. We will apply momentum conservation for the collision, then energy conservation as the kinetic energy of the two carts is converted to spring potential energy to determine the amplitude. The frequency of oscillations is given by Equation 14.26.

Prepare: Let us denote the 250 g and 500 g masses as m_1 and m_2, which have initial velocities $(v_1)_i$ and $(v_2)_i$. After m_1 collides with and sticks to m_2, the two masses move together with velocity v_f. We will first find the final velocity using momentum conservation and then use the mechanical energy conservation equation for the two stuck gliders to determine the amount of compression or the amplitude.

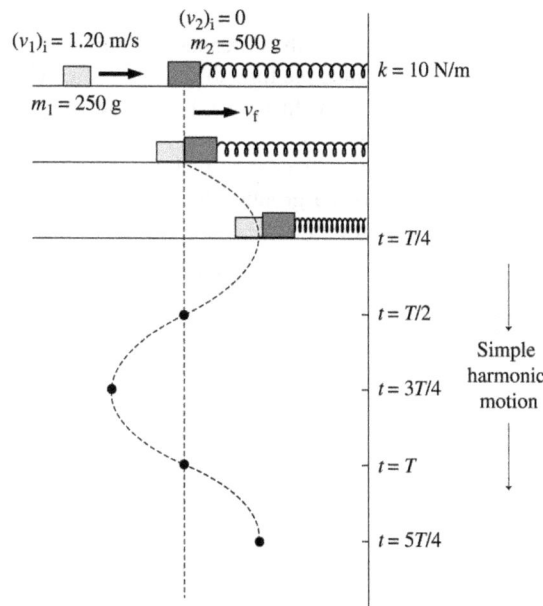

Solve: The momentum conservation equation $p_f = p_i$ for the completely inelastic collision is $(m_1 + m_2)v_f = m_1(v_1)_i + m_2(v_2)_i$. Substituting the given values,

$$(0.750 \text{ kg})v_f = (0.25 \text{ kg})(1.20 \text{ m/s}) + (0.50 \text{ kg})(0 \text{ m/s}) \Rightarrow v_f = 0.40 \text{ m/s}$$

We now use the conservation of mechanical energy equation,

$$(K + U_s)_{compressed} = (K + U_s)_{equilibrium} \Rightarrow 0 \text{ J} + \frac{1}{2}KA^2 = \frac{1}{2}(m_1 + m_2)v_f^2 + 0 \text{ J}$$

$$\Rightarrow A = \sqrt{\frac{m_1 + m_2}{k}}v_f = \sqrt{\frac{0.750 \text{ kg}}{10 \text{ N/m}}}(0.40 \text{ m/s}) = 0.11 \text{ m}$$

The period is

$$T = 2\pi\sqrt{\frac{m_1 + m_2}{k}} = 2\pi\sqrt{\frac{0.750 \text{ kg}}{10 \text{ N/m}}} = 1.7 \text{ s}$$

Assess: The magnitudes of both the amplitude and the time period are physically reasonable.

P14.67. Strategize: For the first part, we simply apply Equation 14.27A. For the second part, we consider the fact that a passenger is moving in circular motion. Although the speed is obviously changing, let us assume it is changing slowly enough that we can treat the motion near the lowest point as uniform circular motion.

Prepare: Equation 14.27A tells us $v_{max} = \sqrt{gL}\,\theta_{max}$, where $\theta_{max} = 40° = 0.698$ rad, and this angle must be expressed in radians in order to apply 14.27A. For the second part, we apply Newton's second law to the lowest point in the motion: $\sum F_r = n - mg = \dfrac{mv_{max}^2}{r} = \dfrac{mv_{max}^2}{L} \Rightarrow n = mg + \dfrac{mv_{max}^2}{L}$.

Solve: (a) Inserting the given values, we have $v_{max} = \sqrt{gL}\,\theta_{max} = \sqrt{(9.8 \text{ m/s}^2)(13 \text{ m})}\,(0.698 \text{ rad}) = 7.88 \text{ m/s}$. Here, we have included an additional significant digit for use in part (b). Our correctly-rounded answer to (a) is 7.9 m/s.

(b) Inserting the given values, we have $n = m\left(g + \dfrac{v_{max}^2}{L}\right) = (55 \text{ kg})\left((9.8 \text{ m/s}^2) + \dfrac{(7.88 \text{ m/s})^2}{(13.0 \text{ m})}\right) = 8.0 \times 10^2 \text{ N}$.

Assess: The actual force of gravity on a 55 kg person would be 540 N, so the apparent weight felt at the bottom of the path is significantly higher than this. That is consistent with our understanding that here the normal force must counteract gravity, as well as provide some centripetal force toward the center of the circular path. So our answer is reasonable.

P14.69. Strategize: The rear leg swings forward with a frequency equivalent to the natural frequency of the pendular swing of the leg. From this we can determine the time for one forward swing, and we will add 0.20 s to that for the effective time of the 2.0 m swing. Let us assume that when the rear leg lifts off the ground, the center of mass of the person is directly above the forward leg.

Prepare: Consider the figure below. With the above assumption that, for stability, the center of mass of the person is above the forward leg when the rear leg lifts off the ground, then it is clear that the center of mass moves one amplitude (half the 2.0 m distance of the swing) for each step.

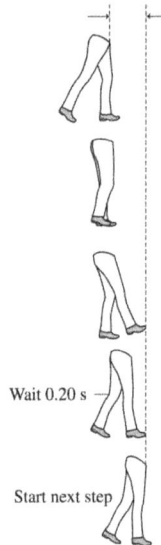

Wait 0.20 s

Start next step

Modeling the leg as a physical pendulum: a rod swinging around the hip, we can write using Equation 14.28:

$f_0 = \dfrac{1}{2\pi}\sqrt{\dfrac{mgd}{I}}$. Inserting the moment of inertia of a rod spinning around one end, we find

$f_0 = \dfrac{1}{2\pi}\sqrt{\dfrac{mg(L/2)}{\left(\dfrac{1}{3}mL^2\right)}} = \dfrac{1}{2\pi}\sqrt{\dfrac{3g}{2L}}$. Equivalently, the time required for half an oscillation is $\dfrac{T}{2} = \dfrac{1}{2f_0} = \pi\sqrt{\dfrac{2L}{3g}}$.

In part (b), we can start with Equation 14.27A: $v_{max} = \sqrt{gL}\theta_{max}$ and use that $\theta_{max}L = \Delta x_{max} = A$ to write $v_{max} = \sqrt{g/L}\,A$. But we note that this the maximum speed relative to the pivot (the hip). The hip itself is also moving forward.

Solve: (a) The speed of the walk is $\Delta x/\Delta t = \dfrac{A}{T/2+t_{rest}} = \dfrac{A}{\pi\sqrt{\dfrac{2L}{3g}}+t_{rest}} = \dfrac{(1.0\text{ m})}{\pi\sqrt{\dfrac{2(0.90\text{ m})}{3(9.8\text{ m/s}^2)}}+(0.20\text{ s})} = 1.02$ m/s.

Here we have reported the speed to one additional significant digit for use in (b). Our correctly-rounded answer to (a) is 1.0 m/s.

(b) The ratio $\dfrac{v_{max,ground}}{v_{walk}} = \dfrac{\sqrt{g/L}\,A+v_{walk}}{v_{walk}} = \dfrac{\sqrt{(9.8\text{ m/s}^2)/(0.90\text{ m})}\,(1.0\text{ m})+(1.02\text{ m/s})}{(1.02\text{ m/s})} = 4.2$, so the maximum swing speed is 3.2 times greater than the walking speed.

Assess: It is certainly reasonable that the maximum swing speed is greater than the walking speed.

P14.71. Strategize: We use Equation 14.29 to relate the loss of amplitude to the time constant. We then use the fact that total energy depends on the amplitude squared to determine the percentage of the pendulum's energy lost each period.

Prepare: Equation 14.29 tells us that $x_{max}(t) = Ae^{-t/\tau}$, and we know that when $t = T = 2.0$ s, $x_{max}/A = 99.47\%$. We insert this information to determine the time constant τ. We know that when the pendulum is at the bottom of its swing, all its energy is kinetic, such that we can write $E = K = \frac{1}{2}mv_{max}^2 = \frac{1}{2}mgL\theta_{max}^2$, and noting that $L\theta_{max} = x_{max}$, we have $E = \frac{1}{2}m\frac{g}{L}x_{max}^2$. We can write this for the initial pendulum swing and for the pendulum after one oscillation to determine the fractional loss in energy.

Solve: (a) Rearranging Equation 14.29, we have $\tau = \dfrac{-t}{\ln(x_{max}/A)}$, and inserting given values, we obtain

$$\tau = \dfrac{-t}{\ln(x_{max}/A)} = \dfrac{-(2.00\text{ s})}{\ln(0.9947)} = 376\text{ s}.$$

(b) We determine the ratio of $\dfrac{E(t=T)}{E(t=0)} = \dfrac{E=\frac{1}{2}m\frac{g}{L}x_{max}^2}{E=\frac{1}{2}m\frac{g}{L}A^2} = \left(\dfrac{x_{max}}{A}\right)^2 = (0.9947)^2 = 0.989428$, where we have kept

several additional significant digits for this intermediate step. The loss is then $1-\dfrac{E(t=T)}{E(t=0)} = 1-0.989428 = 1.06\%$

Assess: Because the loss of amplitude is so small, we expect a relatively small loss of energy in each oscillation. So around 1% is reasonable. Note that a small loss of amplitude in each oscillation means it should take a long time for the amplitude to decrease by a factor of 0.37. Thus, a time constant of a few hundred seconds is also reasonable.

P14.73. Strategize: The maximum speed occurs at the equilibrium position. We will use this maximum initial speed as the launch speed for an object in freefall.

Prepare: We will use $v_{max} = \dfrac{2\pi}{T}A$ for the moment the toy detaches, and then use $(v_f)_y^2 = (v_i)_y^2 + 2a_y\Delta y$ to determine the maximum change in height.

Solve:

$$v_{max} = \frac{2\pi}{T}A = \frac{2\pi}{0.50 \text{ s}}(0.30 \text{ m}) = 0.377 \text{ m/s}$$

After it detaches it is in free fall.

$$\Delta y_{max} = \frac{v_0^2}{2g} = \frac{(3.77 \text{ m/s})^2}{2(9.8 \text{ m/s}^2)} = 0.7251 \text{ m} \approx 73 \text{ cm}$$

Assess: This is a safe and reasonable number.

P14.75. Strategize: The oscillator is in simple harmonic motion and is damped, so we will use Equation 14.29.
Prepare: We will use $x_{max}(t) = Ae^{-t/\tau}$ once using the given information to determine the time constant, and a second time to determine the time required for the additional reduction in amplitude.
Solve: We know $x_{max}/A = 0.60$ at $t = 50$ s, so we can find τ as follows:

$$-\frac{t}{\tau} = \ln\left(\frac{x_{max}(t)}{A}\right) \Rightarrow \tau = \frac{50 \text{ s}}{\ln(0.60)} = 97.88 \text{ s}$$

Now we can find the time t_{30} at which $x_{max}/A = 0.30$

$$t_{30} = -\tau\ln\left(\frac{x_{max}(t)}{A}\right) = -(97.88 \text{ s})\ln(0.30) = 118 \text{ s}$$

The undamped oscillator has a frequency $f = 2$ Hz $= 2$ oscillations per second. Then the number of oscillations before the amplitude decays to 30% of its initial amplitude is $N = f \cdot t_{30} = (2 \text{ oscillations/s}) \cdot (118 \text{ s}) = 236 \text{ oscillations}$ or 240 oscillations to two significant figures.

P14.77. Strategize: We will use Hooke's law and Newton's second law to determine the spring constant.
Prepare: Newton's second law tells us $\sum F_y = -k\Delta y - mg = 0$, since the woman is in static equilibrium.
Solve: We are given $m = 61$ kg and $\Delta y = -2.5$ mm.

$$k = \frac{F}{-\Delta y} = \frac{mg}{-\Delta y} = \frac{(61 \text{ kg})(9.8 \text{ m/s}^2)}{-(-2.5 \text{ mm})} = 239\,000 \text{ N/m} \approx 2.4 \times 10^5 \text{ N/m}$$

The correct choice is D.
Assess: This is a very large spring constant, but the tendon is tough.

P14.79. Strategize: Model the situation as simple harmonic motion with a spring, and use Equation 14.26.
Prepare: The period is given by $T = 2\pi\sqrt{\frac{m}{k}}$.
Solve: Inserting the known values, we have

$$T = 2\pi\sqrt{\frac{m}{k}} = 2\pi\sqrt{\frac{61 \text{ kg}}{239\,000 \text{ N/m}}} = 0.10 \text{ s}$$

So the answer is choice A.
Assess: This the contact time given in the problem.

P14.81. Strategize: We will apply Newton's second law and insert Hooke's law for the restoring force of the web.
Prepare: Since the web is horizontal, it will sag until the effective spring force of the web is equal to the weight of the fly. That is $-k\Delta y = mg$.
Solve: Solving the above expression for k, obtain

$$k = -mg/\Delta y = -(12\times10^{-6} \text{ kg})(9.80 \text{ m/s}^2)/\left(-3\times10^{-3} \text{ m}\right) = 0.039 \text{ N/m}$$

The correct choice is A.
Assess: Since the web is very delicate, we expect a small effective spring constant.

P14.83. Strategize: We can examine Equation 14.26 for an oscillation.
Prepare: Changing the orientation of the web (but otherwise leaving the web alone) changes nothing in the expression for the period or frequency.
Solve: The frequency remains the same. The answer is C.
Assess: We've seen that a mass on a vertical spring is analyzed the same way as a mass and spring on a horizontal frictionless surface.

TRAVELING WAVES AND SOUND

Q15.1. Reason: (a) In a transverse wave, the thing or quantity that is oscillating, such as the particles in a string, oscillates in a direction that is transverse (perpendicular) to the direction of the propagation of the wave.
(b) Vibrations of a bass guitar string are a form of transverse wave. You can see that the oscillation is perpendicular to the string.
Assess: The plucking action makes the segment of the string move perpendicular to the string, but the disturbance is propagated along the string.

Q15.3. Reason: Wave speed is independent of wave amplitude, $v_1 = v_2 = v_3$.

Wave speed for mechanical waves depends on the properties of the medium, not the amplitude of the vibration.
Assess: If the wave speed were dependent on the amplitude then it might be the case that a later shout could overtake an earlier whisper.

Q15.5. Reason: Equation 15.2, $v_{string} = \sqrt{T_s/\mu}$, gives the wave speed on a stretched string with tension T_s and linear mass density $\mu = m/L$. We will investigate how T_s and μ are changed in each case below and how that affects the wave speed. Use a subscript 1 for the original string, and a subscript 2 for the altered string.
We are given that $(v_{string})_1 = 200$ cm/s.

(a) $(T_s)_2 = 2(T_s)_1 \quad \mu_2 = \mu_1$

$$\frac{(v_{string})_2}{(v_{string})_1} = \frac{\sqrt{(T_s)_2/\mu_2}}{\sqrt{(T_s)_1/\mu_1}} = \sqrt{\frac{(T_s)_2/\mu_2}{(T_s)_1/\mu_1}} = \sqrt{\frac{2(T_s)_1/\mu_1}{(T_s)_1/\mu_1}} = \sqrt{2}$$

$$(v_{string})_2 = \sqrt{2}(v_{string})_1 = \sqrt{2}(200 \text{ cm/s}) = 280 \text{ cm/s}$$

(b) $(T_s)_2 = (T_s)_1 \qquad m_2 = 4m_1 \Rightarrow \mu_2 = 4\mu_1$

$$\frac{(v_{string})_2}{(v_{string})_1} = \frac{\sqrt{(T_s)_2/\mu_2}}{\sqrt{(T_s)_1/\mu_1}} = \sqrt{\frac{(T_s)_2/\mu_2}{(T_s)_1/\mu_1}} = \sqrt{\frac{(T_s)_1/4\mu_1}{(T_s)_1/\mu_1}} = \frac{1}{2}$$

$$(v_{string})_2 = \frac{1}{2}(v_{string})_1 = \frac{1}{2}(200 \text{ cm/s}) = 100 \text{ cm/s}$$

(c) $(T_s)_2 = (T_s)_1 \qquad L_2 = 4L_1 \Rightarrow \mu_2 = \frac{1}{4}\mu_1$

$$\frac{(v_{string})_2}{(v_{string})_1} = \frac{\sqrt{(T_s)_2/\mu_2}}{\sqrt{(T_s)_1/\mu_1}} = \sqrt{\frac{(T_s)_2/\mu_2}{(T_s)_1/\mu_1}} = \sqrt{\frac{(T_s)_1/\frac{1}{4}\mu_1}{(T_s)_1/\mu_1}} = 2$$

$$(v_{string})_2 = 2(v_{string})_1 = 2(200 \text{ cm/s}) = 400 \text{ cm/s}$$

(d) $(T_s)_2 = (T_s)_1$ $\quad m_2 = 4m_1$ $\quad$ and $\quad L_2 = 4L_1 \Rightarrow \mu_2 = \mu_1$

$$\frac{(v_{\text{string}})_2}{(v_{\text{string}})_1} = \frac{\sqrt{(T_s)_2/\mu_2}}{\sqrt{(T_s)_1/\mu_1}} = \sqrt{\frac{(T_s)_2/\mu_2}{(T_s)_1/\mu_1}} = \sqrt{\frac{(T_s)_1/\mu_1}{(T_s)_1/\mu_1}} = 1$$

$$(v_{\text{string}})_2 = (v_{\text{string}})_1 = 200 \text{ cm/s}$$

Assess: Notice as in part (d) that if both the mass and length of the string are increased by the same factor, then μ is not changed, so the speed is the same (with no change in tension).

Q15.7. Reason: The speed of a sound wave in air increases with temperature $v_{\text{sound}} \sim \sqrt{T}$. So on a hot day the sound pulse will strike the wall and return to the detector in a shorter time than on a cold day. If the machine is calibrated for a cold day (or just a day not as hot as the one in question) the machine will think the short time is due to the walls being closer than they are. Thus, the room will be measured to be shorter than it actually is.
Assess: Similarly, on an extremely cold day, the room will be measured as being longer than it actually is.

Q15.9. Reason: Since the wave is traveling to the left, the snapshot has the same shape as the history graph. To understand why, consider that the history graph tells us about the displacement at one point in space. As the wave moves to the left, that point witnesses spots on the wave further to the right. Thus increasing t by Δt on the history graph has the same effect as increasing x by $v\Delta t$ on the snapshot graph, where v is the speed of the wave. Consequently the history graph and snapshot graph have the same shape. The only difference is that in going from the history graph to the snapshot graph, the horizontal axis is scaled by a factor of v.

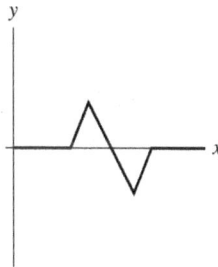

Assess: A similar argument can be used to show that if a wave is traveling to the right, the snapshot graph can be obtained from the history graph by reflecting the graph about the y-axis and scaling the horizontal axis by a factor of v.

Q15.11. Reason: The relationship between wavelength and frequency is $\lambda = v/f$. This tells us that the wavelength is inversely proportional to the frequency—the greater the frequency then, the smaller the wavelength. Since $f_1 < f_2 < f_3$, we can write $\lambda_1 > \lambda_2 > \lambda_3$.
Assess: Since the speed of sound is a constant for any given situation, the wavelength depends only on the frequency.

Q15.13. Reason: The bat's forward motion means the reflected pulse will be Doppler shifted to a higher frequency; consequently, the bat should decrease the frequency of the emitted pulse so the higher reflected pulse will be in the correct frequency range.
Assess: If the bat increases its forward speed, then it should further decrease the emitted frequency.

Q15.15. Reason: The speed of the wave's propagation depends only on the medium (the string). Specifically, $v = \sqrt{\dfrac{T_s}{\mu}}$. The speed of a piece of string moving up and down is different, and one obvious reason why it must be is that it depends on the amplitude of the motion (the distance it must cross) and on the frequency or period of the oscillation (which determines the time in which the piece of string must move up and down). In the special case of a sinusoidal disturbance, we can recall from Chapter 14 that the speed would be given by $v_y(t) = 2\pi fA\sin(2\pi ft)$. So, not only does this up and down speed depend on different things compared to the propagation speed, this one is also not constant in time. There is no fixed relationship between these two speeds.

Assess: The propagation speed depends on properties of the medium, whereas the speed of particles on the string depends on the properties of the wave.

Q15.17. Reason: Because the bullet is traveling faster than the speed of sound it creates a shock wave—a little sonic boom. This is what people at a distance from the gun may hear louder.
Assess: Even the sharp crack of a bull whip is a sonic boom, as the tip of the whip is traveling faster than the speed of sound in air.

Q15.19. Reason: The guns are calibrated for motion directly toward or away from the gun. If a car is moving perpendicular to the line of sight of the gun then it has no motion in the line of sight and there will be no Doppler shift. If the car is coming or going at some other angle then there will be a Doppler shift, but not as big as if the car were traveling along the line of sight of the gun at the same speed. So, if the radar gun reports that a car is moving at a certain speed, it is moving at least that fast, and faster if the motion isn't along the line of sight.
Assess: The police must calibrate their radar guns periodically, but if they have done so and it shows you were speeding, don't try to argue out of it by saying you were going at an angle to the line of sight.

Q15.21. Reason: From Table 15.1 the speed of sound in human tissue is 1540 m/s. We will assume normal room temperature, such that the speed in air is 343 m/s. We know the frequency of the sound is fixed between the two media (the number of compressions cannot be different at the interface), so we can equate

$$\frac{v_{air}}{\lambda_{air}} = f = \frac{v_{human}}{\lambda_{human}} \Rightarrow \lambda_{human} = \frac{v_{human}}{v_{air}}\lambda_{air} = \frac{(1540 \text{ m/s})}{(343 \text{ m/s})}(0.12 \text{ mm}) = 0.54 \text{ mm}. \text{ So the correct answer is D.}$$

Assess: Since the frequency is the same in the two media, and sound moves about four times faster in human tissue than in air, it is reasonable that the wavelength must be about four times longer in human tissue than in air.

Q15.23. Reason: The fundamental relationship for periodic waves is given in Equation 15.10, $v = \lambda f$.
We are told that $\lambda = 32$ cm. We'll compute the frequency from $f = 1/T = 1/0.20 \text{ s} = 5.0 \text{ Hz}$, $v = \lambda f = (32 \text{ cm})$ $(5.0 \text{ Hz}) = 160$ cm/s.
The correct choice is D.
Assess: Equation 15.9 is even more directly applicable, but the customary form (that you should remember) is Equation 15.10. The amplitude was unneeded information.

Q15.25. Reason: When the measured frequency is lower than f_s it is because the source is moving away from you. This is the case from $t = 0$ s to $t = 2$ s. After that the measured frequency is higher than f_s so the source is moving toward you.
The correct choice is D.
Assess: In a Doppler shift situation the measured frequency is constant as long as the source's velocity toward you (or away from you) is constant. The measured frequency doesn't keep increasing as the source gets closer and closer.

Problems

P15.1. Strategize: The wave is a traveling wave on a stretched string. As the tension is changed, let us assume that the string does not stretch appreciably, such that its linear mass density is approximately constant.
Prepare: We will use Equation 15.2 to find the wave speed.
Solve: The wave speed on a stretched string with linear density μ is $v_{string} = \sqrt{T_S/\mu}$. The wave speed if the tension is doubled will be

$$v'_{string} = \sqrt{\frac{2T_S}{\mu}} = \sqrt{2}v_{string} = \sqrt{2}(200 \text{ m/s}) = 280 \text{ m/s}$$

Assess: Wave speed increases with increasing tension.

P15.3. Strategize: Assume the speed of sound is the 20°C value 343 m/s.
Prepare: A pictorial representation of the problem is given.

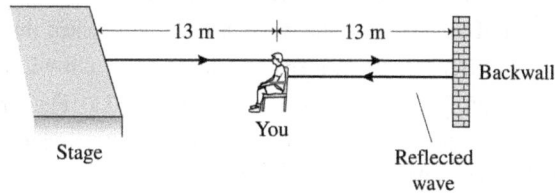

Solve: The time elapsed from the stage to you in the middle row is the distance divided by the speed of sound.

$$\frac{13 \text{ m}}{343 \text{ m/s}} = 0.0379 \text{ s}$$

The time elapsed for the same sound to get reflected from the back wall and reach your ear is

$$\frac{26 \text{ m} + 13 \text{ m}}{343 \text{ m/s}} = 0.114 \text{ s}$$

The difference in the two times is thus $0.114 \text{ s} - 0.038 \text{ s} = 0.076 \text{ s}$.
Assess: Due to the relatively large speed of sound, the observed small time difference would be expected.

P15.5. Strategize: We use Equation 15.3 describing the speed of sound in ideal gases, and solve for the molar mass.
Prepare: Equation 15.3 tells us $v_{\text{sound}} = \sqrt{\dfrac{\gamma RT}{M}}$, where $\gamma = 1.67$ for a monatomic ideal gas. The temperature must be in Kelvin: $T = 293$ K.

Solve: Rearranging and inserting values, we find $M = \dfrac{\gamma RT}{v_{\text{sound}}^2} = \dfrac{(1.67)(8.31 \text{ J/mol} \cdot \text{K})(293 \text{ K})}{(449 \text{ m/s})^2} = 20.2$ g/mol.

Consulting a periodic table, we find that this is the molar mass of neon (Ne), which is, indeed a monatomic gas that is ideal to a very good approximation.
Assess: Since air has a molar mass of about 29 g/mol, it is reasonable that this lighter gas would yield a larger speed of sound.

P15.7. Strategize: We will compute the time it takes for each of the waves to travel 45 km and then subtract to get the difference.
Prepare: Each wave moves at a constant speed, so we can write $\Delta t = \Delta x / v_x$.
Solve:

$$\Delta t_{\text{P}} = \frac{\Delta x}{v_{\text{P}}} = \frac{45 \text{ km}}{5000 \text{ m/s}} = 9.0 \text{ s}$$

$$\Delta t_{\text{S}} = \frac{\Delta x}{v_{\text{S}}} = \frac{45 \text{ km}}{3000 \text{ m/s}} = 15.0 \text{ s}$$

The difference in arrival times between the P and S waves is $15.0 \text{ s} - 9.0 \text{ s} = 6.0 \text{ s}$.
Assess: This seems to be on the order of how quickly things happen in earthquakes.

P15.9. Strategize: We sketch axes and consider what happens at the position $x = 6$ m after various time intervals.

Prepare: This is a wave traveling at constant speed. The pulse moves 1 m to the right every second.

Solve: The snapshot graph shows the wave at all points on the x-axis at $t = 0$ s. You can see that nothing is happening at $x = 6$ m at this instant of time because the wave has not yet reached this point. The leading edge of the wave is still 1 m away from $x = 6$ m. Because the wave is traveling at 1 m/s, it will take 1 s for the leading edge to reach $x = 6$ m. Thus, the history graph for $x = 6$ m will be zero until $t = 1$ s. The first part of the wave causes an upward displacement of the medium. The rising portion of the wave is 2 m wide, so it will take 2 s to pass the $x = 6$ m point. The constant part of the wave, whose width is 2 m, will take 2 seconds to pass $x = 6$ m and during this time the displacement of the medium will be a constant ($\Delta y = 1$ cm). The trailing edge of the pulse arrives at $t = 5$ s at $x = 6$ m. The displacement now becomes zero and stays zero for all later times.

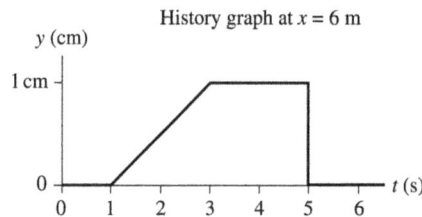

History graph at $x = 6$ m

P15.11. Strategize: The figure shows the waveform, meaning a given point on the graph represents a piece of string/medium as it move up and down. Clearly, the speed of such a piece of string will depend on the slope of this graph and the given propagation speed.

Prepare: The largest slope on the graph corresponds to moving 1.0 cm upward as the wave propagates 1.0 m. We can express this as a speed using the fact that the wave is propagating at 1.0 m/s.

Solve: Given the propagation speed, moving 1.0 m forward takes 1.0 s. Thus the maximum vertical speed of a piece of string is $\left(\dfrac{\Delta y}{\Delta t} \right)_{max} = \left(\dfrac{1.0 \text{ cm}}{1.0 \text{ s}} \right) = 1.0$ cm/s.

Assess: This is a fairly low speed, but the wave is propagating slowly and has a small amplitude, so it is reasonable.

P15.13. Strategize: We will consider how far from the point $x = 2$ m a given wave feature must have been at time $t = 0$ s in order for it to arrive at $x = 2$ m at the time indicated in the given figure.

Prepare: This is a wave traveling at constant speed to the left at 1 m/s.

Solve: This is the history graph of a wave at $x = 2$ m. Because the wave is moving to the left at 1 m/s, the wave passes the $x = 2$ m position a distance of 1 m in 1 s. Because the flat part of the history graph takes 2 s to pass the $x = 2$ m position, its width is 2 m. Similarly, the width of the linearly increasing part of the history graph is 2 m. The center of the flat part of the history graph corresponds to both $t = 0$ s and $x = 2$ m.

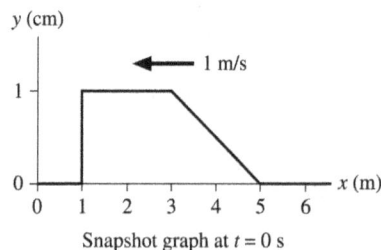

Snapshot graph at $t = 0$ s

P15.15. Strategize: We will use Equation 15.9 to find the wave speed.

Prepare: We are given all necessary quantities.

Solve: The wave speed is

$$v = \frac{\lambda}{T} = \frac{2.0 \text{ m}}{0.20 \text{ s}} = 10 \text{ m/s}$$

P15.17. Strategize: We will apply Equation 15.10 to determine the wavelength. The time is trivially determined by distance over speed, since the wave moves at a constant speed.
Prepare: Since the lab temperature is 20°C, we know the speed of sound is 343 m/s, as given in Table 15.1.
Solve: (a) $f = 40$ kHz

$$\lambda = \frac{v}{f} = \frac{343 \text{ m/s}}{40\times10^3 \text{ Hz}} = 8.6 \text{ mm}$$

(b) For the round trip the distance is 5.0 m.

$$\Delta t = \frac{\Delta x}{v_x} = \frac{5.0 \text{ m}}{343 \text{ m/s}} = 0.015 \text{ s} = 15 \text{ ms}$$

Assess: 15 ms is quick by human reaction time standards, but it is easy to have electronics in the detector record time intervals such as this.

P15.19. Strategize: We will compare this displacement formula to that given in Equation 15.7, such that we can read off properties of the wave. These will assist us in drawing the snapshots.
Prepare: This is a sinusoidal wave with amplitude 2.0 cm. If we rewrite the equation as

$$y(x,t) = (2.0 \text{ cm})\cos\left[2\pi\left(\frac{x}{1.0} - \frac{t}{0.50}\right)\right]$$

then we can see that the wavelength is 1.0 cm and the period is 0.50 s. The minus sign in the formula tells us that the wave is traveling to the right.
Solve: (a) At $t = 0$ s, the wave has a crest at the origin.
(b) At $t = 1/8$ s, one fourth of a period has elapsed since $t = 0$ s, so we redraw the wave one-quarter of a wavelength to the right of its original position.
(c) The speed of the wave is the wavelength divided by the period, $v = (1.0 \text{ cm})/(0.50 \text{ s}) = 2.0$ cm/s.

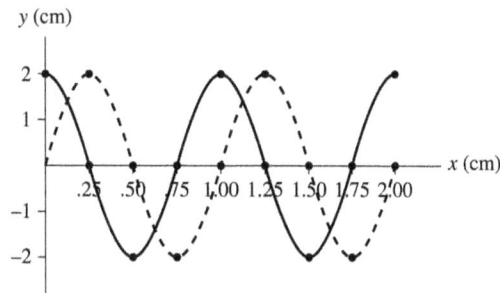

Assess: The minus sign in the formula for the displacement tells us that the wave is moving to the right, which is evidenced by the fact that the snapshot at $t = 1/8$ s, is to the right of the snapshot at $t = 0$ s.

P15.21. Strategize: We apply the definition of the amplitude and period (which gives us the frequency). We can then use the given speed and Equation 15.10 to determine the wavelength.
Prepare: This is a history graph.
Solve: The amplitude of the wave is the maximum displacement which is 6.0 cm.
The period of the wave is 0.60 s, so the frequency $f = 1/T = 1/0.60$ s $= 1.667$ Hz, or 1.7 Hz to two significant figures.
The wavelength, using Equation 15.10, is

$$\lambda = \frac{v}{f} = \frac{2 \text{ m/s}}{1.667 \text{ Hz}} = 1.2 \text{ m}$$

Assess: It is important to know how to read information from graphs.

P15.23. Strategize: We determine the initial position of the peak that will reach the point $x = 0.0$ cm first, and then use the wave speed to determine how long this will take.

Prepare: The wave is moving to the left, so the first crest to reach the point $x = 3.0$ cm will be the one that is initially at $x = 5.0$ cm (one wavelength from the crest at $x = 0.0$ cm). So the wave must propagate 2.0 cm before this first crest reaches the specified point. The speed can be determined from $v = \lambda f$ and this can be equated to $\Delta x / \Delta t$.

Solve: Combining equation and inserting the given values, we have

$$v = \lambda f = \Delta x / \Delta t \Rightarrow \Delta t = \frac{\Delta x}{\lambda f} = \frac{(2.0 \text{ cm})}{(5.0 \text{ cm})(50 \text{ Hz})} = 8.0 \text{ ms}.$$

Assess: This is a short time, which is to be expected since the wave has a speed of 250 cm/s, and it only has to propagate 2 cm.

P15.25. Strategize: We use Equation 15.10 to relate the frequency, wavelength, and speed.

Prepare: We must first look up the speed of sound in water; Table 15.1 says it is 1480 m/s.

Solve:

$$\lambda = \frac{v}{f} = \frac{1480 \text{ m/s}}{100 \times 10^3 \text{ Hz}} = 15 \text{ mm} = 1.5 \text{ cm}$$

Assess: Because the speed of sound in water is over four times the speed of sound in air, dolphins must be quick to process the sonar information.

P15.27. Strategize: We can apply Equation 15.10, in the special case of light.

Prepare: Light is an electromagnetic wave that travels with a speed of 3×10^8 m/s. The frequencies of the blue and red light are 450 nm and 650 nm, respectively.

Solve: (a) The frequency of the blue light is

$$f_{\text{blue}} = \frac{c}{\lambda} = \frac{3.0 \times 10^8 \text{ m/s}}{450 \times 10^{-9} \text{ m}} = 6.7 \times 10^{14} \text{ Hz}$$

(b) The frequency of the red light is

$$f_{\text{red}} = \frac{3.0 \times 10^8 \text{ m/s}}{650 \times 10^{-9} \text{ m}} = 4.6 \times 10^{14} \text{ Hz}$$

Assess: A higher wavelength for the red compared to the blue light means that the frequency for the red light is smaller than the blue light.

P15.29. Strategize: We determine the time required for the bullet to reach the target, and then for the sound to reach the target and we find the difference.

Prepare: We can write $v_{\text{sound}} = \Delta x / \Delta t_{\text{sound}}$ and $v_{\text{bullet}} = \Delta x / \Delta t_{\text{bullet}}$, and we assume $v_{\text{sound}} = 343$ m/s .

Solve: Rearranging, we find $\Delta t_{\text{sound}} - \Delta t_{\text{bullet}} = \Delta x \left(\dfrac{1}{v_{\text{sound}}} - \dfrac{1}{v_{\text{bullet}}} \right) = (500 \text{ m}) \left(\dfrac{1}{343 \text{ m/s}} - \dfrac{1}{1,000 \text{ m/s}} \right) = 0.96$ s .

Assess: Note that the time required for sound to reach the target is larger than the time for the bullet to reach the target. So the bullet strikes the target and almost one second later, you would hear the shot at the target's location.

P15.31. Strategize: We apply the definition of intensity to this case in which the relevant area is the surface area of the ear drum.

Prepare: The power (or energy/time) is the intensity multiplied by the area. The intensity is $I = 1.0 \times 10^{-6}$ W/m². We can deduce from the information given that the area of the eardrum is $A = \pi R^2 = \pi (4.2 \text{ mm})^2 = 5.54 \times 10^{-5}$ m².

Solve:

$$P = Ia = (1.0 \times 10^{-6} \text{ W/m}^2)(5.54 \times 10^{-5} \text{ m}^2) = 5.5 \times 10^{-11} \text{ W}$$

The energy delivered to your eardrum each second is 55 pJ.

Assess: This is an incredibly tiny amount of energy per second; you should be impressed that your ear can detect such small signals! Your eardrum moves back and forth about the width of 100 atoms in such cases!

P15.33. Strategize: We will use the intensity to determine the power delivered to the surface area of the back, and then we will use the duration of the tanning session to determine the total energy.
Prepare: We are asked to find the energy received by your back of area 30 cm×50 cm in 1.0 h if the electromagnetic wave intensity is 1.4×10^3 W/m^2. The energy delivered to your back in time t is $E = Pt$, where P is the power of the electromagnetic wave. The intensity of the wave is $I = P/a$ where a is the area of your back.
Solve: The energy received by your back is

$$E = Pt = Iat = (0.80)(1400 \text{ W/m}^2)(0.30 \times 0.50 \text{ m}^2)(3600 \text{ s}) = 6.0 \times 10^5 \text{ J}$$

Assess: This is equivalent to receiving approximately 170 J of energy per second by your back. This energy is relatively large and will certainly lead to tanning.

P15.35. Strategize: We will use Equation 15.13 to relate the intensities at different distances from the source.
Prepare: If the solar panel produces electrical power with efficiency ε, then the power received from the sun is $P_E = (1.0 \text{ kW})/\varepsilon$. This is because if we multiply the power from the sun by ε, we should get $\varepsilon P_E = 1.0$ kW. If the area of the solar panel is a, then the intensity of radiation from the sun is $(1.0 \text{ kW})/\varepsilon a$. To find the intensity at some other distance from the sun, we use $I_1/I_2 = r_2^2/r_1^2$.

Known
$r_S/r_E = 9.5$
$\varepsilon P_E = 1.0$ kW

Find
εP_S

Solve: Using the previous formula, we can find the intensity of sunlight at the location of Saturn's orbit.

$$I_S = I_E \left(\frac{r_E}{r_S} \right)^2 = \left(\frac{1.0 \text{ kW}}{\varepsilon a} \right) \left(\frac{1}{9.5} \right)^2 = \frac{11 \text{ W}}{\varepsilon a}$$

Then using $P = Ia$, we find that $P_S = (11 \text{ W})/\varepsilon$, that is the solar panel would receive $(11 \text{ W})/\varepsilon$, of power from the sun. Finally, since the panels work with efficiency ε, the power which would be produced if the spacecraft were in orbit around Saturn is $\varepsilon P_S = 11$ W.

Assess: The reduction from 1000 W to 11 W corresponds to a factor of about 100. This is reasonable since intensity is inversely proportional to distance squared. At Saturn's orbit, the satellite is about 10 times farther away from the sun so the intensity of light it receives is reduced by a factor of about 100.

P15.37. Strategize: We use the definition of power for part (a) and the definition of intensity for part (b).
Prepare: To find the power of a laser pulse, we need the energy it contains, U, and the time duration of the pulse, Δt. Then to find the intensity, we need the area of the pulse. Its radius is 0.50 mm.
Solve: (a) Using $P = U/\Delta t$, we find the following:

$$P = (1.0 \times 10^{-3} \text{ J})/(15 \times 10^{-9} \text{ s}) = 6.67 \times 10^4 \text{ W} \approx 6.7 \times 10^4 \text{ W}$$

(b) Then from $I = P/a$, we obtain

$$I = \frac{(6.67 \times 10^4 \text{ W})}{\pi (5.0 \times 10^{-4} \text{ m})^2} = 8.5 \times 10^{10} \text{ W/m}^2$$

Assess: This is very intense light. Using the data from Problem 15.36, the laser light is about 400 million times as intense as the energy from the sun.

P15.39. Strategize: We apply Equation 15.11, which is an effective definition of intensity.
Prepare: Equation 15.11 tells us that $I = P/a$, and here the area is given by $a = \pi r^2$.

Solve: Inserting the given values, we have $I = P/\pi r^2 = \dfrac{(2 \times 10^{15} \text{ W})}{\pi (15 \times 10^{-6} \text{ m})^2} = 2.8 \times 10^{24} \text{ W/m}^2$.

Assess: This is an astronomical intensity, as expected from the problem statement.

P15.41. Strategize: We consult Table 15.3 and use Equation 15.12 to relate the intensity to the distance from the source.
Prepare: Table 15.3 tells us that the intensity of a whisper at one meter is $I_1 = 1.0 \times 10^{-10} \text{ W/m}^2$. Equation 15.12 reads $I = \dfrac{P_{\text{source}}}{4\pi r^2}$, and we will apply it twice: once to the intensity at 2.0 m, and once at 1.0 m.

Solve:

$$\frac{I_2}{I_1} = \frac{\dfrac{P_{\text{source}}}{4\pi r_2^2}}{\dfrac{P_{\text{source}}}{4\pi r_1^2}} = \frac{r_1^2}{r_2^2} = \frac{(1.0 \text{ m})^2}{(2.0 \text{ m})^2} = 0.25$$

$$I_2 = (0.25)I_1 = (0.25)(1.0 \times 10^{-10} \text{ W/m}^2) = 0.25 \times 10^{-10} \text{ W/m}^2 = 2.5 \times 10^{-11} \text{ W/m}^2$$

The sound intensity level is given by Equation 15.14 where $I_0 = 1.0 \times 10^{-12} \text{ W/m}^2$.

$$\beta = (10 \text{ dB}) \log_{10}\left(\frac{I}{I_0}\right) = (10 \text{ dB}) \log_{10}\left(\frac{2.5 \times 10^{-11} \text{ W/m}^2}{1.0 \times 10^{-12} \text{ W/m}^2}\right) = 14 \text{ dB}$$

Assess: Table 15.3 gives the sound intensity level for a whisper at 1 m as 20 dB; we expect it to be less at 2 m, and 14 dB is just about what we expect.

P15.43. Strategize: We apply Equation 15.13 to relate the intensities to the distances from the source for part (a). For part (b), we use Equation 15.14 to determine the sound level.

Prepare: Equation 15.13 tells us $\dfrac{I_1}{I_2} = \dfrac{r_2^2}{r_1^2}$. Once we have determined the intensity, we can apply $\beta = (10 \text{ dB}) \log_{10}\left(\dfrac{I}{I_0}\right)$ to determine the sound level in decibels.

Solve: (a) If a source of spherical waves radiates uniformly in all directions, the ratio of the intensities at distances r_1 and r_2 is

$$\frac{I_1}{I_2} = \frac{r_2^2}{r_1^2} \Rightarrow \frac{I_{50 \text{ m}}}{I_{2 \text{ m}}} = \left(\frac{2 \text{ m}}{50 \text{ m}}\right)^2 = 1.6 \times 10^{-3} \Rightarrow I_{50 \text{ m}} = I_{2 \text{ m}}(1.6 \times 10^{-3}) = (2.0 \text{ W/m}^2)(1.6 \times 10^{-3}) = 3.2 \times 10^{-3} \text{ W/m}^2$$

(b) The sound intensity level is given from Equation 15.14.

$$\beta = (10 \text{ dB}) \log_{10}\left(\frac{I}{I_0}\right) = (10 \text{ dB}) \log_{10}\left(\frac{3.2 \times 10^{-3} \text{ W/m}^2}{1.0 \times 10^{-12} \text{ W/m}^2}\right) = 95 \text{ dB}$$

Assess: The power generated by the sound source is $P = I_{2\,m} [4\pi(2\text{ m})^2] = (2.0\text{ W/m}^2)(50.27) = 101\text{ W}$. This is a significant amount of power.

P15.45. Strategize: The power is fixed, but the intensity and sound level depend on the position of the observer. We can use Equation 15.14 to write the sound level in terms of intensity, and then use Equation 15.12 to relate the intensities at difference distances.

Prepare: Equation 15.14 can be rearranged to yield $I = I_0 10^{\left(\frac{\beta}{10\text{ dB}}\right)}$. This can then be used in Equation 15.12, which we write twice: once for each position described: $I_1 4\pi r_1^2 = P_{\text{source}} = I_2 4\pi r_2^2$, to determine P_{source}.

Solve: Combining the above equations, we have $I_2 = I_1 \left(\dfrac{r_1}{r_2}\right)^2 = I_0 10^{\left(\frac{\beta}{10\text{ dB}}\right)} \left(\dfrac{r_1}{r_2}\right)^2$. Inserting given values, we find

$$I_2 = \left(10^{-12}\text{ W/m}^2\right) 10^{\left(\frac{107\text{ dB}}{10\text{ dB}}\right)} \left(\frac{0.50\text{ m}}{3.0\text{ m}}\right)^2 = 1.4\text{ mW/m}^2.$$

Assess: This is a reasonable intensity for the sound of a very loud insect.

P15.47. Strategize: We will use Equation 15.14 to relate the given sound level to an intensity in W/m^2. We can then use the ratios of distances to determine the intensity 35 m from the speaker using Equation 15.12. We can then apply Equation 15.14 a second time to determine the required sound level.

Prepare: Knowing the sound intensity level $\beta_1 = 120\text{ dB}$ at the point $r_1 = 5\text{ m}$, we can determine the intensity I_1 at this point. Since the power output of the speaker is a constant, knowing the intensity I_1 and distance r_1 from the speaker at one point, we can determine the intensity I_2 at a second point r_2. Knowing the intensity of the sound I_2 at this second point r_2, we can determine the sound intensity level β_2 at the second point.

Solve: The sound intensity may be determined from the sound intensity level as follows: $\beta_1 = (10\text{ dB}) \log_{10} (I_1/I_0)$.

Inserting numbers: $120\text{ dB} = (10\text{ dB}) \log_{10} [I_1/(10^{-12}\text{ W/m}^2)]$ or $12 = \log_{10} [I_1/(10^{-12}\text{ W/m}^2)]$

Taking the antilog of both sides obtains $10^{12} = I_1/(10^{-12}\text{ W/m}^2)$ which may be solved for I_1 to obtain $I_1 = (10^{-12}\text{ W/m}^2)(10^{12}) = 1\text{ W/m}^2$.

Knowing how the sound intensity is related to the power and that the power does not change, we can determine the sound intensity at any other point.

Inserting $a = 4\pi r^2$, into $P = I_1 a_1 = I_2 a_2$ obtain $I_2 = I_1 (r_1/r_2)^2 = (1\text{ W/m}^2)(5/35)^2 = 2.04\times 10^{-2}\text{ W/m}^2$.

Finally, knowing the sound intensity at the second point r_2, we can determine the sound intensity level at this point by

$$\beta_2 = (10\text{ dB}) \log_{10} (I_2/I_0) = (10\text{ dB}) \log_{10} (2.04\times 10^{-2}/1\times 10^{-12}) = (10\text{ dB}) \log_{10} (2.04\times 10^{10})$$

continuing

$$\beta_2 = 10\text{ dB}[\log_{10} (2.04) + \log_{10}(10^{10})] = 10\text{ dB}[0.31 + 10] = 103\text{ dB} \approx 100\text{ dB}$$

Assess: This is still as loud as a pneumatic hammer. Either take ear protection or move farther away from the speaker.

P15.49. Strategize: We will use Equation 15.14 to relate the intensity level to the intensity. Solve this in two steps; first find the intensity that corresponds to 94 dB, then solve for the distance r.

Prepare: We note that $I_0 = 1.0\times 10^{-12}\text{ W/m}^2$, and we will use that $\beta = (10\text{ dB}) \log_{10}\left(\dfrac{I}{I_0}\right)$.

Solve:

$$\beta = (10\text{ dB}) \log_{10}\left(\frac{I}{I_0}\right) \Rightarrow I = I_0 10^{\beta/10\text{ dB}} = (1.0\times 10^{-12}\text{ W/m}^2) 10^{94\text{ dB}/10\text{ dB}} = 2.51\text{ mW/m}^2$$

The distance r is the radius of the spherical wave.

$$I = \frac{P_{source}}{4\pi r^2} \Rightarrow r = \sqrt{\frac{P_{source}}{4\pi I}} = \sqrt{\frac{50 \text{ W}}{4\pi (2.51 \text{ mW/m}^2)}} = 40 \text{ m}$$

Assess: This is a **minimum** distance you should stand from such a speaker for the time of a concert.

P15.51. Strategize: The frequency of the opera singer's note is altered by the Doppler effect. Since the source is moving, we will use Equations 15.16.
Prepare: The frequency is f_+ as the car approaches and f_- as it moves away. f_s is the frequency of the source. The speed of sound in air is 343 m/s.
Solve: (a) Using 90 km/hr = 25 m/s, the frequency as her convertible approaches the stationary person is

$$f_+ = \frac{f_s}{1 - v_s/v} = \frac{600 \text{ Hz}}{1 - \dfrac{25 \text{ m/s}}{343 \text{ m/s}}} = 650 \text{ Hz}$$

(b) The frequency as her convertible recedes from the stationary person is

$$f_- = \frac{f_s}{1 + v_s/v} = \frac{600 \text{ Hz}}{1 + \dfrac{25 \text{ m/s}}{343 \text{ m/s}}} = 560 \text{ Hz}$$

Assess: As would have been expected, the pitch is higher in front of the source than it is behind the source.

P15.53. Strategize: Sound frequency is altered by the Doppler effect. The frequency increases for an observer approaching the source and decreases for an observer receding from the source. You need to ride your bicycle away from your friend to lower the frequency of the whistle. We will use Equation 15.17, since the observer is moving.
Prepare: Since you will ride away from your friend, we will use $f_- = \left(1 - \dfrac{v_s}{v}\right) f_s$.

Solve: The minimum speed you need to travel is calculated as follows:

$$f_- = \left(1 - \frac{v_s}{v}\right) f_s \Rightarrow 20 \text{ kHz} = \left(1 - \frac{v_s}{343 \text{ m/s}}\right)(21 \text{ kHz}) \Rightarrow v_s = 16 \text{ m/s}$$

Assess: A speed of 16 m/s corresponds to approximately 35 mph. This is a possible but very fast speed on a bicycle.

P15.55. Strategize: We will use Equation 15.16 for part (a), because the source is moving. For part (b), you become the source, and it is the observer that is moving. So, for part (b) we will use Equation 15.17.
Prepare: Your friend's frequency is altered by the Doppler effect. The frequency of your friend's note increases as he races toward you (moving source and a stationary observer). The frequency of your note for your approaching friend is also higher (stationary source and a moving observer). In both parts, we will use the expressions in which the frequency is increased due to the source and observer approaching each other: $f_+ = \dfrac{f_s}{1 - \frac{v_s}{v}}$ and $f_+ = f_s\left(1 + \frac{v_s}{v}\right)$.

Solve: (a) The frequency of your friend's note as heard by you is

$$f_+ = \frac{f_s}{1 - \frac{v_s}{v}} \Rightarrow \frac{400 \text{ Hz}}{1 - \frac{25.0 \text{ m/s}}{343 \text{ m/s}}} = 431 \text{ Hz}$$

(b) The frequency heard by your friend of your note is

$$f_+ = f_s\left(1 + \frac{v_s}{v}\right) = (400 \text{ Hz})\left(1 + \frac{25.0 \text{ m/s}}{343 \text{ m/s}}\right) = 429 \text{ Hz}$$

Assess: As would have been expected, the pitch is higher in front of the source than it is behind the source.

P15.57. Strategize: The frequency will be shifted due to the Doppler effect. We will use Equation 15.18, because this involves a reflection of a sound wave.

Prepare: The frequency shift of the ultrasound reflected from blood moving in the artery may be determined by $\Delta f = \pm 2 f_s(v_s/v)$. Here f_s is the frequency of the ultra sound, v is the speed of the ultrasound in human tissue (1540 m/s) and v_s is the speed of the blood cell reflecting the ultrasound.

Solve:

$$\Delta f = \pm 2 f_s(v_s/v) = \pm 2(5.0 \times 10^6 \text{ Hz})[(0.20 \text{ m/s})/(1540 \text{ m/s})] = 1.3 \text{ kHz}$$

Assess: Compared to examples in the text, this is a reasonable number.

P15.59. Strategize: We determine the speed of sound in air at the given temperature.

Prepare: Equation 15.3 describes the speed of sound and how it depends on different properties of a medium: $v_{sound} = \sqrt{\dfrac{\gamma R T}{M}}$. We do not need to determine the values of γ and M if we simply assume that γ is the same at $-57°$ C and $20°$ C. In that case, we can simply write this expression at each temperature and use the ratio to determine the speed of sound at the colder temperature.

Solve: Writing Equation 15.3, we have

$$\frac{v_{sound}\left(-57° \text{ C}\right)}{v_{sound}\left(20° \text{ C}\right)} = \frac{\sqrt{\dfrac{\gamma R T_{cold}}{M}}}{\sqrt{\dfrac{\gamma R T_{room\ temp}}{M}}} = \sqrt{\frac{T_{cold}}{T_{room\ temp}}}$$

$$v_{sound}\left(-57° \text{ C}\right) = v_{sound}\left(20° \text{ C}\right)\sqrt{\frac{216 \text{ K}}{293 \text{ K}}} = (343 \text{ m/s})\sqrt{\frac{216 \text{ K}}{293 \text{ K}}} = 294.5 \text{ m/s}$$

Since the Concorde flew at twice this speed, its speed was 589 m/s.

Assess: We know the speed of sound is lower at lower temperatures, so we expect an answer that is less than twice the speed of sound at room temperature. So, the fact that our answer is less than 686 m/s makes sense.

P15.61. Strategize: To get the time needed for a wave to travel down a string, we need the length of the string and the speed of the wave. The speed comes from Equation 15.2. We are given the mass and length of the string from which we can find the mass density.

Prepare: The diagram below will help us organize our efforts.

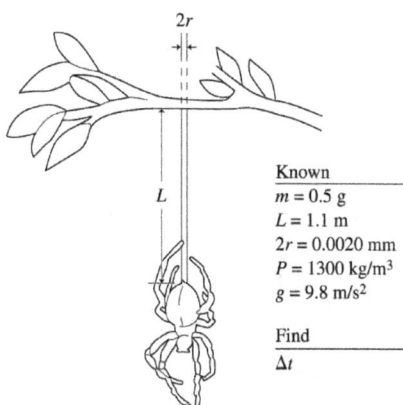

Known
m = 0.5 g
L = 1.1 m
2r = 0.0020 mm
P = 1300 kg/m³
g = 9.8 m/s²

Find
Δt

Solve: Since the silk is cylindrical, its volume is given by $V = \pi r^2 L$. From this equation we can find the linear density as follows:

$$\mu = \frac{m}{L} = \frac{\pi r^2 m}{\pi r^2 L} = \frac{\pi r^2 m}{V} = \pi r^2 \rho = \pi (1.0 \times 10^{-6} \text{ m})^2 (1300 \text{ kg/m}^3) = 4.08 \times 10^{-9} \text{ kg/m}$$

The speed of the wave on the string is $v = \sqrt{\dfrac{T}{\mu}} = \sqrt{\dfrac{4.9 \times 10^{-3} \text{ N}}{4.08 \times 10^{-9} \text{ kg/m}}} = 1.10 \times 10^3$ m/s. The time needed for a pulse to travel the length of the string is just the length of the string divided by the speed of waves on the string.

$$\Delta t = \frac{L}{v} = \frac{1.1 \text{ m}}{1.10 \times 10^3 \text{ m/s}} = 1.0 \text{ ms}$$

Assess: It is good for the spider that very little time is needed for it to receive this information. Notice that the speed of the wave is greater than the speed of sound in air.

P15.63. Strategize: The time lapse is calculated trivially, since we assume the waves propagated at a constant speed. We can answer part (b) by simply considering the direction of propagation and of oscillation. Equation 15.10 can be used to find the wavelength from the speed and frequency. We will use Equation 14.17 to relate the maximum acceleration to the frequency and amplitude.

Prepare: We collect together the relevant equations: $v = \Delta x / \Delta t$, $v = f\lambda$, and $a_{max} = (2\pi f)^2 A$.

Solve: **(a)** To find the time between the earthquake and first detection we simply divide the distance traveled by the speed of the wave, $\Delta t = \dfrac{\Delta x}{v} = \dfrac{200 \text{ km}}{7.0 \text{ km/s}} = 29$ s.

(b) This was a longitudinal wave since it traveled from east to west and caused the ground to vibrate in an east–west direction.

(c) The wavelength was $\lambda = \dfrac{v}{f} = \dfrac{7.0 \text{ km/s}}{1.1 \text{ Hz}} = 6.4$ km.

(d) The maximum horizontal displacement of the ground, that is, A, can be obtained by solving Equation 14.17 for A:

$$A = \frac{a_{max}}{(2\pi f)^2} = \frac{0.25(9.8 \text{ m/s}^2)}{(2\pi(1.1 \text{ Hz}))^2} = 0.051 \text{ m} = 5.1 \text{ cm}.$$

Assess: It is interesting to note that even though the wave traveled incredibly fast—7.0 km/s is about 16 000 mph, the ground moved *much* slower. The maximum speed of the ground, using Equation 14.15, is given by the following:

$$v_{max} = (2\pi f)A = 2\pi(1.1 \text{ Hz})(5.5 \times 10^{-2} \text{ m}) = 0.38 \text{ m/s} \left(\frac{1 \text{ mi}}{1600 \text{ m}}\right)\left(\frac{3600 \text{ s}}{1 \text{ h}}\right) = 0.85 \text{ mph},$$

which is a snail's pace.

P15.65. Strategize: We will use Equation 15.2 to relate the speed of a wave on a string to the linear mass density of the string. We assume that the tension is the same throughout the rope and we assume material is the same regardless of diameter.

Prepare: We start by noting that if $\mu = m/L$ and $\rho = m/Vol$, then $\mu = \rho A = \rho \pi r^2$. Since we assume the material is the same, the volume density is the same throughout the rope; the linear mass density only increases because the radius increases. We can then write $v = \sqrt{\dfrac{T_s}{\mu}}$ twice (once for each radius) and use this to compare the speeds in the thin (t) segment and the fat segment (f).

Solve: $\dfrac{v_f}{v_t} = \dfrac{\sqrt{\dfrac{T_s}{\rho \pi r_f^2}}}{\sqrt{\dfrac{T_s}{\rho \pi r_t^2}}} \Rightarrow v_f = v_t \left(\dfrac{r_t}{r_f} \right) = (22 \text{ m/s}) \left(\dfrac{1}{2} \right) = 11 \text{ m/s}.$

Assess: We expect the thicker rope to have a greater linear mass density; we therefore expect the speed to be lower than 22 m/s, which it is.

P15.67. Strategize: We will use Equation 15.10 to relate the frequency and wavelength to the speed.
Prepare: We know that $v = \lambda f$, and we are given both wavelength and frequency.
Solve: (a) We are given that $\lambda = 30$ m and $v = \lambda f = (30 \text{ m})(0.30 \text{ Hz}) = 9.0 \text{ m/s}.$
(b) The frequency of the wave is $f = v/\lambda$ where the speed is the speed of the waves **plus** the speed of the boat. $f = v/\lambda = (14.0 \text{ m/s})/(30 \text{ m}) = 0.47 \text{ Hz}.$
Assess: Horizontal oscillations can also contribute to motion sickness, as can rolling oscillations. In fact, motion sickness can even be produced without physical oscillation of the body, but by moving (including oscillating) visual stimuli such as flight trainers.

P15.69. Strategize: We use the facts that the wave is cyclic in a spatial translation of $\Delta x = \lambda$ or a time translation of $\Delta t = T$ to relate the arguments inside the cosine function to the wavelength and period of this wave.

Prepare: The argument inside the cosine function must be of the form $\dfrac{2\pi x}{\lambda}$ and $\dfrac{2\pi t}{T}$ in order for the cosine function to repeat itself under the above-mentioned translations. We equate these to the terms in the given expression, and finally use $v = \lambda f = \lambda / T$ to determine the wave speed.

Solve: For the wavelength, we know $\dfrac{2\pi x}{\lambda} = 2.4x \Rightarrow \lambda = \dfrac{2\pi}{2.4} = 2.62$ m, and for the period, we know

$\dfrac{2\pi t}{T} = \left(1.4 \times 10^4 \text{ rad/s} \right) t \Rightarrow T = \dfrac{2\pi}{\left(1.4 \times 10^4 \text{ rad/s} \right)} = 4.49 \times 10^{-4}$ s, or equivalently $f = 1/T = 2,230$ Hz.

Thus, $v = \lambda f = (2.62 \text{ m})(2,230 \text{ Hz}) = 5.8 \times 10^3$ m/s.

Assess: Table 15.1 does not give the speed of sound in steel, but it gives the speed of sound in other solids like aluminum and granite. The speed of the compression wave we calculated here lies directly between the speeds of sound in aluminum and granite, which is very encouraging.

P15.71. Strategize: We will insert the given quantities into Equation 15.8. Equation 15.9 will be used to determine the period.
Prepare: All quantities, except the period, needed to write the y-equation for a wave traveling in the negative x-direction are given. We match the format: $y(x,t) = A \cos \left(2\pi \left(\dfrac{x}{\lambda} + \dfrac{t}{T} \right) \right)$, and use the fact that $T = \dfrac{\lambda}{v}$.

Solve: The period is calculated as follows: $T = \dfrac{\lambda}{v} = \dfrac{0.50 \text{ m}}{4.0 \text{ m/s}} = 0.125$ s.

The displacement equation for the wave is

$$y(x,\ t) = (5.0 \text{ cm}) \cos \left[2\pi \left(\dfrac{x}{50 \text{ cm}} + \dfrac{t}{0.125 \text{ s}} \right) \right]$$

Assess: The positive sign in the cosine function's argument indicates motion along the $-x$ direction.

P15.73. Strategize: We will reach the sound level of each threshold from the graph, and then find use Equation 15.14 to relate these sound levels to the intensity in W/m^2.

Prepare: We are asked for the ratio of the intensities: $I = I_0 10^{\beta/(10\ dB)}$. We use this equation for the threshold at 1kHz (which is 25 dB) and for the threshold at 6 kHz (which is 80 dB).

Solve: The ratio of the two intensities is

$$\frac{I(1\ kHz)}{I(6\ kHz)} = \frac{I_0 10^{(25\ dB)/(10\ dB)}}{I_0 10^{(80\ dB)/(10\ dB)}} = 10^{-5.5} = 3.2 \times 10^{-6}.$$

Assess: Clearly the faintest sound that can be heard at 1 kHz is much quieter than the faintest sound that can be heard at 6 kHz, which is consistent with the graph.

P15.75. Strategize: This integrated problem will bring back some concepts from thermodynamics. We will first determine the power incident on the cross-sectional area of the bottle, and from that determine the total energy input over the 5.0 minute period. We will then use the specific heat of water to determine what temperature change this increase in thermal energy could cause.

Prepare: We will need $P = Ia$, $E = P\Delta t$, and $E = cm\Delta T$. The mass of the water is $m = \rho V = \rho \pi r^2 L$. The cross sectional area of the bottle lying on its side is the length times the diameter. The specific heat of water is $c = 4190\ J/kg \cdot K$.

Solve: Set the two equations for energy equal to each other.

$$P\Delta t = cm\Delta T$$
$$Ia\Delta t = c(\rho \pi r^2)\Delta T$$

Solve for the temperature change.

$$\Delta T = \frac{Ia\Delta t}{c(\rho \pi r^2)L} = \frac{(1000\ W/m^2)(0.60)(0.22\ m \times 0.070\ m)(300\ s)}{(4190\ J/kg \cdot K)\left((1000\ kg/m^3)\pi(0.035\ m)^2\right)(0.22\ m)} = 0.78\ K$$

Assess: This is not a very large temperature rise, due to the large specific heat capacity of water.

P15.77. Strategize: Assume the wave spreads out in a spherical pattern, so the area goes with the square of the radius. Then Equation 15.13 allows us to determine the intensity farther from the siren. Equation 15.14 then allows us to express this as a sound level in decibels.

Prepare: The ratio of intensities is related to the ratio of distances through $\frac{I_1}{I_2} = \left(\frac{r_2}{r_1}\right)^2$, and the sound level is given by $\beta = (10\ dB) \log_{10}\left(\frac{I}{I_0}\right)$.

Solve: Since 300 m is six times farther away than 50 m is, the intensity drops off by a factor of $\left(\frac{1}{6}\right)^2$.

$$0.10\ W/m^2 \left(\frac{1}{6}\right)^2 = 2.778\ mW/m^2$$

The sound intensity level of the source is calculated as follows:

$$\beta = (10\ dB) \log_{10}\left(\frac{I}{I_0}\right) = (10\ dB) \log_{10}\left(\frac{.2778\ mW/m^2}{1.0 \times 10^{-12}\ W/m^2}\right) = 94\ dB$$

Assess: The sound intensity level (compare with values in Table 15.3) is quite loud, as you would expect.

P15.79. Strategize: We will use Equation 15.14 to determine the intensity of the mouse's threshold of hearing, and then we will use Equation 15.13 to determine the distance based on the ratio of intensities.

Prepare: We are given the intensity level of the sound but we need the intensity so we solve Equation 15.14, $I = I_0 \times 10^{(\beta/10 \text{ dB})} = 1.0 \times 10^{-13} \text{ W/m}^2$. Let r_1 and I_1 be the distance to and intensity experienced by you and let r_2 and I_2 be the distance to and intensity experienced by the mouse. Then we can use Equation 15.13 to get the distance from the leaf to the mouse, r_2.

Known
$\beta_1 = 0 \text{ dB}$
$\beta_2 = -10 \text{ dB}$
$r_1 = 1.5 \text{ m}$

Find
r_2

Solve: Solving Equation 15.13 for r_2, we obtain the following:

$$r_2 = \sqrt{r_1^2 \left(\frac{I_1}{I_2}\right)} = \sqrt{(1.5 \text{ m})^2 \left(\frac{1.0 \times 10^{-12} \text{ W/m}^2}{1.0 \times 10^{-13} \text{ W/m}^2}\right)} = 4.7 \text{ m}$$

Assess: To decrease β by 10 dB, the intensity must be decreased by a factor of 10. We know that intensity is inversely proportional to the distance to the source squared, so a reduction in the intensity by a factor of 10 means an increase in r by a factor of $\sqrt{10}$, which is about 3. So it makes sense that the mouse is about three times farther away from the leaf than you are.

P15.81. Strategize: The sound generator's frequency is altered by the Doppler effect. According to Equations 15.16, the frequency increases as the generator approaches the student, and it decreases as the generator recedes from the student.

Prepare: At all times it is the source that is moving; the only thing that changes is that the generator is sometimes moving toward the students and sometimes away from them. We will therefore use $f_+ = \dfrac{f_S}{1 - v_s/v}$ and $f_- = \dfrac{f_S}{1 + v_s/v}$.

Convert rpm into SI units. Use 343 m/s for the speed of sound.

Solve: The generator's speed is

$$v_s = r\omega = r(2\pi f) = (1.0 \text{ m})2\pi\left(\frac{100}{60} \text{ rev/s}\right) = 10.47 \text{ m/s}$$

The frequency of the approaching generator is

$$f_+ = \frac{f_S}{1 - v_s/v} = \frac{600 \text{ Hz}}{1 - \frac{10.47 \text{ m/s}}{343 \text{ m/s}}} = 620 \text{ Hz}$$

Doppler effect for the receding generator, on the other hand, is

$$f_- = \frac{f_S}{1 + v_s/v} = \frac{600 \text{ Hz}}{1 + \frac{10.47 \text{ m/s}}{343 \text{ m/s}}} = 580 \text{ Hz}$$

Thus, the highest and the lowest frequencies heard by the student are 620 Hz and 580 Hz.

P15.83. Strategize: The time can be trivially calculated using the speed of sound, since it is assumed to be constant.

Prepare: Knowing that $v = d/t$ and that the distance d includes out and back, we can determine the time after a pulse is emitted that a bat is ready to detect its echo. We assume the speed of sound is 343 m/s.

Solve:

$$t = d/v = 2\Delta x/v = 2(1 \text{ m})/343 \text{ m/s} = 6 \text{ ms}$$

The correct choice is D.

Assess: This time will allow the bat to keep track of insects flying at standard insect speeds. Blind humans have learned to click with their tongues and use echolocation also.

P15.85. Strategize: Some bats emit ultrasonic pulses through their nostrils into a parabolic nose reflector that concentrates the pulse in the forward direction. As a result, the reflected signal from the prey (usually in the forward direction) is much stronger than the signal from other directions. Furthermore, since the pulse is concentrated in a small area, it is more intense, which makes it easier to detect prey.

Prepare: If we could hear the ultrasound produced by the bat, it would be as loud as a jet engine (100 dB), so the bat does need some ear protection from the ultrasound. However this is provided by another mechanism (see Problem P15.83).

Solve: Choice B is correct.

Assess: We can learn a lot from bats. They might hold the key to detecting stealth bombers.

16

SUPERPOSITION AND STANDING WAVES

Q16.1. Reason: Where there is a change in medium—in particular a change in the wave speed—then reflection can occur.
Assess: Light travels at different speeds in water and air, and so some is reflected at a water-air interface.

Q16.3. Reason: It is possible, especially if the loudspeaker is playing only a single frequency. This would require the two speakers to be coherent (making compressions/expansions in sync or with a fixed time lag between them). If those criteria are met, then at some locations throughout the room, a compression from one speaker will reach your ear at the same time as a rarification from the other speaker. This can cause a reduction in the overall amplitude and the sound can be quieter. This is destructive interference.
Assess: If music involving many simultaneous and varying tones were being played, it would be more difficult to notice this effect, since the positions at which this reduction occurs depend on the wavelength of the sound. Note that being different distances from the two speakers means the intensity from each speaker will be different. So total cancellation of the sound is unlikely.

Q16.5. Reason: The frequency of vibration of your vocal cords is related to their linear mass density by $f_m = (m/(2L))\sqrt{T/\mu}$. Due to the inflammation, the vocal cords are more massive and hence have a greater linear mass density. Since the inflamed vocal cords have a greater linear mass density, the frequency of vibration of the cords and hence the frequency of the sound generated will decrease.
Assess: You are no doubt aware that the frequency of your voice lowers with certain illnesses.

Q16.7. Reason: (a) When standing waves are set up in a tube that is open at both ends, the length of the tube is an integral number of half wavelengths $L = m\lambda/2$. Looking at the figure we see four half wavelengths, hence $m = 4$, and the air column is vibrating in the fourth harmonic. **(b)** Since this is sound, we have a longitudinal wave and the air molecules are vibrating horizontally, parallel to the tube.
Assess: The wave diagram superimposed on top of the open-ended tube sketch is a representation of the pressure at different points in the gas. It is not an air molecule displacement sketch.

Q16.9. Reason: The advantage of having low-frequency organ pipes closed at one end is that they will sound an octave lower without being twice as long as an open-open pipe. Because a pipe closed on one end contains only 1/4 of a wavelength in the lowest mode, the wavelength is twice as long as for an open-open pipe of the same length. If the wavelength is twice as long then the frequency is half, and that corresponds to a musical interval of an octave.
Assess: Pipes closed on one end also have a different sound (timbre) than open-open pipes sounding the same note.

Q16.11. Reason: You would initially hear a loud tone, then a fairly soft tone, then a loud tone again, and so on. As you get very near one speaker, this loud-soft-loud effect would become less noticeable. The waves from each speaker are identical, so if you are equidistant from the speakers the compressions (and rarifications) from each speaker reach you at the same time and cause constructive interference. As you move toward one speaker, the distance the sound must cross to reach you is different for each speaker, so the compressions will reach you at different times. When a compression from speaker A reaches you just as the rarification from speaker B, there is destructive interference. As you move further, one compression from speaker A will reach you at the same time as a different compression (emitted at a different time) from speaker B, and you will experience constructive inference again.

Assess: Note that as you move, the intensities of the sound from each speaker change. So the sounds never cancel perfectly.

Q16.13. Reason: The introduction of helium into the mouth allows harmonics of higher frequencies to be excited more than in the normal voice. The fundamental frequency of the voice is the same but the quality has changed. Our perception of the quality is a function of which harmonics are present.

Assess: The fundamental frequency of a complex tone from the voice is determined by the vibration of the vocal cords and depends on the tension and linear mass density of the cords, not the gas in the mouth.

Q16.15. Reason: The settings on the synthesizer change the amount and proportion of higher harmonics that are mixed in (added to) the fundamental frequency.

Assess: Adding a lot of 2nd and 4th harmonics makes it sound more like a flute.

Q16.17. Reason: The pitch may be the same, but the vocal tract is considerably different. The vocal tract determines the formants that emphasize different harmonics and allow us to distinguish different vocalizations.

Assess: Due to the formants, it is easy to distinguish the voice of a child from the voice of an adult.

Q16.19. Reason: The earliest (and only) time that y will equal 2 mm at the point $x = 3$ cm is when the top of the triangular peak on the left moves right to the point $x = 3$ cm. That wave pulse is moving at 6 cm/s, so it will take 0.5 s to move 3 cm.

The correct choice is A.

Assess: Because the pulse moving from right to left is a negative pulse, it will make $y = -2$ mm at $t = 3$ s, but not a positive $+2$ mm.

Q16.21. Reason: The maximum displacement will be the sum of the contributions from the two traveling waves at each point and at each time; however, you are not guaranteed to be watching a point where a crest will meet a crest. It is true that you might be watching an antinode where the maximum displacement would be $2A$, but it is also possible that you might be watching a node that doesn't move at all—or any place in between.

The correct choice is D.

Assess: If you watch the whole string, you will see points whose maximum displacement is $2A$ and other points whose maximum displacement is 0 and other points in between.

Q16.23. Reason: We know that the speed, frequency, and wavelength of a traveling wave are related by $f = v / \lambda$. As the air and pipe warm from $20°C$ to $25°C$, there is an insignificant expansion of the pipe, hence the resonant wavelengths remain the same. However, the change in temperature of the air is significant compared to the temperature of the air. Since the speed of sound increases with temperature, the frequency will increase. The correct response is A.

Assess: Your ability to look at an expression that describes a situation and predict what will happen under certain circumstances will be enhanced by your study of physics.

Q16.25. Reason: For the lowest standing wave mode on a string, the wavelength is twice the length of the string. If this is the case the speed of the disturbance is determined by

$$v = f_1 \lambda_1 = (20 \text{ Hz})(2.0 \text{ m}) = 40 \text{ m/s}$$

The correct choice is D.

Assess: For the lowest standing wave mode on a string, one-half a wavelength fits into the string.

Problems

P16.1. Strategize: We will translate the waveforms different distances to the left and right, respectively, for different times. Whenever there is overlap of the waves, we will use the superposition principle.

Prepare: The waves are approaching each other at a speed of 1 m/s, that is, each part of each wave is moving 1 m every second. For each snapshot time Δt, we will translate the waveforms a distance $v\Delta t$.

Solve: The graph at $t = 1\,\text{s}$ differs from the graph at $t = 0\,\text{s}$ in that the left wave has moved to the right by 1 m and the right wave has moved to the left by 1 m. This is because the distance covered by the wave pulse in 1 s is 1 m. The snapshot graphs at $t = 2\,\text{s}$, $3\,\text{s}$, and $4\,\text{s}$ are a superposition of the left- and the right-moving waves. The overlapping parts of the two waves are shown by the dotted lines.

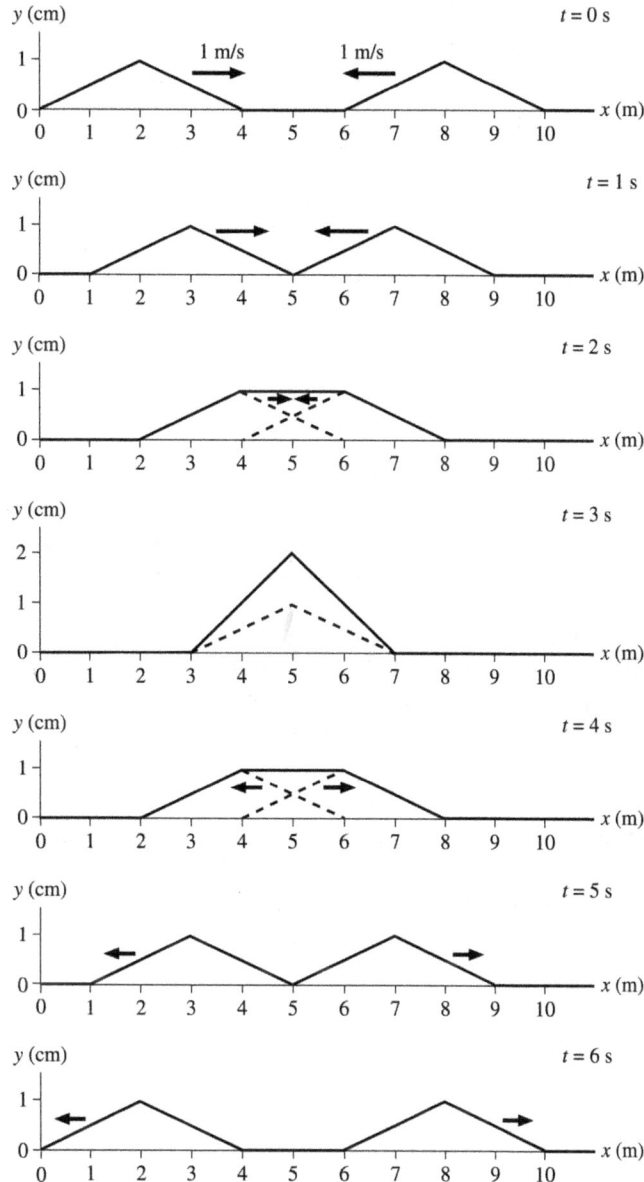

Assess: This is an excellent problem because it allows you to see the progress of each wave and the superposition (addition) of the waves. As time progresses, you know exactly what has happened to each wave and to the superposition of these waves.

P16.3 Strategize: We will determine the displacement at $x = 4$ cm due to each wave separately at various times, then use the superposition principle to determine the net displacement.

Prepare: One way of organizing the displacements is to make a table. We can fill in the values by translating each waveform to the left or right by a distance given by $\Delta x = v_x \Delta t$ and determining the displacement in the y direction at each time.

t	y displ. 1	y displ. 2	net displ.
0	0	0	0
1	0.5	0	0.5
2	1	0	1
3	0	−0.5	−0.5
4	0	−1	−1
5	0	0	0
6	0	0	0
7	0	0	0
8	0	0	0

Solve: Plotting these numerical values, we obtain

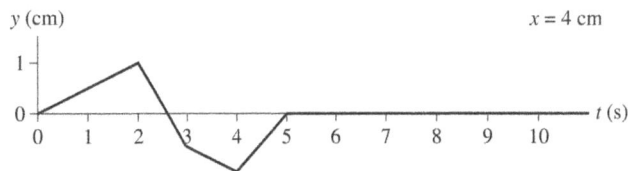

Assess: Since the wave that is above the axis reaches the $x = 4$ cm point before the wave that is below the axis, it is reasonable that our history graph shows a positive displacement at earlier times than the negative displacement.

P16.5. Strategize: We will translate each waveform to the left or right, respectively by a distance equal to $\Delta x = v \Delta t$. At each time step the net displacement will be determined by the superposition of the two waves.

Prepare: The principle of superposition tells us the net displacement is the sum (or difference) of the displacements from each wave separately. The waves are approaching each other at a speed of 1 m/s, that is, each part of each wave is moving 1 m every second.

Solve: The graph at $t = 1$ s differs from the graph at $t = 0$ s in that the left wave has moved to the right by 1 m and the right wave has moved to the left by 1 m. This is because the distance covered by the wave pulse in 1 s is 1 m. The snapshot graphs at $t = 2$ s, 3 s, 4 s, 5 s, and 6 s are a superposition of the left- and the right-moving waves. The overlapping parts of the two waves are shown by the dotted lines.

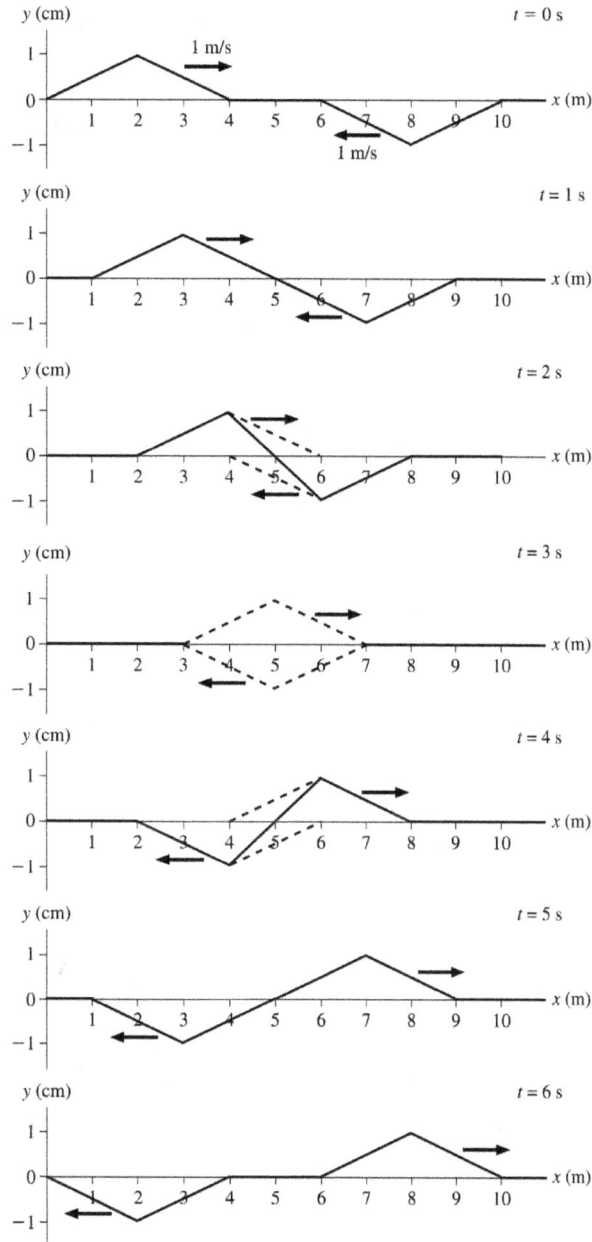

From the figure, the values of the displacement at $x = 5.0$ cm are

t (s)	y (cm)
0	0
1	0
2	0
3	0
4	0
5	0
6	0

Assess: The point at $x = 5.0$ m is a special point.

P16.7. Strategize: The pulse will have to travel to the end of the string (4.0 m), be reflected back to the other end of the string (10.0 m), and then be reflected again and travel to the same position (6.0 m) in order to have the same appearance as at $t = 0$ s.

Prepare: The total distance traveled is 20.0 m and the pulse is traveling at a speed of 4.0 m/s.

Solve: The time required is $\Delta t = \Delta x / v_x = 20.0$ m/(4.0 m/s) = 5.0 s.

Assess: This is basically a straightforward kinematics problem, with the slight complication of the pulse being inverted upon reflection.

P16.9. Strategize: We will determine the wavelength from the figure and then use the given velocity to determine the frequency.

Prepare: Reflections at both ends of the string cause the formation of a standing wave. Figure P16.9 indicates that there are three full wavelengths on the 2.0-m-long string and that the wave speed is 40 m/s. We will use Equation 15.10 to find the frequency of the standing wave.

Solve: The wavelength of the standing wave is $\lambda = \frac{1}{3}(2.0 \text{ m}) = 0.667$ m. The frequency of the standing wave is thus

$$f = \frac{v}{\lambda} = \frac{40 \text{ m/s}}{0.667 \text{ m}} = 60 \text{ Hz}$$

Assess: The units are correct and this is a reasonable frequency for a vibrating string.

P16.11. Strategize: We will use our understanding of standing waves to determine the wavelength, and then relate the speed to the frequency, and wavelength.

Prepare: We assume that the string is tied down at both ends so there are nodes there. This means the length of the string is $L = \frac{1}{2}\lambda$ in the fundamental mode (there are no nodes between the ends). $\lambda = 2L = 2(0.89 \text{ m}) = 1.78$ m.

Solve: Use the fundamental relationship for periodic waves: $v = \lambda f = (1.78 \text{ m})(30 \text{ Hz}) = 53$ m/s.

Assess: Remember, we are talking about the speed of the wave on the string, not the speed of sound in air. These numbers are reasonable for bass guitar strings.

P16.13. Strategize: For the first part, we simply use our understanding of a standing wave on a string with two fixed ends, as discussed in the chapter. For the second part, we recognize that the wave speed relates the frequency and wavelength. Since the wave speed is determined by the tension and linear mass density of the string, the wave speed should be the same for any harmonic.

Prepare: A string fixed at both ends supports standing waves. A standing wave can exist on the string only if its wavelength is given by Equation 16.1, that is, $\lambda_m = \frac{2L}{m}, m = 1, 2, 3 \ldots$ The length L of the string is 240 cm.

Solve: (a) The three longest wavelengths for standing waves will therefore correspond to $m = 1, 2,$ and 3. Thus,

$$\lambda_1 = \frac{2(2.40 \text{ m})}{1} = 4.80 \text{ m} \qquad \lambda_2 = \frac{2(2.40 \text{ m})}{2} = 2.40 \text{ m} \qquad \lambda_3 = \frac{2(2.40 \text{ m})}{3} = 1.60 \text{ m}$$

(b) Because the wave speed on the string is unchanged from one m value to the other,

$$f_2\lambda_2 = f_3\lambda_3 \Rightarrow f_3 = \frac{f_2\lambda_2}{\lambda_3} = \frac{(50.0 \text{ Hz})(2.40 \text{ m})}{1.60 \text{ m}} = 75.0 \text{ Hz}$$

Assess: The units on each determination are correct and the values are reasonable. The maximum wavelength of a standing wave in a string is twice the length of the string and all other possible wavelengths are fractions of this value.

P16.15. Strategize: We will use our understanding of standing waves on a string with two fixed ends to determine the wavelength. We will then determine the speed using the given frequency and the wavelength. Since the tension and linear mass density are not changed, this speed will be the same for the fifth harmonic, allowing us to use speed and wavelength to determine the frequency of the that harmonic.

Prepare: A string fixed at both ends forms standing waves. Three antinodes means the string are vibrating as the $m = 3$ standing wave. The wavelengths of standing wave modes of a string of length L are given by Equation 16.1.

Solve: (a) The frequency is $f_3 = 3f_1$, so the fundamental frequency is $f_1 = \frac{1}{3}(420\ \text{Hz}) = 140\ \text{Hz}$. The fifth harmonic will have the frequency $f_5 = 5f_1 = 700\ \text{Hz}$.

(b) The wavelength of the fundamental mode is $\lambda_1 = 2L = 1.20\ \text{m}$. The wave speed on the string is $v = \lambda_1 f_1 = (1.20\ \text{m})(140\ \text{Hz}) = 168\ \text{m/s}$. Alternatively, the wavelength of the $n = 3$ mode is $\lambda_3 = \frac{1}{3}(2L) = 0.40\ \text{m}$, from which $v = \lambda_3 f_3 = (0.40\ \text{m})(420\ \text{Hz}) = 168\ \text{m/s}$. The wave speed on the string, given by Equation 15.2, is

$$v = \sqrt{\frac{T_S}{\mu}} \Rightarrow T_S = \mu v^2 = (0.0020\ \text{kg/m})(168\ \text{m/s})^2 = 56\ \text{N}$$

Assess: You must remember to use the linear density in SI units of kg/m. Also, the speed is the same for all modes, but you must use a matching λ and f to calculate the speed.

P16.17. Strategize: We know that in normal operation, the frequency excited on a musical instrument is the fundamental frequency. The wave speed is determined by the tension and linear mass density of the string, and the speed is related to frequency and wavelength. So we can relate these quantities to determine the tension.

Prepare: Reflections at the string boundaries cause a standing wave on a stretched string. The wavelengths of standing wave modes of a string of length L are given by Equation 16.1, so we can easily determine the wavelength from the vibrating length of the string, which is 1.90 m. With a known frequency of 27.5 Hz we can find the wave speed using Equation 15.10. Equation 15.2 will now allow us to find the tension in the wire. Mass density μ of the wire is equal to the ratio of its mass and length.

Solve: Because the vibrating section of the string is 1.9 m long, the two ends of this vibrating wire are fixed, and the string is vibrating in the fundamental harmonic. The wavelength is

$$\lambda_m = \frac{2L}{m} \Rightarrow \lambda_1 = 2L = 2(1.90\ \text{m}) = 3.80\ \text{m}$$

The wave speed along the string is $v = f_1 \lambda_1 = (27.5\ \text{Hz})(3.80\ \text{m}) = 104.5\ \text{m/s}$. The tension in the wire can be found as follows:

$$v = \sqrt{\frac{T_S}{\mu}} \Rightarrow T_S = \mu v^2 = \left(\frac{\text{mass}}{\text{length}}\right) v^2 = \left(\frac{0.400\ \text{kg}}{2.00\ \text{m}}\right)(104.5\ \text{m/s})^2 = 2180\ \text{N}$$

Assess: You must remember to use the linear density in SI units of kg/m. Also, the speed is the same for all modes, but you must use a matching λ and f to calculate the speed.

P16.19. Strategize: Antinodes in standing waves are separated by half a wavelength (as are antinodes). We will use this for part (a). For the second part, we determine the period from the frequency.

Prepare: Assuming we want the next antinode closest to the one at the wall, it is ½ wavelength away.

Solve: The antinodes are a half-wavelength apart, or in this case $\lambda/2 = 26\ \text{m}/2 = 13\ \text{m}$. The period of her motion is related to the speed and wavelength.

$$T = \frac{1}{f} = \frac{\lambda}{v} = \frac{26\ \text{m}}{4.4\ \text{m/s}} = 5.9\ \text{s}$$

Assess: These seem like reasonable answers.

P16.21 Strategize: This is equivalent to an open-closed tube.

Prepare: For the first part, we use $f_1 = \dfrac{v}{4L}$, where $v_{\text{sound,water}} = 1480\ \text{m/s}$ (from Table 15.1). For the second part, we note that the lowest frequency corresponds to the largest wavelength. So we wish to find the largest wavelength such that the distance from the antinode to a node is 0.28 m. We note that the distance from one antinode to the nearest node is ¼ of a wavelength.

Solve: (a) $f_1 = \dfrac{v}{4L} = \dfrac{(1480 \text{ m/s})}{4(1.4 \text{ m})} = 2.6 \times 10^2$ Hz.

(b) We require $\lambda / 4 = 0.28 \text{ m} \Rightarrow \lambda = 1.12$ m. Then $f = v / \lambda = (1480 \text{ m/s}) / 1.12 \text{ m} = 1.3$ kHz.

Assess: This is within the range of human hearing, whether or not it is in the range of frog hearing could be determined by the frog's behavior.

P16.23 Strategize: This is an open-open pipe, meaning we simply apply Equation 16.6.
Prepare: We will assume room temperature, such that $v = 343$ m/s . We must convert to SI units:

$$(64 \text{ ft})\left(\frac{12 \text{ in}}{1 \text{ ft}}\right)\left(\frac{2.54 \text{ cm}}{1 \text{ in}}\right)\left(\frac{1 \text{ m}}{100 \text{ cm}}\right) = 19.51 \text{ m}.$$

Solve: Inserting the given values, we have $f_1 = \dfrac{(343 \text{ m/s})}{2(19.51 \text{ m})} = 8.8$ Hz.

Assess: Note that the lowest frequency humans can hear is around 20 Hz. So it is very reasonable that the longest organ pipe would correspond to a frequency right around the lower frequency limit of human hearing. Why build a longer one, if no one can hear it?

P16.25. Strategize: The length of the tube is unchanged, so the wavelength is also unchanged. We will use ratios.
Prepare: We assume that (as in normal operation of a musical instrument) the fundamental frequency will be excited. Thus, we know $m = 1$ in $\lambda_m = 2L/m$.
Solve: Using Equation 15.10,

$$\frac{f_{\text{He}}}{f_{\text{air}}} = \frac{v_{\text{He}}/\lambda}{v_{\text{air}}/\lambda} \Rightarrow f_{\text{He}} = \frac{v_{\text{He}}}{v_{\text{air}}} f_{\text{air}} = \frac{1010 \text{ m/s}}{343 \text{ m/s}} (315 \text{ Hz}) = 928 \text{ Hz}$$

Assess: Note that the length of the tube is one-quarter the wavelength, whether the tube is filled with helium or air.

P16.27. Strategize: We will use Equation 16.6 to calculate the different harmonics. For the second part, we will use our knowledge that humans can hear sounds with frequencies as low as 20 Hz. Finally, we know that the speed of sound decreases as temperature decreases.
Prepare: The frequencies at which resonance will occur in an open-open pipe are determined by $f_m = m(v/(2L))$.
Solve: (a) Knowing the speed of sound, the fundamental frequency is determined by

$$f_1 = (1)(v/(2L)) = (340 \text{ m/s})/(2(30.0 \text{ m})) = 5.67 \text{ Hz}$$

(b) The lowest frequency we can hear is about 20 Hz. For an open-open tube the frequency of the harmonics are related to the fundamental frequency by $f_m = mf_1$, hence the lowest harmonic that would be audible to the human ear is

$$m = f_m/f_1 = 20 \text{ Hz}/(5.67 \text{ Hz}) = 3.5$$

But m must be an integer, so we must take $m = 4$ and calculate the frequency.

$$4f_1 = 4(5.67 \text{ Hz}) = 22.7 \text{ Hz}$$

(c) As the air cools the speed of sound will decrease. Inspecting the function given in the previous Prepare step, we see that this would in turn result in a decrease in frequency.
Assess: This is a straightforward example of resonance in an open-open pipe.

P16.29. Strategize: We can use Equation 16.7 to determine the fundamental frequency of an open-closed tube.
Prepare: Equation 16.7 reads

$$f_m = m\frac{v}{4L}$$

where we are given that $m = 1$ (for the fundamental frequency) and that $f_1 = 200$ Hz. We are also given that $v = 350$ m/s.

Solve: Solve the equation for L.

$$L = m\frac{v}{4f_m} = (1)\frac{350 \text{ m/s}}{4(200 \text{ Hz})} = 0.44 \text{ m} = 44 \text{ cm}$$

Assess: Since the 200 Hz is a "typical" fundamental frequency we don't really expect the length obtained to be the exact length of *your* vocal tract; but we do note that it is in the right ballpark (put a meter stick next to your stretched neck and guesstimate the distance from your mouth to your diaphragm).

P16.31. Strategize: We use the expression for the harmonics of an open-closed tube to determine what sounds resonate well in a cavity of the given dimensions.
Prepare: Follow Example 16.6 very closely. Assume the ear canal is an open-closed tube, for which we need Equation 16.7

$$f_m = m\frac{v}{4L} \qquad m = 1, 3, 5, 7 \dots$$

where we are given that $L = 1.3 \text{ cm} = 0.013 \text{ m}$. We assume that $v = 350 \text{ m/s}$ in the warm ear canal.
Take the audible range as 20 Hz–20 kHz.
Solve: Plug in various values of m and obtain the corresponding frequencies.

$$f_1 = (1)\frac{350 \text{ m/s}}{4(0.013 \text{ m})} = 6730 \text{ Hz} \approx 6700 \text{ Hz}$$

Higher frequencies are odd multiples of this fundamental.

$$f_3 = 3(6730 \text{ Hz}) = 20\,200 \text{ Hz}$$

Already f_3 is out of the audible range. So the only one in the audible range is $f_1 = 6700 \text{ Hz}$.

P16.33. Strategize: For the first part, we simply consult Figure 16.23. We will use the known expressions for the harmonics of an open-closed tube to relate this to the length of the formant.

Prepare: We are considering the fundamental frequency. Equation 16.7 tells us $f_1 = \frac{v}{4L}$.

Solve: Notice in the graph that the first 1000 Hz is divided into seven equal sections and the desired frequency is 2/7 of the way to 1000 Hz or $(2/7)(1000) = 290 \text{ Hz}$. The relationship between the speed of sound, the frequency of the sound, and the length of the open-closed pipe is

$$v = f\lambda = f(4L/m) \quad \text{or} \quad L = mv/4f = v/4f = \frac{343 \text{ m/s}}{4(290 \text{ Hz})} = 0.30 \text{ m}$$

Assess: This is a reasonable length for the vocal system.

P16.35. Strategize: Because the structure of the mouth does not change, the wavelengths of the harmonics will not change. This will allow us to relate the frequencies and speeds under the different conditions.
Prepare: First assume the speed of sound is 350 m/s in the vocal tract. The relationship between speed, wavelength, and frequency for a traveling wave disturbance in any medium is $v = f\lambda$. The frequency of vibration in air is caused by and is the same as the frequency of vibration of the vocal cords. The length of the vocal tract is an integral number of half-wavelengths $L = m\lambda/2$. The length of the vocal tract and hence the wavelengths that cause standing wave resonance do not change as the diver descends. However, since the speed of the sound waves changes, the frequency will also change.
Solve: When a sound of frequency 270 Hz is coming out of the vocal tract, the wavelength of standing waves established in the vocal tract associated with this frequency is

$$\lambda = v/f = (350 \text{ m/s})/(270 \text{ Hz}) = 1.296 \text{ m}$$

As the diver descends, the vocal tract does not change length and hence this wavelength for standing wave resonance will not change. However since the sound is now travelling through a helium-oxygen mixture with a speed of 750 m/s, the frequency of the sound will change to

$$f = v/\lambda = (750\,\text{m/s})/(1.296\,\text{m}) = 580\,\text{Hz}$$

Going through the same procedure for sound at a frequency of 2300 Hz we get the frequency in the helium-oxygen mixture to be $f = 4900\,\text{Hz}$.

Assess: We are aware that the sound should be at a higher frequency and the frequencies obtained have higher values.

P16.37. Strategize: Interference occurs as a result of the path difference between the path lengths of the sound from the two speakers, which we are given. We will use Equation 16.9 to relate the path length difference to the wavelength.

Prepare: A separation of 20 cm between the speakers leads to maximum intensity on the x-axis, but a separation of 30 cm leads to zero intensity.

Solve: (a) When the waves are in phase and lead to constructive interference, $(\Delta d_1) = m\lambda = 20$ cm. For destructive interference, $(\Delta d)_2 = (m + \frac{1}{2})\lambda = 30$ cm. Thus, for the same value of m

$$(\Delta d)_2 - (\Delta d)_1 = \frac{\lambda}{2} \Rightarrow \lambda = 2(30\,\text{cm} - 20\,\text{cm}) = 20\,\text{cm}$$

(b) If the distance between the speakers continues to increase, the intensity will again be a maximum when the separation between the speakers that produced a maximum has increased by one wavelength. That is, when the separation between the speakers is $20\,\text{cm} + 20\,\text{cm} = 40\,\text{cm}$.

Assess: The distances obtained are reasonable. As a check on our work we might want to determine the frequency of sound associated with the wavelength. A wavelength of 40 cm is associated with a frequency of 860 Hz, which is in the audible range.

P16.39 Strategize: As the frequency is tripled, the difference between the number of full oscillations to Victor's right and left changes by 3 (since there are three full oscillations from quiet through loud and back to quiet).

Prepare: We will describe the initial path length difference for sound from speaker A and B using Equation 16.9 as $\Delta d = \left(m + \dfrac{1}{2}\right)\lambda$. Finally, the path length difference has not changed, but the number of wavelengths has: $\Delta d = \left(m + 3 + \dfrac{1}{2}\right)\lambda'$. We know that $\lambda' = \lambda/3$.

Solve: Dividing the equations for the initial and final conditions, we find

$$\frac{\Delta d}{\Delta d} = \frac{\left(m + \dfrac{1}{2}\right)}{\left(m + 3 + \dfrac{1}{2}\right)} \frac{\lambda}{\lambda'} \Rightarrow 1 = \frac{\left(m + \dfrac{1}{2}\right)}{\left(m + 3 + \dfrac{1}{2}\right)} 3 \Rightarrow \left(m + \frac{7}{2}\right) = \left(3m + \frac{3}{2}\right)$$

$$2m = 2 \Rightarrow m = 1$$

Reinserting this, we have $\lambda = \dfrac{\Delta d}{\left(m + \dfrac{1}{2}\right)} \Rightarrow f = \dfrac{v}{\lambda} = \dfrac{v\left(m + \dfrac{1}{2}\right)}{\Delta d} = \dfrac{(343\ \text{m/s})(3/2)}{(1.0\ \text{m})} = 515\ \text{Hz}$.

Assess: This is in the range of human hearing.

P16.41. Strategize: We assume that the speakers are identical and that they are emitting in phase. Since you don't hear anything, the separation between the two speakers corresponds to the condition of destructive interference.

Prepare: Equation 16.9 for destructive interference is $\Delta d = \left(m + \dfrac{1}{2}\right)\lambda$. We will combine this with $v = \lambda f$.

Solve: Considering the first few values, we have

$$\Delta d = \left(m + \frac{1}{2}\right)\lambda \Rightarrow \Delta d = \frac{\lambda}{2}, \frac{3\lambda}{2}, \frac{5\lambda}{2}$$

Since the wavelength is

$$\lambda = \frac{v}{f} = \frac{340 \text{ m/s}}{170 \text{ Hz}} = 2.0 \text{ m}$$

three possible values for d are 1.0 m, 3.0 m, and 5.0 m.

Assess: The units worked out and these are reasonable distances.

P16.43. Strategize: The two speakers are identical, and so they are emitting circular waves in phase. The overlap of these waves causes interference. We will find the path lengths from each speaker to the point described and determine what type of interference occurs here.

Prepare: An overview of the problem follows.

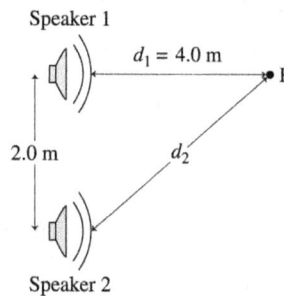

Solve: The wavelength of the sound waves is

$$\lambda = \frac{v}{f} = \frac{340 \text{ m/s}}{1800 \text{ Hz}} = 0.1889 \text{ m}$$

From the geometry of the figure,

$$d_2 = \sqrt{d_1^2 + (2.0 \text{ m})^2} = \sqrt{(4.0 \text{ m})^2 + (2.0 \text{ m})^2} = 4.472 \text{ m}$$

So, $\Delta d = d_2 - d_1 = 4.472 \text{ m} - 4.0 \text{ m} = 0.472 \text{ m}$.

Because $\Delta d / \lambda = 0.472 \text{ m}/0.1889 \text{ m} = 2.5 = 5/2 = 2 + \frac{1}{2}$ or $\Delta d = 2 + \frac{1}{2}\lambda$, the interference is perfectly destructive.

Assess: Destructive interference (for two waves in phase) will occur when the path difference is an integral number of half wavelengths.

P16.45. Strategize: The beat frequency is the difference of the two frequencies.

Prepare: We know the flat flute's frequency is lower.

Solve: $f_{\text{untuned}} = f_{\text{tuned}} - f_{\text{beat}} = 440 \text{ Hz} - 2 \text{ Hz} = 438 \text{ Hz}$.

Assess: Trained ears are quite sensitive to such differences in frequency.

P16.47. Strategize: The beats occur because the two strings have slightly different frequencies. We can determine which string has the higher frequency, and equate the beat frequency to the difference of the two fundamental frequencies of the strings.

Prepare: Knowing the following relationships for a vibrating string, $v = \sqrt{T/\mu}$, $v = f\lambda$, and $L = m\lambda/2$, we can establish what happens to the frequency as the tension is increased. Knowing the relationship $f_{beat} = f_1 - f_2$, we can determine the frequency of the string with the increased tension.

Solve: Combining the first three expressions above, obtain $f = \dfrac{m}{2L}\sqrt{\dfrac{T}{\mu}}$, which allows us to determine that the frequency will increase as the tension increases. Using the expression for the beat frequency, we know that the difference frequency between the two frequencies is 3 Hz. Combining this with our knowledge that the frequency increases, we obtain

$$f_{beat} = f_1 - f_2 \Rightarrow 3\,\text{Hz} = f_1 - 200\,\text{Hz} \Rightarrow f_1 = 203\,\text{Hz}$$

Assess: f_1 is larger than f_2 because the increased tension increases the wave speed and hence the frequency.

P16.49. Strategize: Knowing the expression for the beat frequency, we can determine by which the amount the frequency will change.
Prepare: At this point we don't know if the frequency increases or decreases. Examining the expression for the frequency of a flute (modeled as an open-open pipe) as a function of its length, we can establish if the frequency increases or decreases when the "tuning joint" is removed.
Solve: Using the expression for the beat frequency, the flute player's initial frequency is either $523\,\text{Hz} + 4\,\text{Hz} = 527\,\text{Hz}$ or $523\,\text{Hz} - 4\,\text{Hz} = 519\,\text{Hz}$. Modeling the flute as an open-open pipe we see that $v = f\lambda$ and $L = m\lambda/2$, which may be combined to obtain $f = mv/(2L)$

This expression allows us to see that as the length increases, the frequency decreases. As a result we know that the initial frequency of sound from the flute was 527 Hz.
Assess: Since she matches the tuning fork's frequency by lengthening her flute, she is increasing the wavelength of the standing wave in the flute. A wavelength increase means a decrease of frequency because $v = f\lambda$. This confirms that the initial frequency was greater than the frequency of the tuning fork.

P16.51. Strategize: Since when a sound wave hits the boundary between soft tissue and air, or between soft tissue and bone, most of the energy is reflected, the situation is like a vibrating string with reflections at both ends. If that is the case, standing wave resonance will be established (and heating will occur) when twice the thickness of the soft tissue is an integral number of wavelengths.
Prepare: The speed of ultrasound in soft tissue is 1540 m/s. The frequency of standing waves in this scenario is $f_m = m\left(\dfrac{v}{2L}\right)$.
Solve: Inserting the given numbers, we have

$$f_m = m\left(\frac{v}{2L}\right) \Rightarrow m = \frac{f_m(2L)}{v} = \frac{(0.70 \times 10^6\,\text{Hz})(2)(0.0055\,\text{m})}{1540\,\text{m/s}} = 5.0$$

Since this is an integer, then yes, there will be standing waves and heating of the tissue.
Assess: Since the frequency of the ultrasound and the thickness of the soft tissue are reasonable, we expect heating to occur.

P16.53. Strategize: We will use the density and cross sectional area to determine the linear mass density. This can be used (with the tension) to determine the velocity of a wave on the tendon. The harmonics follow directly.
Prepare: The relationship between the velocity, frequency, and wavelength of a traveling wave disturbance is $v = f\lambda$. The relationship between the velocity, tension, and linear mass density for a traveling wave disturbance in a string is $v = \sqrt{T/\mu}$. The relationship that must be satisfied to create standing wave resonance in a stretched tendon is $L = m\lambda/2$. Finally, the relationship between linear and volume mass density is $\mu = \rho A$.

Solve: Start with $v = \sqrt{T/\mu}$ and insert $\mu = \rho A$ to obtain $v = \sqrt{T/(\rho A)}$. Express the wavelength as $\lambda = 2L/m$. Insert both v and λ into $v = f\lambda$ and solve for the frequency.

$$\sqrt{T/(\rho A)} = f_m(2L/m) \Rightarrow f_m = (m/(2L))\sqrt{T/(\rho A)} \quad \text{where } m = 1, 2, 3, \ldots$$

Inserting values, obtain the fundamental frequency:

$$f_1 = (1/(2(0.20\,\text{m})))\sqrt{500\,\text{N}/((1100\,\text{kg/m}^3)(1.00\times10^{-4}\,\text{m}^2))} = 160\,\text{Hz}$$

Other possible frequencies are multiples of the fundamental

$$f_2 = 2f_1 = 320\,\text{Hz}, \quad f_3 = 3f_1 = 480\,\text{Hz}, \ldots \text{ etc.}$$

Assess: These are reasonable frequencies for the vibration of a tendon.

P16.55. Strategize: We can determine the necessary wave speed using the fundamental wavelength and the frequency. We will use the density and cross sectional area to determine the linear mass density. This will allow us to determine the required tension.

Prepare: The relationship between the velocity, frequency, and wavelength of a traveling wave disturbance is $v = f\lambda$. The relationship between the velocity, tension and linear mass density for a traveling wave disturbance in a string is $v = \sqrt{T/\mu}$. The relationship that must be satisfied to create standing wave resonance in a stretched tendon is $L = m\lambda/2$. Finally, the relationship between linear and volume mass density is $\mu = \rho A$.

Solve: First determine the velocity of the traveling wave disturbance of the struggling insect

$$v = f\lambda = f(2L/m) = 2fL = 2(100\,\text{Hz})(0.12\,\text{m}) = 24\,\text{m/s}$$

Next determine the linear mass density of the web

$$\mu = \rho A = \rho\pi d^2/4 = (1.300\times10^3\,\text{kg/m}^3)(\pi/4)(2.0\times10^{-6})^2 = 4.08\times10^{-9}\,\text{kg/m} \approx 4.1\times10^{-9}\,\text{kg/m}$$

Finally, determine the tension in the web

$$T = v^2\mu = (24\,\text{m/s})^2(4.08\times10^{-9}\,\text{kg/m}) = 2.4\times10^{-6}\,\text{N}$$

Assess: This is a very small tension; however, it is several orders of magnitude greater than the weight of a small insect that the spider might wish to capture.

P16.57. Strategize: We can determine the fundamental frequency by using the length of the wire and the speed of a wave on the wire. The speed of the wave depends on the linear mass density and the tension.

Prepare: For the stretched wire vibrating at its fundamental frequency, the wavelength of the standing wave from Equation 16.1 is $\lambda_1 = 2L$. From Equation 15.2, the wave speed is equal to $\sqrt{T_S/\mu}$, where $\mu = \text{mass/length} = 5.0\times10^{-3}\,\text{kg}/0.90\,\text{m} = 5.555\times10^{-3}\,\text{kg/m}$. The tension T_S in the wire equals the weight of the sculpture or Mg.

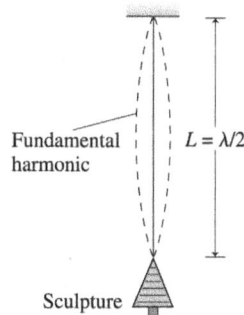

Fundamental harmonic $L = \lambda/2$

Sculpture

Solve: The wave speed on the steel wire is

$$v_{wire} = \sqrt{\frac{T_s}{\mu}} = \sqrt{\frac{Mg}{\mu}} = \sqrt{\frac{(12 \text{ kg})(9.8 \text{ m/s}^2)}{5.55 \times 10^{-3} \text{ kg/m}}} = 145.6 \text{ m/s}$$

Now we can solve for frequency.

$$v_{wire} = f\lambda \Rightarrow f = \frac{v_{wire}}{\lambda} = \frac{v_{wire}}{2L} = \frac{145.6 \text{ m/s}}{2(0.90 \text{ m})} = 81 \text{ Hz}$$

Assess: A frequency of 81 Hz for the wire is reasonable.

P16.59 Strategize: We will use the initial (unbent) case and compare it to the "bent" case. Use Newton's second law to determine the tension in the string when it is "bent", and we calculate the new frequency.

Prepare: We will $f = v/2L = \dfrac{1}{2L}\sqrt{\dfrac{T_s}{\mu}}$ once for the initial straight string and once for the "bent" string. We note

that Newton's second law in the direction of the push from the fingers is $\sum F_x = F_{push} - 2T_s \sin(\theta) = ma_x = 0$, where

$$\theta = \tan^{-1}\left(\frac{\Delta x}{(L/2)}\right) = \tan^{-1}\left(\frac{8.0 \times 10^{-3} \text{ m}}{0.32 \text{ m}}\right) = 1.43°.$$

Solve: The new tension in the string is $T_s = \dfrac{F_{push}}{2\sin(\theta)} = \dfrac{(4.0 \text{ N})}{2\sin(1.43°)} = 80.0 \text{ N}$. Writing the expression for the

fundamental frequency for the initial and final cases and taking their ratio, we have

$$\frac{f_{1,f}}{f_{1,i}} = \frac{(2L)^{-1}\sqrt{T_{s,f}/\mu}}{(2L)^{-1}\sqrt{T_{s,i}/\mu}} \Rightarrow f_{1,f} = f_{1,i}\sqrt{\frac{T_{s,f}}{T_{s,i}}} = (392 \text{ Hz})\sqrt{\frac{80.0 \text{ N}}{74.0 \text{ N}}} = 408 \text{ Hz}.$$

Assess: Since the tension is increased, we expect the frequency to increase. Our answer is consistent with this.

P16.61. Strategize: We rearrange the condition for standing waves to give the length as a function of the order of the harmonic. Then we can take the difference between the lengths of tube that yield two adjacent harmonics and relate it to the wavelength.

Prepare: The nodes of a standing wave are spaced $\lambda/2$ apart. The wavelength of the mth mode of an open-open tube from Equation 16.6 is $\lambda_m = 2L/m$. Or, equivalently, the length of the tube that generates the mth mode is $L = m(\lambda/2)$. Here λ is the same for all modes because the frequency of the tuning fork is unchanged.

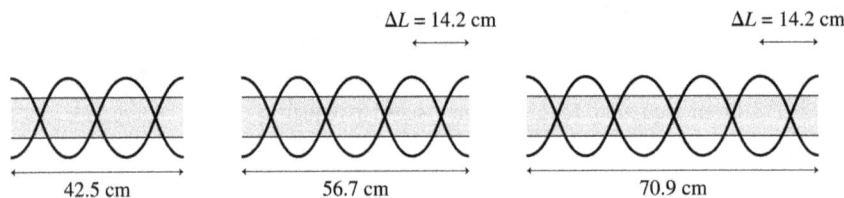

Solve: Increasing the length of the tube to go from mode m to mode $m+1$ requires a length change:

$$\Delta L = (m+1)(\lambda/2) - m\lambda/2 = \lambda/2$$

That is, lengthening the tube by $\lambda/2$ adds an additional antinode and creates the next standing wave. This is consistent with the idea that the nodes of a standing wave are spaced $\lambda/2$ apart. This tube is first increased by $\Delta L = 56.7 \text{ cm} - 4.25 \text{ cm} = 14.2 \text{ cm}$, then by $\Delta L = 70.9 \text{ cm} - 56.7 \text{ cm} = 14.2 \text{ cm}$. Thus $\lambda/2 = 14.2 \text{ cm}$ and $\lambda = 28.4 \text{ cm} = 0.284 \text{ m}$. Therefore, the frequency of the tuning fork, using Equation 15.10, is

$$f = \frac{v}{\lambda} = \frac{343 \text{ m/s}}{0.284 \text{ m}} = 1210 \text{ Hz}$$

Assess: This is a reasonable value for the frequency of a tuning for k in the audible range and the units are correct.

P16.63. Strategize: We use the known condition for constructive interference to relate the path length difference to the wavelength. The frequency is trivially related to the wavelength and the speed of sound.

Prepare: The waves constructively interfere when speaker 2 is located at 0.75 m and 1.00 m, but not in between. Assume the two speakers are in phase (helpful for visualization, but the result will be generally true as long as the two frequencies are the same). For constructive interference the path length difference must be an integer number of wavelengths, $0.75 \text{ m} = n\lambda$, and $1.00 \text{ m} = (n+1)\lambda$. Subtracting the two Equations gives $\lambda = 0.25 \text{ m}$.

Solve:

$$f = \frac{v}{\lambda} = \frac{340 \text{ m/s}}{0.25 \text{ m}} = 1360 \text{ Hz} \approx 1400 \text{ Hz}$$

Assess: 1400 Hz is near the "middle" of the range of human hearing, so it is probably right.

P16.65. Strategize: The changing sound intensity is due to the interference of two overlapped sound waves. Minimum intensity implies destructive interference. We determine the path length difference and use the condition for destructive interference.

Prepare: Destructive interference occurs where the path length difference for the two waves is $\Delta d = (m + \frac{1}{2})\lambda$.

Solve: The wavelength of the sound is $\lambda = v_{\text{sound}}/f = (343 \text{ m/s})/(686 \text{ Hz}) = 0.500 \text{ m}$. Consider a point that is a distance d in front of the top speaker. Let d_1 be the distance from the top speaker to the point and d_2 the distance from the bottom speaker to the point. We have

$$d_1 = x \quad d_2 = \sqrt{x^2 + (3.00 \text{ m})^2}$$

Destructive interference occurs at distances d such that

$$\Delta d = \sqrt{x^2 + 9 \text{ m}^2} - x = \left(m + \frac{1}{2}\right)\lambda$$

To solve for x, isolate the square root on one side of the equation and then square:

$$x^2 + 9 \text{ m}^2 = \left[x + \left(m + \frac{1}{2}\right)\lambda\right]^2 = x^2 + 2\left(m + \frac{1}{2}\right)\lambda x + \left(m + \frac{1}{2}\right)^2 \lambda^2 \Rightarrow x = \frac{9 \text{ m}^2 - (m + \frac{1}{2})^2 \lambda^2}{2(m + \frac{1}{2})\lambda}$$

Evaluating x for different values of m:

m	x (m)
0	17.88
1	5.62
2	2.98
3	1.79

Because you start at $x = 2.5$ m and walk *away* from the speakers, you will only hear minima for values $x > 2.5$ m. Thus, minima will occur at distances of 2.98 m, 5.62 m, and 17.88 m.

Assess: These are reasonable distances and the units are correct.

P16.67. Strategize: We calculate the frequencies of the two different harmonics. Their difference will equal the beat frequency.

Prepare: The superposition of two slightly different frequencies gives rise to beats.

Solve: The third harmonic of note A and the second harmonic of note E are

$$f_{3A} = 3f_{1A} = 3(440 \text{ Hz}) = 1320 \text{ Hz} \quad f_{2E} = 2f_{1E} = 2(659 \text{ Hz}) = 1318 \text{ Hz} \Rightarrow f_{3A} - f_{2E} = 1320 \text{ Hz} - 1318 \text{ Hz} = 2 \text{ Hz}$$

The beat frequency between f_{3A} and f_{2E} is 2 Hz. It therefore emerges that the tuner looks for a beat frequency of 2 Hz.

Assess: It would be impossible to tune a piano without a good understanding of beat frequency and harmonics.

P16.69. Strategize: We determine the Doppler shifted frequency in the case of radar reflected off a vehicle driving at 55 mph. The difference between this shifted frequency and the initial emitted frequency will be the beat frequency emulated by the tuning fork.

Prepare: Frequencies for the Doppler shift in the microwave range may be summarized by:

$$f_\pm = f_s \frac{(c \pm v_0)}{(c \mp v_s)}$$

Where c = the speed of light, v_s = speed of the source and v_0 = speed of the observer. In the numerator and the denominator, the top sign is used when the observer and source are moving toward each other and the bottom sign is used if the observer and source are moving away from each other. A speed of 55 mph is approximately 25 m/s.

Solve: The frequency sent out by the radar unit is $f_s = 10.5 \times 10^9$ Hz.

The frequency of waves observed by the moving car is

$$f_c = f_s \frac{(c + v_c)}{c} = 10.5 \times 10^9 \text{ Hz} \frac{(3.00 \times 10^8 + 25) \text{ m/s}}{3.00 \times 10^8 \text{ m/s}} = 10.50000088 \times 10^9 \text{ Hz}$$

Now the car acts like a source of this frequency and since it is moving towards the police unit, the frequency of the reflected waves arriving at the unit may be determined by

$$f_u = f_c \frac{c}{c - v_c} = (10.50000088 \times 10^9 \text{ Hz}) \frac{3.00 \times 10^8 \text{ m/s}}{(3.00 \times 10^8 - 25) \text{ m/s}} = 10.500001750 \times 10^9 \text{ Hz}$$

As a result, the beat frequency determined by the unit is 1750 Hz.

When the unit is switched to calibration mode, a tuning fork vibrating in front of the unit at a frequency of 1750 Hz will register on the unit as 55 mph.

Assess: A frequency of 1750 Hz for a tuning fork is in the range that we hear very well. According to the operations manual that comes with the police radar unit, to ensure accuracy every unit is supposed to be calibrated with tuning forks before each working shift. Using several different tuning forks would ensure that the unit is accurate over a range of speeds. Finally, the problem could be solved in one step by inserting values into the first equation, but it is instructive to solve the problem in two steps as shown previously.

P16.71. Strategize: From the beat frequency, we determine the frequency of the Doppler-shifted reflected wave. From that, we determine the maximum speed of the heart tissue.

Prepare: We will need concepts from Chapter 15 to solve this problem. The speed of ultrasound waves in human tissue is given in Table 15.1 as $v = 1540$ m/s. The frequency of the reflected wave must be 2.0 MHz $\pm$ 520 Hz, that is, $f_+ = 2\,000\,520$ Hz and $f_- = 1\,999\,480$ Hz.

There are a couple of mathematical paths we could take, but it is probably easiest to carefully review Example 15.13 and use the mathematical result there, as it is very similar to our problem and gives an expression for the speed of the source v_s.

Solve:

$$v_s = \frac{f_+ - f_-}{f_+ + f_-} v = \frac{2\,000\,520 \text{ Hz} - 1\,999\,480 \text{ Hz}}{2\,000\,520 \text{ Hz} + 1\,999\,480 \text{ Hz}} 1540 \text{ m/s} = \frac{1040 \text{ Hz}}{4\,000\,000 \text{ Hz}} 1540 \text{ m/s} = 0.40 \text{ m/s}$$

Assess: We may not have a good intuition about how fast heart muscles move, but 0.40 m/s seems neither too fast nor too slow for a maximum speed.

P16.73. Strategize: We calculate the higher harmonics of the G-flat and compare them with the harmonics of C.

Prepare: The harmonics of the G-flat will be the integer multiples of 370 Hz: 370 Hz, 740 Hz, 1110 Hz, 1480 Hz, 1850 Hz, etc.

Solve: Comparing these harmonics with those given for C in the figure shows that some of them have differences in the range that produces dissonance. For example, 740 Hz (the second harmonic of G-flat) is 46 Hz away from 786 Hz (the third harmonic of C), and 1110 Hz is 62 Hz (the third harmonic of G-flat) away from 1048 Hz (the fourth harmonic of C). Those differences may not be quite *maximally* dissonant, but most people agree that that musical interval is a dissonant one.

The correct answer is B.

Assess: It should be obvious that G-flat and G sounded together would also be dissonant since each corresponding harmonic is just different enough from the corresponding harmonic for the other note to produce dissonance.

Acoustics in general and music in particular are fascinating areas of physics. It should in no way reduce our enjoyment of the aesthetic pleasures of music to understand the physics of it all; much rather the understanding should increase our appreciation for the beauty of the music.

P16.75. Strategize: We use our knowledge of allowed harmonics in the different tube types.

Prepare: A pipe that is open on one end and closed on the other only produces the odd harmonics. See Equation 16.7 and the preceding discussion.

Solve: The odd harmonics are 262 Hz, 786 Hz, and 1310 Hz, corresponding to $m = 1$, $m = 3$, and $m = 5$.

The correct answer is B.

Assess: Because only the odd harmonics are present in an open-closed pipe, it sounds different from an open-open pipe sounded at the same fundamental frequency.

OSCILLATIONS AND WAVES

PptIV.1. Reason: Nothing goes faster than light in a vacuum, so it is first. Waves through the solid earth are next fastest, followed by sound in air. According to the passage, tsunamis travel at hundreds of kilometers per hour, (about 200 m/s), but not quite as fast as the speed of sound in air.

$$v_{light} > v_{earthquake} > v_{sound} > v_{tsunami}$$

Assess: This ranking matches expectations.

PptIV.3. Reason: Use $T = 1/f$ in $v = \lambda f$.

$$T = \frac{\lambda}{v} = \frac{150\,000 \text{ m}}{200 \text{ m/s}} = 750 \text{ s} = 12.5 \text{ min} \approx 15 \text{ min}$$

The answer is D.
Assess: This is a long period, but it makes sense given the long wavelength.

PptIV.5. Reason: A tsunami is not a standing wave, but it does reflect as shown in the simulation. The reflected wave can interfere with the primary wave to make extra large crests.
The answer is B.
Assess: A standing wave would need a continuous source of new pulses.

PptIV.7. Reason: Use $T = 1/f$ in $v = \lambda f$ and the given equation for v in deep water. The effective speed with which the waves pass the ship is the speed of the waves plus the speed of the ship toward the waves.

$$T = \frac{\lambda}{v} = \frac{\lambda}{\sqrt{\dfrac{g\lambda}{2\pi}} + v_{ship}} = \frac{75 \text{ m}}{\sqrt{\dfrac{(9.8 \text{ m/s}^2)(75 \text{ m})}{2\pi}} + 4.5 \text{ m/s}} = 4.9 \text{ s} \approx 5 \text{ s}$$

The answer is B.
Assess: We expect the waves to reach the ship more frequently if the ship is sailing toward them.

PptIV.9. Reason: We need to know how far 40 wavelengths is, so we find λ. In Table 15.1 find the speed of ultrasound in human tissue: $v = 1540 \text{ m/s}$.

$$\lambda = \frac{v}{f} = \frac{1540 \text{ m/s}}{1.0 \text{ MHz}} = 0.00154 \text{ m}$$

Now find the number of wavelengths in 12 cm.

$$12\,\text{cm} = 12\,\text{cm}\left(\frac{1\,\text{wavelength}}{0.00154\,\text{m}}\right) = 78 \text{ wavelengths.}$$

This is about twice the 40-wavelength half-distance, so the intensity is halved twice. In other words, the new intensity is 250 W/m^2. The answer is C.

Assess: The penetration depth depends on the frequency, but this is reasonable for 1 MHz ultrasound.

PptIV.11. Reason: Higher frequency provides less penetration, as we saw in the previous problem, but with smaller wavelength it provides better resolution. See Example 15.7. The correct answer is B.
Assess: There are tradeoffs in deciding which frequency to use.

PptIV.13. Reason: There are three full oscillations in 0.004 s so the period is 0.00133 s. The frequency is the inverse of this.

$$f = \frac{1}{T} = \frac{3}{0.004\ \text{s}} = 750\,\text{Hz}$$

The answer is C.
Assess: All of the choices are plausible answers, so we must have faith in the math.

PptIV.15. Reason: The wave pulse travels down the tube and back in 0.008 s

$$\text{dist} = \text{speed} \times \text{time} \quad \Rightarrow \quad 2L = (343\,\text{m/s})(0.008\text{s}) \quad \Rightarrow \quad L = 1.4\,\text{m}$$

The answer is C.
Assess: This is a reasonable length for a tube for this measurement.

PptIV.17. Reason: It will take half a period to return to the starting spot.

$$\frac{T}{2} = \pi\sqrt{\frac{L}{g}} = \pi\sqrt{\frac{5.0\ \text{m}}{9.8\ \text{m/s}^2}} = 2.2\ \text{s}$$

The answer is C.
Assess: This seems about right for a 5.0 m long swing.

PptIV.19. Reason: At the farthest point you are 1.0 m higher than at the bottom. See figure.

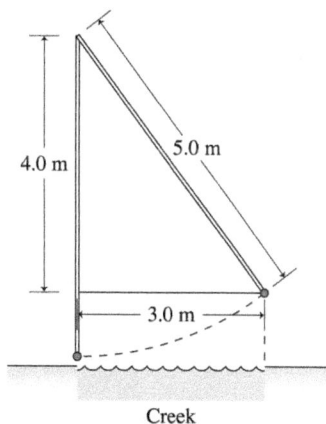

Creek

We now use conservation of energy to find the speed at the bottom.

$$\frac{1}{2}mv^2 = mgh \quad \Rightarrow \quad v = \sqrt{2gh} = \sqrt{2(9.8 \text{ m/s}^2)(1.0 \text{ m})} = 4.4 \text{ m/s}$$

The correct answer is C.

Assess: This seems like a reasonable speed for a big swing.

PptIV.21. Reason: It is a logarithmic scale.

$$\beta(\text{dB}) = 10 \log \frac{I}{I_0} - 27 \text{dB} \quad \Rightarrow \quad \frac{I}{I_0} = 10^{2.7} = 500$$

Assess: The answer may be surprising to those who don't realize the decibel scale is logarithmic.